5.4	Ratchet	109
5.5	Electron correlation and nonreciprocal transport	112
5.6	Shift current	115
5.7	Discussion and conclusions	120
	Acknowledgments	122
	References	122

6. **Spintronics Meets Topology**   125

Shuichi Murakami

6.1	Introduction	125
6.2	Intrinsic spin Hall effect	126
6.3	Topological insulators	128
6.4	Topological semimetals	129
6.5	Material realization of topological phases	133
6.6	Non-hermitian systems	139
6.7	Summary	143
	Acknowledgments	143
	References	143

7. **Thermalization and Its Absence within Krylov Subspaces of a Constrained Hamiltonian**   147

Sanjay Moudgalya, Abhinav Prem, Rahul Nandkishore, Nicolas Regnault and B. Andrei Bernevig

7.1	Introduction	147
7.2	Model and its symmetries	151
7.3	Hamiltonian at 1/2 filling	154
7.4	Krylov fracture	157
7.5	Integrable subspaces	160
7.6	Nonintegrable subspaces and Krylov-restricted ETH	170
7.7	Quasilocalization from thermalization	174
7.8	Connections with Bloch MBL	177
7.9	Conclusions and open questions	181
	Acknowledgments	206
	References	206

8. **Classification of Strongly Disordered Topological Wires Using Machine Learning**   211

Ye Zhuang, Luiz H. Santos and Taylor L. Hughes

8.1	Introduction	211

8.2	Model	213
8.3	Results	214
8.4	Summary	221
	Acknowledgments	222
	References	222

9. Topological Physics with Mercury Telluride   225

Saquib Shamim, Hartmut Buhmann and Laurens W. Molenkamp

9.1	Introduction	225
9.2	Bandstructure of HgTe	226
9.3	HgTe as 2D topological insulator: Realization of quantum spin Hall effect	228
9.4	HgTe as 3D topological insulator	234
9.5	Conclusion and outlook	238
	References	239

10. Topology and Interactions in InAs/GaSb Quantum Wells   243

Rui-Rui Du

10.1	Introduction	244
10.2	Quantum spin hall effect	246
10.3	Topological excitonic insulator	251
10.4	1D helical Luttinger liquid	253
10.5	Future perspectives	255
	Acknowledgments	257
	References	257

11. First Principle Calculation of the Effective Zeeman's
    Couplings in Topological Materials   263

Zhida Song, Song Sun, Yuanfeng Xu, Simin Nie,
Hongming Weng, Zhong Fang and Xi Dai

11.1	Introduction	263
11.2	Theory	264
11.3	First principle calculations	266
11.4	Discussions	273
11.5	Summary	275
	Acknowledgments	277
	References	278

12. Anomaly Inflow and the η-Invariant 283

Edward Witten and Kazuya Yonekura

12.1 Introduction 283
12.2 A precise formula for anomaly inflow 286
12.3 The anomaly 318
12.4 Examples in dimensions $d = 1, 2, 3, 4$ 329
Acknowledgments 349
References 349

13. Detection of the Orbital Hall Effect by the Orbital–Spin Conversion 353

Jiewen Xiao, Yizhou Liu and Binghai Yan

13.1 Introduction 353
13.2 Results and discussions 354
13.3 Summary 363
Acknowledgments 363
References 363

14. Non-Bloch Band Theory and Beyond 365

Zhong Wang

14.1 Introduction 365
14.2 Band theory and topology 366
14.3 Non-Hermitian physics 368
14.4 Non-Bloch band theory 371
14.5 Applications of non-Bloch band theory 379
14.6 Outlooks 383
Acknowledgments 384
References 384

15. Quantum Anomalous Hall Effect in Magnetic Topological Insulators 389

Yayu Wang, Ke He and Qikun Xue

15.1 Introduction 389
15.2 Experimental realization of the quantum anomalous
Hall effect 391

15.3   Recent progresses on the quantum anomalous
       Hall effect   398
References   400

16.   SciviK: A Versatile Framework for Specifying and
      Verifying Smart Contracts   403

Shaokai Lin, Xinyuan Sun, Jianan Yao, and Ronghui Gu

16.1   Introduction   403
16.2   Overview   406
16.3   The annotation system   411
16.4   Generating annotated IR   412
16.5   Translating annotated IR into WhyML   415
16.6   Generating verification conditions   421
16.7   Evaluation   424
16.8   Related work   432
16.9   Conclusion   433
Acknowledgments   434
References   434

Appendix: Schedule of the Shoucheng Zhang Memorial Workshop   439

Foreword

As the family of our beloved Shoucheng Zhang — physicist, educator, collaborator, husband, father, and friend — it is with deep gratitude that we introduce this volume. From his childhood years, Shoucheng was an intensely avid reader, and few things brought him such peace and contentment as nestling into his armchair with a book as companion, letting his thoughts flow with its poetic lines or scientific ideas. To have been gifted a book not only so full of ideas, but written by his life's closest colleagues and students — that would have been pure delight. Imagining Shoucheng's enthusiasm filling our home, we share his joy and excitement as well that this beautiful book has come to fruition.

If Shoucheng were unwrapping this volume, it would certainly have been within his study — between two ceiling-high shelves of books, overlooking Stanford's Hoover Tower on one side, and with Raphael's inspirational painting *The School of Athens* hanging on the opposite wall. Shoucheng chose *The School of Athens* as the centerpiece of his study years ago, and it is a vivid illustration of his life's endeavors. Just like the ancient Greek philosophers and mathematicians — Socrates, Plato, Euclid, Pythagoras, and many more — Shoucheng was fascinated by the natural world and the human quest for knowledge. Yet his pursuits were never individual, but were fostered within a flourishing community of other academics — not just within one ancient empire, but reaching across the entire globe.

Amidst our sadness of losing our beloved Shoucheng, we knew Shoucheng had touched many lives worldwide and wanted them to be able to grieve and remember him with us. We were overwhelmed by the outpouring of love towards him and our family as dozens of his worldwide collaborators traveled from China, Germany, Japan, South Korea, the United Kingdom, Canada, and across the US for a set of memorial events in May 2019. These included a service in Stanford's Memorial Church, a physics and science frontiers workshop, and a dinner banquet for the workshop attendees. Through the intimate and touching stories that were shared, we were reminded that

Shoucheng was truly fortunate to have been part of such a close-knit, caring community of physicists during his lifetime.

This volume of tributes captures the spirit of worldwide collaboration that Shoucheng embodied and sought to champion throughout his career. Most of the contributions stem out of talks that were delivered in the memorial workshop. This volume contains contributions from Shoucheng's long term collaborators Daniel Arovas, Xi Dai, Shuichi Murakami, and Naoto Nagaosa, and his former advisees Chao-Xing Liu, Zhong Wang, and Binghai Yan. We are particularly pleased to have the contributions of several of Shoucheng's former PhD students at Stanford: Andrei Bernevig, Jiangping Hu, Taylor Hughes, Leonid Pryadko, and Congjun Wu. Each of these collaborators continue to carry on a piece of Shoucheng's legacy in their research and teaching at universities around the world.

Even though Shoucheng was primarily a condensed-matter theorist, this volume reminds us of his wide-ranging interests throughout his life. He sought scientific beauty in the intersection of elegant mathematical theory with real-world experiments, and through his career developed many key collaborations with experimentalists. These include those with Hartmut Buhmann, Rui-Rui Du, He Ke, Laurens Molenkamp, Yayu Wang, and Qikun Xue, who have all shared their reflections on the thrills of doing science with Shoucheng. On the theoretical side, Edward Witten, one of the preeminent mathematical physicists of our time, describes how ideas from Shoucheng's work on topological insulators have a deep connection to fundamental particle physics.

Shoucheng aspired to be an interdisciplinary scientist in the model of da Vinci. As a cofounder of Danhua Venture Capital (DHVC) from 2013 onwards, he enjoyed the challenge of applying principles from theoretical physics — simplicity and universality — to assess emerging technologies from blockchain to genome editing to AI. At the May memorial workshop, we invited several speakers to present on Shoucheng's broader interests in hopes of inspiring other researchers with Shoucheng's vision. Within these pages, Ronghui Gu, a computer science professor, describes his research on formal verification for proving the correctness of blockchain smart contracts, a topic that he and Shoucheng frequently discussed.

Even if the foreground figures of Shoucheng's "School of Athens" are physics faculty around the world, the reality is that such a thriving community requires many supporting figures working behind the scenes. Our family would like to thank all the staff members of Stanford's Physics Department and the Geballe Laboratory for Advanced Materials who worked with us to organize the memorial events for Shoucheng, and especially Noelle

Rudolph, Cynthia Sanchez, and Rosenna Yau. Their tireless efforts working around the clock made these events all we had hoped for and brought us peace in our memory of Shoucheng.

Thanks to Eugene Demler, Steve Kivelson, Biao Lian, Chao-Xing Liu, Xiao-Liang Qi, and Xiaoqi Sun, all dear friends of our family, for convening and organizing the memorial workshop, as well as for writing the overview article on Shoucheng's scientific life. Our family has been moved time and time again by all the memories of Shoucheng that you have helped us preserve.

Our thanks goes to the publishing team at World Scientific Publishing Company for editing this volume. Lastly, we once again thank all the attendees of the memorial workshop, for being with us to remember Shoucheng and for helping us to glimpse the fullness of his life, even as it was cut short. In our darkest season, each of the visits and notes we received helped us to hold on and find hope in our loss.

And to the reader of this volume, we hope that you may be inspired by the pieces here and that it will help guide you on a journey of knowledge towards deeper and broader horizons. May you read it with the words of Stanford University's motto in mind: "May the winds of freedom blow!"

Brian, Ruth, Stephanie, Charlie, and Barbara Zhang
December 2020

Attendees of the Shoucheng Zhang Memorial Workshop in May 2019

Shoucheng with his students and postdocs at a group hike at Russian Ridge, Palo Alto, 2008

A group photo at Shoucheng's 50th birthday party, 2013

Shoucheng at Stanford campus, 2013

Shoucheng in a physics discussion, Tsinghua University, 2012

The Zhang family with Prof Chen-Ning Yang in Prof Yang's office, 2008

Foreword by the Editors

On May 2–4, 2019, several members of the Stanford University Physics Department organized a memorial workshop for the late professor Shoucheng Zhang. Shoucheng had been an intellectual leader in theoretical physics. He was also renowned as an outstanding advisor and mentor who influenced many scientists deeply with his great enthusiasm and unique talent for discovering the beauty and simplicity in physics. The speakers at the memorial workshop were Shoucheng's colleagues, friends and current and former advisees, who honored Shoucheng by presenting and discussing the latest developments in many areas of research including condensed matter physics, high energy physics, quantum information science, computer science, biology, etc. (A schedule of the workshop is included as an appendix of this memorial volume.)

This memorial volume contains 16 research articles by speakers from the memorial workshop, and cover almost all the fields discussed during the workshop. From these articles, the reader can not only learn about exciting frontiers of science, but also see how deep and broad the influence of Shoucheng's ideas remain. We greatly appreciate the efforts of all the authors to this unique collection. We are particularly pleased that Prof. Chen-Ning Yang agreed to include his beautiful summary of Shoucheng's scientific life as a preface for this volume. Prof. Yang was Shoucheng's mentor and role model. In 2019, he also organized a memorial workshop for Shoucheng at Tsinghua University. We are honored to include his reflections on Shoucheng in this book.

In addition, we have included an overview of Shoucheng's scientific life, written by the five of us together with Xiaoqi Sun, one of Shoucheng's last Ph.D. students. During the memorial workshop, we presented a series of 12 posters summarizing Shoucheng's scientific contributions and some significant events in his life. This overview article was based on the posters. In this article, we have included the images of some selected handwritten notes of Shoucheng's, which he preserved in a very organized way. These notes are precious historical documents which allow us to gain some insight into how some of his groundbreaking ideas took shape. We are grateful to the Zhang family for permitting us to access and publish this invaluable document.

We would like to acknowledge the strong support from the Department of Physics and the Geballe Laboratory of Advanced Materials (GLAM) at Stanford University, without which the memorial workshop would not have been possible. We would especially like to recognize the contributions of the GLAM and Physics staff members and student/postdoc volunteers, in particular Noelle Rudolph. We appreciate the efforts of World Scientific in publishing this volume, especially Prof. Kok Khoo Phua, Dr. Sun Han and Ms. Cheryl Heng.

Last but not least, our special thanks go to the Zhang family members, Brian Zhang, Ruth Fong, Stephanie Zhang, Charlie Yang, and Barbara Zhang, for sharing their precious memories of Shoucheng, for writing a beautiful foreword for this memorial volume, and for being so caring and inspiring to all of us, the members of Shoucheng's extended scientific family.

Eugene Demler, Steven Kivelson, Biao Lian, Chao-Xing Liu, Xiao-Liang Qi

Shoucheng Zhang — A physicist of the first rank*

C. N. Yang

Institute for Advanced Study, Tsinghua University, P. R. China

Published 12 February 2019

In condensed matter physics, researchers investigate properties of different kinds of matter: why copper can conduct electricity, while rubber cannot; why water can freeze into ice, but also evaporate into steam; etc. This area of physics has deep relations with applications, with the economic development of the world, and with daily lives of mankind. Thus, it is an especially important area of research. S. C. Zhang has led several revolutionary advances in this area of physics, of which the most important is in Quantum Spin Hall Effect and related areas.

In 2005–2006, C. L. Kane of the University of Pennsylvania and Zhang independently published theories pointing to the possible existence, in composite matter under proper conditions, of surface electric conductivity, an entirely new phenomenon. These two papers immediately caught the attention of all condensed matter physicists. But which composite matter, under what conditions, would exhibit this remarkable new phenomenon remained a very difficult problem.

Zhang told me that together with several semiconductor collaborators, he next made calculations for specially designed quantum wells in composite semiconductors, and published on 15 December 2006, a paper predicting that in a Hg-Te-Cd semiconductor quantum well there could be surface electric conductivity plus other new important phenomena. Now, there are many types of semiconductors, and quantum wells can have many different shapes and dimensions, how did they choose to focus on their specific Hg-Te-Cd

---

*An earlier version of this paper, in Chinese, had appeared in GuangMing Daily-net in December 2018.

quantum well? I think the only answer is: Zhang had deep intuition in the quantum mechanics of semiconductors.

In 2007, Molenkamp of Würzburg University and his experimental group, following Zhang's prediction, tried several kinds of HgTe/(HgCd)Te quantum wells, and created the first example of surface electric conductivity and related phenomena. In November that year, they published, together with Zhang and his student X. L. Qi, a paper announcing this great success. This paper is one of the most important papers in physics in recent years.

It has been the general opinion of all physicists that Kane, Zhang and Molenkamp will receive the Nobel prize in time. Now that Zhang has passed away, I firmly believe Kane and Molenkamp will one day be awarded the Nobel prize.

To See a World in a Grain of Sand

— The Scientific Life of Shoucheng Zhang

Biao Lian, Chao-Xing Liu, Xiao-Qi Sun, Steven Kivelson,
Eugene Demler and Xiao-Liang Qi

Our friend and colleague, Prof. Shoucheng Zhang, passed away in 2018, which was a great loss for the entire physics community. For all of us who knew Shoucheng, it is difficult to overcome the sadness and shock of his early departure. However, we are very fortunate that Shoucheng has left us such a rich legacy and so many memories in his 55 years of life as a valuable friend, a world-leading physicist, a remarkable advisor, and a great thinker. On May 2–4, 2019, a memorial workshop for Shoucheng was organized at Stanford University, where we displayed a small exhibition of 12 posters, as a brief overview of Shoucheng's wonderful scientific life. This article is prepared based on those posters.

1 Early experience

1.1 *Childhood and school age*

Shoucheng Zhang (张首晟) was born in 1963 in Shanghai, China. In his childhood, Shoucheng already showed exceptional talent and a strong interest in various fields of knowledge. In the attic of Shoucheng's home, there were many books on art, history, philosophy, and science left by his grandfather and others of his parents' generation. These included books on the philosophy of Russell and Kant, and the art of da Vinci and Rodin. Shoucheng's favorite activity after school was to read books in the attic. In an era when educational resources were scarce, these books opened a new world to him. In 1976, Shoucheng's father bought Shoucheng a set of high school textbooks on mathematics, physics, and chemistry. He was immediately attracted by the amazing beauty of science.

1.2 *College time in Shanghai and Berlin*

In 1977, the National Higher Education Entrance Exam of China was restarted after the end of the Cultural Revolution. Without attending high school, Shoucheng took the first exam and got admitted to the Physics

Fig. 1.   Shoucheng at age two.

Fig. 2.   Shoucheng in front of the grave of nuclear physicist Otto Hahn in Göttingen, 1981.

Department of Fudan University in Shanghai in 1978. At the age of 15, he was the youngest student in his class. One year later, in recognition of his excellent academic performance, Shoucheng was selected for an exchange program to study abroad at the Free University of Berlin, where he received his *Diplom-Physiker* (Bachelor of Science degree) in 1983.

During his college time in Germany, besides studying physics, Shoucheng also had rich exposure to German culture. On a trip back from Bonn to Berlin in 1981, he and some friends visited the Stadtfriedhof cemetery in Göttingen that houses the graves of many scientists. Even after many years, Shoucheng often remembered this visit as a source of inspiration. Upon observing the equations on the tombstones of Max Planck, Otto Hahn, and Max Born, Shoucheng wrote that he would "spend my energy on the pursuit of science, hoping that I too would leave behind a life's work that could be summed up in a simple equation."

Fig. 3.   Shoucheng (second from the right in the second row) and his roommates at Fudan University, 1978.

1.3   *Ph.D. at Stony Brook University*

Shoucheng began his Ph.D. studies on supergravity at the State University of New York at Stony Brook in 1983, advised by Peter van Nieuwenhuizen. In the final year of Shoucheng's Ph.D. (1986–1987), following Prof. Chen-Ning Yang's suggestion, he started shifting his research direction to condensed matter physics. He began a collaboration with Steven Kivelson, a faculty member at Stony Brook who later became Shoucheng's colleague and lifelong friend at Stanford.

1.4   *Santa Barbara, IBM and Stanford*

After receiving his Ph.D. in 1987 from Stony Brook, Shoucheng became a postdoctoral fellow at the Institute of Theoretical Physics (ITP) in UC Santa Barbara. In 1987, he married his childhood sweetheart Barbara Yu. He then joined IBM Almaden Research Center as a Research Staff Member from 1989 to 1993. Thereafter, he joined the faculty of Stanford University, which remained his academic home for the rest of his life.

During his career, Shoucheng has worked on many different directions in physics, such as the theory of fractional quantum Hall states and

Fig. 4.   Barbara, Shoucheng, Chen-Ning Yang, and Shoucheng's father Hongfan, 1987.

Fig. 5.   Shoucheng with his advisor Peter van Nieuwenhuizen (middle) and his future wife Barbara (left) at their graduation ceremony, 1987.

high temperature superconductivity, topological insulators and topological superconductors, spintronics, etc. In Table 1, we provide a sketch of some key events in his life and career. His scientific achievements in different areas of physics will be overviewed in the rest of this article.

Table 1.   Some key events in Shoucheng's career and life.

Year	Events
1963	Shoucheng was born in Shanghai, China to Manfan Ding and Hongfan Zhang
1980	Began studying at Free University of Berlin
1983	Graduated from Free University of Berlin
	Started Ph.D. at Stony Brook University
1986	Started working on condensed matter physics
1987	Received Ph.D. and started postdoc at ITP, UC Santa Barbara
	Married his childhood sweetheart, Barbara Yu
	Chern-Simons theory of fractional quantum Hall states
1989	Started at IBM Almaden Research Center, San Jose
	Global phase diagram of fractional quantum Hall states
1993	Joined the faculty of Stanford University
	Shoucheng's son Brian was born
1996	Shoucheng's daughter Stephanie was born
1997	Proposed the SO(5) theory of high Tc superconductors
2001	Generalized quantum Hall effect to 4 dimensions
2003	Proposed the intrinsic spin Hall effect
2004–2005	Early models of quantum spin Hall effect (2004–2005)
2006	Predicted the quantum spin Hall effect in HgTe
2007	Quantum spin Hall effect realized in HgTe
2008-2009	Topological magneto-electric effect
	Prediction and realization of Bi_2Te_3 family of topological insulators
	Prediction of new topological superconductors
2010	Prediction of quantum anomalous Hall effect
2013	Quantum anomalous Hall effect realized
	Founded venture capital firm DHVC
	Celebrated 50th birthday with his friends, current and past group members.
2017	Experimental evidence of chiral topological superconductor reported
2018	Founded the Stanford Center for Topological Quantum Physics
	Passed away on December 1, 2018

2   Theory of fractional quantum Hall effect

2.1   *The fractional quantum Hall effect*

When Shoucheng began to study condensed matter physics in the late 1980s, one of the first topics he became interested in was the fractional quantum Hall effect (FQHE).

The FQHE, experimentally discovered in 1982 by D. C. Tsui, H. L. Stormer, and A. C. Gossard [41], is a remarkable quantum phenomenon of two dimensional (2D) metals near the absolute zero temperature in a strong magnetic field B, where the Hall resistance R_{xy} is quantized at value

$$R_{xy} = h/\nu e^2$$

Fig. 6.   A group photo of Shoucheng and his students after a class at Stanford in 1994.

Fig. 7.   Shoucheng and J. Robert Schrieffer at Santa Barbara, 1989.

Here ν is one of a particular set of rational fractions, h is Planck's constant, and e is the electron charge.

The understanding of the FQHE fundamentally challenged conventional condensed matter theory. In 1983, Robert B. Laughlin [43] proposed the Laughlin wavefunction which successfully described the ground state of

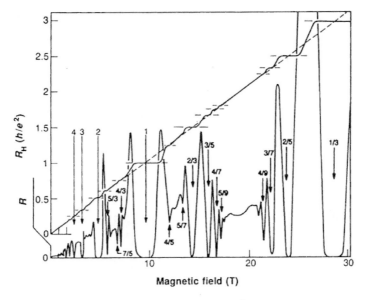

Fig. 8.   The FQHE with $R_{xy} = h/\nu e^2$, from Ref. [10]

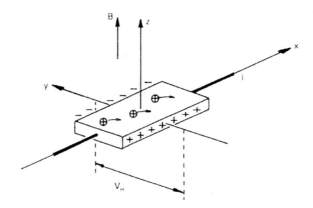

Fig. 9.   The experiment of Hall resistance $R_{xy} = V_y/I_x$ (Kosmos 1986).

fractional quantum Hall fluids with $\nu = \frac{1}{2k-1}$ ($k = 1, 2, \ldots$); this laid the foundation for the theory of the FQHE.

2.2   *Chern-Simons Ginzburg-Landau Theory and the Global phase diagram*

Shoucheng was interested in looking for a quantum field theory (QFT) for the FQHE. The presence of the background magnetic field B implies that this field theory must be quite different from those that were familiar in high energy physics. The field theory needed to capture essential features of the fractional quantum Hall state such as the quantized Hall conductance, fractional charged excitations, etc.

In 1989, Shoucheng and his collaborators Thors Hans Hansson and Steven Kivelson [44] proposed the Chern-Simons-Landau-Ginzburg field theory of the FQHE:

$$\mathcal{L} = \mathcal{L}_{GL}\left[\phi, \left(i\partial_\mu - A_\mu + a_\mu\right)\phi\right] + \left(\frac{e\pi}{2\theta\Phi_0}\right)\epsilon^{\mu\nu\lambda}a_\mu\partial_\nu a_\lambda.$$

The key idea of this theory is **flux attachment**, which is achieved by introducing a gauge field with a Chern-Simons term. The dynamics of this gauge field attaches $2k-1$ magnetic fluxes to each electron, which transmutes them into bosons. The FQHE is then interpreted as the condensate of this boson.

In an interview with Stanford News after Shoucheng's death, Steven Kivelson described this discovery: "One day Shoucheng came to visit and he said, 'Look what I figured out.' He then sketched out a basic idea of the theory and all of these mysterious features of the fractional quantum Hall effect just dropped in your lap incredibly simply," Kivelson said. "That's not how physics usually works. You usually slave away at things. But on that day Shoucheng's idea was just so focused and so perfect."

Along the same direction, in 1992 Shoucheng collaborated with Steven Kivelson and Dung-Hai Lee to propose a global phase diagram of the FQHE,

Fig. 10.   Shoucheng with James H. Simons (middle), one of the founders of Chern-Simons theory, and Edward Witten (right) at Stanford in 2010.

Fig. 11.   Shoucheng with Per Bak, Steven Kivelson, and Steven's wife Pamela Davis at Stony Brook in 1987, celebrating Shoucheng's thesis defense.

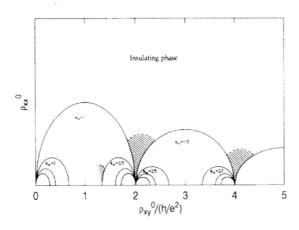

Fig. 12.   Global phase diagram of FQHE, from Ref. [6].

which derived a set of interrelations among various FQHE states [6]. The Chern-Simons effective theory and the global phase diagram also revealed the relation between fractional quantum Hall physics and various forms of **particle-vortex duality**, which still remains an active research topic after 30 years.

3 SO(5) Theory of high temperature superconductivity

3.1 *High T_c cuprates and the SO(5) theory*

The discovery of high temperature superconductivity in copper oxides (known as cuprates) in 1987 was one of the most important breakthroughs in condensed matter physics. Despite enormous theoretical and experimental efforts, many questions about cuprates remain open today.

Shoucheng started to work in this field in 1990, when he collaborated with C. N. Yang to point out that the Hubbard model, a model that is widely believed to capture the essential physics of cuprate superconductors, has an enhanced SO(4) symmetry [45]. Starting from this work, Shoucheng and collaborators explored the possibility of a (broken) enhanced symmetry in high T_c superconductors and its physical consequences. For example, in 1995, Eugene Demler and Shoucheng studied the relation of a neutron scattering feature in cuprates with enhanced symmetry [46].

Fig. 13.   A levitated magnet above superconducting $YBa_2Cu_3O_7$, from Science photo library.

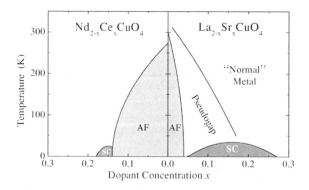

Fig. 14.   Schematic phase diagram of Cuprates, from Ref. [19].

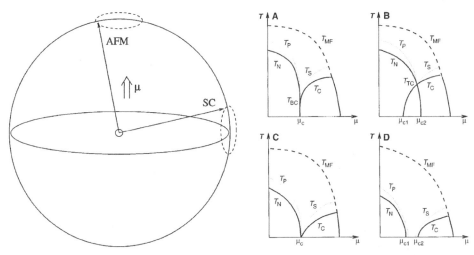

Fig. 15. (Left) The SO(5) superspin, from Ref. [4]. (Right) The SO(5) phase diagram, from Ref. [35].

Based on these works, in 1997 Shoucheng proposed a new effective field theory for high T_c superconductivity [35]. As an elegant application of symmetry principles, he proposed a five-component vector order parameter that incorporates both antiferromagnetism and superconductivity, the main two competing orders in the cuprates. This theory is known as the SO(5) theory, in which spin rotation symmetry SO(3) and charge conservation symmetry U(1) $\simeq$ O(2) are considered as subgroups of SO(5). This theory offered a simple possible framework for interpreting many of the rich phenomena in cuprate superconductors, and it clearly reflected Shoucheng's unique style: relating simple and universal principles to experimental reality in condensed matter physics.

3.2 *Further developments*

Although the SO(5) theory was proposed as a phenomenological effective field theory, Shoucheng and collaborators also identified and studied microscopic models with an exact SO(5) symmetry [47, 48]. The latter paper further generalized the model to even larger symmetry, which is realizable in large-spin ultra-cold fermion systems [49].

A single electron carries the spinor representation of SO(5), in a manner analogous to the case of the spatial rotation group. Interestingly, the four-component SO(5) spinors are intrinsically related to the Dirac equation. Surprisingly, starting from these works, the SO(5) Clifford algebra and SO(5) spinors formed a theme that appeared again and again in Shoucheng's significant contributions in seemingly unrelated topics, including the four-dimensional quantum Hall effect, the intrinsic spin Hall effect, the quantum spin Hall effect, and topological insulators.

4   The four-dimensional quantum Hall effect

4.1   *From quaternions to quantum Hall effect in 4D*

Shoucheng was deeply committed to the pursuit of mathematical beauty in physical theories, and his theory of four dimensional (4D) quantum

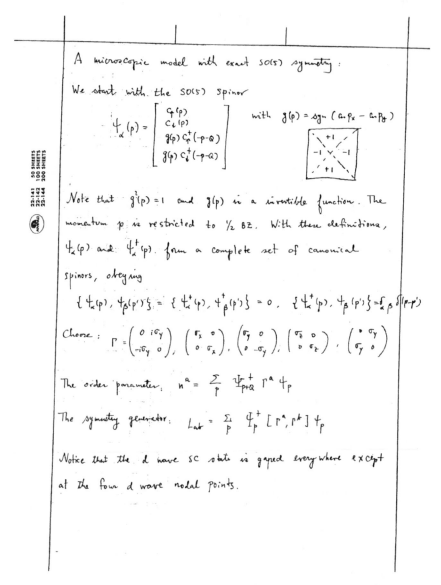

Fig. 16.   Shoucheng's note about an SO(5) model in 1997, with Dirac Γ matrices.

Fig. 17. William Rowan Hamilton invented quaternions.

Hall (QH) effect [50] is one of the most vivid examples of the power of this approach.

A remarkable feature of the two-dimensional (2D) QH effect is that the QH wave functions can be written as functions of complex numbers $z = x + iy$, where (x, y) represents the 2D coordinate of an electron. This motivated Shoucheng to ask: is there a QH wave function of quaternions instead of complex numbers?

In mathematics, in addition to the real numbers R, and complex numbers C, there are also quaternions H invented by William Rowan Hamilton, which have the form:

$$u = a + b\boldsymbol{i} + c\boldsymbol{j} + d\boldsymbol{k}$$

with $\boldsymbol{i}^2 = \boldsymbol{j}^2 = \boldsymbol{k}^2 = -1$. Since a quaternion consists of 4 components, Shoucheng was led to consider the 4D QH effect with 4 coordinates. In theoretical physics, $\boldsymbol{i}, \boldsymbol{j}, \boldsymbol{k}$ can be written as the 2×2 Pauli matrices $i\sigma_1, i\sigma_2, i\sigma_3$, which are the generators of the spin-1/2 representation of the SU(2) group. Therefore, the 4D QH effect quaternion wave function couples to an SU(2) Yang-Mills gauge field, instead of the U(1) gauge field in 2D quantum Hall states.

The original paper of Shoucheng and his student Jiangping Hu on the 4D QH effect [50] (2001) is a beautiful synthesis of mathematics and physics. They employed the 2nd Hopf map, a topological map from the 7D sphere to the 4D sphere, to construct the 4D QH effect quaternion wave function. This revealed that the 4D QH state is a topological state with a nonzero 2nd Chern number, which leads to an SU(2) quantized Hall effect.

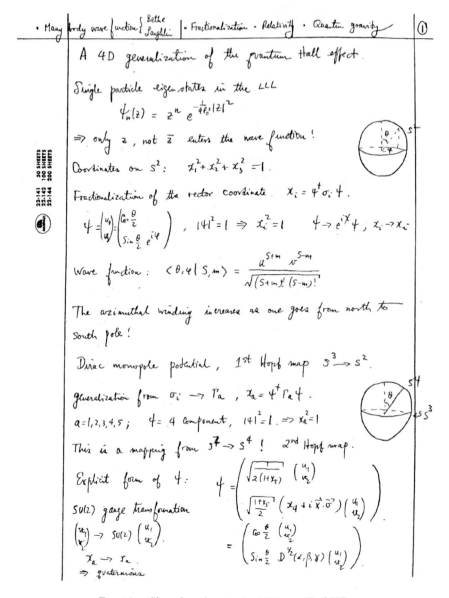

Fig. 18.   Shoucheng's note in 2001 on 4D QHE.

4.2   Relation to more recent developments

Noting that the 4D QH state has edge states consisting of relativistic particles with different spins living in 3D space, Zhang and Hu proposed it as a candidate "unified theory" of fundamental interactions and particles. This theory is also one of the earliest studies of topological states of

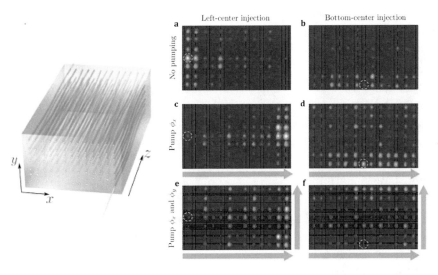

Fig. 19.   4D QHE edge state pumping observed in the photonic crystal experiment [14].

matter in high dimensions, and its idea has been generalized in many directions, including in numerous studies of other high dimensional topological states.

The 4D QH effect theory partially motivated Shoucheng's later work on the quantum spin Hall (QSH) effect and topological insulators (TI). Zhang, Hu (2001) [50] proposed the following equation of a spin Hall current, as a consequence of the 4D QHE:

$$\dot{X}_a = \frac{R^4 \partial V}{I^2 \partial X_b} F^i_{ab} I_i, \ \dot{I}_i = \epsilon_{ijk} A^j_\mu \dot{X}_\mu I_k$$

The spin Hall current of electrons corresponds to the SU(2) Hall current of the 4D. In 2008, the work from his group clarified that the 3D TI and 2D QSH can be obtained from a 4D QH state (though a different one from the original Zhang-Hu proposal) via dimensional reduction [3].

4.3   *Experimental realization*

Since we only have 3 spatial dimensions in reality, one might not expect a 4D QH effect to be experimentally accessible. However, in 2018 experiments on photonic crystal [14] and cold atom systems [51] effectively "realized" the 4D QH effect. The key idea is to use control parameters to create synthetic dimensions, much in the way Thouless pumping in 1D is related to the 2D QH effect. These experiments effectively constructed a 4D system with a non-zero 2nd Chen number by introducing two pumping parameters tunable by laser beams, together with two spatial dimensions.

5 Spin Hall effect

5.1 Intrinsic spin Hall effect

Shoucheng's work on the 4D QH effect implied the possibility of topological and dissipationless spin transport in 3D systems. Although the original 4D QH model is not directly related to any experimental system, this connection motivated Shoucheng to start thinking about spin-orbit coupling and its role in dissipationless transport. Near the end of the Zhang-Hu paper, the authors commented that "The single-particle states also have a strong gauge coupling between iso-spin and orbital degrees of freedom, which is ultimately responsible for the emergence of the *relativistic helicity* of the collective modes." In 2003, Shuichi Murakami, Naoto Nagaosa and Shoucheng predicted an intrinsic spin Hall effect in semiconductors [8], which referred to a transverse dissipationless spin current induced by an electric field in semiconductor compounds with strong spin-orbit coupling. The spin Hall current is described by the following formula:

$$j_j^i = \sigma_s \epsilon^{ijk} E_k.$$

As was stated at the beginning of their paper, this work "is driven by the confluence of the important technological goals of quantum spintronics with the quest of generalizing the QH effect to higher dimensions."

Different from charge current, spin current is even under time reversal, which allows this effect to occur in materials without a magnetic field. Spin-orbit coupling replaced the role of magnetic field in QH effect and leads to

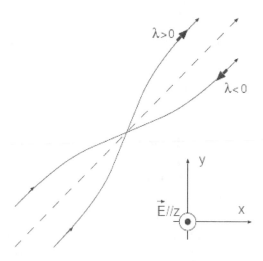

Fig. 20.   Schematic illustration of the mechanism of intrinsic spin Hall effect in Ref. [8].

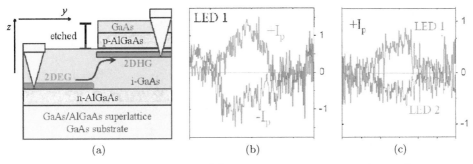

(a)   (b)   (c)

Fig. 21.   (a) Experimental device and (b) observed edge spin accumulation (by measurement of light polarization) due to spin Hall effect in a 2D hole gas system from Ref. [28].

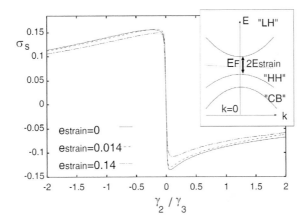

Fig. 22.   Spin Hall conductivity and band structure of the spin Hall insulator [7].

the nontrivial Berry's phase that is the essential reason of the spin Hall effect. The 4D QH and intrinsic spin Hall effect are the first proposals of dissipationless transport in dimensions higher than two, which is the overture of the upcoming breakthrough of new topological insulators and topological superconductors. (Intrinsic spin Hall effect of a different type was also proposed in the concurrent work of [52])

Soon after its theoretical proposal, the spin Hall effect was observed experimentally in hole doped semi-conductors [28].

5.2 *Spin Hall insulators*

In 2004, Murakami, Nagaosa, and Shoucheng generalized their results to a family of materials that they named "spin Hall insulators," which are systems with zero charge conductivity but finite spin Hall conductivity [7]. The materials mentioned in this paper are zero gap and narrow gap

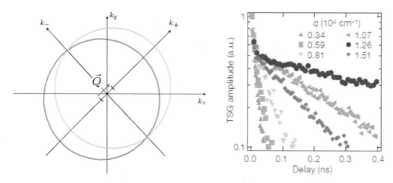

Fig. 23.   (Left) Theoretical prediction of the special Fermi surface that leads to persistent spin helix. (Right) Experimental realization of the spin helix in Ref. [25].

semiconductors, including HgTe, HgSe, β-HgS, α-Sn, PbTe, PbSe, PbS. Many of the materials mentioned here were later identified as topological insulators, although the connection to topological physics had not yet been uncovered in this work.

5.3   *Other works in spintronics*

In addition to the spin Hall effect, Shoucheng also worked more broadly in the field of spintronics. For example, in his collaboration with B. A. Bernevig and J. Orenstein [53], they predicted, and then subsequently experimentally realized, a "persistent spin helix" which supported a long-lived helical spin configuration due to an emergent SU(2) symmetry. Shoucheng's works brought many ideas from fundamental physics into the field of semiconductors and spintronics.

6   Quantum spin Hall effect

6.1   *Early models of quantum spin Hall effect*

The 4D QH and the intrinsic spin Hall effect are both proposed as generalizations of the QH effect. In 2004, Shoucheng started to consider a more direct generalization of the QH effect. B. A. Bernevig and Shoucheng proposed that in a semiconductor with a spatially inhomogeneous strain, the spin-orbit coupling can realize a uniform magnetic field that is opposite for spin up and spin down electrons. This manifests as a quantized spin Hall (QSH) effect preserving time reversal symmetry [15].

Concurrently, a different QSH model was independently proposed by Charles L. Kane and J. Eugene Mele, who studied spin-orbit coupling in graphene [31]. This model is a time-reversal invariant generalization of a model proposed by F. D. M. Haldane in 1988 [54]. In another work soon

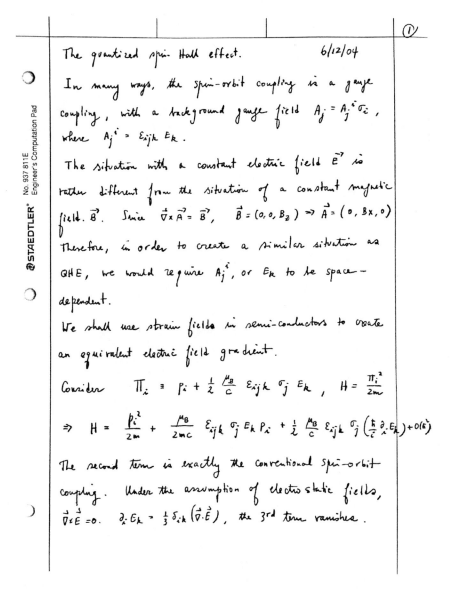

Fig. 24.   Shoucheng's note in 2004 on quantized spin Hall effect.

afterwards, Kane and Mele pointed out that the QSH state is classified by Z_2, which means that the states with odd pairs of edge states are topologically robust in the presence of time-reversal symmetry [55].

Shoucheng and his students Congjun Wu and B. Andrei Bernevig studied the stability of QSH edge states in the presence of electron interaction. They pointed out that the interacting edge states are qualitatively different

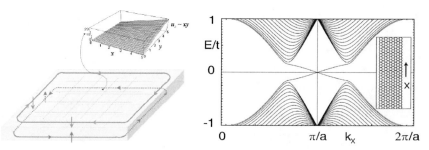

Fig. 25. The two earliest models of the quantum spin Hall effect. (Left) The illustration of the Bernevig-Zhang proposal [15]. (Right) The energy spectrum of the Kane-Mele model [31].

from the regular Luttinger liquid, and named it the "helical liquid" [56] (see also [57]).

6.2 *HgTe theory and experiments*

The earliest models reviewed above set up the theoretical foundation of QSH, but they are difficult to realize experimentally for various reasons. In 2006, Shoucheng and his students B. A. Bernevig and T. L. Hughes made the first realistic proposal of a QSH material: HgTe/CdTe quantum wells [5]. They obtained the low energy effective theory of this quantum well (known as the Bernevig-Hughes-Zhang (BHZ) model) and predicted that the QSH phase could be identified by a topological phase transition controlled by the thickness of the quantum well. The BHZ work not only proposed the first realistic QSH material, but also proposed a general mechanism to identify topological materials, which they named "band inversion."

Soon after the proposal, the QSH effect in HgTe quantum wells was verified in Laurens Molenkamp's group [16]. They indeed observed a $2e^2/h$ conductance for $d > d_c$ while the resistance is much higher when $d < d_c$, agreeing well with theoretical predictions.

6.3 *Other quantum spin Hall materials*

The discovery of HgTe QSH heralded the beginning of a new era with tremendous theoretical and experimental developments in topological states of matter. Since then Shoucheng and collaborators have also predicted

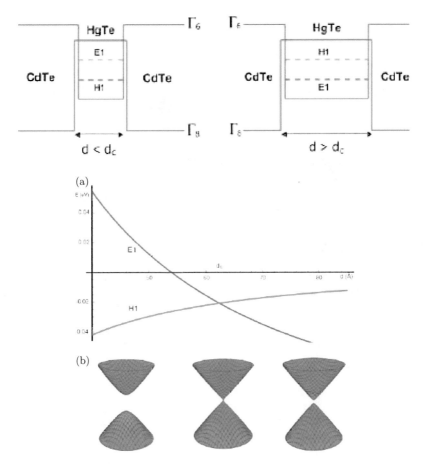

Fig. 26.   The band edge energy of narrow and wide quantum wells [5].

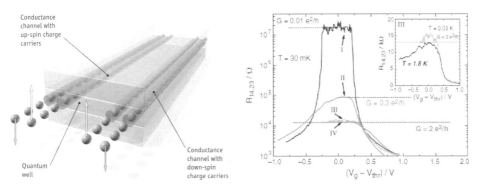

Fig. 27.   (Left) Illustration of the quantum spin Hall effect. (Right) Transport measurement in HgTe quantum wells, from Ref. [16].

Fig. 28.   Shoucheng (5th from left), Charles L. Kane (2nd) and Laurens Molenkamp (7th) at the Oliver E. Buckley Prize ceremony (2012), together with Shoucheng's son Brian (6th) and colleagues Eric Fullerton (1st), Stuart S. P. Parkin (3rd), and Dimitri Basov (4th).

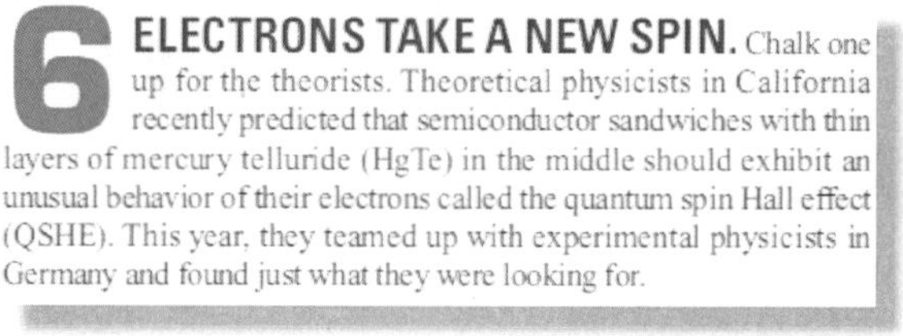

6 **ELECTRONS TAKE A NEW SPIN.** Chalk one up for the theorists. Theoretical physicists in California recently predicted that semiconductor sandwiches with thin layers of mercury telluride (HgTe) in the middle should exhibit an unusual behavior of their electrons called the quantum spin Hall effect (QSHE). This year, they teamed up with experimental physicists in Germany and found just what they were looking for.

Fig. 29.   Dicovery of the quantum spin Hall effect was listed by Science magazine as one of the top 10 Breakthroughs of Year in 2007.

several other QSH materials, such as InAs/GaSb quantum wells [9] and stanene, a single-layer thin film of Sn [13].

Shoucheng's group has also investigated many topological effects in QSH systems, such as electron interaction effects on the edge, and fractional charge on an edge magnetic domain wall, spin-charge separation induced by a π flux, etc.

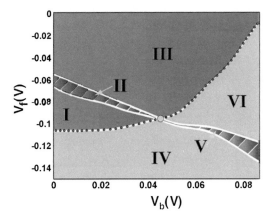

Fig. 30.   Phase diagram of the InAs/GaSb quantum wells with QSH effect, from Ref. [9].

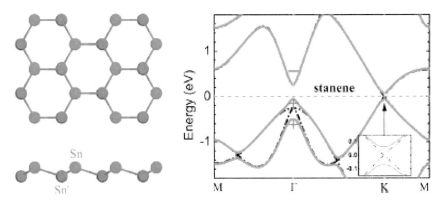

Fig. 31.   Stanene and its non-trivial band structure, from Ref. [13].

7   Three dimensional topological insulators

7.1   *Generalization of quantum spin Hall effect to three dimensions*

The understanding of the QSH and Z_2 topological protection implied that the QSH could be generalized to 3D insulators, with robust two-dimensional surface states. The surface electrons are "helical", with their spin direction locked with the momentum. This new state, named 3D topological insulators (TI), was proposed in 2006 by three papers [58–60].

The first 3D TI was the alloy of Bi and Sb, proposed by L. Fu and C. L. Kane [61], and realized by M. Z. Hasan's group in Princeton [62].

In 2008, Shoucheng and collaborators proposed a new family of topological insulators, including three materials: Bi_2Te_3, Bi_2Se_3, and Sb_2Te_3 [63]. These materials are described by a simple effective model similar to the

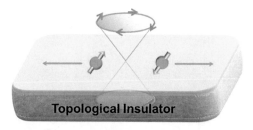

Fig. 32.   An illustration of the spin-momentum locking of topological insulator surface states.

(a)

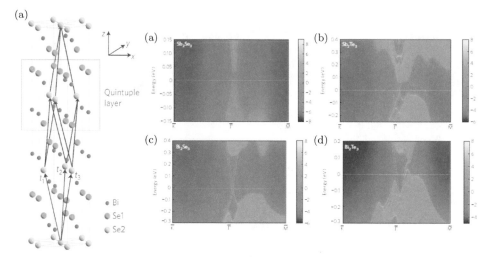

Fig. 33.   (Left) Crystal structure of the Bi$_2$Se$_3$ family of TI. (Right) Theoretical prediction of the surface state dispersion relation based on ab initio calculation. From Ref. [17].

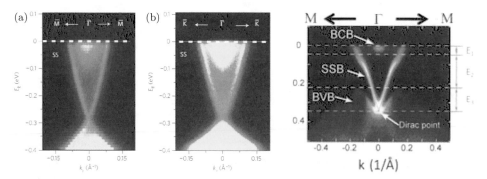

Fig. 34.   Experimental observation of topological surface states in angle-resolved photoemission. (Left) Bi$_2$Se$_3$ from Ref. [29]. (Right) Bi$_2$Te$_3$ from Ref. [33].

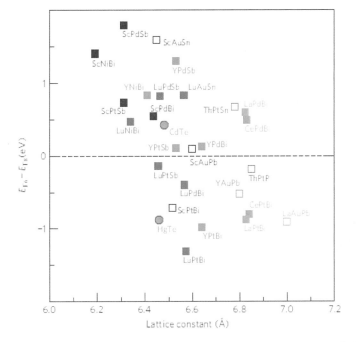

Fig. 35.   Half-Heusler family of topological insulators. From Ref. [21].

BHZ model. This family of TI was realized experimentally in 2009 [33]. The Bi_2Se_3 family of TI is easy to grow and has a large gap and simple surface states, which immediately became the "gold standard" of TI. They remain the most widely studied TI materials to this day.

7.2   *Other 3D TI materials*

Since 3D TI's are single crystals rather than quantum wells, it is easier to realize experimentally. Shoucheng and collaborators have predicted many of the early 3D TI materials. This includes strained HgTe [64] (Xi Dai *et al.*, *Phys. Rev. B* **77**, 125319 (2008), realized in L. Molenkamp's group), half-Heusler compounds [21], $TlBiSe_2$ family [65], filled skutterudites [66], Actinide compounds [67], etc.

7.3   *TI and axion electrodynamics*

Similar to the Chern-Simons theory that Shoucheng investigated in the FQHE, the new states QSH and 3D TI should be described by some topological field theories, which characterize their universal features. In early 2008, Xiao-Liang Qi, Taylor L. Hughes and Shoucheng developed the topological field theory of 3D TI [3]. Surprisingly, the field theory describing

TI is related to electromagnetic duality and has appeared before in high
energy physics as "axion electrodynamics" (For an introduction of the
relation of TI and axion physics, see Frank Wilczek, Nature **458** 129 (2009)).
The axion is a hypothetical particle in high energy physics with a particular
coupling with the electromagnetic field $\theta E \cdot B$ [68]. In TI, θ is an angle
determined by the material, with $\theta = 0, \pi$ corresponding to trivial insulator
and TI. The effect of θ is a modification of the constituent equations:

$$D = E + 4\pi P - \alpha \frac{\theta}{\pi} B$$

$$H = B - 4\pi M + \alpha \frac{\theta}{\pi} E$$

The physical consequence of the axion coupling is a topological *magnetoelectric* effect, which means an electric field induces magnetization, while a
magnetic field induces charge polarization. The axion angle θ determines the
amplitude of this effect. Different consequences of the topological magnetoelectric effect have been proposed, including an image monopole induced
by a charge near the surface [2] and a topological contribution to the
Faraday and Kerr rotation of linear polarized light [3, 69]. The latter has
been experimentally realized recently in 2016–2017 [36, 70, 71].

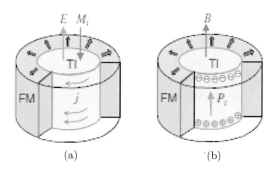

Fig. 36.   Topological magneto-electric effect. From Ref. [3].

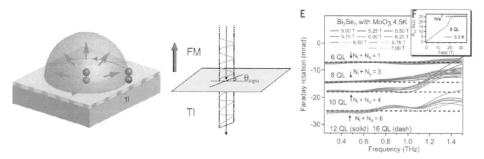

Fig. 37.   (Left) Image magnetic monopole effect [2]. (Middle) Illustration of the topological
Faraday rotation [3]. (Right) Faraday rotation experiment from Ref. [36].

7.4   *Further generalization*

The topological field theory approach also provided a unified view to topological states in different dimensions and symmetry classes. The Qi-Hughes-Zhang paper proposed a family tree of topological insulators, in which the 4D QH state (determined by the second Chern number, a topological invariant) is identified as the "parent state" from which 3DTI and QSH are obtained by dimensional reduction. A different approach of classification was independently proposed by two other works [72, 73], which was based on non-interacting fermions but covered a larger family of topological states than [3]. Starting from these works, the family of topological states has greatly expanded, by considering new symmetry classes, interaction effects and gapless systems (topological semimetals). Shoucheng has made very broad contributions to many different aspects of this recent progress. In the next two sections we review two of these major directions, the quantum anomalous Hall (QAH) effect and topological superconductors (TSC).

8   Quantum anomalous Hall effect

8.1   *Quantum anomalous Hall effect*

The QAH effect refers to QH effect in a band insulator without an external magnetic field. Topologically it belongs to the same phase as integer QH effect, but it has dispersing bands and no Landau levels. The first model for QAH was proposed by F. D. M. Haldane in 1988 [54]. Although in principle QAH does not require QSH effect and spin-orbit coupling, it is only after the proposal of QSH and TI that a path to realize the QAH effect was found. In a 2006 paper Shoucheng collaborated on with X. L. Qi and Y. S. Wu about a QSH model [74], the authors also proposed a new model of QAH effect. (This paper is possibly the origin of the name "quantum anomalous Hall effect".)

QAH effect can be realized by making use of TI because the latter already has nontrivial Berry curvature that contributes to the Hall conductance. One just needs to break time-reversal in an appropriate way to avoid the cancellation.

Based on further developments in TI, Shoucheng and collaborators made two more realistic proposals of QAH: one in Mn doped HgTe/CdTe quantum wells, making use of QSH effect [75]; the other in magnetically doped Bi_2Se_3 (or Sb_2Te_3, Bi_2Te_3) thin films, making use of the surface state of 3D TI [30]. The Hg(Mn)Te material is paramagnetic and therefore requires an external field, while the magnetically doped Bi_2Se_3 film was proposed to be

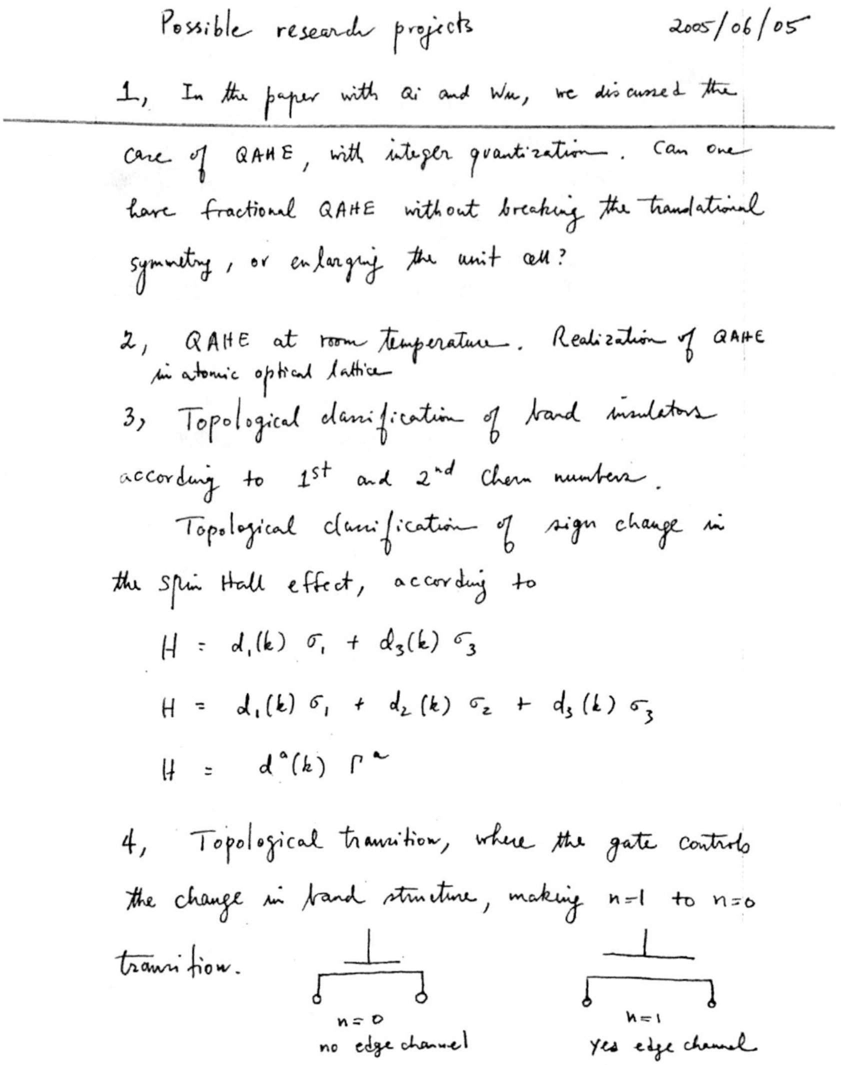

Possible research projects   2005/06/05

1, In the paper with Qi and Wu, we discussed the

case of QAHE, with integer quantization. Can one

have fractional QAHE without breaking the translational

symmetry, or enlarging the unit cell?

2, QAHE at room temperature. Realization of QAHE
in atomic optical lattice

3, Topological classification of band insulators

according to 1st and 2nd Chern numbers.

  Topological classification of sign change in

the spin Hall effect, according to

$$H = d_1(k)\, \sigma_1 + d_3(k)\, \sigma_3$$

$$H = d_1(k)\, \sigma_1 + d_2(k)\, \sigma_2 + d_3(k)\, \sigma_3$$

$$H = d^a(k)\, \Gamma^a$$

4, Topological transition, where the gate controls

the change in band structure, making $n=1$ to $n=0$

transition.

$n = 0$
no edge channel

$n = 1$
yes edge channel

Fig. 38.   Shoucheng's note on a list of possible research projects related to QAH effect in June 5, 2005. The first project is on fractional QAH effect, which became an active research field later in 2011. The second project of QAH effect at room temperature is still a major goal of the field.

ferromagnetic, which was experimentally realized in 2013 [18]. More recently, Shoucheng and his collaborators also identified a new family of intrinsic magnetic topological materials, $MnBi_2Te_4$, for the realization of the QAH effect [76]. The quantization of Hall resistance in this system has also been experimentally observed [77].

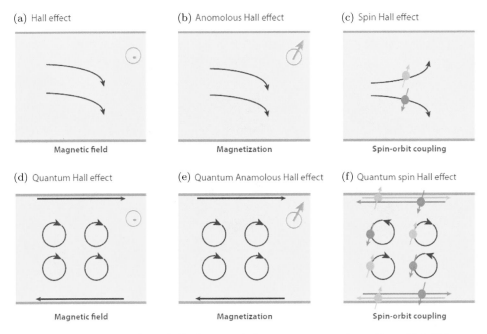

Fig. 39.   Table of three Hall effects and their quantum versions. From Ref. [24].

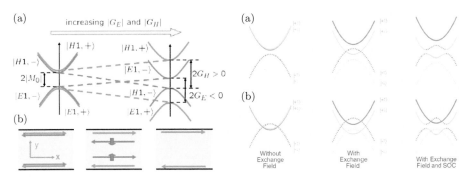

Fig. 40.   Band inversion mechanism for QAH effect in Mn doped HgTe (left) and magnetically doped $Bi_2 Se_3$ film. From Ref. [12] and Ref. [30].

8.2   *Experimental realization*

The QAH effect in magnetically doped Bi_2Se_3 family TI was first realized in Cr-doped $(Bi,Sb)_2Te_3$ thin films in Qikun Xue's group in Tsinghua University [18]. Shoucheng played an important role in this collaboration. Since then, the QAH state has been firmly verified in more and more studies. Many unique features of the QAH state, such as zero Hall plateau [78],

Fig. 41.   Qikun Xue and Shoucheng in a discussion at Stanford campus, 2014.

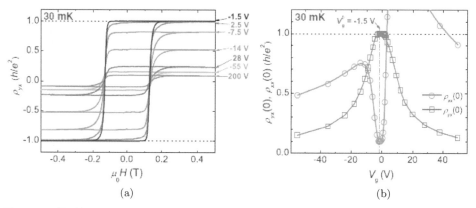

(a)   (b)

Fig. 42.   (Left) Hall resistivity vs magnetic field. (Right) Gate voltage dependence of Hall resistivity and longitudinal resistivity. From Ref. [18].

universal scaling behavior at the plateau transition [79], QAH state with higher plateaus [80], and anomalous edge transport due to coexisting chiral and helical modes [81], were studied by Shoucheng's group. These works guided the rapid development of this field and many of the theoretical predictions have been demonstrated in experiments [82–86].

9   Topological superconductors and more

9.1   *Topological superconductors*

The Bardeen–Cooper–Schrieffer (BCS) mean-field theory of superconductivity describes quasiparticles in a similar way as electrons in band insulators.

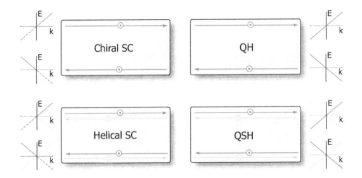

Fig. 43. Comparison between 2D time-reversal invariant TSC and QSH, from Ref. [23]

When the superconductor has a full gap, the classification is very similar to that of insulators, which suggested the concept of topological superconductors (TSC). (This is analogous to how Ettore Majorana generalized the Dirac equation to his Majorana equation in 1937.) Historically, the first topologically nontrivial superconductors discovered were 2D p+ip superconductors [87] and 1D p-wave superconductors [88].

Following the discovery of time reversal invariant (TRI) TI, new TSC with time reversal symmetry were proposed in 2008–2009, by independent works from Shoucheng's group and other groups [23,72,73] Interestingly, the B phase of ^{3}He superfluid, known from many years ago, was proposed as a 3D TRI topological superfluid.

9.2 *2D chiral TSC from QAH*

The 2D p+ip TSC (also known as chiral TSC) is an interesting system for quantum computation, because of non-Abelian statistics of vortices with Majorana zero modes. A candidate material for chiral TSC is Sr_2RuO_4, but it has not been confirmed. In 2010, Shoucheng's group proposed a new mechanism for realizing chiral TSC, by making use of QAH effect [1]. The proposal is that in general a QH phase transition will be broadened into a TSC phase when superconducting proximity effect is introduced. If realized, this proposal also enables new way to probe TSC through the QAH edge states. Shoucheng's group proposed that the chiral Majorana fermions yield a half QH plateau in a hybrid device of QAH and TSC [32,89]. Experimental evidence of half QH plateau was reported in 2017 [90], although the interpretation remains controversial [91,92]. Based on this proposal, in 2018 Shoucheng's group further proposed the possibility of realizing non-Abelian quantum gates with chiral Majorana fermions [11].

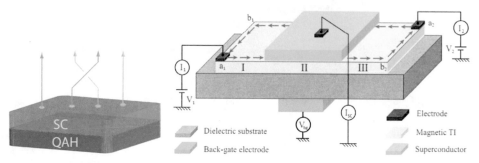

Fig. 44. Shoucheng's proposals for 2D chiral TSC [1] and device for detecting chiral Majorana fermion [32].

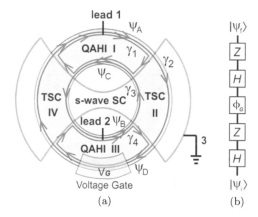

Fig. 45. A single qubit quantum gate of chiral Majorana fermions [11].

10   Other areas of condensed matter physics

Besides the directions discussed so far, Shoucheng also conducted research in a wide variety of other topics in physics, such as topological semimetals, interacting topological states, Fe-pnictide superconductors, the role of symmetry in quantum Monte Carlo method, etc. For example, in Ref. [42], Shoucheng and Congjun Wu proposed a new idea in dealing with the long-standing sign problem of quantum Monte Carlo method. The latter is a widely used numerical method in quantum many-body physics and statistical mechanics, which allows an efficient sampling of a probability distribution. A generic quantum problem cannot be reduced to evaluating the expectation value of simple physical quantity in a probability distribution, due to the complex phase of the wavefunction. This is known as the sign problem. Ref. [42] pointed out how the sign problem can be avoided in some systems by making use of discrete symmetries such as time reversal symmetry and particle-hole symmetry. This method has been applied to a wide variety of problems (see e.g. Z.-C. Wei, Congjun Wu, Yi Li, Shiwei Zhang, and T. Xiang Phys. Rev. Lett. 116, 250601) As another example, at the early stage of the discovery of Fe-pnictide high-temperature superconductors, in Ref. [40], Shoucheng and collaborators proposed the first minimal model for this system, which has influenced a lot of later works in this direction.

Shoucheng has also collaborated with high-energy physicists on interdisciplinary problems related to topological physics. For example, in Ref. [38] in collaboration with Edward Witten and Xiao-Liang Qi, they related the physics of three-dimensional TRI TSC to anomaly inflow and axion theory studied in high energy physics. In Ref. [39], in collaboration with Biao Lian, Cumrun Vafa and Farzan Vafa, they showed that the low-energy field theory

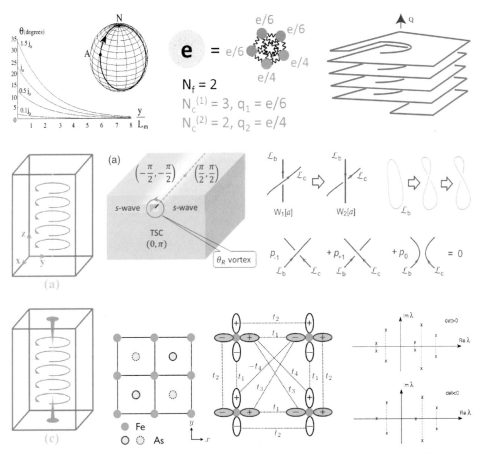

Fig. 46. Snapshots of Shoucheng's works on (starting from top left) giant-magnetoresistive materials [22], fractional TI [26, 27], Weyl semimetals and axion strings [34], Fe-pnictide topological vortex [37], axion theory of TSC [38], nodal semimetals and Chern-Simons theory [39], Fe-pnictide effective model [40], time reversal and quantum Monte Carlo [42].

of topological nodal line superconductors is equivalent to Wilson loops in the Chern-Simons theory, which relates the linking invariants of nodal lines to thermal magnetoelectric effect in a superconductor. We end this section with a selection of figures from Shoucheng's papers (Figure), which provide a glance into his diverse research interests.

11 Exploration beyond physics

Shoucheng always believed that science should have no disciplinary borders. He often told his students to spend 80% of their time on their research focus, and 20% time exploring broader directions and thinking about big questions. In recent years, Shoucheng investigated many directions beyond physics such as artificial intelligence (AI), blockchain, distributed computing, bioinformatics, and more. Just like the previous years, Shoucheng kept a collection of well-organized handwritten notes in 2018. In Appendx A, we display a few pages of this precious record, from which we can see how many different directions he was exploring actively until the last days of his life.

Shoucheng's ideas often uncovered new connections between fundamental physics principles and various other fields. In 2018, Shoucheng and collaborators used AI to study chemical compounds [20] (Q. Zhou *et. al.*, *PNAS* **115**, 6411 (2018)). They developed an "Atom2Vec" algorithm and showed that the AI did not only rediscover the periodic table but was also able to predict new chemical compounds from the properties of elements it discovered.

Another area of Shoucheng's recent interest was blockchain technology. Blockchain (and its generalizations) is a mathematical mechanism to reach consensus on a record without requiring any centralized agency. In analogy with the concept of information entropy, Shoucheng emphasized that consensus has an intrinsic value, and envisioned that "humanity is now reaching a new era where trust and consensus is built upon math," a viewpoint that he summarized in the motto "in math we trust." Shoucheng believed that new consensus mechanisms would help to build a fairer, more efficient, and more diverse society.

Shoucheng was also interested in many other areas related to how mathematics and algorithms could improve human society. An example is the application of secure multi-party computation in bioinformatics and medical sciences. Shoucheng's son Brian Zhang suggested a common theme behind the different areas of Shoucheng's latest interests, which is "algorithms for the future."

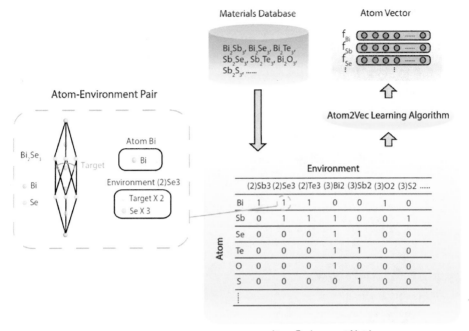

Fig. 47. Illustration of the Atom2Vec algorithm. From Ref. [20].

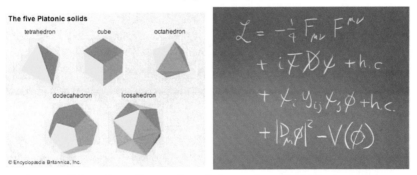

- Among all branches of human knowledge, we have the highest consensus on mathematics.
- The deepest truth of our physical universe is revealed to us by the most profound equations of theoretical physics, a beautiful application of mathematics to the natural world.
- Humanity is now reaching a new era where trust and consensus is built upon math!
- Math is the language of God.

Fig. 48. A slide from Shoucheng's talk "In Math We Trust — Foundations of Crypto-economic Science" in Sep. 2018.

12   To see a world in a grain of sand

There are many aspects of Shoucheng's legendary scientific life that we are not able to cover in our brief overview. Despite his abrupt departure, his groundbreaking scientific works; his dreams and vision for science, life, and humanity; and his inspiration to all of us will forever be preserved as an eternal monument to him. We would like to end this article by his favorite quote from William Blake's poem:

> *To see a World in a Grain of Sand*
> *And a Heaven in a Wild Flower*
> *Hold Infinity in the palm of your hand*
> *And Eternity in an hour*

Acknowledgments

We would like to thank the Zhang family members Barbara, Stephanie, Brian, and Ruth for sharing with us the precious photos and notes of Shoucheng, and for providing a lot of helpful suggestions. We would like to thank Yaroslaw Bazaliy, B. Andrei Bernevig, Leonid Pryadko, and Congjun Wu for helpful suggestions. We would like to thank all authors of this memorial volume, and all other speakers and participants of the Shoucheng Zhang Memorial Workshop.

References

[1] X.-L. Qi, T. L. Hughes and S.-C. Zhang, Chiral topological superconductor from the quantum Hall state, *Physical Review B* **82**(18), 184516 (2010).

[2] X.-L. Qi, *et al.*, Inducing a magnetic monopole with topological surface states, *Science* **323**(5918), 1184–1187 (2009).

[3] X.-L. Qi, T. L. Hughes and S.-C. Zhang, Topological field theory of time-reversal invariant insulators, *Physical Review B* **78**(19), 195424 (2008).

[4] E. Demler, W. Hanke and S.-C. Zhang, SO (5) theory of antiferromagnetism and superconductivity, *Reviews of Modern Physics* **76**(3), 909 (2004).

[5] B. A. Bernevig, T. L. Hughes and S.-C. Zhang, Quantum spin Hall effect and topological phase transition in HgTe quantum wells, *Science* **314**(5806), 1757–1761 (2006).

[6] S. Kivelson, D.-H. Lee and S.-C. Zhang, Global phase diagram in the quantum Hall effect, *Physical Review B* **46**(4), 2223 (1992).

[7] S. Murakami, N. Nagaosa and S.-C. Zhang, Spin-hall insulator, *Physical Review Letters* **93**(15), 156804 (2004).

[8] S. Murakami, N. Nagaosa and S.-C. Zhang, Dissipationless quantum spin current at room temperature, *Science* **301**(5638), 1348–1351 (2003).

[9] C. Liu, *et al.*, Quantum spin hall effect in inverted type-II semiconductors, *Physical Review Letters* **100**(23), 236601 (2008).

[10] J. P. Eisenstein and H.L. Stormer, The Fractional quantum hall effect, *Science* **248**(4962), 1510–1516 (1990).

[11] B. Lian, *et al.*, Topological quantum computation based on chiral Majorana fermions, *Proceedings of the National Academy of Sciences* **115**(43), 10938–10942 (2018).

[12] C.-X. Liu, *et al.*, Quantum anomalous Hall effect in Hg 1-y Mn y Te quantum wells, *Physical Review Letters* **101**(14), 146802 (2008).

[13] Y. Xu, *et al.*, Large-gap quantum spin Hall insulators in tin films, *Physical review letters* **111**(13), 136804 (2013).

[14] O. Zilberberg, *et al.*, Photonic topological boundary pumping as a probe of 4D quantum Hall physics, *Nature* **553**(7686), 59–62 (2018).

[15] B. A. Bernevig and S.-C. Zhang, Quantum spin Hall effect, *Physical review letters* **96**(10), 106802 (2006).

[16] M. König, *et al.*, Quantum spin Hall insulator state in HgTe quantum wells, *Science* **318**(5851), 766–770 (2007).

[17] H. Zhang, *et al.*, Topological insulators in Bi_2Se_3, Bi_2 Te_3 and Sb_2 Te_3 with a single Dirac cone on the surface, *Nature Physics* **5**(6), 438–442 (2009).

[18] C.-Z. Chang, *et al.*, Experimental observation of the quantum anomalous Hall effect in a magnetic topological insulator, *Science* **340**(6129), 167–170 (2013).

[19] A. Damascelli, Z. Hussain and Z.-X. Shen, Angle-resolved photoemission studies of the cuprate superconductors, *Reviews of Modern Physics* **75**(2), 473–541 (2003).

[20] Q. Zhou, *et al.*, Learning atoms for materials discovery, *Proceedings of the National Academy of Sciences* **115**(28), E6411–E6417 (2018).

[21] S. Chadov, *et al.*, Tunable multifunctional topological insulators in ternary Heusler compounds, *Nature Materials* **9**(7), 541–545 (2010).

[22] Y. B. Bazaliy, B. A. Jones and S.-C. Zhang, Modification of the Landau-Lifshitz equation in the presence of a spin-polarized current in colossal- and giant-magnetoresistive materials, *Physical Review B* **57**(6), R3213–R3216 (1998).

[23] X.-L. Qi, *et al.*, Time-reversal-invariant topological superconductors and superfluids in two and three dimensions, *Physical Review Letters* **102**(18), 187001 (2009).

[24] C.-X. Liu, S.-C. Zhang and X.-L. Qi, The quantum anomalous hall effect: theory and experiment, *Annual Review of Condensed Matter Physics* **7**(1), 301–321 (2016).

[25] J. D. Koralek, *et al.*, Emergence of the persistent spin helix in semiconductor quantum wells, *Nature* **458**(7238), 610–613 (2009).

[26] J. Maciejko, *et al.*, Fractional topological insulators in three dimensions, *Physical Review Letters* **105**(24), 246809 (2010).

[27] S.-C. Zhang, A unified theory based on SO(5) symmetry of superconductivity and antiferromagnetism, *Science* **275**(5303), 1089–1096 (1997).

[28] J. Wunderlich, *et al.*, Experimental observation of the spin-hall effect in a two-dimensional spin-orbit coupled semiconductor system, *Physical Review Letters* **94**(4), 047204 (2005).

[29] Y. Xia, *et al.*, Observation of a large-gap topological-insulator class with a single Dirac cone on the surface, *Nature Physics* **5**(6), 398–402 (2009).

[30] R. Yu, *et al.*, Quantized anomalous Hall effect in magnetic topological insulators, *Science* **329**(5987), 61–64 (2010).

[31] C. L. Kane and E. J. Mele, Quantum spin Hall effect in graphene, *Physical Review Letters* **95**(22), 226801 (2005).

[32] J. Wang, *et al.*, Chiral topological superconductor and half-integer conductance plateau from quantum anomalous Hall plateau transition. *Physical Review B* **92**(6), 064520 (2015).

[33] Y. Chen, *et al.*, Experimental realization of a three-dimensional topological insulator, Bi_2Te_3, *Science* **325**(5937), 178–181 (2009).

[34] Z. Wang and S.-C. Zhang, Chiral anomaly, charge density waves, and axion strings from Weyl semimetals, *Physical Review B* **87**(16), 161107 (2013).

[35] S.-C. Zhang, A unified theory based on SO (5) symmetry of superconductivity and antiferromagnetism, *Science* **275**(5303), 1089–1096 (1997).

[36] L. Wu, *et al.*, Quantized Faraday and Kerr rotation and axion electrodynamics of a 3D topological insulator, *Science* **354**(6316), 1124–1127 (2016).

[37] G. Xu, *et al.*, Topological Superconductivity on the Surface of Fe-Based Superconductors. *Physical Review Letters* **117**(4), 047001 (2016).

[38] X.-L. Qi, E. Witten and S.-C. Zhang, Axion topological field theory of topological superconductors, *Physical Review B* **87**(13), 134519 (2013).

[39] B. Lian, *et al.*, Chern-Simons theory and Wilson loops in the Brillouin zone, *Physical Review B* **95**(9), 094512 (2017).

[40] S. Raghu, *et al.*, Minimal two-band model of the superconducting iron oxypnictides, *Physical Review B* **77**(22), 220503 (2008).

[41] D. C. Tsui, H. L. Stormer and A. C. Gossard, Two-dimensional magnetotransport in the extreme quantum limit, *Physical Review Letters* **48**(22), 1559–1562 (1982).

[42] C. Wu and S.-C. Zhang, Sufficient condition for absence of the sign problem in the fermionic quantum Monte Carlo algorithm, *Physical Review B* **71**(15), 155115 (2005).

[43] R. B. Laughlin, Anomalous Quantum Hall Effect: An Incompressible Quantum Fluid with Fractionally Charged Excitations, *Physical Review Letters* **50**(18), 1395–1398 (1983).

[44] S. C. Zhang, T. H. Hansson and S. Kivelson, Effective-field-theory model for the fractional quantum Hall effect, *Physical Review Letters* **62**(1), 82 (1989).

[45] C. N. Yang and S. Zhang, SO_4 symmetry in a Hubbard model, *Modern Physics Letters B* **4**(11), 759–766 (1990).

[46] E. Demler and S.-C. Zhang, Theory of the resonant neutron scattering of high-T c superconductors, *Physical Review Letters* **75**(22), 4126 (1995).

[47] S. Rabello, *et al.*, Microscopic electron models with exact SO (5) symmetry, *Physical Review Letters* **80**(16), 3586 (1998).

[48] C. Wu, J.-p. Hu and S.-c. Zhang, Exact SO (5) symmetry in the spin-3/2 fermionic system, *Physical Review Letters* **91**(18), 186402 (2003).

[49] S. Taie, *et al.*, Realization of a SU (2) × SU (6) system of fermions in a cold atomic gas, *Physical Review Letters* **105**(19), 190401 (2010).

[50] S.-C. Zhang and J. Hu, A four-dimensional generalization of the quantum Hall effect, *Science* **294**(5543), 823–828 (2001).

[51] M. Lohse, *et al.*, Exploring 4D quantum Hall physics with a 2D topological charge pump, *Nature* **553**(7686), 55–58 (2018).

[52] J. Sinova, *et al.*, Universal intrinsic spin Hall effect, *Physical Review Letters* **92**(12), 126603 (2004).

[53] B. A. Bernevig, J. Orenstein and S.-C. Zhang, Exact SU (2) symmetry and persistent spin helix in a spin-orbit coupled system, *Physical Review Letters* **97**(23), 236601 (2006).

[54] F. D. M. Haldane, Model for a Quantum Hall Effect without Landau Levels: Condensed-Matter Realization of the "Parity Anomaly", *Physical Review Letters* **61**(18), 2015–2018 (1988).

[55] C. L. Kane and E. J. Mele, Z_2 Topological Order and the Quantum Spin Hall Effect, *Physical Review Letters* **95**(14), 146802 (2005).

[56] C. Wu, B. A. Bernevig and S.-C. Zhang, Helical Liquid and the Edge of Quantum Spin Hall Systems, *Physical Review Letters* **96**(10), 106401 (2006).

[57] C. Xu and J.E. Moore, Stability of the quantum spin Hall effect: Effects of interactions, disorder, and Z_2 topology, *Physical Review B* **73**(4), 045322 (2006).

[58] J. E. Moore and L. Balents, Topological invariants of time-reversal-invariant band structures, *Physical Review B* **75**(12), 121306 (2007).

[59] L. Fu, C. L. Kane and E. J. Mele, Topological Insulators in Three Dimensions. *Physical Review Letters* **98**(10), 106803 (2007).

[60] R. Roy, Topological phases and the quantum spin Hall effect in three dimensions, *Physical Review B* **79**(19), 195322 (2009).

[61] L. Fu and C. L. Kane, Topological insulators with inversion symmetry, *Physical Review B* **76**(4), 045302 (2007).

[62] D. Hsieh, *et al.*, A topological Dirac insulator in a quantum spin Hall phase, *Nature* **452**(7190), 970–974 (2008).

[63] H. Zhang, *et al.*, Topological insulators in Bi_2Se_3, Bi_2Te_3 and Sb_2Te_3 with a single Dirac cone on the surface, *Nature Physics* **5**(6), 438–442 (2009).

[64] X. Dai, *et al.*, Helical edge and surface states in HgTe quantum wells and bulk insulators, *Physical Review B* **77**(12), 125319 (2008).

[65] K. Kuroda, *et al.*, Experimental realization of a three-dimensional topological insulator phase in ternary chalcogenide $TlBiSe_2$, *Physical Review Letters* **105**(14), 146801 (2010).

[66] B. Yan, *et al.*, Topological insulators in filled skutterudites, *Physical Review B* **85**(16), 165125 (2012).

[67] X. Zhang, *et al.*, Actinide topological insulator materials with strong interaction, *Science* **335**(6075), 1464–1466 (2012).

[68] F. Wilczek, Two applications of axion electrodynamics, *Physical Review Letters* **58**(18), 1799–1802 (1987).

[69] J. Maciejko, *et al.*, Topological quantization in units of the fine structure constant, *Physical Review Letters* **105**(16), 166803 (2010).

[70] K. N. Okada, *et al.*, Terahertz spectroscopy on Faraday and Kerr rotations in a quantum anomalous Hall state, *Nature Communications* **7**(1), 12245 (2016).

[71] V. Dziom, *et al.*, Observation of the universal magnetoelectric effect in a 3D topological insulator, *Nature Communications* **8**(1), 15197 (2017).

[72] A. P. Schnyder, *et al.*, Classification of topological insulators and superconductors in three spatial dimensions, *Physical Review B* **78**(19), 195125 (2008).

[73] A. Kitaev, V. Lebedev and M. Feigel'man, Periodic table for topological insulators and superconductors, *AIP Conference Proceedings* **1134**(1), 22–30 (2009).

[74] X.-L. Qi, Y.-S. Wu and S.-C. Zhang, Topological quantization of the spin Hall effect in two-dimensional paramagnetic semiconductors, *Physical Review B* **74**(8), 085308 (2006).

[75] C.-X. Liu, *et al.*, Quantum anomalous Hall effect in $Hg_{1-y} Mn_y Te$ quantum wells, *Physical Review Letters* **101**(14), 146802 (2008).

[76] J. Li, *et al.*, Intrinsic magnetic topological insulators in van der Waals layered MnBi2Te4-family materials, *Science Advances* **5**(6), eaaw5685 (2019).

[77] Y. Deng, *et al.*, Quantum anomalous Hall effect in intrinsic magnetic topological insulator MnBi2Te4, *Science* **367**(6480), 895–900 (2020).

[78] J. Wang, *et al.*, Quantized topological magnetoelectric effect of the zero-plateau quantum anomalous Hall state, *Physical Review B* **92**(8), 081107 (2015).

[79] J. Wang, B. Lian and S.-C. Zhang, Universal scaling of the quantum anomalous Hall plateau transitio, *Physical Review B* **89**(8), 085106 (2014).

[80] J. Wang, *et al.*, Quantum anomalous Hall effect with higher plateaus, *Physical Review Letters* **111**(13), 136801 (2013).

[81] J. Wang, *et al.*, Anomalous edge transport in the quantum anomalous Hall state, *Physical Review Letters* **111**(8), 086803 (2013).

[82] Y. Feng, *et al.*, Observation of the zero Hall plateau in a quantum anomalous Hall insulator, *Physical Review Letters* **115**(12), 126801 (2015).

[83] X. Kou, *et al.*, Metal-to-insulator switching in quantum anomalous Hall states, *Nature Communications* **6**(1), 1–8 (2015).

[84] C.-Z. Chang, *et al.*, Observation of the quantum anomalous Hall insulator to Anderson insulator quantum phase transition and its scaling behavior, *Physical Review Letters* **117**(12), 126802 (2016).

[85] Y.-F. Zhao, *et al.*, Tuning Chern Number in Quantum Anomalous Hall Insulators. arXiv preprint arXiv:2006.16215, 2020.

[86] C.-Z. Chang, *et al.*, Zero-field dissipationless chiral edge transport and the nature of dissipation in the quantum anomalous Hall state, *Physical Review Letters* **115**(5), 057206 (2015).

[87] N. Read and D. Green, Paired states of fermions in two dimensions with breaking of parity and time-reversal symmetries and the fractional quantum Hall effect, *Physical Review B*, **61**(15), 10267–10297 (2000).

[88] A. Y. Kitaev, Unpaired Majorana fermions in quantum wires, *Physics-Uspekhi*, **44**(10S), 131–136 (2001).

[89] S. B. Chung, *et al.*, Conductance and noise signatures of Majorana backscattering, *Physical Review B* **83**(10), 100512 (2011).

[90] Q. L. He, *et al.*, Chiral Majorana fermion modes in a quantum anomalous Hall insulator–superconductor structure, *Science* **357**(6348), 294–299 (2017).

[91] M. Kayyalha, *et al.*, Absence of evidence for chiral Majorana modes in quantum anomalous Hall-superconductor devices, *Science* **367**(6473), 64–67 (2020).

[92] M. Kayyalha, *et al.*, Non-Majorana origin of the half-quantized conductance plateau in quantum anomalous Hall insulator and superconductor hybrid structures, arXiv preprint arXiv:1904.06463, (2019).

Appendix A. Shoucheng's notes in 2018

In this appendix, we include the snapshot of a few pages from Shoucheng's handwritten notes in 2018 on various topics. These are a small part of the many topics that he was learning about and/or brainstorming about. Starting from the first page, it covers the following topics:

1. CRISPR (clustered regularly interspaced short palindromic repeats, CAS9), which is a simple powerful tool for editing genomes.

2. Vector clock and causality, which is a mathematical data structure for determining the partial ordering of events (i.e., determining time) in distributed systems.

3. CAS12 based molecular diagnosis, which is a method similar to CAS9 for editing genomes.

4. DNA finger printing, the technique for identifying an individual from a sample of DNA by looking at unique patterns in their DNA.

5. DAG (directed acyclic graph), which is a mathematical concept in graph theory. Shoucheng was interested in its application in the generalization of the block chain.

6. A simple review of ZKP (zero knowledge proofs), which is a method by which one can prove to another person that he/she knows some knowledge without conveying any information apart from the fact that they know the knowledge.

7. Maxwell's demon. A thought experiment proposed by James C. Maxwell about a hypothetical way of violating the second law of thermodynamics. The note is about one way of understanding the resolution of the paradox.

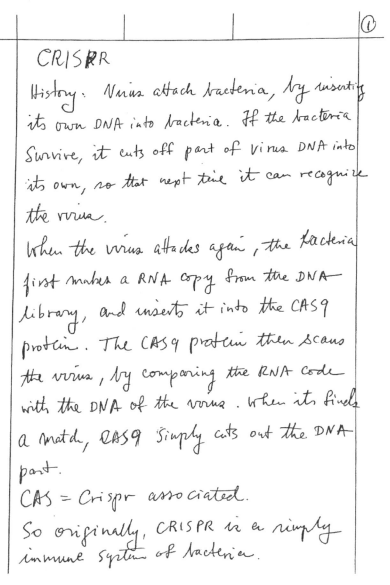

Fig. 49. Shoucheng's note on CRISPER (clusters of regularly interspaced short palindromic repeats)

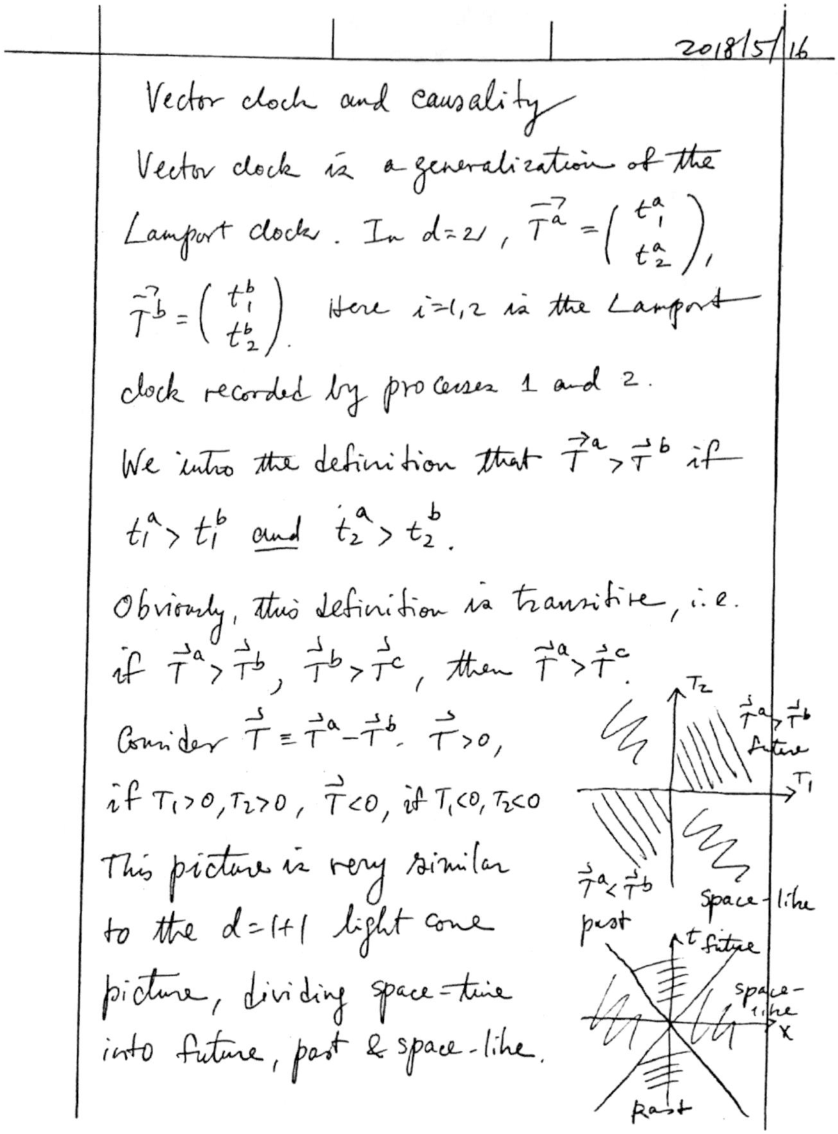

2018/5/16

Vector clock and causality

Vector clock is a generalization of the Lamport clock. In $d=2$, $\vec{T}^a = \begin{pmatrix} t_1^a \\ t_2^a \end{pmatrix}$, $\vec{T}^b = \begin{pmatrix} t_1^b \\ t_2^b \end{pmatrix}$. Here $i=1,2$ is the Lamport clock recorded by processes 1 and 2.

We intro the definition that $\vec{T}^a >_{>} \vec{T}^b$ if $t_1^a > t_1^b$ and $t_2^a > t_2^b$.

Obviously, this definition is transitive, i.e. if $\vec{T}^a >_{>} \vec{T}^b$, $\vec{T}^b >_{>} \vec{T}^c$, then $\vec{T}^a >_{>} \vec{T}^c$.

Consider $\vec{T} = \vec{T}^a - \vec{T}^b$. $\vec{T} > 0$, if $T_1 > 0, T_2 > 0$, $\vec{T} < 0$, if $T_1 < 0, T_2 < 0$

This picture is very similar to the $d = 1+1$ light cone picture, dividing space-time into future, past & space-like.

Fig. 50.   Shoucheng's note on vector clock and causality.

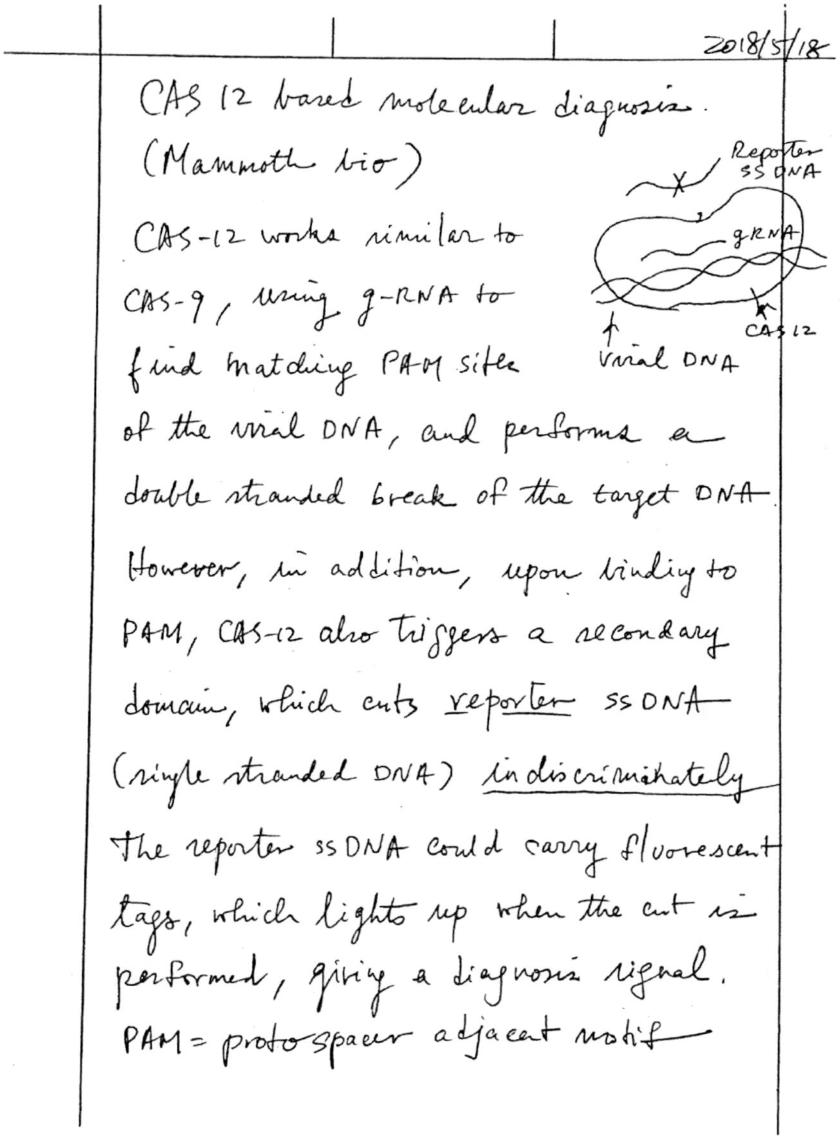

2018/5/18

CAS 12 based molecular diagnosis.
(Mammoth bio)

CAS-12 works similar to
CAS-9, using g-RNA to
find matching PAM sites
of the viral DNA, and performs a
double stranded break of the target DNA.
However, in addition, upon binding to
PAM, CAS-12 also triggers a secondary
domain, which cuts <u>reporter</u> ss DNA
(single stranded DNA) <u>indiscriminately</u>
The reporter ss DNA could carry fluorescent
tags, which lights up when the cut is
performed, giving a diagnosis signal.
PAM = protospacer adjacent motif

Fig. 51.   Shoucheng's note on CAS12 based molecular diagnosis.

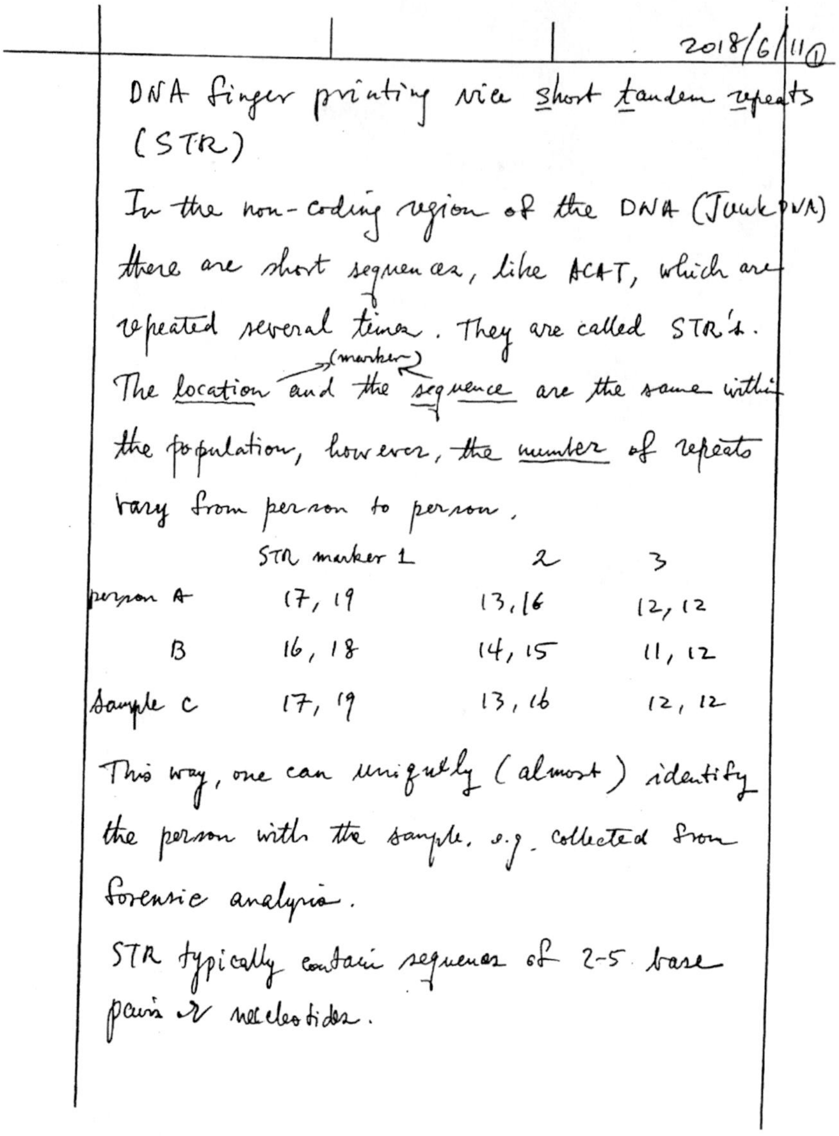

DNA finger printing via short tandem repeats (STR)

In the non-coding region of the DNA (Junk DNA) there are short sequences, like ACAT, which are repeated several times. They are called STR's. The location and the sequence (marker) are the same within the population, however, the number of repeats vary from person to person.

	STR marker 1	2	3
person A	17, 19	13, 16	12, 12
B	16, 18	14, 15	11, 12
Sample C	17, 19	13, 16	12, 12

This way, one can uniquely (almost) identify the person with the sample. e.g. collected from forensic analysis.

STR typically contain sequences of 2-5 base pairs of nucleotides.

Fig. 52. Shoucheng's note on DNA finger printing.

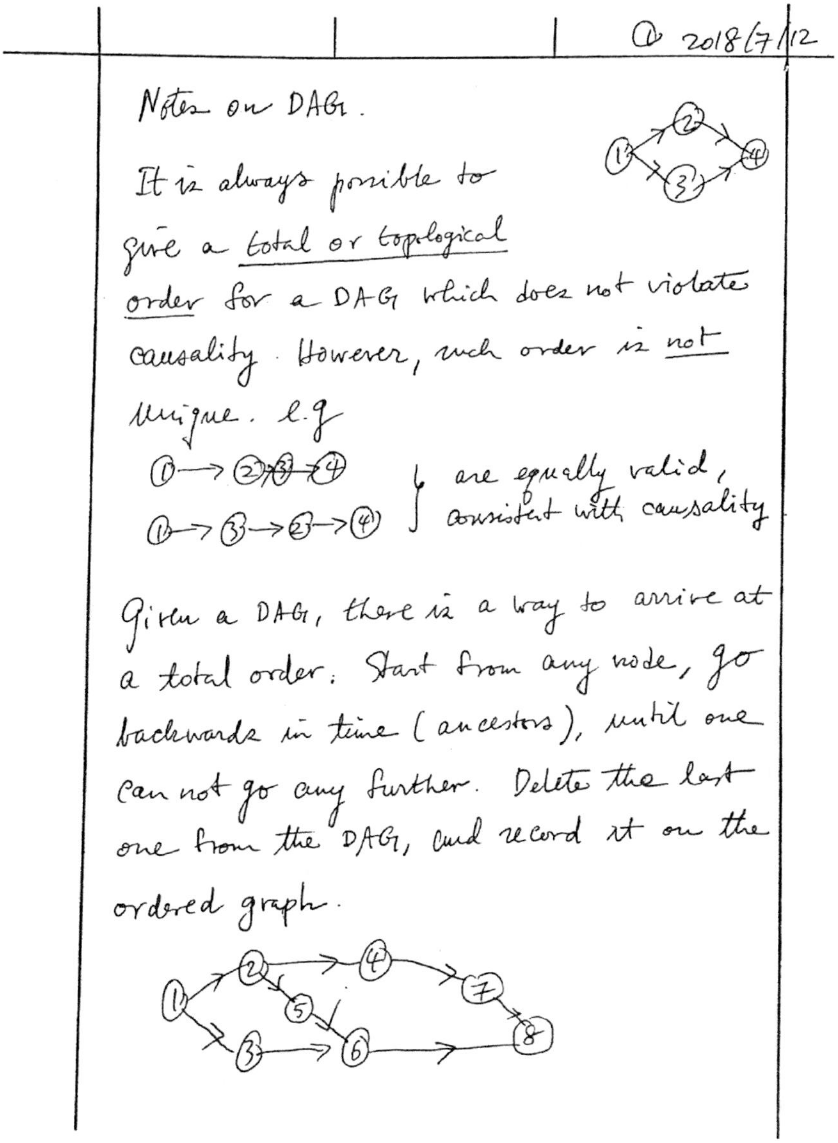

Notes on DAG.

It is always possible to give a total or topological order for a DAG which does not violate causality. However, such order is not unique. e.g

①→②⟍③ ④ } are equally valid,
①→③→②→④ } consistent with causality.

Given a DAG, there is a way to arrive at a total order: Start from any node, go backwards in time (ancestors), until one can not go any further. Delete the last one from the DAG, and record it on the ordered graph.

Fig. 53. Shoucheng's note on DAG (directed acyclic graph) and its application in block chain.

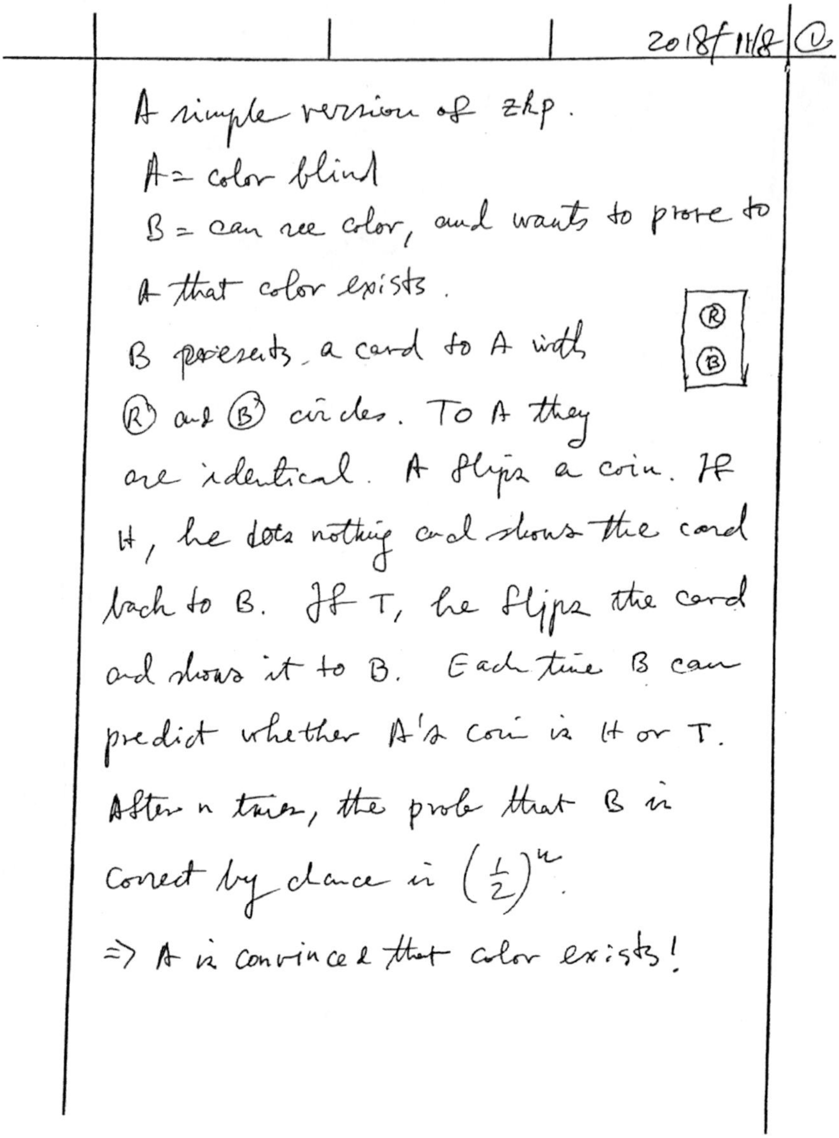
2018年11/8 ①

A simple version of ZKP.

A = color blind

B = can see color, and wants to prove to A that color exists.

B presents a card to A with Ⓡ and Ⓑ circles. To A they are identical. A flips a coin. If H, he does nothing and shows the card back to B. If T, he flips the card and shows it to B. Each time B can predict whether A's coin is H or T. After n tries, the prob that B is correct by chance is $\left(\frac{1}{2}\right)^n$.

$\Rightarrow$ A is convinced that color exists!

Fig. 54. Shoucheng's note on zero knowledge proofs.

Maxwell's demon

We consider a container
with only one particle,
in contact with a thermal
bath at T. The demon is trying to
extract some useful work out of the
random motion of the particle.

① The demon inserts a wall at the middle.

② If the demon "knows" that the particle
is on the left, he attaches a weight on
the left side. If he knows that the
particle is on the right, he attaches the
weight on the right.

③ The wall can move without friction,
and will expand isothermally, doing work

$$W = \int A\, p(x)\, dx = \int A\, \frac{k_B T}{A X}\, dx = k_B T \ln x \Big|_L^{2L}$$

$$= k_B T \ln 2$$

Fig. 55.   Shoucheng's note on Maxwell's demon.

Chapter 1

Topological Defects in General Quantum LDPC Codes

Pak Kau Lim, Kirill Shtengel and Leonid P. Pryadko

Department of Physics & Astronomy, University of California, Riverside, CA, 92521 USA

We consider the structure of defects carrying quantum information in general quantum low-density parity-check (LDPC) codes. These generalize the corresponding constructions for topological quantum codes, without the need for locality. Relation of such defects to (generalized) topological entanglement entropy is also discussed.

1.1 Introduction

Quantum computation offers exponential algorithmic speed-up for some classically hard problems. It relies on multi-particle quantum-correlated states which are very fragile and are destroyed rapidly in the presence of errors — noise, environment, or just random control errors. Quantum error correction gives a unique way of dealing with such a fragility. Selective measurements are performed to separate just one of exponentially many possible outcomes. As a result, instead of the net error probability which grows linearly or faster with the number of qubits n and saturates rapidly, one only has to worry about the error probability per qubit, p. This leads to the threshold theorem [1, 2], basically stating that an arbitrarily long quantum computation is possible when p is below certain threshold, $p_c > 0$.

Most important class of quantum error-correcting codes are the stabilizer codes [3,4]. Of these, most studied, and closest to a practical implementation, are the surface codes [5,6]. A surface code can be viewed as the degenerate ground-state manifold of certain quantum spin model with the Hamiltonian formed by a sum of commuting terms local on a two-dimensional lattice. One of the many advantages of surface codes is the flexibility they offer in the code parameters and the structure of logical operators. To add an extra qubit, one may simply create a hole in the surface. A larger hole, well separated from other defects, offers better protection (larger minimal code distance). Pairs of such holes can be moved around to perform encoded Clifford gates, etc. [6–8].

On the other hand, a substantial disadvantage of surface codes, or any stabilizer code with generators local on a D-dimensional Euclidean lattice, is that such codes necessarily have small rates $R = k/n$ whenever the code distance d gets large [9, 10]. Here k is the number of encoded qubits and n is the block length of the code. To get a finite asymptotic rate, one needs more general quantum codes. In particular, any family of *w-bounded quantum LDPC codes* with stabilizer generators of weight not exceeding $w > 0$ and distances divergent as a logarithm or a power of n has a nonzero asymptotic error correction threshold even in the presence of measurement errors [11,12]. Several families of bounded-weight quantum LDPC codes with finite rates have been constructed. Best-known constructions are quantum hypergraph-product (qHP) and related codes [13–15], and various hyperbolic codes [16–19].

The biggest obstacle to practical use of finite-rate quantum LDPC codes is that their stabilizer generators must include far separated qubits, regardless of the qubit layout in a D-dimensional space [9, 10]. Error correction requires frequent measurement of all stabilizer generators, and measuring such non-local generators is not practical if the hardware only allows local measurements. Nevertheless, there is a question of whether other advantages of surface codes, e.g., the ability to perform protected Clifford gates by code deformations, can be extended to more general quantum LDPC codes.

Such a construction generalizing the surface-code defects and gates by code deformations to the family of qHP codes [13] has been recently proposed by Krishna and Poulin [20]. However, their defect construction is very specific to qHP codes. Second, Krishna and Poulin do not discuss the distance of the defect codes they construct, even though it is important for the accuracy of the resulting gates. Indeed, since gates by code deformation are relatively slow, the distance has to be large enough to suppress logical errors.

The purpose of this work is to give a general defect construction applicable to any stabilizer code. In the simplest form, one may just remove a stabilizer generator which produces an additional logical qubit, $k \to k+1$. However, the distance d' of such a code will not exceed the maximum stabilizer generator weight, $d' \leq w$. Given a degenerate quantum LDPC code with the stabilizer generator weights bounded by w and a distance $d > w$, we would actually like to construct a related code encoding more qubits but retaining degeneracy, i.e., with a distance $d' > w$. We propose a three-step defect construction: remove qubits in an erasable region to obtain a subsystem code, do gauge-fixing to obtain a stabilizer code with some generators of weight exceeding w, and promote one or more such generators of the resulting code to logical operators. The choice of the gauge-fixing prescription is easier in the case of Calderbank, Shor, and Steane (CSS) codes [21, 22], which makes the construction more explicit. For such codes, with some additional assumptions, we give a lower bound on the distance of the defect code. This shows that defect codes with unbounded distances can be constructed, as is also the case with surface codes.

An interesting and a rather unexpected application of this analysis is the relation of qubit-carrying capacity of a defect to its (generalized) topological entanglement entropy [23–25] (TEE), denoted γ. Namely, a degenerate defect code with distance $d' > w$ can only be created when $\gamma > 0$. Further, when distance d' is large, the TEE γ acquires stability: it remains nonzero whenever the defect is deformed within certain bounds.

1.2 Defect construction

Generally, an n-qubit quantum code is a subspace of the n-qubit Hilbert space $\mathbb{H}_2^{\otimes n}$. A quantum $[[n, k, d]]$ stabilizer code is a 2^k-dimensional subspace $\mathcal{Q} \subseteq \mathbb{H}_2^{\otimes n}$ specified as a common $+1$ eigenspace of all operators in an Abelian *stabilizer* group $\mathcal{S} \in \mathcal{P}_n$, $-1 \notin \mathcal{S}$, where $\mathcal{P}_n$ denotes the n-qubit Pauli group generated by tensor products of single-qubit Pauli operators. The stabilizer is typically specified in terms of its generators, $\mathcal{S} = \langle S_1, \dots, S_r \rangle$. If the number of independent generators is $r \equiv \operatorname{rank} \mathcal{S}$, the code encodes $k = n - r$ qubits. The weight of a Pauli operator is the number of qubits that it affects. The distance d of a quantum code is the minimum weight of a Pauli operator $L \in \mathcal{P}_n$ which commutes with all operators from the stabilizer $\mathcal{S}$, but is not a part of the stabilizer, $L \notin \mathcal{S}$. Such operators act nontrivially in the code and are called logical operators.

An n-qubit CSS stabilizer code $\mathcal{Q} \equiv \mathtt{CSS}(P, Q)$ is specified in terms of two n-column binary stabilizer generator matrices $H_X \equiv P$ and $H_Z \equiv Q$. Rows of the matrices correspond to stabilizer generators of X- and Z-type, respectively, and the orthogonality condition $PQ^T = 0$ is required to ensure commutativity. The code encodes $k = n - \operatorname{rank} P - \operatorname{rank} Q$ qubits, and has the distance $d = \min(d_X, d_Z)$,

$$d_X = \min_{b \in \mathcal{C}_Q^\perp \backslash \mathcal{C}_P} \operatorname{wgt}(b), \quad d_Z = \min_{c \in \mathcal{C}_P^\perp \backslash \mathcal{C}_Q} \operatorname{wgt}(c). \tag{1.1}$$

Here $\mathcal{C}_Q \in \mathbb{F}_2^{\otimes n}$ is the binary linear code (linear space) generated by the rows of Q, and $\mathcal{C}_Q^{\perp}$ is the corresponding dual code formed by all vectors in $\mathbb{F}_2^{\otimes n}$ orthogonal to the rows of Q. Matrix Q is the parity-check matrix of the code $\mathcal{C}_Q^{\perp}$. A generating matrix of $\mathcal{C}_Q^{\perp}$, Q^*, has rank $Q^* = n - \text{rank} Q$ and is called *dual* to Q. Also, if $V = \{1, \ldots, n\}$ is the set of indices and $B \subset V$ its subset, for any vector $b \in \mathbb{F}_2^{\otimes n}$, we denote $b[B]$ the corresponding *punctured* vector with positions outside B dropped. Similarly, $Q[B]$ (with columns outside of B dropped) generates the code $\mathcal{C}_Q$ *punctured* to B. We will also use the notion of a binary code $\mathcal{C}$ *shortened* to B, which is formed by puncturing only vectors in $\mathcal{C}$ supported inside B,

$$\text{Code } \mathcal{C} \text{ shortened to } B = \{c[B] : c \in \mathcal{C} \wedge \text{supp}(c) \in B\}.$$

We will denote Q_B a generating matrix of the code $\mathcal{C}_Q$ shortened to B. If G and $H = G^*$ is a pair of mutually dual binary matrices, i.e., $GH^T = 0$ and $\text{rank} G + \text{rank} H = n$, then H_B is a parity-check matrix of the punctured code $\mathcal{C}_{G[B]}$, and [26]

$$\text{rank} G[B] + \text{rank} H_B = |B|. \tag{1.2}$$

The distance d of a linear code $\mathcal{C}$ is the minimal Hamming weight of a nonzero vector in $\mathcal{C}$. In general puncturing reduces the code distance. More precisely, if d and d' are the distances of the original and the punctured code, respectively, they satisfy $d - |A| \le d' \le d$, where $A = V \backslash B$ is the complement of B. On the other hand, the minimum distance d'' of a shortened code is not smaller than that of the original code, $d'' \ge d$.

For a quantum code, if A is a set of qubits and $B = V \backslash A$ its complement, the stabilizer group $\mathcal{S}$ can also be punctured to B, by dropping all positions outside B. With the exception of certain special cases [27, 28], the resulting group $\mathcal{G} \equiv \mathcal{S}[B]$ will not be Abelian, and can be viewed as a gauge group of a subsystem code [29, 30] called the *erasure* code. A stabilizer code can be obtained by removing some of the generators from $\mathcal{G}$ to make it Abelian; such a procedure is called gauge-fixing. In the case of a CSS code with stabilizer generator matrices $H_X = P$ and $H_Z = Q$, the punctured group has generators $P[B]$ and $Q[B]$, while a gauge-fixed stabilizer code can be obtained, e.g., by replacing punctured matrix $Q[B]$ with the corresponding shortened matrix, Q_B. This latter construction can be viewed as a result of measuring qubits outside B in the X-basis. Qubits in an erasable set A can be removed without destroying quantum information. In this case, according to the cleaning lemma [10], the logical operators of the original code can all be chosen with the support outside A. From here, with the help of Eqs. (1.1) and (1.2), we obtain the following statement (see Appendix A for all proofs).

Statement 1. *Consider a CSS code $Q \equiv \mathtt{CSS}(P, Q)$ on qubit set V of cardinality $|V| = n$, encoding k qubits and with the CSS distances d_X, d_Z. Let $A \subset V$ be an erasable in Q set of qubits, and $B \equiv V \setminus A$ its complement. Then, the length-$|B|$ code $Q' \equiv \mathtt{CSS}(P[B], Q_B)$ encodes the same number of qubits, $k' = k$, and has the CSS distances d'_X, d'_Z such that:*

$$d_X - |A| \le d'_X \le d_X, \quad d'_Z \ge d_Z. \tag{1.3}$$

The statement about the number of encoded qubits is true in general: an erasure code and any of the corresponding gauge-fixed codes encode the same number of qubits as the original code as long as the set A of removed qubits is erasable. (And, of course, we want to stick to erasable sets since we do not want to lose quantum information). To construct a code that encodes $k'' > k$ qubits, it is not sufficient to just remove some qubits, one has to also remove some group generators. If we do not care about the weight of stabilizer generators and start with a generic stabilizer code, a code with a decent distance may be obtained simply by dropping one of the existing stabilizer generators. Our general construction below is focused on quantum LDPC codes with weight-limited stabilizer generators:

Construction 1. *Given an original $[[n, k, d]]$ degenerate code with stabilizer generator weights bounded by some $w < d$, in order to create a degenerate "defect" code with $k' > k$ and $d' > w$, (**i**) remove some qubits in an erasable set, (**ii**) gauge fix the resulting subsystem code, and then (**iii**) drop one or more stabilizer generators with weights bigger than w.*

The gauge group $\mathcal{G} = \mathcal{S}[B]$ of the erasure code in step (**i**) has generators of weights w or smaller; generators of weight greater than w are obtained after gauge fixing in step (**ii**). This construction does not guarantee whether we get a degenerate code or not. Below, with the help of some additional assumptions, we prove several inequalities that guarantee the existence of not only degenerate defect codes with $d' > w$, but also highly-degenerate defect codes with unbounded distances.

1.3 Distance bounds for a defect in a CSS code

First, let us get general expressions for the distances d'_X, d'_Z of a CSS code with a removed Z-type generator. Given the original code $\mathtt{CSS}(P, Q)$, we choose a linearly-independent row of Q, u_0, as the additional type-Z logical operator, and denote Q' the corresponding matrix with the row dropped (and of the rank reduced by one). Denote

$$d_Z^{(0)} = \min_\alpha \text{wgt}(u_0 + \alpha Q'), \tag{1.4}$$

the minimum weight of a linear combination of u_0 with the rows of Q'. Then, Eq. (1.1) gives

$$d_Z' = \min(d_Z, d_Z^{(0)}). \tag{1.5}$$

The additional type-X logical operator has to be taken from the set of detectable errors of the original code. Specifically, it has to anticommute with the element of the stabilizer being removed, but commute with the remaining operators in the stabilizer and all logical operators of the original code. In addition to the X-type logical operators of the original code, the logical operators of the new code include all errors with the same syndrome as the chosen canonical operator. Respectively, the expression for the distance reads:

$$d_X' = \min(d_X, d_X^{(0)}), \quad d_X^{(0)} = \min_{b\,:\,u_0 b^T = 1 \wedge Q' b^T = 0} \text{wgt}(b). \tag{1.6}$$

The lower bounds constructed in the following two subsections both rely on geometry in a bipartite (Tanner) graph associated with the Z-type generator matrix $H_Z = Q$. Namely, given its row-set U (check-nodes) and column-set V (value-nodes), the Tanner graph has the union $U \cup V$ as its vertex set, and an undirected edge $(u, v) \in U \times V$ for each nonzero matrix element Q_{uv}. On a graph there is a natural notion of the distance between a pair of nodes, the number of edges in the shortest path between them; a ball $\Omega_R(u_0)$ of radius R centered around u_0 is the set of all vertices at distance R or smaller from u_0. Then, an erasable region $A = \Omega_R(u_0) \cap V$ is chosen as a set of value nodes within the radius R from a check node $u_0 \in U$, subject to the condition that a row of the shortened matrix Q_B contains u_0 in its expansion over the rows of Q.

The condition is not a trivial one, as it is actually equivalent to region A being erasable in the code $\text{CSS}(P, Q')$ formed by the original matrix $H_X \equiv P$ and the matrix Q', the original matrix Q with the row u_0 (considered linearly independent) dropped, same code as in Eqs. (1.4)–(1.6). As an equivalent but easier to check condition, one may request that row $u_0[A]$ be a linear combination of the rows of the punctured matrix $Q'[A]$ (remember that the support of u_0 is a subset of A, while u_0 is linearly independent from the rows of Q'). In addition, we use a corresponding sufficient condition as a part of lower X-distance bound in Statement 2, and formulate a related necessary condition in terms of the topological entanglement entropy associated with the defect A in Sec. 1.4.

1.3.1 *Code with locally linearly-independent generators*

We need a condition to guarantee a lower bound on the weight of the operator conjugate to the row u_0 removed from the matrix Q_B, see Eq. (1.6). Here, we will assume that the set of Z-type stabilizer generators forming the rows of the matrix $H_Z = Q$ be overcomplete. That is, there be one or more linear relations between the rows of Q, and that we start with a row u_0 which takes part in such a relation.

In the case of the toric code (or any surface code on a locally planar graph without boundaries), see Fig. 1.1(a), the linear relation is simply the statement that the sum of all rows of H_Z be zero (necessarily so since each column has weight two). Such a relation exists for any matrix with even column weights, e.g., qHPs from (ℓ, m)-regular binary codes with both ℓ and m even. Further, many such linear relations exist for CSS codes forming chain complexes of length 3 or more, e.g., the D-dimensional hyperbolic [18, 31] and higher-dimensional qHP codes [15] with $D > 2$.

Statement 2. *Given a CSS code* $\mathsf{CSS}(P, Q)$ *and a natural* R_1, *consider the bipartite Tanner graph associated with the matrix* Q, *and a ball* $W = \Omega_{2R_1}(u_0)$ *of radius* $2R_1$ *centered around the row* $u_0 \in U$. *Assume* **(a)** *that the row* u_0 *is involved in at least one linear relation with other rows of* Q, *and* **(b)** *there exists* $R_2 > R_1$ *such that all rows within radius* $2R_2$ *from the center be linearly independent of each other. Let* Q_1 *denote a full-row-rank matrix obtained from* Q *by removing some (linearly-dependent) rows outside* W. *Then weight of any* $b \in \mathbb{F}_2^{\otimes n}$ *such that the syndrome* $Q_1 b^T$ *has*

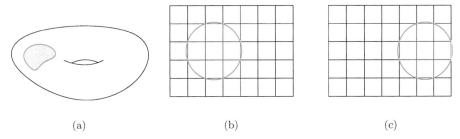

(a)   (b)   (c)

Fig. 1.1. (a) Homologically trivial hole on a torus. (b) Surface code with a smooth boundary. Removing qubits (edges) inside of the circle we get a nontrivial defect. (c) This circle contains a boundary edge with no neighboring plaquette; removing the corresponding edges we again get a trivial defect.

the only nonzero bit at the check node u_0 satisfies wgt$(b) \geq R_2$, *and for the complement $B = V \setminus A$ of any region $A \subseteq W$,* wgt$(b[B]) \geq R_2 - R_1$.

The punctured vector $b[B]$ is a representative of the new X-type codeword in the defect code CSS$(P[B], Q'_B)$, where Q' is obtained from Q_1 by removing u_0; the constructed bound gives $d_X^{(0)} \geq R_2 - R_1$ for the distance in Eq. (1.6).

We also note that additional, linearly-dependent with u_0, rows in Q need not have bounded weight, as long as on the Tanner graph they are located outside the ball W_2. The corresponding requirement is of course equivalent to any of the two conditions discussion above this subsection title, with $A = W_2 \cap V$. In the case of a surface code with smooth boundary, see Figs. 1.1(b) and 1.1(c), the extra row may be chosen as the product of all plaquette generators, with the support along the actual boundary. In such a case, the lower distance bound in Statement 2 is saturated.

1.3.2   *Stabilizer group with an expansion*

Here we construct a simple lower bound on the Z-distance of the defect, in essence, relying on the monotonicity of the distance d_Z with respect to X-basis measurement of qubits in an erasable set, see Statement 1. To make it nontrivial, we assume that Z-type stabilizer generators of the original code satisfy an expansion condition, namely, there exists an increasing real-valued function f such that a product Π_m of any m distinct generators has weight bounded by $f(m)$,

$$\text{wgt}(\Pi_m) \geq f(m), \quad f(m+1) > f(m). \tag{1.7}$$

Such a *global* condition on code generators guarantees that the boundary condition is good for the defect we are trying to construct. For example, in case of the toric code on an $L \times L$ square lattice with periodic boundary conditions, there are L^2 plaquette generators but only $L^2 - 1$ of them are independent. Namely, the product of all plaquette generators is an identity, so $m \to L^2 - m$ is a symmetry of the weight distribution. Necessarily, the function f in Eq. (1.7) has a trivial maximum, $f(L^2) \leq 0$. Respectively, a single hole in Fig. 1.1(a) has a homologically trivial boundary — meaning that it can be pushed out and eventually contracted to nothing by a sequence of single-plaquette steps. On the other hand, for a planar smooth-boundary surface code configuration as in Fig. 1.1(b), one gets $f(m)$ scaling as a perimeter of m plaquettes with a non-trivial maximum.

Generally, as one increases the set A of removed qubits, there will be rows in the shortened matrix Q_B formed as linear combinations of increasing numbers of rows of the original matrix Q. The expansion condition (1.7) with $\max_m f(m) > 0$ guarantees that the corresponding rows cannot be contracted to nothing. For example, when we remove a single qubit corresponding to a weight-ℓ column of Q, if the corresponding adjacent rows all have weights w and do not overlap (in the case of a surface code $\ell \le 2$ and w is the number of sides in the corresponding plaquette), the shortened matrix Q_B necessarily has $\kappa = \ell - 1$ rows of weight $2w - 2$. Assuming $f(2) = 2w - 2$, at any $w > 2$ this is already sufficient to guarantee the existence of a degenerate defect code with $d_Z' > w$.

With larger defects, combinations of larger numbers of rows may become necessary. If so, the expansion condition (1.7) will also guarantee that codes with Z-type distances (1.5) much greater than w can be constructed (assuming big enough original code distance d_Z).

Statement 3. *Given a code* CSS(P, Q) *and a natural* R_1, *consider the bipartite Tanner graph associated with the matrix* Q, *and a ball* $W = \Omega_{2R_1}(u_0)$ *of radius* $2R_1$ *centered around the row* $u_0 \in U$. *Denote* Q' *the matrix obtained by removing* u_0 *from* Q. *Assume* **(a)** *that* $A \equiv W \cap V$ *is erasable in the code* CSS(P, Q'), *and* **(b)** *that the set of* Z-*generators defined by the rows of matrix* Q *satisfies the expansion condition* (1.7) *with* $f(2) \ge 1$. *Then, weight of any linear combination of* u_0 *with rows of the matrix* Q' *supported on the complement* $B \equiv V \setminus A$ *satisfies* $d_Z^{(0)} \ge f(R_1)$.

Notice that the condition **(a)** here is the same as discussed above Sec. 1.3.1 title; the corresponding sufficient condition is a part of Statement 2, where any $R_2 > R_1$ will do.

1.3.3 *Defect codes with arbitrary large distances*

Notice that Statement 2 requires a linear *dependence* between generators of Q, while Statement 3 requires the expansion condition (1.7) with nontrivial f which is stronger than just linear *independence*. Nevertheless, these conditions are not necessarily incompatible. The condition in Statement 2 only needs to be satisfied for some parent code. For example, in the case of a toric code, *two* distinct holes are needed in order to create a defect code with distance $d' = \min(d_Z', d_X') > 4$. Here a linear dependence between plaquette operators can be found in the parent toric code, one hole is needed to satisfy the conditions of Statement 3, while the other one is the actual erasable set in Construction 1.

Generally, suppose we have a parent CSS code CSS(H_X, H_Z) with bounded-weight generators, sufficiently large distance, and matrix H_Z with even-weight columns so that the sum of all rows be zero. Such a pair of matrices satisfies conditions of Statement 2 but not of Statement 3. Similar to the toric code, where one needs two holes to create a single-qubit defect, here we also may need to take an erasable set A formed by two or more disjoint erasable defects, e.g., balls as in Statement 3; that the set A be erasable can be guaranteed by the union lemma (Lemma 2 in Ref. [9]). Then, one (or more if needed) balls can be used to ensure the existence of the function f in Eq. (1.7) with sufficiently large $\max_m f(m)$, while the qubits in the last remaining ball would be used as the erasable set.

Explicitly, as a parent code family, one can use, e.g., qHP codes [13] created from random matrices with even-valued row and column weights. For example, $(4,6)$-regular random matrices would do well, leading to qHP codes with asymptotically finite rates, whose CSS generator matrices have column weights 4 and 6, regular row weights $w = 10$, and $\mathcal{O}(n^{1/2})$ linear relations between the rows of generator matrices with number of nonzero coefficients in each linear relation scaling linearly with block length n of the resulting code. The distance of such parent codes grows as $\mathcal{O}(n^{1/2})$; this is sufficient to ensure that for any $d_0 > 0$ one can choose n large enough so that sufficiently large erasable balls exist to guarantee the existence of defect codes with $d' \equiv \max(d'_X, d'_Z) \geq d_0$.

1.4   Relation with topological entanglement entropy

There exists a suggestive parallel between the structure of a large-distance qubit-carrying defect we discussed, and (generalized) topological entanglement entropy (TEE) which can be associated with such a defect [23,24]. The latter may be defined in terms of the usual entanglement entropy (EE), which characterizes what happens when some of the qubits carrying a normalized quantum state $|\psi\rangle \in \mathcal{H}_2^{\otimes n}$ are erased (traced over). Namely, if the set of qubits is decomposed into A and its complement $B = V \setminus A$, one considers the binary von Neumann entropy $\Upsilon(A; B) \equiv -\operatorname{tr}_B \rho_B \log_2 \rho_B$, where the density matrix $\rho_B = \operatorname{tr}_A |\psi\rangle\langle\psi|$ is obtained by tracing over the qubits in A. The definition is actually symmetric with respect to interchanging A and B, $\Upsilon(A; B) = \Upsilon(B; A)$.

The entanglement entropy has a particularly simple form when $|\psi\rangle$ is a stabilizer state [32]. Such a state is just a stabilizer code encoding no qubits, so that its dimension is $2^0 = 1$. With $n = |V|$ total qubits, this requires a stabilizer group with n independent generators. According to Fattal *et al.* [32], the EE of any stabilizer state $|\psi\rangle \in \mathcal{Q}$ is uniquely determined by the decomposition of the stabilizer group $\mathcal{S} = \mathcal{S}_A \times \mathcal{S}_B \times \mathcal{S}_{AB}$, where nontrivial elements of subgroups $\mathcal{S}_A$ and $\mathcal{S}_B$ are supported only on A and only on B, respectively, and those of $\mathcal{S}_{AB}$ are necessarily split between A and B. Namely, rank $\mathcal{S}_{AB} = 2p$ is always even, and it is this p (or, equivalently, the number of EPR pairs split between A and B) that determines the entanglement entropy,

$$\Upsilon(A; B) = p \equiv \tfrac{1}{2} \operatorname{rank} \mathcal{S}_{AB}. \tag{1.8}$$

Given a stabilizer code $\mathcal{Q}$ with parameters $[[n, k, d]]$ and a stabilizer group $\mathcal{S}$ of rank $n - k$, a stabilizer state $|\psi\rangle \in \mathcal{Q}$ can be formed by adding any k mutually commuting logical Pauli operators to the stabilizer group. Then, if set $A \subset V$ is erasable, according to the cleaning lemma [10], we can select all logical operators with the support in $B = V \setminus A$. With the logical operators in $\mathcal{S}_B$, both $\mathcal{S}_A$ and $\mathcal{S}_{AB}$ are subgroups of the stabilizer group $\mathcal{S}$ of our original code, and the entanglement entropy is given by Eq. (1.8). The same quantity p can also be expressed in terms of the punctured stabilizer group $\mathcal{S}[B] = \mathcal{S}_B \times \mathcal{S}_{AB}[B]$ (gauge group of the subsystem erasure code), written as a product of its center, the (shortened) stabilizer group $\mathcal{S}_B$, and p pairs of canonically conjugated "gauge" qubits which generate the (punctured) subgroup $\mathcal{S}_{AB}[B]$.

In the case of a CSS code, such a decomposition exists for both X-type and Z-type subgroups of the stabilizer, e.g., $\mathcal{S}^{(Z)} = \mathcal{S}_A^{(Z)} \times \mathcal{S}_B^{(Z)} \times \mathcal{S}_{AB}^{(Z)}$, with rank $\mathcal{S}_{AB}^{(Z)} = \operatorname{rank} \mathcal{S}_{AB}^{(X)} = p$. These p independent generators are obtained from the rows of the original generator matrices that are split between A and B. In a weight-limited LDPC code, the total number of such rows, e.g., in H_Z, can be called the perimeter $L(A; B)$ of the cut. However, the number of generators of $\mathcal{S}_{AB}^{(Z)}$ can actually be smaller than $L(A; B)$ since some linear combination(s) of the generators split between A and B combined with other generators may form an element of $\mathcal{S}_A$ or $\mathcal{S}_B$. Thus, we can write EE as

$$\Upsilon(A; B) = L - \gamma, \tag{1.9}$$

with $L = L(A; B)$ the perimeter of the cut and some integer $\gamma \equiv \gamma(A; B) \geq 0$. While this expression strongly resembles the Kitaev–Preskill definition of TEE [23, 24], for now γ is just a parameter associated with the particular cut.

Let us now consider weights of generators of $\mathcal{S}_B^{(Z)}$. These correspond to the rows of Q_B. Clearly, each row of the original matrix Q may be supported in A, or in B, or be split between the two sets. Rows already supported in B can be moved directly to Q_B and preserve their original weights. Thus no more than $\gamma \geq 0$ generators of $\mathcal{S}_B^{(Z)}$ may need to have larger weights. Necessarily, if we want to construct a defect forming a degenerate code with the distance $d' > w$, the additional number of qubits is bounded by

$$\kappa \equiv k' - k \leq \gamma. \tag{1.10}$$

Thus, with $\gamma = 0$, the defect cannot support a degenerate code with $\kappa > 0$. However, whether or not a particular defect does, in fact, support $\kappa > 0$, also depends on the global structure of the code, e.g., the boundary conditions.

Now, let us imagine that we have a defect code with a sufficiently large distance d. Then, such a defect is also *stable* to small deformations, e.g., when B is changed to some B' as a result of up to $M < d$ steps, where at each step a single position is added or removed from the set. That is, our defect code retains the same number κ of additional qubits when we change the set B to a set B', $|B \triangle B'| \leq M$, where $B \triangle B' = (B \setminus B') \cup (B' \setminus B)$ is the symmetric set difference. For deformations such that $M + w < d$, the inequality $\gamma \geq \kappa$ must be satisfied in the course of deformations.

Now, TEE is normally considered a property of ground-state wavefunction of some many-body Hamiltonian, while our focus was on quantum LDPC codes with bounded-weight but not necessarily local generators. Different terms in a Hamiltonian can be viewed as generators of the code. However, in the absence of locality, why would we care about weights of terms in a quantum spin Hamiltonian?

In a physical system, multi-qubit Pauli operators may appear as terms in an n-spin quantum Hamiltonian, e.g.,

$$H_0 = -A \sum_a P_a - B \sum_b Q_b, \tag{1.11}$$

where and $A > 0$ and $B > 0$ are the coupling constants, and, to connect with our discussion of CSS codes, P_a and Q_b could be Pauli operators of X- and Z-type, respectively, specified by rows of the binary matrices P and Q. Then, if all terms in the Hamiltonian commute, i.e., $PQ^T = 0$, the ground-state space of H_0 is exactly the code with the stabilizer group generated by these operators.

Any simple spin Hamiltonian (1.11) is usually just the leading-order approximation to a real problem. Even at zero temperature, additional interaction terms are virtually always present. Such terms may break the degeneracy of the ground-state of the Hamiltonian H_0. The effect is weak if the code has a large distance, while perturbations be small and local. The standard example is the effect of an external magnetic field $\mathbf{h} = (h_x, h_y, h_z)$, which can be introduced as an additional perturbation Hamiltonian

$$H_1 = -\frac{1}{2} \sum_i (h_x X_i + h_y Y_i + h_z Z_i). \tag{1.12}$$

For a code with distance d, only a Pauli operator of weight d or larger may act within the code. Respectively, assuming the magnetic field small, degenerate perturbation theory gives the ground state subspace energy splitting scaling as $\mathcal{O}(h^d)$, where $h = |\mathbf{h}|$ is the field magnitude.

However, the code distance d gives only a part of the story. Large-weight operators appearing in H_0 make the ground-state order particularly susceptible to local perturbations such as the magnetic field. In this case, the relevant scale for the magnetic field is $Wh \sim \max(A, B)$, that is, the effect of the magnetic field may be magnified by the operator weight W. Indeed, if we start with the spin-polarized ground state of H_1, a weight-W Pauli operator will generically flip W spins, producing a state with the energy increased by $\mathcal{O}(Wh)$. The effect of such a perturbation will be small as long as the corresponding coefficient, A or B in Eq. (1.11), remains small compared to Wh. Thus, with W large, the ground state of the spin Hamiltonian H_0 gets destroyed already with very small $h \sim \max(A, B)/W$. The same estimate can be also obtained with the help of an exact operator map similar to that used by Trebst *et al.* [33].

1.5   Conclusions

To summarize, we discussed a general approach to adding logical qubits to an existing quantum stabilizer code, with the focus on quantum LDPC codes with weight-limited stabilizer generators. In short, a stabilizer generator needs to be promoted to a logical operator, which puts a bound on the distance of the obtained code in terms of the generator weight w. As in a surface code, a degenerate code can be obtained by removing some qubits in an erasable set, and gauge-fixing the resulting subsystem code in such a way as to ensure that stabilizer generators of sufficiently large weight be created. We also constructed some lower bounds on the distance of thus obtained defect codes which show that construction can in principle be used to obtain highly degenerate codes with distances much larger than w.

An interesting observation is a relation between the ability of a particular defect (erasable set of qubits) to support an additional logical qubit in a degenerate code, and a quantity analogous to TEE, γ. A degenerate defect code can be only created with $\gamma > 0$. Further, when a defect code has a large distance d', a lower bound on $\gamma > 0$ is maintained in the course of deformations, not unlike for the conventionally defined TEE.

Many open problems remain. First, our lower distance bounds are constructed by analogy with surface codes. In particular, the lower bound in Statement 2 applies only for a single qubit. In addition, we do not have good lower distance bounds for defects in non-CSS codes.

Second, the notion of generalized TEE γ in Eq. (1.9) needs to be cleaned up. Here we are working with lattice systems, not necessarily local, and the usual expansions in term of $1/L$ do not necessarily help. Further, as defined, γ certainly depends of the chosen set of generators. Redundant sets of small-weight generators imply the existence of higher homologies, as in higher-dimensional toric codes; it would be nice to be able to interpret values of γ, as, e.g., was done by Grover *et al.* in a field theory setting [25].

Third, if we start with a finite-rate family of codes, are there defects of size $|A|$ with $\gamma = \mathcal{O}(|A|)$? Coming back to defect codes, it appears that a typical defect with large γ would generically lead to an entire spectrum of operator weights in the generators of $\mathcal{S}_B$. Is there a situation when there is a large gap in this weight distribution, as in the surface codes with $\gamma = 1$, where only one high-weight operator may exist?

Appendix: All the proofs

Proof of Statement 1. The number of encoded qubits follows from the identity (1.2). Namely, the exact dual Q^* of the matrix Q can be obtained from P by adding k rows corresponding to inequivalent codewords $b \in \mathcal{C}_Q^\perp \setminus \mathcal{C}_P$. According to the cleaning lemma [10], these can be chosen with the support outside of an erasable set A. Dropping these k rows from $Q^*[B]$ recovers the punctured matrix $P[B]$ with the correct rank to ensure $k' = k$. The distance inequalities are obtained from Eq. (1.1) by considering removal of a single qubit at a time.

Proof of Statement 2. Indeed, since Q' differs from Q only by some rows outside the ball $W_2 \equiv \Omega_{2R_2}(u_0)$ which are linearly dependent with u_0, the corresponding full-matrix syndrome Qb^T must have nonzero bits outside $W_2 \cap U$. With the exception of u_0, any row in $W_2 \cap U$ must be incident on an even number of set bits in b, and there must be a continuous path on the graph from u_0 to outside W_2 formed by pairs of set bits in b (otherwise b could be separated into a pair of vectors with nonoverlapping supports, $b = b_1 + b_2$, such that $Qb_1^T = 0$ outside W_2 and $Qb_2^T = 0$ inside W_2, which would contradict the assumptions). This guarantees that at any odd distance from u_0 (up to $2R_2 - 1$), b has at least one set bit, which recovers the two lower bounds.

Proof of Statement 3. The inequality follows from the locality of row operations on the Tanner graph: a nonzero bit v in some vector $c \in \mathbb{F}_2^{\otimes n}$

can only be removed by adding a row $u \in U$ neighboring with v. The lower bound on f equivalent to linear independence of rows of Q guarantees that rows of Q' be linearly independent from u_0, thus weight must remain nonzero at every step. The condition (**a**) guarantees the existence of a linear combination in question.

Acknowledgment

This work was supported in part by the NSF Division of Physics via grant No. 1820939.

References

[1] P. W. Shor, Fault-tolerant quantum computation, in *Proc. 37th Ann. Symp. Fundamentals of Comp. Sci.*, Los Alamitos, IEEE (IEEE Computer Society Press, 1996), pp. 56–65. URL http://arxiv.org/abs/quant-ph/9605011v2.

[2] J. Preskill, Fault-tolerant quantum computation, in *Introduction to Quantum Computation*, eds. H.-K. Lo, S. Popescu and T. P. Spiller (World Scientific, 1998).

[3] D. Gottesman, *Stabilizer codes and quantum error correction*, Ph.D. thesis, Caltech (1997); URL http://arxiv.org/abs/quant-ph/9705052.

[4] A. R. Calderbank, E. M. Rains, P. M. Shor and N. J. A. Sloane, Quantum error correction via codes over GF(4), *IEEE Trans. Inform. Theory* **44**, 1369 (1998); URL http://dx.doi.org/10.1109/18.681315.

[5] A. Y. Kitaev, Fault-tolerant quantum computation by anyons, *Ann. Phys.* **303**, 2 (2003); URL http://arxiv.org/abs/quant-ph/9707021.

[6] E. Dennis, A. Kitaev, A. Landahl and J. Preskill, Topological quantum memory, *J. Math. Phys.* **43**, 4452 (2002); URL http://dx.doi.org/10.1063/1.1499754.

[7] H. Bombin and M. A. Martin-Delgado, Quantum measurements and gates by code deformation, *J. Phys. A* **42**(9), 095302 (2009); doi:10.1088/1751-8113/42/9/095302.

[8] H. Bombin, Topological subsystem codes, *Phys. Rev. A* **81**, 032301 (2010); doi:10. 1103/PhysRevA.81.032301; URL http://link.aps.org/doi/10.1103/PhysRevA.81.032 301.

[9] S. Bravyi and B. Terhal, A No-Go theorem for a two-dimensional self-correcting quantum memory based on stabilizer codes, *New J. Phys.* **11**(4), 043029 (2009); URL http://stacks.iop.org/1367-2630/11/i=4/a=043029.

[10] S. Bravyi, D. Poulin and B. Terhal, Tradeoffs for reliable quantum information storage in 2D systems, *Phys. Rev. Lett.* **104**, 050503 (2010); doi:10.1103/PhysRevLett. 104.050503; URL http://link.aps.org/doi/10.1103/PhysRevLett.104.050503.

[11] A. A. Kovalev and L. P. Pryadko, Fault tolerance of quantum low-density parity check codes with sublinear distance scaling, *Phys. Rev. A* **87**, 020304(R) (2013); doi:10.1103/PhysRevA.87.020304; URL http://link.aps.org/doi/10.1103/PhysRevA. 87.020304.

[12] I. Dumer, A. A. Kovalev and L. P. Pryadko, Thresholds for correcting errors, erasures, and faulty syndrome measurements in degenerate quantum codes, *Phys. Rev. Lett.* **115**, 050502 (2015); doi:10.1103/PhysRevLett.115.050502; URL http://link.aps.org/ doi/10.1103/PhysRevLett.115.050502.

[13] J.-P. Tillich and G. Zemor, Quantum LDPC codes with positive rate and minimum distance proportional to $\sqrt{n}$, in *Proc. IEEE Int. Symp. Inf. Theory (ISIT)* (June, 2009); pp. 799–803; doi:10.1109/ISIT.2009.5205648.

[14] A. A. Kovalev and L. P. Pryadko, Quantum Kronecker sum–product low-density parity-check codes with finite rate, *Phys. Rev. A* **88** 012311 (2013); doi:10.1103/PhysRevA.88.012311; URL http://link.aps.org/doi/10.1103/PhysRevA.88.012311.

[15] W. Zeng and L. P. Pryadko, Higher-dimensional quantum hypergraph-product codes with finite rates, *Phys. Rev. Lett.* **122**, 230501 (2019); doi:10.1103/PhysRevLett.122.230501; URL https://link.aps.org/doi/10.1103/PhysRevLett.122.230501.

[16] G. Zémor, On Cayley graphs, surface codes, and the limits of homological coding for quantum error correction, in *Proc. Coding and Cryptology: Second International Workshop, IWCC 2009*, eds. Y. M. Chee, C. Li, S. Ling, H. Wang and C. Xing, pp. 259–273 (Springer, 2009); doi:10.1007/978-3-642-01877-0_21; URL http://dx.doi.org/10.1007/978-3-642-01877-0_21.

[17] N. Delfosse, Tradeoffs for reliable quantum information storage in surface codes and color codes, in *2013 IEEE Int. Symp. Information Theory Proceedings (ISIT)*, pp. 917–921; (IEEE, 2013) pp. 917–921; doi:10.1109/ISIT.2013.6620360.

[18] L. Guth and A. Lubotzky, Quantum error correcting codes and 4-dimensional arithmetic hyperbolic manifolds, *J. Math. Phy.* **55**(8), 082202, (2014); doi:http://dx.doi.org/10.1063/1.4891487; URL http://scitation.aip.org/content/aip/journal/jmp/55/8/10.1063/1.4891487.

[19] N. P. Breuckmann and B. M. Terhal, Constructions and noise threshold of hyperbolic surface codes, *IEEE Trans. Inform. Theory* **62**(6), 3731–3744 (2016); doi:10.1109/TIT.2016.2555700.

[20] A. Krishna and D. Poulin, Fault-tolerant gates on hypergraph product codes, To be published in Phys. Rev. X (2021).

[21] A. R. Calderbank and P. W. Shor, Good quantum error-correcting codes exist, *Phys. Rev. A* **54**(2), 1098 (1996); doi:10.1103/PhysRevA.54.1098.

[22] A. M. Steane, Simple quantum error-correcting codes, *Phys. Rev. A* **54**, 4741 (1996); URL http://dx.doi.org/10.1103/PhysRevA.54.4741.

[23] M. Levin and X.-G. Wen, Detecting topological order in a ground state wave function, *Phys. Rev. Lett.* **96**, 110405 (2006); doi:10.1103/PhysRevLett.96.110405; URL https://link.aps.org/doi/10.1103/PhysRevLett.96.110405.

[24] A. Kitaev and J. Preskill, Topological entanglement entropy, *Phys. Rev. Lett.* **96**, 110404 (2006); doi:10.1103/PhysRevLett.96.110404; URL https://link.aps.org/doi/10.1103/PhysRevLett.96.110404.

[25] T. Grover, A. M. Turner and A. Vishwanath, Entanglement entropy of gapped phases and topological order in three dimensions, *Phys. Rev. B* **84**, 195120 (2011); doi:10.1103/PhysRevB.84.195120; URL https://link.aps.org/doi/10.1103/PhysRevB.84.195120.

[26] F. J. MacWilliams and N. J. A. Sloane, *The Theory of Error-Correcting Codes.* (North-Holland, 1981).

[27] E. M. Rains, Nonbinary quantum codes, *IEEE Trans. Inform. Theory* **45**(6), 1827 (1999); doi:10.1109/18.782103.

[28] P. K. Sarvepalli, *Quantum stabilizer codes and beyond*, PhD thesis, Texas A&M University (2008); URL http://hdl.handle.net/1969.1/86011.

[29] D. Poulin, Stabilizer formalism for operator quantum error correction, *Phys. Rev. Lett.* **95**, 230504 (2005); doi:10.1103/PhysRevLett.95.230504.

[30] D. Bacon, Operator quantum error-correcting subsystems for self-correcting quantum memories, *Phys. Rev. A* **73**, 012340 (2006); doi:10.1103/PhysRevA.73.012340.

[31] N. P. Breuckmann, *Homological quantum codes beyond the toric code*, PhD thesis, RWTH Aachen University (2017).

[32] D. Fattal, T. S. Cubitt, Y. Yamamoto, S. Bravyi and I. L. Chuang. Entanglement in the stabilizer formalism, preprint (2004); URL http://arXiv.org/abs/quant-ph/0406 168; arXiv:quant-ph/0406168.

[33] S. Trebst, P. Werner, M. Troyer, K. Shtengel and C. Nayak, Breakdown of a topological phase: Quantum phase transition in a loop gas model with tension, *Phys. Rev. Lett.* **98**, 070602 (2007); doi:10.1103/PhysRevLett.98.070602; URL http://link. aps.org/doi/10.1103/PhysRevLett.98.070602.

Chapter 2

Quantum Nucleation of Skyrmions in Magnetic Films by Inhomogeneous Fields

Sebastián A. Díaz[*,‡] and Daniel P. Arovas[†,§]

*Department of Physics, University of Basel
Klingelbergstrasse 82, CH-4056 Basel, Switzerland
†Department of Physics, University of California
San Diego, La Jolla, CA 92093, USA
‡s.diaz@unibas.ch
§darovas@ucsd.edu

We show that in magnetic ultrathin films with interfacial Dzyaloshinskii–Moriya interaction, single skyrmions can be nucleated by creating a local distortion in the magnetic field. In our study, we have considered zero temperature quantum nucleation of a single skyrmion from a ferromagnetic phase. The physical scenario we model is one where a uniform field stabilizes the ferromagnet, and an opposing local magnetic field, generated by the tip of a local probe, drives the skyrmion nucleation. Using spin path integrals and a collective coordinate approximation, the tunneling rate from the ferromagnetic to the single skyrmion state is computed as a function of the tip's magnetization and height above the sample surface. Suitable parameters for the experimental observation of the quantum nucleation of single skyrmions are identified.

2.1 Introduction

Magnetic skyrmions (also known as "baby skyrmions") are two-dimensional topological configurations in which the direction $\hat{n}(r)$ of magnetization field $M(r)$ wraps around the unit sphere. More precisely, $M(r)$ supports a single skyrmion when its integer-valued topological charge, or Pontrjagin index,

$$Q = \frac{1}{4\pi} \int d^2 r\, \hat{n} \cdot \frac{\partial \hat{n}}{\partial x} \times \frac{\partial \hat{n}}{\partial y}, \qquad (2.1)$$

is equal to -1, with $\hat{n}(r) = M(r)/|M(r)|$. Skyrmions are stable against smooth variations of the magnetization since Q cannot jump continuously from one integer value to another. Because skyrmions are localized in space and are topologically stable, they behave as particles.

In certain chiral magnets, the competition between local exchange, Dzyaloshinskii–Moriya interactions (DMI), and external magnetic field stabilizes a skyrmion crystal phase, in which Q is thermodynamically large and given by the number of magnetic unit cells of the structure. Such configurations were first observed in bulk MnSi by neutron scattering [1], and real space observation of skyrmions was achieved in a thin film of $Fe_{0.5}Co_{0.5}Si$ using Lorentz transmission electron microscopy (TEM) [2]. Owing to their topological nature, systems supporting skyrmions exhibit novel properties such as emergent magnetic monopoles [3] and electromagnetic fields [4], as well as the topological [5] and skyrmion [6] Hall effects. Their microscopic size, topologically-protected stability and effective coupling to electric currents make skyrmions attractive for applications. Their small depinning current densities [7], some six orders of magnitude smaller than those needed for domain walls, and their ability to move around obstacles [8,9] make them promising information carriers in magnetic storage and logic devices, where their motion can be controlled by currents or electric fields.

Recent experiments have demonstrated the ability to nucleate skyrmions in a controlled fashion. In thin magnetic films, skyrmions can be nucleated at the sample edge by spin-polarized electric currents [10]. Using spin-polarized tunneling from the tip of a scanning tunneling microscope (STM), skyrmions have been written and erased from magnetic ultrathin films [11]. Another experiment uses an in-plane electric current to force magnetic stripe domains through a geometrical constriction to nucleate skyrmions in a process resembling soap bubble blowing [12]. Finally, time-dependent magnetic fields generated by sending current pulses down a microcoil were used to nucleate skyrmions in magnetic disks [13].

Topological charge is created during the skyrmion nucleation process. Although smooth deformations of the magnetization field cannot make Q jump from one integer value to another in the continuum, this restriction does not apply to spins on a lattice [14]. Here, we shall derive the instanton paths corresponding to skyrmion nucleation in a lattice model, using a collective coordinate approach. From an application-oriented perspective, understanding and controlling the nucleation of skyrmions should be of importance in exploiting these textures as information carriers. Thus far, experimental and theoretical studies have relied on electric currents, time-dependent magnetic fields, and local heating to provide the energy injection necessary to overcome the energy barrier preventing the system from reaching the magnetic skyrmion texture. However, even in the absence of thermal fluctuations or external perturbations, the system has a nonzero probability to escape from a metastable state by quantum tunneling through the energy barrier.

Here we model skyrmion nucleation in magnetic ultrathin films with interfacial Dzyaloshinskii–Moriya interaction (DMI). In the continuum limit, the energy density is [2] $\mathcal{E}(\boldsymbol{r}) = \frac{1}{2}J(\boldsymbol{\nabla M})^2 + D[M^z\boldsymbol{\nabla} \cdot \boldsymbol{M} - (\boldsymbol{M} \cdot \boldsymbol{\nabla})M^z] - BM^z$, with $B(\boldsymbol{r})$ initially uniform. At zero temperature, the low B phase is a helical structure (H) and the high B phase is a uniformly magnetized ferromagnet (FM). Interpolating these phases is the skyrmion crystal (SkX). We consider the effect of a local reduction in B close to the SkX–FM boundary, as described in Sec. 2.2 below. If the local field is sufficiently reduced over a sufficiently large spot, we find that the lowest energy state is one accommodating a single skyrmion (SSk). In Sec. 2.3, we introduce a lattice-based version of this model in which the individual spins are endowed with quantum dynamics. Then using spin path integrals and a collective coordinate approximation, we compute the rate at which single skyrmions are nucleated out of the metastable FM configuration. Section 2.4 contains the results and discussion of the tunneling rate calculation, followed by conclusions.

2.2   Setup and system preparation

Since the nucleation rate is expected to scale exponentially with the number of spins involved in the tunneling process, here we are interested in fairly compact skyrmions of microscopic dimensions. The experimental setup we envision is depicted in Fig. 2.1. A uniform magnetic field, B_{ext}, is applied perpendicular to the film ($\hat{\boldsymbol{z}}$) to control its global magnetic phase. A local probe, such as a magnetized tip of an STM or magnetic force microscope (MFM), is then invoked to provide an additional opposing local field B_{tip}. We adopt a model for the magnetic field of the tip that has been recently used to analyze MFM imaging data of individual Néel skyrmion in Ir/Fe/Co/Pt multilayer films [15], that had also been previously used with superconducting vortices [16]. The model describes the tip as a magnetized truncated cone. Setting the origin of coordinates on the surface of the thin film, when the tip is at a height, h, right above the origin, the magnetic field it produces on the film surface ($z = 0$) has the form

$$B_{\text{tip}}(\boldsymbol{r}) = \frac{\mu_0 \widetilde{m}}{4\pi} \left\{ \left[\frac{h_0\,\rho}{[\rho^2 + h^2]^{3/2}} + \frac{1}{\rho}\left(1 - \frac{h}{\sqrt{\rho^2 + h^2}} \right) \right] \hat{\boldsymbol{\rho}} \right.$$

$$\left. - \left[\frac{1}{\sqrt{\rho^2 + h^2}} + \frac{h_0\,h}{[\rho^2 + h^2]^{3/2}} \right] \hat{\boldsymbol{z}} \right\}, \tag{2.2}$$

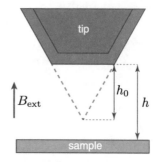

Fig. 2.1.   Schematic of the setup to study quantum nucleation of a single skyrmion. A uniform field, B_{ext}, and an opposing local magnetic field, generated by the tip of a local probe, $\boldsymbol{B}_{\text{tip}}$ (not shown), are simultaneously applied to the sample. Thus, the net local magnetic field is $\boldsymbol{B}(\boldsymbol{r}) = B_{\text{ext}}\hat{\boldsymbol{z}} + \boldsymbol{B}_{\text{tip}}(\boldsymbol{r})$. The tip is modeled as a truncated cone, with truncation height h_0. A coating layer of a magnetic material (blue) generates the magnetic field of the tip.

where μ_0 is the vacuum permeability, $\rho = |\boldsymbol{r}|$, and h_0 is the truncation height (see Fig. 2.1). Details of the magnetic properties of the tip such as the cone angle, thickness of the coating layer, and magnetization of the coating are encoded in $\widetilde{m}$. Thus, the net local magnetic field is $\boldsymbol{B}(\boldsymbol{r}) = B_{\text{ext}}\hat{\boldsymbol{z}} + \boldsymbol{B}_{\text{tip}}(\boldsymbol{r})$. The quantities B_{ext}, $\widetilde{m}$, h, and h_0 are all adjustable parameters within our model.

To prepare the system for the quantum nucleation of a single skyrmion (SSk), a uniform magnetic field is first applied to bring the sample to the ferromagnetic state, close to the boundary separating it from the skyrmion crystal phase. Then, the local probe field is applied. Varying the probe field strength, it can be shown that an SSk can become energetically more favorable than the FM. The local distortion in the magnetic field produced by the field of the local probe is crucial because if we tried not to use it and we simply decreased the applied, uniform magnetic field the system would at some point undergo a phase transition into the SkX, instead of the SSk. Thus, under these conditions, the system is still ferromagnetically ordered, but now in a metastable state. At zero or low enough temperature, thermal fluctuations cannot overcome the energy barrier separating the FM and SSk states. However, quantum tunneling renders the FM state unstable, hence there is a nonzero probability to decay to the SSk state.

2.3   Theoretical model

2.3.1   *Discretization and quantum mechanical action*

Since the Pontrjagin index cannot change under a smooth deformation of the field $\hat{\boldsymbol{n}}(\boldsymbol{r})$, in order to accommodate quantum tunneling between topological sectors in our model, we must account for the underlying discrete lattice on which the spins are situated [14]. We discretize on a square lattice of lattice constant a, and henceforth we measure all lengths in units of a. We assume the magnitude $|\boldsymbol{M}(\boldsymbol{r})|$ of the local magnetization is fixed at M_0, which is set by a consideration of interaction effects [1]. Assuming all the spins have magnetic moment μ, then $M_0 = \mu/(a^2 t)$, with t the thickness of the film. After introducing a uniaxial anisotropy term $-K(M^z)^2$ and identifying an energy scale $E_0 = tJM_0^2$, the discretized dimensionless energy, $\bar{H} = E/E_0$, with $E = t \int d^2 r \mathcal{E}(\boldsymbol{r})$, takes the form

$$\bar{H} = -\sum_{\boldsymbol{r}} \{ \hat{\boldsymbol{n}}_{\boldsymbol{r}} \cdot (\hat{\boldsymbol{n}}_{\boldsymbol{r}+\hat{\boldsymbol{x}}} + \hat{\boldsymbol{n}}_{\boldsymbol{r}+\hat{\boldsymbol{y}}}) + \alpha[\hat{\boldsymbol{x}} \cdot \hat{\boldsymbol{n}}_{\boldsymbol{r}} \times \hat{\boldsymbol{n}}_{\boldsymbol{r}+\hat{\boldsymbol{y}}} - \hat{\boldsymbol{y}} \cdot \hat{\boldsymbol{n}}_{\boldsymbol{r}} \times \hat{\boldsymbol{n}}_{\boldsymbol{r}+\hat{\boldsymbol{x}}}]$$

$$+ \kappa(n_{\boldsymbol{r}}^z)^2 + b_{\boldsymbol{r}} n_{\boldsymbol{r}}^z \}, \tag{2.3}$$

where $\alpha = aD/J$ is a ratio of the lattice constant to the length scale $R_0 = J/D$ set by the competition between DMI and exchange terms, $\kappa = Ka^2/J$, and $b_{\boldsymbol{r}} = B(\boldsymbol{r})/B_0$ with $B_0 = E_0/\mu$ a magnetic field scale.

To endow our model with quantum dynamics, we extend each $\hat{\boldsymbol{n}}_{\boldsymbol{r}}$ to a function of imaginary time, and write the quantum action[a],

$$\mathcal{A}_{\mathrm{E}}[\{\hat{\boldsymbol{n}}_{\boldsymbol{r}}(\tau)\}] = S \int_0^{\beta \tilde{E}_0} d\tau \left\{ i \sum_{\boldsymbol{r}} \frac{d\omega_{\boldsymbol{r}}(\tau)}{d\tau} + \bar{H}[\hat{\boldsymbol{n}}_{\boldsymbol{r}}(\tau)] \right\}, \tag{2.4}$$

where β is the inverse temperature, $\tilde{E}_0 = E_0/S$, and $d\omega_{\boldsymbol{r}}(\tau)/d\tau$ is the rate at which solid angle is swept out by the evolution of $\hat{\boldsymbol{n}}_{\boldsymbol{r}}(\tau)$. The quantized spin value, S, is related to the magnetic moment of the spins by $\mu = g\mu_{\mathrm{B}}S$, with g the gyromagnetic ratio. Note that we have rescaled imaginary time by units of $\hbar S/E_0$ in order to render it dimensionless.

2.3.2 *Phase diagram*

In order to identify parameters of our model suitable for the potential experimental observation of quantum nucleation of skyrmions, we have borne in mind the following two considerations. First, as anticipated above, the tunneling rate is expected to scale exponentially with the number of spins that participate in the tunneling process. Since the size of skyrmions is controlled by the length scale $R_0 = J/D$, [17, 18], and we are interested in skyrmions that extend over a few lattice sites, we set $R_0 = a$, hence

---

[a]In Eq. 2.3, the value of S is absorbed into the model parameters. For convenience we then rescale the imaginary time τ so that a common factor of S multiplies the Lagrangian density, hence the upper limit on τ is $\beta \tilde{E}_0 = \beta E_0/S$.

$\alpha = 1$. Noting that the uniaxial anisotropy is typically small compared to the exchange interaction and DMI, we will neglect it in our study, hence $\kappa = 0$. For these values of α and κ, in the absence of the local probe, we have determined that the zero temperature SkX–FM boundary is crossed when the external magnetic field is swept between $0.65\,B_0$ and $0.7\,B_0$. Second, for the parameters describing the magnetic field generated by the local probe we will take as reference the values reported by the experiments in [16]. In turn, these guided us in choosing the parameters that fix B_0. We have determined that for $E_0 = 0.5$ meV, $a = 0.3$ nm, and $\mu = \mu_B$, with μ_B the Bohr magneton, the magnetic field scale of our system is $B_0 = 8.34$ T.

Employing a combination of Monte Carlo and numerical relaxation methods we computed the energy of the FM and SSk states on a square lattice of 30×30 spins with periodic boundary conditions, using the Hamiltonian $\bar{H}(\{\hat{n}_r\})$ in Eq. (2.3). To ensure that the SSk state could be achieved for reasonable experimental configurations, the truncation height was chosen as $h_0 = 200$ nm; $\tilde{m}$, which encodes the geometric and magnetic properties of the tip of the local probe, was allowed to vary between $0.5\,\tilde{m}_0$ and $1.5\,\tilde{m}_0$, with $\tilde{m}_0 = 0.027$ A/m; and the local probe height above the sample surface, h, was swept between $60\,a$ and $100\,a$. Figure 2.2 shows the zero temperature phase diagram, with $E_0 = 0.5$ meV, $\alpha = 1.0$, $\kappa = 0$,

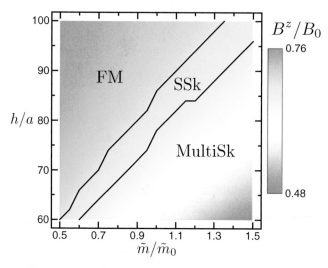

Fig. 2.2.   Phase diagram as a function of the local probe parameters $\tilde{m}$ and h. Zero temperature phase diagram of the model Hamiltonian from Eq. (2.3), with $E_0 = 0.5$ meV, $\alpha = 1.0$, $\kappa = 0$, $\mu = \mu_B$, $a = 0.3$ nm, $h_0 = 200$ nm, and $\tilde{m}_0 = 0.027$ A/m, for a 30×30 square lattice of spins with periodic boundary conditions. The overlaid density plot corresponds to the vertical component of the net magnetic field on the surface of the sample and below the center of the tip, B^z, in units of the magnetic field scale B_0.

$\mu = \mu_B$, and $a = 0.3$ nm, indicating regions where the FM and SSk configurations are the ground states. Also shown in the phase diagram is a region we define as MultiSk, where the ground state accommodates more than one skyrmion. We should point out that the FM state referred to in this phase diagram deviates slightly from the configuration where all the spins point along $\hat{z}$ because of the in-plane component of the local probe field. The phase diagram is overlaid with a density plot of B^z, defined as the vertical component of the net magnetic field on the surface of the sample and below the center of the local probe. We have confirmed that the FM–SSk boundary lies between the level curves $B^z = 0.69 \, B_0$ and $B^z = 0.70 \, B_0$, in the range where the FM–SkX boundary was obtained in the absence of the local probe field. The overall structure of the phase diagram can be understood by noting that B^z, for a given $\tilde{m}$ and h, is the maximum value attained by the vertical component of the net magnetic field at the surface of the sample. Either increasing $\tilde{m}$ or decreasing h results in a smaller B^z. Therefore, starting in the FM state, as B^z is reduced, it will first become energetically more favorable for the system to accommodate a single skyrmion (SSk), and then multiple ones (MultiSk).

2.3.3 *Tunneling rate and collective coordinates*

According to a classical description, once in the FM metastable state, in the absence of thermal fluctuations, it is impossible for the system to overcome the energy barrier that separates it from the SSk state. However, within a quantum mechanical description, the FM state is rendered unstable due to quantum tunneling. This process is analyzed by looking at the survival probability amplitude that the system remains in the FM state after a long time T, i.e., $\langle \text{FM} \, | \, e^{-iHT/\hbar} \, | \, \text{FM} \rangle$, where H is the quantum Hamiltonian. The magnitude of the tunneling amplitude decays for large T as $\exp(-\Gamma T/2)$, where Γ is the inverse lifetime of the metastable FM state. We will calculate Γ using path integrals and the standard technique of instantons [19]. To that end we first write the above probability amplitude as the following multi-spin coherent state path integral [20] in Euclidean time,

$$\langle \text{FM} \, | \, e^{-\beta H} \, | \, \text{FM} \rangle = \int \mathcal{D}\hat{n}(\tau) \, e^{-\mathcal{A}_{\text{E}}[\hat{n}(\tau)]}, \qquad (2.5)$$

where $\hbar\beta = iT$, $\hat{n} = \{\hat{n}_r\}$, and the dimensionless Euclidean action $\mathcal{A}_{\text{E}}[\hat{n}(\tau)]$ is given in Eq. (2.4). Naïvely, the boundary conditions on the path integral are $\hat{n}(0) = \hat{n}(\beta E_0/S) = \hat{n}_{\text{FM}}$, but as the action is linear in time derivatives, supplying initial and final conditions is problematic. However, for long-time bounce paths as we shall consider below, the instantons come very close to satisfying these conditions. A careful discussion of boundary conditions on the spin path integral is provided in the work of Braun and Garg [21].

Rather than expressing the path integral in terms of the unit vectors $\hat{n}_r$, we find it useful to instead use their stereographic projections $w_r = v_r/u_r$, where $u_r = \cos(\theta_r/2)$ and $v_r = \sin(\theta_r/2)\exp(i\phi_r)$ are spinor coordinates for the spin at site r. Let $(u_r^{\mathrm{Sk}}, v_r^{\mathrm{Sk}})$ be the spinor coordinates corresponding to a static single skyrmion configuration $\hat{n}_r^{\mathrm{Sk}}$ which extremizes the energy function $\bar{H}$ in Eq. (2.3). Rather than attempting to solve for the full instanton, we adopt here a simplifying collective coordinate description, parameterized by a complex scalar $\lambda(\tau)$, and we write

$$w_r(\tau) = \frac{\lambda(\tau)\, v_r^{\mathrm{Sk}}}{1 + \lambda(\tau) u_r^{\mathrm{Sk}}}, \qquad \bar{w}_r(\tau) = \frac{\bar{\lambda}(\tau)\, \bar{v}_r^{\mathrm{Sk}}}{1 + \bar{\lambda}(\tau)\, \bar{u}_r^{\mathrm{Sk}}}. \qquad (2.6)$$

For $\lambda \to 0$, we have that w_r describes an FM configuration, while for $\lambda \to \infty$, w_r corresponds to a single skyrmion. For $\lambda = \pm 1$, we encounter a singularity in the continuum limit, where the Pontrjagin index changes discontinuously. On the lattice, however, the singularity is avoided by discretizing in such a way that no lattice point lies at the origin. The central plaquette then lies at the spatial center of the space–time hedgehog defect responsible for the change in topological index [14]. Our collective coordinate description provides us with a rather simple (and perhaps simplistic) description of the topology change via quantum nucleation. One can envisage a more complete description of our instanton, accounting for the full dynamics of the spin field, such as in [22].

By investing the interpolation parameters $\{\lambda, \bar{\lambda}\}$ with time dependence, the evolution of the set $\{w_r(\tau), \bar{w}_r(\tau)\}$ now depends on the evolution of these new collective coordinates. Our approach here parallels that taken in [23], where a collective coordinate path integral approach was applied to analyze metastable Bose–Einstein condensates. The situation is depicted in Fig. 2.3. As a function of real λ, the energy $\bar{H}(\lambda)$ has a local minimum at the FM state $\lambda = 0$, and a global minimum at the skyrmion state $\lambda = \infty$ (blue curve). In the quantum tunneling process, the collective coordinate moves under the barrier, emerging at the point marked X in the figure, at which point it may "roll downhill" toward $\lambda = \infty$. At the conclusion of the tunneling process, the Pontrjagin index of the X state will be $Q = -1$ or $Q = 0$, if $\lambda_X > 1$ or $\lambda_X < 1$, respectively.

Within the collective coordinate approximation, the spin path integral becomes

$$\langle \mathrm{FM}\,|\,e^{-\beta H}\,|\,\mathrm{FM}\rangle \approx \int \mathcal{D}[\lambda, \bar{\lambda}]\, e^{-A_{\mathrm{E}}^{\mathrm{eff}}[\lambda, \bar{\lambda}]}, \qquad (2.7)$$

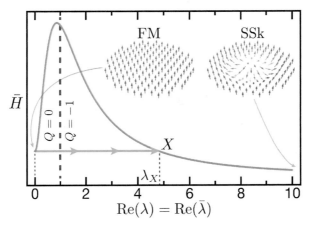

Fig. 2.3.   Quantum tunneling process. Energy landscape as a function of the collective coordinates along the line $\mathrm{Re}(\lambda) = \mathrm{Re}(\bar{\lambda})$, with $\mathrm{Im}(\lambda) = 0 = \mathrm{Im}(\bar{\lambda})$, for the model parameters given in the caption to Fig. 2.2, as well as $\tilde{m}/\tilde{m}_0 = 0.85$ and $h/a = 70$. The classically metastable FM state, at $\lambda = 0 = \bar{\lambda}$, is rendered unstable due to quantum tunneling. The system can tunnel through the energy barrier to the X state, at λ_X, and then reach the SSk state, at $\lambda, \bar{\lambda} \to \infty$, by classical evolution. The topological charge of the X state is -1 or 0, for $\lambda_X > 1$ or $\lambda_X < 1$, respectively.

where

$$\mathcal{A}_{\mathrm{E}}^{\mathrm{eff}}[\lambda, \bar{\lambda}] = S \int_0^{\beta \tilde{E}_0} d\tau \left\{ \sum_r \frac{\bar{w}_r \partial_\lambda w_r \partial_\tau \lambda - w_r \partial_{\bar{\lambda}} \bar{w}_r \partial_\tau \bar{\lambda}}{1 + \bar{w}_r w_r} + \bar{H}(\lambda, \bar{\lambda}) \right\}, \quad (2.8)$$

where $\bar{H}(\lambda, \bar{\lambda})$ is obtained from $\bar{H}(\{\hat{\boldsymbol{n}}_r\})$ by substituting

$$n_r^z = \frac{1 - \bar{w}_r w_r}{1 + \bar{w}_r w_r}, \quad n_r^x + i n_r^y = \frac{2 w_r}{1 + \bar{w}_r w_r}. \quad (2.9)$$

This reduced path integral over collective coordinates is dominated by the paths that extremize $\mathcal{A}_{\mathrm{E}}^{\mathrm{eff}}$, the so-called bounce instantons. Summing over all the multi-bounce instantons [19] and the quadratic fluctuations about them, the tunneling rate from the FM to the SSk state can be approximated as $\Gamma = C \exp(-\Delta \mathcal{A}_{\mathrm{E}}^{\mathrm{eff}})$, where C is the fluctuation determinant prefactor, and the reduced effective action,

$$\Delta \mathcal{A}_{\mathrm{E}}^{\mathrm{eff}} = S \int_0^{\beta \tilde{E}_0} d\tau \sum_r \frac{\bar{w}_r \partial_\lambda w_r \partial_\tau \lambda - w_r \partial_{\bar{\lambda}} \bar{w}_r \partial_\tau \bar{\lambda}}{1 + \bar{w}_r w_r}, \quad (2.10)$$

is evaluated in the single-bounce instanton; here we focus on the computation of $\Delta \mathcal{A}_{\mathrm{E}}^{\mathrm{eff}}$. The calculation of the tunneling rate has now been reduced to solving the Euler–Lagrange (EL) equations of motion

$$\frac{d\lambda}{d\tau} = -\frac{1}{M}\frac{\partial \bar{H}}{\partial \bar{\lambda}}, \quad \frac{d\bar{\lambda}}{d\tau} = +\frac{1}{M}\frac{\partial \bar{H}}{\partial \lambda}, \tag{2.11}$$

where $\bar{H}$ and M are functions of both λ and $\bar{\lambda}$, with

$$M(\lambda, \bar{\lambda}) = 2\sum_r \frac{1}{(1 + w_r \bar{w}_r)^2}\frac{\partial w_r}{\partial \lambda}\frac{\partial \bar{w}_r}{\partial \bar{\lambda}}. \tag{2.12}$$

Owing to the relative minus sign in Eq. (2.11), the EL equations for $\lambda(\tau)$ and $\bar{\lambda}(\tau)$ are not complex conjugates of each other. Nor, since the EL equations are first order in time, are we permitted to impose boundary conditions on both λ and $\bar{\lambda}$ at $\tau = 0$ and $\tau = \beta E_0/S$. Rather, $\lambda(\tau)$ is to be evaluated forward from initial data $\lambda(0) = \lambda_0$ (with $\lambda_0 = 0$ in our case, corresponding to the metastable FM state) and $\bar{\lambda}(\tau)$ is to be evaluated backward from final data $\bar{\lambda}(\beta E_0/S) = \lambda_0^*$, where star denotes complex conjugation. Thus, during the bounce path, $\lambda(\tau)$ and $\bar{\lambda}(\tau)$ are generally not complex conjugates, and thus the components n_r^α of local spin field, obtained in Eq. (2.9) from $(w_r, \bar{w}_r)$, are not always real [22,24]. Indeed the bounce instanton equations imply $\bar{\lambda}(\tau) = \lambda^*(\beta E_0/S - \tau)$, hence $\bar{w}_r(\tau) = w_r^*(\tau)$ only on three time slices: $\tau = 0$ and $\tau = \beta E_0/S$, corresponding to the FM state, and $\tau = \frac{1}{2}\beta E_0/S$, where the field emerges from the barrier and the topological charge has been nucleated. In fact, for finite $\beta E_0/S$, there are exponentially small differences between $\lambda(\beta E_0/S)$ and λ_0, and between $\bar{\lambda}(0)$ and λ_0^*. These differences can be made arbitrarily small by increasing the value of $\beta E_0/S$.

That λ and $\bar{\lambda}$ are not related by complex conjugation during the bounce instanton can be illustrated within the familiar context of quantum tunneling in a potential $V(x)$. Using a phase space path integral, the Euclidean Lagrangian in this case reads: $L_{\rm E} = ip\dot{x} + H(p, x)$. The Euler–Lagrange equations of motion are

$$i\dot{x} = -\frac{\partial H}{\partial p}, \quad i\dot{p} = \frac{\partial H}{\partial x}. \tag{2.13}$$

Since H is real and remains constant during the bounce instanton, we see that the particle has to travel under the energy barrier, therefore p must become imaginary. Employing coherent states, the equations of motion are $\partial z/\partial \tau = -\partial H/\partial \bar{z}$ and $\partial \bar{z}/\partial \tau = \partial H/\partial z$, with $z = (x + ip)/\sqrt{2}$ and $\bar{z} = (x - ip)/\sqrt{2}$. Given that we already established that p is imaginary, it follows that $\bar{z} \neq z^*$.

For a continuous family of magnetization fields $\boldsymbol{M}(\boldsymbol{r}, u)$ parameterized by a real number u, a topology change between different Pontrjagin number sectors is possible only via Bloch points [25–27], which are configurations where $\boldsymbol{M}(\boldsymbol{r}, u)$ vanishes at some location $\boldsymbol{r}$. This occurs for a critical value of u, since this state of affairs is nongeneric, and thereby corresponds to a three-dimensional singularity such as a hedgehog. In our tunneling formalism, the fields $w_{\boldsymbol{r}}(\tau)$ and $\bar{w}_{\boldsymbol{r}}(\tau)$ are in general *not* complex conjugates of each other, and thus there is no corresponding field $\boldsymbol{M}(\boldsymbol{r}, \tau)$ which has a classical interpretation except at the initial and final (imaginary) times, and at the midpoint where the fields emerge from the tunneling barrier. Nevertheless, if one defines the fields $n_{\boldsymbol{r}}^z \equiv (1 - w_{\boldsymbol{r}}^* w_{\boldsymbol{r}})/(1 + w_{\boldsymbol{r}}^* w_{\boldsymbol{r}})$, $n_{\boldsymbol{r}}^+ \equiv 2w_{\boldsymbol{r}}/(1 + w_{\boldsymbol{r}}^* w_{\boldsymbol{r}})$, $\bar{n}_{\boldsymbol{r}}^z \equiv (1 - \bar{w}_{\boldsymbol{r}} \bar{w}_{\boldsymbol{r}}^*)/(1 + \bar{w}_{\boldsymbol{r}} \bar{w}_{\boldsymbol{r}}^*)$, and $\bar{n}_{\boldsymbol{r}}^+ \equiv 2\bar{w}_{\boldsymbol{r}}^*/(1 + \bar{w}_{\boldsymbol{r}} \bar{w}_{\boldsymbol{r}}^*)$, where bar *does not* signify complex conjugation, one has that $\hat{\boldsymbol{n}}(\boldsymbol{r}, \tau)$ and $\bar{\hat{\boldsymbol{n}}}(\boldsymbol{r}, \tau)$ *each* go though Bloch points at different times, in the continuum limit. Again, the fact that our model is defined on a lattice avoids any actual singularities.

2.4 Results and discussion

We first computed the energy of the system as a function of the collective coordinates along the line $\text{Re}(\lambda) = \text{Re}(\bar{\lambda})$, with $\text{Im}(\lambda) = 0 = \text{Im}(\bar{\lambda})$, in the SSk region of the phase diagram in Fig. 2.2. It was confirmed that the FM state was indeed a metastable state in this region. Then the reduced effective Euclidean action, $\Delta\mathcal{A}_E^{\text{eff}}$, was computed as a function of $\tilde{m}$ and h in the SSk region, for the same model parameters used in the construction of the phase diagram. Our results are presented as a contour plot in Fig. 2.4.

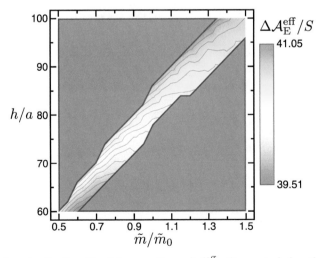

Fig. 2.4.   Reduced effective Euclidean action, $\Delta\mathcal{A}_E^{\text{eff}}$. Computed for the same model parameters in Fig. 2.2, as a function of the local probe parameters $\tilde{m}$ and h in the region where the SSk state is the ground state. S is the spin of the individual magnetic moments in the sample. $\Delta\mathcal{A}_E^{\text{eff}}$ was not computed in the (gray) regions where the ground state is either the FM or the MultiSk state.

Since $\Gamma = C \exp(-\Delta\mathcal{A}_{\mathrm{E}}^{\mathrm{eff}})$, a large/small $\Delta\mathcal{A}_{\mathrm{E}}^{\mathrm{eff}}$ corresponds to a small/large tunneling rate. $\Delta\mathcal{A}_{\mathrm{E}}^{\mathrm{eff}}$ is simply the sum of the (complex) Berry phases accumulated by all the spins in the lattice during a single-bounce instanton. The spin Berry phase has the geometrical interpretation of the solid angle swept by the spin as it evolves in time. Therefore, Γ is strongly dependent on the effective number of spins flipped during the tunneling process. A general feature of our calculation of $\Delta\mathcal{A}_{\mathrm{E}}^{\mathrm{eff}}$ is that it decreases as both $\widetilde{m}$ and h are reduced. We have correlated this behavior with the height of the energy barrier separating the FM from the X state: the smaller the energy barrier the smaller $\Delta\mathcal{A}_{\mathrm{E}}^{\mathrm{eff}}$ gets.

Two representative bounce instantons have been plotted in Fig. 2.5 along with the respective phase portrait of solutions determined by Eq. (2.11).

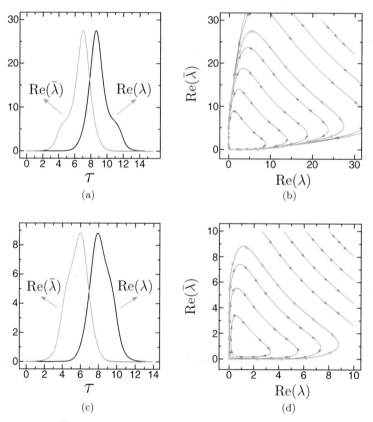

Fig. 2.5.   (a) and (c) Bounce instantons as a function of imaginary time, measured in units of $\hbar S / E_0$. Only the real part of λ and $\bar{\lambda}$ are plotted, their imaginary parts vanished within numerical error. (b) and (d) Phase portraits of the instanton ODEs, Eq. (2.11); the orange contours correspond to the bounce instantons. Local probe parameters $\{\widetilde{m}/\widetilde{m}_0, h/a\} = \{1.3, 98\}$ and $\{\widetilde{m}/\widetilde{m}_0, h/a\} = \{0.85, 70\}$ were used for (a) and (b) and (c) and (d), respectively. For all panels, the same model parameters in Fig. 2.2 were used.

The spin configuration exhibited by the system as it emerges from under the tunneling barrier, the X state, admits a classical interpretation (see Fig. 2.3). Therefore, its topological charge can be computed. For the parameters we have worked with, it was found that the X state always had a topological charge of -1 since $\lambda_X > 1$. Once the X state is reached, the system evolves classically toward the minimum energy state. During this evolution the magnetic texture deforms, but preserves its topological charge of -1.

The inverse of the tunneling rate, $1/\Gamma$, is also the lifetime of the FM, i.e., a measure of the characteristic time that one has to wait to observe the decay of the FM into the SSk state. To complete the calculation of Γ, we also need to compute the fluctuation determinant prefactor. C has units of inverse time, and thus it can be estimated using the energy scale of our model as $C \approx E_0/\hbar$. In the calculations presented here, we assumed the spins had a magnetic moment $\mu = g\mu_B S = \mu_B$, consistent with a system of spins with $S = 1/2$ and a gyromagnetic ratio $g = 2.0$. Therefore, from our computed $\Delta\mathcal{A}_E^{\mathrm{eff}}$, $E_0 = 0.5$ meV, and $S = 1/2$, we obtain that $1/\Gamma$ ranges between 0.5 and 1 ms.

2.5   Conclusions

Quantum nucleation of individual skyrmions in magnetic ultrathin films with interfacial DMI was studied. At zero temperature, a localized magnetic field, generated by the tip of a local probe, applied to a sample in the FM state can render an SSk energetically more favorable. Using spin path integrals and a collective coordinate approximation, the tunneling rate from the metastable FM state to the SSk state was computed as a function of the magnetization of the local probe and its height above the sample surface. Nucleating an SSk from the FM state is unavoidably accompanied by the creation of topological charge, a process forbidden in the continuum. However, since the magnetic state of the system was described by spins on a lattice, we were able to model the quantum nucleation process using the continuous evolution of the spins. Model parameters leading to tunneling rate values that could result in experimentally observable skyrmion nucleation events were determined.

Acknowledgments

SAD acknowledges partial support from the International Fulbright Science and Technology Award. DPA is grateful for support from the UC San Diego Academic Senate. We are grateful to O. Tchernyshyov for helpful comments. This chapter contains material from arXiv:1604.04010, *Quantum Nucleation of Skyrmions in Magnetic Films by Inhomogeneous Fields* and from

Chapter 5 of S. A. Díaz Santiago, *Toward Magnetic Skyrmion Manipulation* (Ph.D. thesis, UC San Diego, 2017).

References

[1] S. Mühlbauer, B. Binz, F. Jonietz, C. Pfleiderer, A. Rosch, A. Neubauer, R. Georgii and P. Böni, *Science* **323**, (915), (2009); http://www.sciencemag.org/content/323/5916/915.abstract.

[2] X. Z. Yu, Y. Onose, N. Kanazawa, J. H. Park, J. H. Han, Y. Matsui, N. Nagaosa and Y. Tokura, *Nature* **465**, 901 (2010); http://dx.doi.org/10.1038/nature09124.

[3] P. Milde, D. Köhler, J. Seidel, L. M. Eng, A. Bauer, A. Chacon, J. Kindervater, S. Mühlbauer, C. Pfleiderer, S. Buhrandt, C. Schütte and A. Rosch, *Science* **340**, 1076 (2013); http://www.sciencemag.org/content/340/6136/1076.abstract.

[4] T. Schulz, R. Ritz, A. Bauer, M. Halder, M. Wagner, C. Franz, C. Pfleiderer, K. Everschor, M. Garst and A. Rosch, *Nat Phys.* **8**, 301 (2012); http://dx.doi.org/10.1038/nphys2231.

[5] A. Neubauer, C. Pfleiderer, B. Binz, A. Rosch, R. Ritz, P. G. Niklowitz and P. Böni, *Phys. Rev. Lett.* **102**, 186602 (2009); http://link.aps.org/doi/10.1103/PhysRevLett.102.186602.

[6] J. Zang, M. Mostovoy, J. H. Han and N. Nagaosa, *Phys. Rev. Lett.* **107**, 136804 (2011); http://link.aps.org/doi/10.1103/PhysRevLett.107.136804.

[7] F. Jonietz, S. Mühlbauer, C. Pfleiderer, A. Neubauer, W. Münzer, A. Bauer, T. Adams, R. Georgii, P. Böni, R. A. Duine, K. Everschor, M. Garst and A. Rosch, *Science.* **330**, 1648 (2010); http://www.sciencemag.org/content/330/6011/1648.abstract.

[8] A. Fert, V. Cros and J. Sampaio, *Nat. Nano.* **8**, 152 (2013); http://dx.doi.org/10.1038/nnano.2013.29.

[9] J. Iwasaki, M. Mochizuki and N. Nagaosa, *Nat. Commun.* **4**, 1463 (2013); http://dx.doi.org/10.1038/ncomms2442.

[10] X. Yu, M. Mostovoy, Y. Tokunaga, W. Zhang, K. Kimoto, Y. Matsui, Y. Kaneko, N. Nagaosa and Y. Tokura, *Proc. Natl. Acad. Sci. USA.* **109**, 8856 (2012); http://www.pnas.org/content/109/23/8856.full.pdf; http://www.pnas.org/content/109/23/8856.abstract.

[11] N. Romming, C. Hanneken, M. Menzel, J. E. Bickel, B. Wolter, K. von Bergmann, A. Kubetzka and R. Wiesendanger, *Science.* **341**, 636 (2013); http://www.sciencemag.org/content/341/6146/636.abstract.

[12] W. Jiang, P. Upadhyaya, W. Zhang, G. Yu, M. B. Jungfleisch, F. Y. Fradin, J. E. Pearson, Y. Tserkovnyak, K. L. Wang, O. Heinonen, S. G. E. te Velthuis and A. Hoffmann, *Science* **349** 283 (2015); http://www.sciencemag.org/content/349/6245/283.abstract.

[13] S. Woo, K. Litzius, B. Kruger, M.-Y. Im, L. Caretta, K. Richter, M. Mann, A. Krone, R. M. Reeve, M. Weigand, P. Agrawal, I. Lemesh, M.-A. Mawass, P. Fischer, M. Klaui and G. S. D. Beach, *Nat Mater.* **15**, 501 (2016); http://dx.doi.org/10.1038/nmat4593.

[14] F. D. M. Haldane, *Phys. Rev. Lett.* **61**, 1029 (1988); http://link.aps.org/doi/10.1103/PhysRevLett.61.1029.

[15] A. Yagil, A. Almoalem, A. Soumyanarayanan, A. K. C. Tan, M. Raju, C. Panagopoulos and O. M. Auslaender, *Appl. Phys. Lett.* **112**, 192403 (2018); https://doi.org/10.1063/1.5027602.

[16] A. Yagil, Y. Lamhot, A. Almoalem, S. Kasahara, T. Watashige, T. Shibauchi, Y. Matsuda and O. M. Auslaender, *Phys. Rev. B.* **94**, 064510 (2016); http://link. aps.org/doi/10.1103/PhysRevB.94.064510.

[17] A. Bogdanov and A. Hubert, *J. Magn. Magn. Mater.* **138**, 255 (1994); http://www. sciencedirect.com/science/article/pii/0304885394900469.

[18] J. H. Han, J. Zang, Z. Yang, J.-H. Park and N. Nagaosa, *Phys. Rev. B.* **82**, 094429 (2010); http://link.aps.org/doi/10.1103/PhysRevB.82.094429.

[19] S. Coleman, *Aspects of Symmetry* (Cambridge University Press, 1985); http://dx.doi. org/10.1017/CBO9780511565045.

[20] A. Auerbach, *Interacting Electrons and Quantum Magnetism*, Graduate Texts in Contemporary Physics, (Springer, 1998); https://books.google.com/books?id=tiQlK zJa6GEC.

[21] C. Braun and A. Garg, *J. Math. Phys.* **48**, 032104 (2007); http://dx.doi.org/ 10.1063/1.2710198; http://scitation.aip.org/content/aip/journal/jmp/48/3/10.1063/ 1.2710198.

[22] J. A. Freire, D. P. Arovas and H. Levine, *Phys. Rev. Lett.* **79**, 5054 (1997); http:// link.aps.org/doi/10.1103/PhysRevLett.79.5054.

[23] J. A. Freire and D. P. Arovas, *Phys. Rev. A* **59**, 1461 (1999); http://link.aps.org/doi/ 10.1103/PhysRevA.59.1461.

[24] J. K. Jain and S. Kivelson, *Phys. Rev. A.* **36**, 3467 (1987); http://link.aps.org/doi/ 10.1103/PhysRevA.36.3467.

[25] S. K. Kim and O. Tchernyshyov, *Phys. Rev. B.* **88**, 174402 (2013); http://link.aps. org/doi/10.1103/PhysRevB.88.174402.

[26] R. Hertel and C. M. Schneider, *Phys. Rev. Lett.* **97**, 177202 (2006); http://link.aps. org/doi/10.1103/PhysRevLett.97.177202.

[27] A. Thiaville, J. M. García, R. Dittrich, J. Miltat and T. Schrefl, *Phys. Rev. B.* **67**, 094410 (2003); http://link.aps.org/doi/10.1103/PhysRevB.67.094410.

Chapter 3

In the Pursuit of Majorana Modes
in Iron-Based High-T_c Superconductors

Xianxin Wu[*,‡‡], Rui-Xing Zhang[†], Gang Xu[‡,§], Jiangping Hu[¶,∥,**]
and Chao-Xing Liu[††,§§]

[*]Institut für Theoretische Physik und Astrophysik,
Julius-Maximilians-Universität Würzburg,
97074 Würzburg, Germany
[†]Condensed Matter Theory Center and Joint Quantum Institute,
Department of Physics, University of Maryland,
College Park, MD 20742-4111, USA
[‡]Wuhan National High Magnetic Field Center,
Huazhong University of Science and Technology,
Wuhan, Hubei 430074, China
[§]School of Physics, Huazhong University of Science and Technology,
Wuhan, Hubei 430074, China
[¶]Beijing National Laboratory for Condensed Matter Physics
and Institute of Physics, Chinese Academy of Sciences,
Beijing 100190, China
[∥]Kavli Institute of Theoretical Sciences,
University of Chinese Academy of Sciences,
Beijing, 100049, China
[**]Collaborative Innovation Center of Quantum Matter,
Beijing 100049, China
[††]Department of Physics, the Pennsylvania State University,
University Park, PA, 16802, USA
[‡‡]Max-Planck-Institut für Festkörperforschung,
Heisenbergstrasse 1, D-70569 Stuttgart, Germany
[§§]cxl56@psu.edu

Majorana zero mode is an exotic quasiparticle excitation with non-Abelian statistics in topological superconductor systems and, can serve as the cornerstone for topological quantum computation, a new type of fault-tolerant quantum computation architecture. This chapter highlights recent progress in realizing Majorana modes in iron-based high-temperature superconductors. We begin with the discussion on topological aspect of electronic band structures in iron-based superconductor compounds. Then we focus on several concrete proposals for Majorana modes in this system, including the Majorana zero modes inside the vortex core on the surface of Fe(Te,Se), helical Majorana modes at the hinge of Fe(Te,Se), the Majorana zero modes at the corner of the Fe(Te,Se)/FeTe heterostructure or the monolayer Fe(Te,Se) under an in-plane magnetic field. We also review the current experimental stage and provide the perspective and outlook for this rapidly developing field.

3.1 Introduction

In conventional solid materials, quasiparticle excitations normally obey either bosonic or fermionic statistics and their wavefunctions will acquire a phase factor of $e^{i\theta}$ ($\theta = 0$ for bosons and $\theta = \pi$ for fermions) upon the exchange (braiding) of two quasiparticles. In two dimensions, collective excitations can in principle be anyons with more exotic statistical behaviors that are fundamentally different from bosonic or fermionic statistics [1, 2]. Exchanging two anyons can give rise to a fractional phase $\theta = \pi p/q$ with integers p and q for Abelian anyons or a unitary transformation among multiple degenerate many-body ground states for non-Abelian anyons [3–8]. In particular, non-Abelian anyons can serve as topological quantum-bits (qubits) and the logic gate operations for these qubits can be implemented by simply braiding these anyons [7–12]. The nonlocality and topological nature of non-Abelian anyons can protect the stored quantum information against weak local perturbations from environments, thus providing a unique route to overcome the major challenge in quantum computation.

Majorana zero mode (MZM) is one example of such quasiparticle excitations, corresponding to a particular type of non-Abelian anyons called "Ising anyons". Here the zero mode refers to the zero-energy midgap excitation which appears inside the superconductor (SC) gap and normally locates at the boundary or inside the vortex core of low-dimensional topological SC (TSC) systems, including 5/2 fractional quantum Hall state (which can be viewed as a TSC of composite fermions) [5], p-wave Sr_2RuO_4 SCs [13, 14] and the heterostructures consisting of SCs and spin–orbit coupled materials [15–34]. Current major experimental efforts focus on the heterostructure approach in a variety of different systems, including semiconductor nanowires in proximity to SCs under magnetic fields [15–18, 21, 26–28], magnetic ion chains on top of SC substrates [19, 20], the surface of topological insulator (TI) compound $(Bi,Sb)_2Te_3$ in proximity to SCs [24, 25, 32–34] and the heterostructure with a quantum anomalous Hall insulator coupled to a SC [23,29,30]. Despite of impressive experimental progress in the heterostructure approach, unambiguous detection and manipulation of MZMs in these heterostructures, however, heavily rely on the SC proximity effect that suffers from the complexity of the interface. Furthermore, the low operation temperature of conventional SC materials further complicates both the realization and manipulation of MZMs. On the other hand, high-temperature SCs, such as cuprate SCs [35] and iron-based SCs [36], are not suitable for the heterostructure approach due to their short coherence length. It is thus desirable to find an intrinsic, robust and controllable Majorana platform that is compatible with existing fabrication and patterning technologies.

During the past few years, significant progress has been made in the search of intrinsic superconductor compounds that host nontrivial topological band structures, termed as *"connate* TSCs" [37]. These compounds include Cu-, Nb- or Sr-doped Bi_2Se_3 [38–40], p-doped $TlBiTe_2$ [41–43], p-doped Bi_2Te_3 under pressure [44], half-Heusler SCs [45–48] and iron-based SCs [37]. Among these SC compounds, iron-based SC family is of particular interest because of their high transition temperature, the abundance of the compounds in this family and the intricate phase diagram with superconducting, nematic and magnetic orders. In this family of materials, nontrivial band topology has been either theoretically proposed or experimentally observed in the normal states of $CaFeAs_2$ [49], monolayer or bulk Fe(Te,Se) (FTS) [50–53], Li(Fe,Co)As [54], $(Li_{1-x}Fe_x)OHFeSe$ [55] and $CaKFe_4As_4$ [56]. These iron-based SC compounds provide ideal high-temperature SC platforms to explore Majorana modes and the relevant topological physics. In this chapter, we will review recent theoretical and experimental progress towards realizing Majorana physics in iron-based SCs. In Sec. 3.2, we will first discuss the underlying physical mechanism of topological electronic band structure in iron-based SCs. Section 3.3 will review several theoretical proposals to achieve MZMs at different locations, including the vortex cores, the hinge and the corner, in iron-based SC systems, as well as recent experimental progress. Conclusion and outlook are given in Sec. 3.4.

3.2 Topological band structure in iron-based superconductors

For most iron-based SCs, the electronic band structure is mainly determined by the Fe–X (X = As, P, Se, Te) tri-layer, as shown in Fig. 3.1(a), in which the primitive unit cell contains two iron and two pnictogen/chalcogen atoms. Due to the existence of the glide symmetry (combination of half translation and reflection in the xy-plane), two iron and pnictogen/chalcogen sites in the primitive unit cell can be related to each other. Consequently, the unit cell can also be redefined to contain only one Fe and one X atoms based on the glide symmetry [57, 58]. Therefore, in literature, the Brillouin zone (BZ) of iron-based SCs can be either defined for the primitive unit cell (blue color region in Fig. 3.1(b)) or for the reduced unit cell (the whole regions

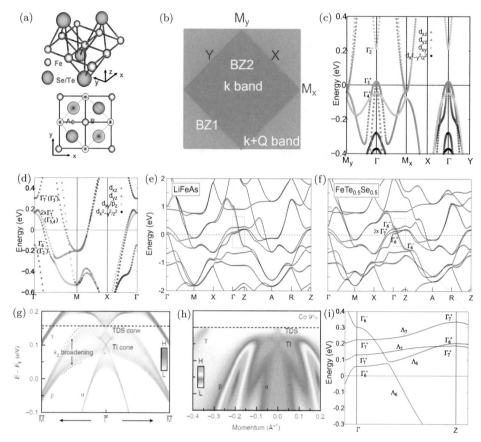

Fig. 3.1.   (a) Crystal structure for Fe(Te,Se) (adapted from Ref. [50]). (b) One-Fe BZ and two-Fe BZ. (c) The typical orbital-resolved band structure near the Fermi level for FeSe. (d) The inverted bands for monolayer Fe(Te,Se) in DFT calculations. Band structures for bulk LiFeAs (e) and FeTe$_{0.5}$Se$_{0.5}$ (f) and zoom-in band structure along Γ–Z (i) for FeTe$_{0.5}$Se$_{0.5}$ [54]. The area of red circles represent the p_z-orbital weight for As/Te,Se atoms. (001) surface spectrum of Li(Fe$_{1-x}$Co$_x$)As: (g) calculated surface spectrum; (h) ARPES intensity plot of Li(Fe$_{1-x}$Co$_x$)As ($x = 9\%$) [54]. The observed two Dirac cones originate from topological insulator phase and topological Dirac semimetal phase.

with both red and blue colors in Fig. 3.1(b)). In this chapter, we will always discuss the electronic band structure in the BZ for two-Fe primitive unit cell.

The energy bands near the Fermi energy are mainly contributed from the d-orbitals of two Fe atoms. Figure 3.1(c) shows the electronic band structure and atomic orbital projections for monolayer FeSe film, from which one can see that the hole pockets around Γ points are mainly attributed from d_{xz}- and d_{yz}-orbitals, while the electron pockets around M are dominated by d_{xz}-, d_{yz}- and d_{xy}-orbitals (here x, y is along Fe–Fe direction). Due to the multiple Fermi pockets, the extended s-wave SC pairing with the opposite sign of pairing gap at the electron and hole pockets becomes possible and this topic has been discussed and reviewed in literature [59, 60]. Besides the superconductivity, an intriguing feature of the band structure is that the valence bands at Γ with d_{xz}-, d_{yz}- and d_{xy}-orbitals are of even parity under inversion (denoted as $\Gamma_{4,5}^+$ bands in Fig. 3.1(c)) while the conduction band at Γ is of d_{xy}-orbital character with odd parity (denoted as Γ_2^- bands in Fig. 3.1(c)). Therefore, according to the Fu–Kane criterion [61], if the odd-parity Γ_2^- bands drop below the even-parity $\Gamma_{4,5}^+$ bands, the system will undergo a topological phase transition and becomes topologically nontrivial. However, the conduction band bottom at Γ has relatively high energy, around 0.2 eV, above the valence band top. Furthermore, spin–orbit coupling (SOC) is negligible for the d-orbitals of Fe atoms. Therefore, at the first sight, it seems hopeless to realize topological states in iron-based SCs.

A crucial theoretical insight comes from the understanding that the energy of the conduction band bottom strongly relies on the hybridization between the d_{xy}-orbital of Fe atoms and the p_z-orbital of anions (X atoms). Although the p_z-orbital band of anions is around $\sim$2 eV below the Fermi energy, its parity (the bonding states of p_z-orbitals between two X-atom sites) is also odd and thus it can hybridize with the odd-parity Γ_2^- bands, but not the even-parity $\Gamma_{4,5}^+$ bands at the Γ point. Due to the level repulsion, this strong hybridization push the Γ_2^- band above the $\Gamma_{4,5}^+$ bands in FeSe films. Therefore, by reducing the hybridization through enlarging the anion height with respect to Fe plane, the Γ_2^- band bottom can sink in energy and eventually have lower energy than the $\Gamma_{4,5}^+$ bands, leading to a band inversion. It was theoretically proposed to enlarge the anion height by either replacing the Se atoms by Te atoms in FTS films or the strain effect from the substrate [51]. Additional effect from the hybridization is to enhance SOC in the bands near the Fermi energy since the SOC strength of p-orbital of anions is normally significant. After including SOC, the Γ_2^- (Γ_4^+) band should be changed to the Γ_6^- (Γ_7^+) bands while the Γ_5^+ bands are split into the Γ_6^+ and Γ_7^+ bands. The new Γ_6^-, Γ_6^+ and Γ_7^+ bands are all spin degenerate due to the inversion and time reversal symmetry. The typical DFT band structure for FTS with inverted bands is shown in Fig. 3.1(d), where the Γ_6^- state sinks below the $\Gamma_{6,7}^+$ states. Since this band inversion mechanism only involves the energy bands around Γ, it can be captured by the well-established Bernevig–Hughes–Zhang (BHZ) model of the quantum spin Hall (QSH) effect [62] on the basis of the odd-parity Γ_6^- bands with the angular momentum $j_z = \pm\frac{1}{2}$ (d_{xy}- and p_z-orbitals) and the even-parity Γ_7^+ bands with the angular momentum $j_z = \pm\frac{3}{2}$ (d_{xz}- and d_{yz}-orbitals). A more sophisticated study on this band inversion mechanism can also be carried out on a tight-binding model with all d-orbitals of Fe atoms and p-orbitals of X atoms and justify this physical picture [51].

Similar physical mechanism also occurs in three-dimensional (3d) bulk iron-based SCs. Different from two-dimensional (2d) monolayer FTS, the Γ_6^- energy bands are highly dispersive along the k_z-direction due to the p_z-orbital nature of X atoms. Figures 3.1(e) and 3.1(f) show the electronic band structure for FeTe$_{0.5}$Se$_{0.5}$ and LiFeAs, respectively. Figure 3.1(i) shows the zoom-in of the bands near the Fermi energy for FeTe$_{0.5}$Se$_{0.5}$ along the Γ–Z line, from which one can see that the Γ_6^- bands are above the Γ_6^+ and Γ_7^+ bands at Γ, but below these bands at Z point in the 3d BZ. An anti-crossing gap is opened between the Γ_6^- and Γ_6^+ bands along the Γ–Z line and leads to nontrivial topological state with surface states within this anti-crossing gap [52]. Figure 3.1(g) shows the direct theoretical calculations of topological surface states coexisting with bulk bands in Li(Fe$_{1-x}$Co$_x$)As, which shares a similar bulk band structure as FeTe$_{0.5}$Se$_{0.5}$ and LiFeAs. It should be mentioned that although the anti-crossing gap between the Γ_6^- and Γ_6^+ bands is ~0.1 eV above the Fermi energy in the DFT calculation, the ARPES measurements show that the Fermi energy exactly lies within this gap for FeTe$_{0.5}$Se$_{0.5}$ and LiFeAs, thus allowing for the direct observation of the topological surface states [53,54]. This discrepancy presumably comes from the inadequate treatment of the strong correlation effect of iron-based SCs in the DFT calculations. In addition to the anti-crossing between the Γ_6^- and Γ_6^+ bands, the band sequence between Γ_6^- and Γ_7^+ bands is also inverted, but these two bands can cross with each other along the Γ–Z path and form a 3d Dirac cone protected by the C_4 rotational symmetry due to the different eigenvalues of C_4 rotation for these two bands ($e^{\pm\frac{\pi}{4}i}$ for Γ_6^- bands and $e^{\pm\frac{3\pi}{4}i}$ for Γ_7^+ bands). A complete eight-band tight-binding model including Γ_6^-, Γ_6^+ and two Γ_7^+ bands has been developed by Xu et $al.$ [63] and all the relevant topological physics of this system can be studied within this model. Further simplified models also exist. The nontrivial anti-crossing between the Γ_6^- and Γ_6^+ bands and the topological surface states can be studied within a four-band model that was first developed for the prototype TI Bi$_2$(Se,Te)$_3$ family of materials [64,65]. On the other hand, the crossing between Γ_6^- and Γ_7^+ bands can be well described by the 3d Dirac Hamiltonian which was first used for 3d Dirac semimetals, such as Na$_3$Bi [66] and Cd$_2$As$_3$ [67].

It should be emphasized that the above physical scenario is not limited for one or two specified compounds, but generally exists in most families of iron-based SCs, including 122, 111 and 11 families [54], thus making iron-based SCs really a fertile platform to explore TSC physics. The physics can be slightly different for different compounds. For example, in most iron-based SC compounds, both topological properties and high T$_c$ superconductivity originate from Fe–X (X = As, Se, Te) tri-layers, but CaFeAs$_2$ is an exception. In CaFeAs$_2$, the superconductivity comes from FeAs layer while the topologically nontrivial bands occur in the CaAs layer. Thus, CaFeAs$_2$ can be regarded as an intrinsic TI-SC hetero-structure [49]. Furthermore, the band inversion may also occur around M point, in additional to the Γ point, as demonstrated in FeSe with the strain effect from the substrate [50].

Exciting experimental progress has been achieved in probing topological electronic bands of iron-based SCs during the past years. The angular resolved photoemission spectroscopy (ARPES) measurements with high energy and momentum resolution were performed for the (001) surface of FTS, which clearly resolves the surface Dirac cone [53]. Moreover, in the same experiment, spin-resolved ARPES measurements directly verify the helical nature of the surface states, confirming their topological origin. Additionally, in the iron pnictide Li(Fe$_{1-x}$Co$_x$)As [54], the chemical potential can be controlled by tuning the Co doping ratio. Therefore, both the topological surface states within the anti-crossing gap between the Γ$_6^-$ and Γ$_6^+$ bands and the 3d bulk Dirac cone due to the crossing between the Γ$_6^-$ and Γ$_7^+$ bands have been simultaneously observed, as shown in Fig. 3.1(h), in good agreement with the theoretical calculations in Fig. 3.1(g).

For 2d FTS thin films, strong evidence for band inversion has been found in ARPES measurements when tuning the Se concentrations of FTS [68]. Very recently, ARPES measurements reveal the topological phase transition process with the variation of Se concentration in monolayer FTS/STO, while in the samples with inverted bands, the scanning tunneling microscopy (STM) measurements provide strong evidence for the existence of nontrivial edge states [69].

3.3 Majorana modes in iron-based superconductors

The generic coexistence of superconductivity and multiple topological states makes a broad class of iron-based SCs a promising high T$_c$ SC platform for exploring topological SCs and Majorana modes. The topics of Majorana physics in topological SCs has been reviewed in a number of papers [10–12, 15], to which the readers can refer for more details. Here we focus on how to achieve Majorana physics in different sample configurations based on iron-based SCs. In particular, it has been theoretically proposed that MZMs can exist inside the vortex cores or at the corners while the helical Majorana modes can appear at the hinge of iron-based SCs [63, 70–72]. The corner and hinge Majorana modes are also relevant to another emergent subfield, called higher order topological state.

3.3.1   *Vortex line transition and Majorana zero modes in the vortex core*

The early studies on MZMs focus on the intrinsic $p + ip$ superconductor, in which the MZMs can be localized in the $h/2e$ vortex core [5] for the 2d case or at the end for the one-dimensional (1d) case [7]. However, the intrinsic $p + ip$ superconductor is rare in nature and the experimental evidence of MZMs in p-wave SC, such as Sr_2RuO_4 [13, 14], is still lacking. In a seminal work [24], Fu and Kane notice that the effective Hamiltonian for the surface states of a strong TI in proximity to a s-wave SC is equivalent to that of a spinless $p + ip$ SC up to a unitary transformation. Consequently, the MZM is also expected to exist at the $h/2e$ vortex core at the surface of a TI with a conventional SC deposited on top. Although the original idea is for a TI-SC heterostructure, the iron-based SCs provide an ideal intrinsic platform to realize this theoretical proposal. As discussed above, the topological surface states have been experimentally observed in several iron-based SC compounds, with the example of FTS [53]. Besides the surface states, bulk electron and hole bands also exist at the Fermi energy and are responsible for the occurrence of superconductivity in these compounds. The intrinsic or self-proximity effect for the superconductivity from the bulk bands to the topological surface bands has also been demonstrated experimentally. For example, a superconducting gap around 1.8 meV has been observed on the topological surface state of $FeTe_{0.55}Se_{0.45}$ in the ARPES measurements [53]. However, there is still one subtle issue. Fu and Kane's original proposal relies on the Fermi energy only crossing the topological surface bands, while in iron-based SCs, the surface bands are normally buried in the bulk bands. Thus, it is natural to ask if the MZMs inside the vortex core at the surface can still survive when both bulk and surface bands appear at the Fermi energy and are strongly hybridized.

This question was first addressed by Hosur *et al.* in a model for doped TIs, which can be applied to several SC compounds based on doped $Bi_2(Se,Te)_3$ family of materials [73]. It was found that there is a critical doping, at which a one-dimensional (1d) topological phase transition can occur along the vortex line. This vortex line phase transition separates a trivial phase with a fully gapped vortex line from a nontrivial phase, in which MZMs are trapped inside the vortex core at the surface of doped TIs (or equivalently at the end of the vortex line), as shown in Figs. 3.2(a)–3.2(d). Compared to doped TI systems, iron-based SCs possess a more complex

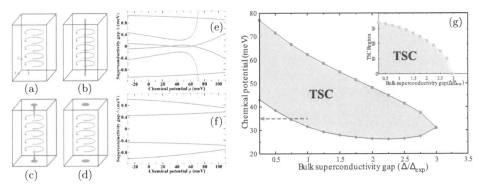

Fig. 3.2.   (a)–(d) The schematic evolution of the surface MZMs in a vortex line. As the chemical potential is tuned from the trivial regime (a) towards the surface TSC regime, two Majorana zero modes arise [(b)] and then become more and more localized at the ends of the vortex line [(c) and (d)]. (e) The energy spectrum at the Γ point of a vortex line along the z-direction as a function of the chemical potential μ. The energy gap closes at $\mu_1 = 31$ meV and $\mu_2 = 62$ meV, respectively, showing the (001) surface is a TSC in the range $\mu \in (\mu_1, \mu_2)$. (f) The energy spectrum at the Z point of the vortex line as a function of the chemical potential μ. There is no gap closing at the Z point, so the phase transitions are solely determined by the gap closing at the Γ point. In (a)–(f), chemical potential $\mu = 0$ corresponds to the Fermi level of the stoichiometric FST. (g) TSC phase space vs. bulk superconducting gap. The red and blue lines are the upper and lower phase boundaries of the TSC phase, respectively. The gray dash at $\mu = 35$ meV indicates the phase transition from a TSC at 0 K ($\Delta = \Delta_{\mathrm{exp}}$) to NSC at $T = T_c$ ($\Delta \sim 0$) with increasing temperature. The inset shows an evolution of the TSC region (red line minus blue line) with respect to the bulk pairing gap. From Ref. [63].

electronic band structure near the chemical potential. Specifically, the Γ_6^- bands are inverted with two Γ_7^+ and one Γ_6^+ bands along the Γ–Z line and all these bands may affect the existence of vortex core MZMs at the surface. Furthermore, it is well accepted that iron-based SCs have an extended s-wave pairing, rather than a simple s-wave pairing and its influence of the surface MZMs is unclear.

To test this idea in bulk $\mathrm{Fe}_{1+y}\mathrm{Se}_{0.5}\mathrm{Te}_{0.5}$, Xu *et al.* [63] built up an eight-band tight-binding model including all the bands that are relevant for the band inversion (Γ_6^-, Γ_6^+ and two Γ_7^+ bands with spin degeneracy). Based on this model, the bulk energy dispersion can fit well with that from the first principles calculations and the topological surface states can be found on the (001) surface, thus confirming the nontrivial topology of this material. By further including the extended s-wave pairing into this model, Xu *et al.* numerically studied the energy levels of a magnetic vortex line in a cylinder geometry for $\mathrm{Fe}_{1+y}\mathrm{Se}_{0.5}\mathrm{Te}_{0.5}$ with the momentum k_z still being a good quantum number along the vortex line direction. Their numerical results show that the SC gap at Γ point along the magnetic vortex line closes at two different chemical potentials μ_1 and μ_2 shown in Fig. 3.2(e), while it remains open at Z point for all the μ values in Fig. 3.2(f). This suggests the vortex line transition occurring at two chemical potentials μ_1 and μ_2 and the system is in the TSC phase with the MZMs localized in the vortex core at the (001) surface when $\mu_1 < \mu < \mu_2$. The phase diagram on the parameter space spanned by the pairing strength Δ and chemical potential μ is also obtained and depicted in Fig. 3.2(g), from which a finite regime of TSC phase is identified. From this phase diagram, a vortex line transition can be induced by varying temperature for a fixed μ. For instance, at $\mu = 35$ meV, the system is in the TSC phase at zero temperature ($\Delta \sim \Delta_{\mathrm{exp}}$), but becomes a trivial SC phase when temperature is close to T_c ($\Delta \to 0$), as indicated by the gray dashed arrow in Fig. 3.2(g).

Recently, several experimental groups have reported the observation of zero-bias peak at the vortex core at the surface of $FeSe_{0.45}Te_{0.55}$ [74–76] and $(Li_{0.84}Fe_{0.16})OHFeSe$ [55] through STM measurements. The spatial profile and the magnetic field dependence of zero-bias peak are consistent with the physical picture of MZM, thus providing strong evidence of its topological origin, while alternative explanation based on the conventional Caroli–de Gennes–Matricon states was still not excluded [77].

Besides MZMs at the end of the vortex line, recent theoretical studies demonstrate that a quasi-1d helical Majorana state protected by C_{4z} rotation symmetry can exist along the vortex line when the chemical potential is tuned close to the 3d Dirac cone formed by the Γ_6^- and Γ_7^+ bands [78–80]. When the bulk state is in the 3d weak TI phase, such as in $(Li_{1-x}Fe_x)OHFeSe$, Qin et $al.$ [81] showed that the vortex phase can also be nodal SC with pairs of helical Majorana modes along the vortex line.

3.3.2 Higher order topology in iron-based superconductors

The past few years have witnessed the rapid developments of "higher order topology" as a direct generalization of TIs [82, 83]. By definition, a D-dimensional topological state has nth-order topology if the system has nontrivial gapless state on some of its $(D - n)$-dimensional boundary manifolds. For example, a second-order TI in 3d is usually featured by gapped 3d bulk and 2d surface states, just like a topologically trivial atomic insulator. However, there exist 1d gapless modes that live on the "hinges" between adjacent surfaces. Experimentally, if we simply probe a second-order TI with ARPES measurement, we will most certainly conclude that it is a trivial band insulator given the gapped surface spectrum. The higher order topology is thus only revealed when the sample hinges are probed by STM, transport studies and other hinge-sensitive measurements. In experiment, phenomena of higher order topological insulators have been observed in photonic systems [84], phononic systems [85], acoustic systems [86] and bismuth [87].

Compared with the higher order TIs, their superconducting counterparts, the higher order TSC, appear to be more exotic for hosting corner or hinge Majorana modes [88–92], which provide a new platform for topological quantum computation. One promising route to achieve higher order TSC is to start with a normal state TI and further introduce bulk unconventional superconductivity. As an example, let us consider a 3d superconducting TI whose 2d Dirac surface states become gapped due to the pairing effects. When the bulk pairing is *not* isotropic s-wave, it is possible that the projected surface pairings will acquire a π-phase difference for two neighboring surfaces. In this case, the hinge between these two surfaces is equivalent to a superconducting domain wall of the surface Dirac fermions, which necessarily binds 1d Majorana modes. As a result, such superconducting TI system realizes 1d hinge Majorana modes along with gapped bulk and surfaces, which manifests itself as a second-order TSC in three dimensions. In principle, we can consider similar physics in a 2d superconducting TI that realizes Majorana zero modes localized on the sample corners. A prerequisite for this simple mechanism is the coexistence of nontrivial band topology and unconventional superconductivity, which has been experimentally demonstrated in a number of iron-based SC compounds. To demonstrate this idea of higher order TSC phase in iron-based SCs, we will first describe how 1d dispersing helical Majorana modes naturally emerge on the hinges of a bulk FTS following Ref. [70]. While a FTS monolayer is unlikely higher order topological by itself, we will show that localized Majorana zero modes will appear on the sample corners, when the monolayer is covered by an additional FeTe monolayer [71] or placed under in-plane magnetic field [72].

3.3.2.1 *1d helical Majorana modes on the hinges*

The idea that an iron-based superconductor can be *intrinsically* a higher order TSC was first proposed and demonstrated by Zhang *et al.* [70]. This proposal is inspired by the ARPES observation of spin-momentum locked topological Dirac surface state in the normal state of bulk $FeTe_xSe_{1-x}$, which is also believed to develop the extended s-wave pairing (also known as $s_\pm$ pairing) below the superconductor transition temperature T_c. Being an unconventionally superconducting TI, the central question for FTS is whether hinge Majorana modes could show up as a result of the in-plane $s_\pm$ pairing.

To model the topological physics in FTS, Zhang *et al.* construct a minimal lattice model that captures the following important properties of this material: (i) a topological band inversion at Z in the Brillouin zone; (ii) bulk $s_\pm$ pairing with $\Delta(\mathbf{k}) = \Delta_0 + \Delta_1(\cos k_x + \cos k_y)$. Here, Δ_0 and Δ_1 denote on-site and nearest-neighbor s-wave pairing, respectively.

To understand the origin of higher order topology in FTS, Zhang *et al.* derived an effective surface theory H_Σ in the continuum limit for an arbitrary surface $\Sigma(\phi, \theta)$. Here the surface $\Sigma(\phi, \theta)$ is characterized by a unit vector $\mathbf{n} = (\sin \theta \cos \phi, \sin \theta \sin \phi, \cos \theta)^T$ in the spherical coordinate and can thus be understood as the tangent surface of a unit sphere defined by $\mathbf{n}$. By solving for the surface theory of $\Sigma(\phi, \theta)$ analytically and project the bulk $s_\pm$ pairing onto the surface state basis, Zhang *et al.* obtain the effective surface theory for $\Sigma(\phi, \theta)$ as

$$H_\Sigma(k_1, k_2) = \begin{pmatrix} k_1\varsigma_2 + k_2\varsigma_1 & -i\varsigma_y\Delta_{\text{eff}}(\theta) \\ i\varsigma_y\Delta_{\text{eff}}(\theta) & -k_1\varsigma_2 + k_2\varsigma_1 \end{pmatrix}. \tag{3.1}$$

Here, the in-plane crystal momenta (k_1, k_2) on the surface $\Sigma(\phi, \theta)$ is related to (k_x, k_y) by two successive Euler rotations $R_Z(-\phi)$ and $R_Y(-\theta)$ around z- and y-axes, respectively. ς_i are defined as the Pauli matrices in the TI surface state bases. Importantly, the effective pairing gap on the surface has a strong θ dependence, which is given by

$$\Delta_{\text{eff}}(\theta) = \Delta_0 + 2\Delta_1 - \Delta_1 \frac{m_0 - 2m_1 + m_2}{m_2 \cos^2 \theta - m_1 \sin^2 \theta} \sin^2 \theta, \tag{3.2}$$

where $m_{0,1,2}$ are model parameters. Note that $\Delta_{\text{eff}}(\theta)$ is isotropic in ϕ, simply because the continuum limit of $H(\mathbf{k})$ has full rotational symmetry around z-axis. With Eq. (3.2), Zhang *et al.* arrive at a simple criterion for higher order TSC in FTS: *Given two surfaces with distinct values of θ (e.g., θ_1 and θ_2), the hinge connecting them will necessarily host a pair of helical Majorana modes if $\Delta(\theta_1)\Delta(\theta_2) < 0$.* In particular, for the band parameters used in their work, the hinge Majorana physics emerges when $-2\Delta_1 < \Delta_0 < -1.5\Delta_1$. Physically, this topological criterion is the condition to form a superconducting mass domain wall betweem the two surfaces. Thus, the helical hinge Majorana modes can be interpreted as the domain wall modes that arise due to the π-phase difference of the effective surface pairing functions.

Zhang *et al.* further provide numerical results to demonstrate the higher order topological physics in this model. As shown in Fig. 3.3(a), they first calculate the surface gap evolution as a function of Δ_0 for both (001) and (010) surfaces. By comparing with the analytical results predicted

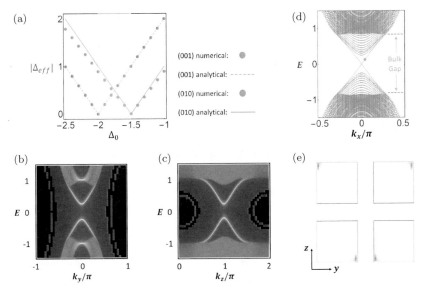

Fig. 3.3.   (a) Evolution of the (001) and (010) surface gaps. The green dashed and red solid lines use Eq. (3.2). Results for both surfaces agree well with numerical calculations on the full lattice model (dots). The red region denotes the phase with helical hinge Majorana modes. In (b) and (c), we plot the surface spectrum of both (001) and (010) surfaces using the iterative Green function method. (d) Energy spectrum in a wire geometry along x with open boundary conditions on both y and z with 20 lattice sites along each direction. The linear modes inside the surface gap are the helical hinge Majorana modes. (e) Spatial profile of the eigenstates at the red dot in (d). From Ref. [70].

by Eq. (3.2), they found a quantitative agreement between these results. At $\Delta_0 = -1.75\Delta_1$, they observed gapped energy spectra for both (001) and (010) surfaces in the semi-infinite geometry, as shown in Figs. 3.3(b) and 3.3(c). To reveal the hinge Majorana modes, they further performed a calculation for the energy spectrum in a infinitely long wire geometry along x-direction, with open boundary conditions imposed for both y- and z-directions. As shown in Fig. 3.3(d), while both bulk and surface gaps are clearly illustrated, there exist four pairs of 1d helical Majorana modes penetrating the surface gap. By plotting the spatial profile of these Majorana modes, Fig. 3.3(e) demonstrates the localization of these modes around four corners of the y–z cross-section. Consequently, for a cubic sample with open boundary conditions in all three spatial directions, the helical Majorana modes will propagate along the hinges between top/bottom and side surfaces, which thus establishes their nature as *helical hinge Majorana modes*.

Zhang *et al.* also studied the robustness of the higher order topology in their model with respect to a finite chemical potential and weak potential disorder effects. As the first material proposal of 3d higher order TSC, Ref. [70]

also inspires experimental efforts to probe the hinge Majorana physics in the iron-based materials. For example, Ref. [93] performed a systematic point contact study on FTS samples and observed interesting zero-bais peaks in the tunneling spectroscopy that suggest possible Majorana-related physics on the sample hinges.

3.3.2.2 *Majorana zero modes at the corners*

Recently, monolayer FTS, a known 2d high-T_c superconductor, is also found to host nontrivial band topology in its normal state. Specifically, topological band gap closing process at Γ point has been observed through ARPES and STM techniques [69]. Thus, the normal state physics resembles that of a 2d TI. Unlike its 3d counterpart, however, the pairing mechanism of monolayer FTS remains mysterious and under debate, which makes it difficult to concretely predict promising topological phenomenon based on this 2d platform. Despite the normal-state band topology, it is nonetheless unlikely for monolayer FTS to be topologically superconducting in an intrinsic way. Therefore, a natural question is that whether and how Majorana physics could emerge in monolayer FTS.

This challenge is resolved simultaneously by Zhang *et al.* [71] and Wu *et al.* [72], where 2d higher order TSC phases with corner Majorana modes are independently proposed by breaking the time-reversal symmetry externally. In particular, Zhang *et al.* proposed to reveal the higher order topology with a hetero-structure combining monolayer FTS and FeTe with anti-ferromagnetic ordering, while Wu *et al.* claimed that applying an in-plane magnetic field to FTS could do the same job. In the following, we will review these two complimentary proposals for realizing Majorana corner modes in monolayer FTS.

The proposal from Zhang *et al.* takes advantageous of the abundant magnetic and superconducting physics in the iron chalcogenides. In particular, the transition between an antiferromagnet (AFM) and an SC is simply tuned by the ratio between Se and Te in FTS. For example, FeSe and FeTe represent examples for a high-Tc superconductor and a bi-collinear AFM [see Fig. 3.4(a) for the magnetic configuration], respectively. Therefore, by growing an extra FeTe layer on top on the existing FTS monolayer, the AFM structure of FeTe will compete with the superconductivity in FTS in an anisotropic way. When projected on the edges, this anisotropy turns out to be the key to drive the system into a higher order TSC.

To demonstrate this idea, Zhang *et al.* considered the bilayer heterostructure that consists of a FTS monolayer and a FeTe monolayer, as shown in Fig. 3.4(b). They wrote down a Bernevig–Hughes–Zhang (BHZ) model to

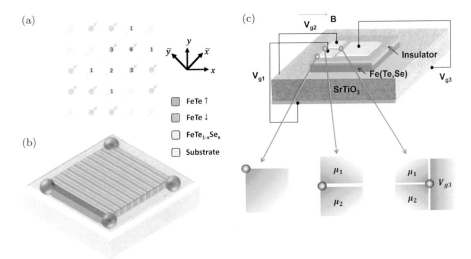

Fig. 3.4. (a) Schematic plot of bicollinear antiferromagnetic order in FeTe. The circle and its arrow represent the Fe atom and its magnetic moment. (b) Schematic plot of the FTS/FeTe heterostructure with corner-localized Majorana modes. (c) Schematics for the Majorana platform based on monolayer FTS under the in-plane magnetic field. MZMs can be found at three different locations: the corner between two perpendicular edges, the chemical potential domain wall along the 1d edge and the tri-junction in the 2d bulk. From Ref. [71, 72].

capture the TI nature of monolayer FTS around Γ point of the Brillouin zone, following Ref. [51]. Introducing bicolinear AFM order M and singlet s-wave pairing Δ enlarges the unit cell to include four inequivalent atom sites, which is shown by the green dashed line in Fig. 3.4(a). The lattice vectors for the enlarged unit cell $\tilde{\mathbf{a}}_{x,y}$ are related to the original lattice vectors $\mathbf{a}_{x,y}$ as $\tilde{\mathbf{a}}_x = 2(\mathbf{a}_x + \mathbf{a}_y)$ and $\tilde{\mathbf{a}}_y = -\mathbf{a}_x + \mathbf{a}_y$.

An interesting observation is that the magnetic configuration in Fig. 3.4(a) is antiferromagnetic along $\tilde{\mathbf{a}}_x$-direction but ferromagnetic (FM) along $\tilde{\mathbf{a}}_y$-direction. To clarify the competition between AFM and superconductivity, Zhang *et al.* chose to first turn off s-wave pairing and the remaining system can be viewed as a 2d TI with additional AFM ordering. On the FM edge along $\tilde{\mathbf{a}}_y$-direction, the TI edge states will develop a Zeeman gap due to the FM exchange coupling effect. In spite of the explicit TRS breaking on the AFM edge along $\tilde{\mathbf{a}}_x$ direction, there exists an AFM TRS symmetry that ensures the gaplessness of the TI edge states, which consists of conventional TRS operation and a half-unit cell translation along $\tilde{\mathbf{a}}_x$.

When superconductivity is turned on, the AFM edge immediately opens up a pairing gap. The higher order topology emerges when the s-wave pairing fails to compete with the Zeeman gap on the FM edge. If this happens, the corner between AFM edge and FM edge will form a zero-dimesional domain wall between FM and superconducting gaps for TI edge states. As shown in Ref. [94] such domain wall will necessarily bind a single localized Majorana zero mode. Mathematically, Zhang *et al.* formulated a simple effective edge theory for the FM edge and analytically identified the condition for corner Majorana physics, which is given by

$$M > \frac{1}{\beta_M} \sqrt{\mu^2 + \Delta^2}. \tag{3.3}$$

Here μ is the chemical potential and β_M is a nonuniversal constant that depends on the details of band parameters. In Fig. 3.5(a), the white dashed line shows the analytical prediction of the topological phase boundary, which agrees well with the numerical results by calculating the FM edge spectrum. When placed on an open boundary geometry with the size of $20\tilde{a}_y \times 10\tilde{a}_x$,

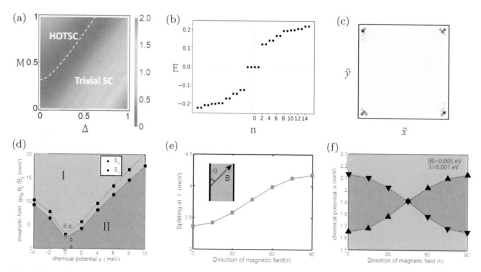

Fig. 3.5.   Figures (a)–(c) and Figs. (d)–(f) discuss the corner Majorana physics for the proposals from Zhang *et al.* [71] and Wu *et al.* [72], respectively. (a) Topological phase diagram for a fixed $\mu = 0.2$. The white dashed line shows the analytical results of Eq. (3.3). (b) Energy spectrum of HFTS with open boundary conditions in both $\tilde{\mathbf{a}}_x$- and $\tilde{\mathbf{a}}_y$-directions, which clearly reveals four Majorana zero modes. (c) Spatial profile of the Majorana zero modes in (b). (d) Topological phase transition as a function of magnetic fields and chemical potentials for the (100) edge. The black circles (squares) correspond to the TPT line for the magnetic field B_X (B_Y) perpendicular (parallel) to the edge. (e) Zeeman splitting of edge states as a function of magnetic field angle with respect to 1d edge (Y-axis), where $g\mu_B|B|$ is fixed to be 5 meV. Inset shows the angle between magnetic field and 1d edge (thick black line). (f) Phase diagram of the existence regime for MZMs at the corner as function of chemical potential μ and magnetic field angle θ. In the pink/blue regimes, the helical edge states at two perpendicular edges have different/the same topological characters, thus can/cannot host MZMs at the corner.

as shown in Fig. 3.5(b), four zero-energy Majorana modes are found inside both the bulk and the edge gaps, with each of them living on one of the four system corners [see Fig. 3.5(c)]. The emergent corner-localized Majorana zero modes unambiguously establish the FTS/FeTe bilayer heterostructure as a 2d higher order TSC. Zhang *et al.* also made estimates on the important energy scales involved, which are based on known experimental results and first-principles calculations. In particular, they found the pairing Δ usually less than 10 meV [95] and a fairly large exchange gap (around 60 meV) induced in the FTS layer. Given $\beta_M \approx 0.5$ and a small chemical potential μ, the corner Majorana physics is experimentally accessible, especially since the growth technique is mature and has been successfully applied to achieve similar heterostructures in iron chalcogenides [96, 97].

The key ingredient for higher order topology in the FTS/FeTe heterostructure is the competing gaps at two orthogonal edges (the magnetic gap at the FM edge and the SC gap at the AFM edge). Similar idea can also be applied to a monolayer FTS under an in-plane magnetic field (Fig. 3.4(c)). Without magnetic field, the helical edge state is expected to open a SC gap while the Zeeman coupling can induce a magnetic gap for a large enough magnetic field. To support this competing edge gap picture, Wu *et al.* [72] numerically studied energy spectrum along the 1d edge of FTS based on both the tight-binding model and the effective BHZ model. Figure 3.5(d) shows the phase diagram of the helical edge states as a function of in-plane magnetic field and chemical potential for a fixed pairing gap. An edge phase transition is found and separates two topologically distinct phases, one with magnetic gap and the other with SC gap. However, to obtain higher order topology, there is a crucial difference for this case compared to the FTS/FeTe heterostructure. In the FTS/FeTe heterostructure, the magnetic gap induced by exchange coupling only exists at the FM edge, but not the AFM edge, while for the in-plane magnetic field, the Zeeman coupling generally exists for the helical edge states along all the edges. While the s-wave SC gap is generally isotropic for any edge, higher order topology *only* appear if the Zeeman coupling is *anisotropic* for the edge states along different edges. Therefore, Wu *et al.* numerically studied the magnetic gap as a function of the angle θ between the edge and in-plane magnetic field and found a significant difference in the magnetic gap (about 80% in Fig. 3.5(e)) when the in-plane magnetic field is parallel or perpendicular to the edge. Due to the anisotropic Zeeman coupling, a unique feature of BHZ model, the topological phase transition lines along one edge for different magnetic field directions (or equivalently for different directional edges with a fixed magnetic field) can be well separated, as shown by the solid and dashed lines for magnetic fields perpendicular or parallel to the edge in Fig. 3.5(d) and by up-pointing and down-pointing triangles for two orthogonal edges, respectively, as a function of the rotating angle θ of the in-plane magnetic field in Fig. 3.5(f). In Fig. 3.5(f), topological properties of two orthogonal edges are the same (distinct) in the blue (pink) regions. Therefore, the MZMs at the corner are expected when μ is tuned to the pink region.

The direct calculation of the system with the open boundaries can verify the existence of the corner MZMs, exactly the same as Figs. 3.5(b) and 3.5(c) for the FTS/FeTe heterostructure. Besides the corner MZMs, Wu *et al.* also predicted that by patterning local electric gates, the MZMs can also be found at the domain wall of chemical potentials at one edge and certain type of tri-junction in the 2d bulk, as depicted in Fig. 3.4(c). Given the high T_c of 40 K [98] and a large in-plane upper critical magnetic field of about 45 T [99], the proposed setup is quite feasible in experiments. Compared to the heterostructure with the fixed AFM direction, utilizing the in-plane magnetic field has the advantage that it can be rotated and thus may provide an efficient approach to perform the braiding operation for the corner MZMs.

3.4   Discussion and perspective

Recently, 112 family of iron pnictides have been predicted to host MZMs owing to the intrinsic TI-SC hetero-structure [100]. Interestingly, the systems are in a boundary-obstructed topological superconducting phase driven by the $s_\pm$ pairing. The experimental detection of MZMs will provide a smoking gun evidence for the extended s-wave pairing in the iron pnictides [100]. Besides the proposals based on topological band structures, recent STM experiments identified a zero-bias peak at the ends of atomic line defects in monolayer Fe(Te,Se)/STO, highly resembling the characteristics of MZMs [101]. They are theoretically interpreted as Majorana ends states of topological Shockley defects [102] or topological magnetic defects [103], waiting for further experimental verifications.

The above discussions have theoretically demonstrated the potential realization of Majorana modes at different locations, including the vortex core at the surface, the hinge of a 3d sample, the corner of a 2d sample and the chemical potential domain walls, in iron-based SCs (monolayer, bulk or heterostructure). However, compared to the heterostructure approach combining the conventional SCs and semiconductors or TIs, both the experimental and theoretical studies in iron-based SCs are still at the infancy stage.

Current experiments in topological aspect of iron-based SCs mainly focus on several 3d bulk iron-based SC compounds, including FTS [53], Li(Fe,Co)As [54], $(Li_{1-x}Fe_x)OHFeSe$ [55] and $CaKFe_4As_4$ [56]. As discussed above, the topological electronic band structure and topological surface states in these compounds have been well established through high-resolution spin-resolved ARPES measurements. The zero-bias peak at the vortex core has been observed at the surface of FTS and $(Li_{1-x}Fe_x)OHFeSe$ through the STM measurements, providing evidence of MZMs [55]. However, it turns out that the selective appearance of zero-bias peaks (around 20% vortex cores during more than 150 measurements) challenges the Majorana interpretation and its physical origin is still under debate. Unlike spin-resolved STM measurement in SC/TI heterostructure [104], spin information of the zero-bias peak has not been extracted yet. A more recent STM study claims to observe a peak value nearly reaching $\frac{2e^2}{h}$ after removing extrinsic instrumental broadening via deconvolution [105]. However, just like the early stage experiments in the heterostructure approach [15], a large subgap density of states still coexists with the zero-bias feature (soft-gap) and these subgap states are detrimental to the topological protection of the coherence of MZMs. Therefore, the next material challenge in iron-based SCs is to improve sample quality and remove these subgap states.

For 2d monolayer FTS, it is still challenging to confirm the topological electronic band structure. As discussed above, a recent ARPES experiment shows the evidence of the gap closing when tuning the ratio between Te and Se compositions, which is supported by the first principle calculations [69]. The STM measurements also show additional density of states at the edge, as compared to that in the bulk, which is attributed to the 1d helical edge state. However, unlike the other QSH systems with the characteristic $\frac{2e^2}{h}$ two-terminal conductance, the existence of 2d bulk Fermi pockets and the high SC transition temperature in FTS prevents the direct observation of the quantized conductance through transport measurements. Therefore, it is still an open question about how to unambiguously demonstrate the helical edge physics in this system. Compared to the 3d case, the 2d monolayer FTS has the advantage of higher T_c, higher upper in-plane critical field and better tunability. Patterning local electric gates or rotating magnetic fields will allow for the manipulation of MZMs and thus may provide a variable route towards topological qubit operations via the braiding and fusion of MZMs. In addition, the phase difference between two SC regions in a Josephson junction provides an additional control knob for tuning Majorana modes. Indeed, recent experimental efforts have demonstrated the feasibility of creating MZMs at the end of a Josephson junction in the semiconductor/SC hybrid system [106, 107]. Such approach has not been explored in iron-based SCs.

In addition to superconductivity, iron-based compounds also exhibit rich phase diagram with other phases, including anti-ferromagnetism and nematicity [108, 109]. A large number of research works focus on the competition between superconductivity, anti-ferromagnetism and nematicity, while their influence on topological states is much less understood. Interestingly, it has been theoretically suggested that nematicity can drive a topological phase transition and lead to the helical edge states in FeSe [110]. Given the rich phase diagram, topological defects, such as domain wall, dislocation, vortex, etc. may also exist and thus iron-based SCs also provide a fertile platform to explore the interplay between topological electronic band structure and topological defects.

Acknowledgments

C.X.L. acknowledges the support of the Office of Naval Research (Grant No. N00014-18-1-2793), and Kaufman New Initiative research grant (Grant No. KA2018-98553) of the Pittsburgh Foundation. R.X.Z. is supported by a JQI Postdoctoral Fellowship. G.X. is supported by the National Key Research and Development Program of China (2018YFA0307000) and the National Natural Science Foundation of China (No. 11874022). J.P. Hu was supported by the Ministry of Science and Technology of China 973 program (No. 2017YFA0303100), National Science Foundation of China (Grant No. NSFC11888101), and the Strategic Priority Research Program of CAS (Grant No.XDB28000000).

References

[1] A. Stern, Anyons and the quantum Hall effect — A pedagogical review, *Ann. Phys.* **323**(1), 204 (2008).

[2] F. Wilczek, *Fractional Statistics and Anyon Superconductivity*, Vol. 5 (World Scientific, 1990).

[3] X.-G. Wen, Non-Abelian statistics in the fractional quantum Hall states, *Phys. Rev. Lett.* **66**(6), 802 (1991).

[4] G. Moore and N. Read, Nonabelions in the fractional quantum Hall effect, *Nucl. Phys. B* **360**(2), 362 (1991).

[5] N. Read and D. Green, Paired states of fermions in two dimensions with breaking of parity and time-reversal symmetries and the fractional quantum Hall effect, *Phys. Rev. B* **61**(15), 10267 (2000).

[6] D. A. Ivanov, Non-abelian statistics of half-quantum vortices in p-wave superconductors, *Phys. Rev. Lett.* **86**(2), 268 (2001).

[7] A. Y. Kitaev, Fault-tolerant quantum computation by anyons, *Ann. Phys.* **303**(1), 2 (2003).

[8] C. Nayak, S. H. Simon, A. Stern, M. Freedman and S. D. Sarma, Non-abelian anyons and topological quantum computation, *Rev. Mod. Phys.* **80**(3), 1083 (2008).

[9] D. Aasen *et al.*, Milestones toward Majorana-based quantum computing, *Phys. Rev. X* **6**(3), 031016 (2016).

[10] J. Alicea, New directions in the pursuit of Majorana fermions in solid state systems, *Rep. Prog. Phys.* **75**(7), 076501 (2012).

[11] C. Beenakker, Search for Majorana fermions in superconductors, *Ann. Rev. Condens. Matter Phys.* **4**(1), 113 (2013).

[12] S. D. Sarma, M. Freedman and C. Nayak, Majorana zero modes and topological quantum computation, *NJP Quantum Inf.* **1**, 15001 (2015).

[13] T. Rice and M. Sigrist, Sr_2RuO_4: an electronic analogue of ^{3}He?, *J. Phys. Condens. Matter* **7**(47), L643 (1995).

[14] S. D. Sarma, C. Nayak and S. Tewari, Proposal to stabilize and detect half-quantum vortices in strontium ruthenate thin films: Non-abelian braiding statistics of vortices in a $p_x + ip_y$ superconductor, *Phys. Rev. B.* **73**(22), 220502 (2006).

[15] R. Lutchyn, E. Bakkers, L. P. Kouwenhoven, P. Krogstrup, C. Marcus and Y. Oreg, Majorana zero modes in superconductor–semiconductor heterostructures, *Nat. Rev. Mater.* **3**(5), 52 (2018).

[16] Y. Oreg, G. Refael and F. von Oppen, Helical liquids and majorana bound states in quantum wires, *Phys. Rev. Lett.* **105**(17), 177002 (2010).

[17] R. M. Lutchyn, J. D. Sau and S. Das Sarma, Majorana fermions and a topological phase transition in semiconductor-superconductor heterostructures, *Phys. Rev. Lett.* **105**(7), 077001 (2010).

[18] J. D. Sau, R. M. Lutchyn, S. Tewari and S. Das Sarma, Generic New Platform for Topological Quantum Computation Using Semiconductor Heterostructures, *Phys. Rev. Lett.* **104**(4), 040502 (2010).

[19] T.-P. Choy, J. Edge, A. Akhmerov and C. Beenakker, Majorana fermions emerging from magnetic nanoparticles on a superconductor without spin–orbit coupling, *Phys. Rev. B.* **84**(19), 195442 (2011).

[20] S. Nadj-Perge, I. K. Drozdov, J. Li, H. Chen, S. Jeon, J. Seo, A. H. MacDonald, B. A. Bernevig and A. Yazdani, Observation of majorana fermions in ferromagnetic atomic chains on a superconductor, *Science.* **346**(6209), 602 (2014).

[21] V. Mourik, K. Zuo, S. M. Frolov, S. R. Plissard, E. P. A. M. Bakkers and L. P. Kouwenhoven, Signatures of majorana fermions in hybrid superconductor-semiconductor nanowire devices, *Science.* **336**(6084), 1003 (2012).

[22] L. P. Rokhinson, X. Liu and J. K. Furdyna, The fractional ac Josephson effect in a semiconductor–superconductor nanowire as a signature of Majorana particles, *Nat. Phys.* **8**(11), 795 (2012).

[23] X.-L. Qi, T. L. Hughes and S.-C. Zhang, Chiral topological superconductor from the quantum Hall state, *Phys. Rev. B.* **82**(18), 184516 (2010).

[24] L. Fu and C. L. Kane, Superconducting proximity effect and Majorana fermions at the surface of a topological insulator, *Phys. Rev. Lett.* **100**(9), 096407 (2008).

[25] M.-X. Wang, C. Liu, J.-P. Xu, F. Yang, L. Miao, M.-Y. Yao, C. L. Gao, C. Shen, X. Ma, X. Chen, Z.-A. Xu, Y. Liu, S.-C. Zhang, D. Qian, J.-F. Jia and Q.-K. Xue, The coexistence of superconductivity and topological order in the Bi_2Se_3 thin films, *Science.* **336**(6077), 52 (2012).

[26] H. Zhang, C.-X. Liu, S. Gazibegovic, D. Xu, J. A. Logan, G. Wang, N. van Loo, J. D. S. Bommer, M. W. A. de Moor, D. Car, R. L. M. Op het Veld, P. J. van Veldhoven, S. Koelling, M. A. Verheijen, M. Pendharkar, D. J. Pennachio, B. Shojaei, J. S. Lee, C. J. Palmstrøm, E. P. A. M. Bakkers, S. D. Sarma, and L. P. Kouwenhoven, Quantized majorana conductance, *Nature.* **556**, 74 (2018).

[27] J. Alicea, Majorana fermions in a tunable semiconductor device, *Phys. Rev. B.* **81**(12), 125318 (2010).

[28] M. Deng, S. Vaitiekėnas, E. B. Hansen, J. Danon, M. Leijnse, K. Flensberg, J. Nygård, P. Krogstrup and C. M. Marcus, Majorana bound state in a coupled quantum-dot hybrid-nanowire system, *Science.* **354**(6319), 1557 (2016).

[29] Q. L. He *et al.*, Chiral majorana fermion modes in a quantum anomalous hall insulator–superconductor structure, *Science* **357**(6348), 294 (2017).

[30] M. Kayyalha, D. Xiao, R. Zhang, J. Shin, J. Jiang, F. Wang, Y.-F. Zhao, R. Xiao, L. Zhang, K. M. Fijalkowski, P. Mandal, M. Winnerlein, C. Gould, Q. Li, L. W. Molenkamp, M. H. W. Chan, N. Samarth and C.-Z. Chang, Absence of evidence for chiral majorana modes in quantum anomalous hall-superconductor devices, *Science* **367**(6473), 64 (2020).

[31] F. Pientka, A. Keselman, E. Berg, A. Yacoby, A. Stern and B. I. Halperin, Topological superconductivity in a planar Josephson junction, *Phys. Rev. X.* **7**(2) 021032, (2017).

[32] X.-L. Qi and S.-C. Zhang, Topological insulators and superconductors, *Rev. Mod. Phys.* **83**(4) 1057, (2011).

[33] M. Z. Hasan and C. L. Kane, Colloquium: topological insulators, *Rev. Mod. Phys.* **82**(4), 3045 (2010).

[34] B. A. Bernevig and T. L. Hughes, *Topological insulators and topological superconductors.* (Princeton university press, 2013).

[35] J. G. Bednorz and K. A. Müller, Possible high T_c superconductivity in the Ba-La-Cu-O system, *Z. Phys. B Condens. Matter.* **64**(2), 189 (1986).

[36] Y. Kamihara, T. Watanabe, M. Hirano and H. Hosono, Iron-based layered superconductor La[$O_{1-x}F_x$]FeAs (x=0.05-0.12) with $T_c = 26K$, *J. Am. Chem. Soc.* **130**(11), 3296 (2008).

[37] N. Hao and J. Hu, Topological quantum states of matter in iron-based superconductors: from concept to material realization, *Nat. Sci. Rev.* **6**(2), 213 (2018).

[38] M. Kriener, K. Segawa, Z. Ren, S. Sasaki and Y. Ando, Bulk superconducting phase with a full energy gap in the doped topological insulator $Cu_xBi_2Se_3$, *Phys. Rev. Lett.* **106**(12), 127004 (2011).

[39] T. Asaba, B. Lawson, C. Tinsman, L. Chen, P. Corbae, G. Li, Y. Qiu, Y. S. Hor, L. Fu and L. Li, Rotational symmetry breaking in a trigonal superconductor nb-doped Bi_2Se_3, *Phys. Rev. X.* **7** (1), 011009 (2017).

[40] Z. Liu, X. Yao, J. Shao, M. Zuo, L. Pi, S. Tan, C. Zhang and Y. Zhang, Superconductivity with topological surface state in $Sr_xBi_2Se_3$, *J. Am. Chem. Soc.* **137**(33), 10512 (2015).

[41] B. Yan, C.-X. Liu, H.-J. Zhang, C.-Y. Yam, X.-L. Qi, T. Frauenheim and S.-C. Zhang, Theoretical prediction of topological insulators in thallium-based III-V-VI$_2$ ternary chalcogenides, *EPL (Europhysics Letters).* **90**(3), 37002 (2010).

[42] Y. L. Chen *et al.*, Single Dirac cone topological surface state and unusual thermoelectric property of compounds from a new topological insulator family, *Phys. Rev. Lett.* **105**(26), 266401 (2010).

[43] H. Lin, R. Markiewicz, L. Wray, L. Fu, M. Hasan and A. Bansil, Single-Dirac-cone topological surface states in the $TlBiSe_2$ class of topological semiconductors, *Phys. Rev. Lett.* **105**(3), 036404 (2010).

[44] J. Zhang, S. Zhang, H. Weng, W. Zhang, L. Yang, Q. Liu, S. Feng, X. Wang, R. Yu, L. Cao, et al., Pressure-induced superconductivity in topological parent compound Bi_2Te_3, *Proc. Nat. Acad. Sci.* **108**(1), 24 (2011).

[45] H. Kim, K. Wang, Y. Nakajima, R. Hu, S. Ziemak, P. Syers, L. Wang, H. Hodovanets, J. D. Denlinger, P. M. Brydon, et al., Beyond triplet: Unconventional superconductivity in a spin-3/2 topological semimetal, *Sci. Adv.* **4**(4), eaao4513 (2018).

[46] B. Yan and A. de Visser, Half-heusler topological insulators, *MRS Bull.* **39**(10), 859 (2014).

[47] T. Bay, T. Naka, Y. Huang and A. de Visser, Superconductivity in noncentrosymmetric YPtBi under pressure, *Phys. Rev. B.* **86**(6), 064515 (2012).

[48] N. P. Butch, P. Syers, K. Kirshenbaum, A. P. Hope and J. Paglione, Superconductivity in the topological semimetal YPtBi, *Phys. Rev. B.* **84**(22), 220504 (2011).

[49] X. Wu, S. Qin, Y. Liang, C. Le, H. Fan and J. Hu, CaFeAs$_2$: A staggered intercalation of quantum spin hall and high-temperature superconductivity, *Phys. Rev. B.* **91**, 081111 (2015).

[50] N. Hao and J. Hu, Topological phases in the single-layer FeSe, *Phys. Rev. X.* **4**(3), 031053 (2014).

[51] X. Wu, S. Qin, Y. Liang, H. Fan and J. Hu, Topological characters in Fe(Te$_{1-x}$Se$_x$) thin films, *Phys. Rev. B.* **93**(11), 115129 (2016).

[52] Z. Wang, P. Zhang, G. Xu, L. K. Zeng, H. Miao, X. Xu, T. Qian, H. Weng, P. Richard, A. V. Fedorov, H. Ding, X. Dai and Z. Fang, Topological nature of the FeSe$_{0.5}$Te$_{0.5}$ superconductor, *Phys. Rev. B.* **92**(11), 115119 (2015).

[53] P. Zhang, K. Yaji, T. Hashimoto, Y. Ota, T. Kondo, K. Okazaki, Z. Wang, J. Wen, G. D. Gu, H. Ding and S. Shin, Observation of topological superconductivity on the surface of an iron-based superconductor, *Science* **360**(6385), 182 (2018).

[54] P. Zhang, Z. Wang, X. Wu, K. Yaji, Y. Ishida, Y. Kohama, G. Dai, Y. Sun, C. Bareille, K. Kuroda, T. Kondo, K. Okazaki, K. Kindo, X. Wang, C. Jin, J. Hu, R. Thomale, K. Sumida, S. Wu, K. Miyamoto, T. Okuda, H. Ding, G. D. Gu, T. Tamegai, T. Kawakami, M. Sato and S. Shin, Multiple topological states in iron-based superconductors, *Nat. Phys.* **15** (1), 41 (2019).

[55] Q. Liu, C. Chen, T. Zhang, R. Peng, Y.-J. Yan, C.-H.-P. Wen, X. Lou, Y.-L. Huang, J.-P. Tian, X.-L. Dong, G.-W. Wang, W.-C. Bao, Q.-H. Wang, Z.-P. Yin, Z.-X. Zhao and D.-L. Feng, Robust and clean majorana zero mode in the vortex core of high-temperature superconductor Li$_{0.84}$Fe$_{0.16}$OHFeSe, *Phys. Rev. X.* **8**, 041056 (2018).

[56] W. Liu, L. Cao, S. Zhu, L. Kong, G. Wang, M. Papaj, P. Zhang, Y. Liu, H. Chen, G. Li, F. Yang, T. Kondo, S. Du, G. Cao, S. Shin, L. Fu, Z. Yin, H.-J. Gao and H. Ding, A new Majorana platform in an Fe-As bilayer superconductor, preprint (2019) arXiv:1907.00904.

[57] P. A. Lee and X.-G. Wen, Spin-triplet p-wave pairing in a three-orbital model for iron pnictide superconductors, *Phys. Rev. B.* **78**, 144517 (2008).

[58] M. Tomić, H. O. Jeschke and R. Valentí, Unfolding of electronic structure through induced representations of space groups: Application to Fe-based superconductors, *Phys. Rev. B.* **90**, 195121 (2014).

[59] G. R. Stewart, Superconductivity in iron compounds, *Rev. Mod. Phys.* **83**, 1589 (2011).

[60] P. J. Hirschfeld, M. M. Korshunov and I. I. Mazin, Gap symmetry and structure of Fe-based superconductors, *Rep. Prog. Phys.* **74**(12), 124508 (2011).

[61] L. Fu and C. L. Kane, Topological insulators with inversion symmetry, *Phys. Rev. B.* **76**, 045302 (2007).

[62] B. A. Bernevig, T. L. Hughes and S.-C. Zhang, Quantum Spin *Hall* Effect and topological phase transition in HgTe quantum wells, *Science.* **314**(5806), 1757 (2006).

[63] G. Xu, B. Lian, P. Tang, X.-L. Qi and S.-C. Zhang, Topological superconductivity on the surface of Fe-based superconductors, *Phys. Rev. Lett.* **117**(4), 047001 (2016).

[64] H. Zhang, C.-X. Liu, X.-L. Qi, X. Dai, Z. Fang and S.-C. Zhang, Topological insulators in Bi_2Se_3, Bi_2Te_3 and Sb_2Te_3 with a single Dirac cone on the surface, *Nat. Phys.* **5**(6), 438 (2009).

[65] C.-X. Liu, X.-L. Qi, H. Zhang, X. Dai, Z. Fang and S.-C. Zhang, Model hamiltonian for topological insulators, *Phys. Rev. B.* **82**(4), 045122 (2010).

[66] Z. Wang, Y. Sun, X.-Q. Chen, C. Franchini, G. Xu, H. Weng, X. Dai and Z. Fang, Dirac semimetal and topological phase transitions in A_3Bi $A = Na, K, Rb$, *Phys. Rev. B.* **85**, 195320 (2012).

[67] Z. Wang, H. Weng, Q. Wu, X. Dai and Z. Fang, Three-dimensional Dirac semimetal and quantum transport in Cd_3As_2, *Phys. Rev. B.* **88**, 125427 (2013).

[68] X. Shi, Z.-Q. Han, P. Richard, X.-X. Wu, X.-L. Peng, T. Qian, S.-C. Wang, J.-P. Hu, Y.-J. Sun and H. Ding, $FeTe_{1-x}Se_x$ monolayer films: towards the realization of high-temperature connate topological superconductivity, *Sci. Bull.* **62**(7), 503 (2017).

[69] X.-L. Peng, Y. Li, X.-X. Wu, H.-B. Deng, X. Shi, W.-H. Fan, M. Li, Y.-B. Huang, T. Qian, P. Richard, J.-P. Hu, S.-H. Pan, H.-Q. Mao, Y.-J. Sun and H. Ding, Observation of topological transition in high-T_c superconducting monolayer $FeTe_{1-x}Se_x$ films on $SrTiO_3(001)$, *Phys. Rev. B.* **100**, 155134 (2019).

[70] R.-X. Zhang, W. S. Cole and S. Das Sarma, Helical hinge majorana modes in iron-based superconductors, *Phys. Rev. Lett.* **122**, 187001 (2019).

[71] R.-X. Zhang, W. S. Cole, X. Wu and S. Das Sarma, Higher-order topology and nodal topological superconductivity in fe(se,te) heterostructures, *Phys. Rev. Lett.* **123**, 167001 (2019).

[72] X. Wu, X. Liu, R. Thomale and C.-X. Liu, High-T_c superconductor Fe (Se, Te) monolayer: an intrinsic, scalable and electrically-tunable majorana platform, Preprint (2019); arXiv:1905.10648.

[73] P. Hosur, P. Ghaemi, R. S. K. Mong and A. Vishwanath, Majorana modes at the ends of superconductor vortices in doped topological insulators, *Phys. Rev. Lett.* **107**, 097001 (2011).

[74] D. Wang, L. Kong, P. Fan, H. Chen, S. Zhu, W. Liu, L. Cao, Y. Sun, S. Du, J. Schneeloch, R. Zhong, G. Gu, L. Fu, H. Ding and H.-J. Gao, Evidence for majorana bound states in an iron-based superconductor, *Science.* **362**(6412), 333 (2018).

[75] T. Machida, Y. Sun, S. Pyon, S. Takeda, Y. Kohsaka, T. Hanaguri, T. Sasagawa, and T. Tamegai, Zero-energy vortex bound state in the superconducting topological surface state of Fe(Se,Te), *Nat. Mater.* **18**(8), 811 (2019).

[76] L. Kong, S. Zhu, M. Papaj, H. Chen, L. Cao, H. Isobe, Y. Xing, W. Liu, D. Wang, P. Fan, Y. Sun, S. Du, J. Schneeloch, R. Zhong, G. Gu, L. Fu, H.-J. Gao and H. Ding, Half-integer level shift of vortex bound states in an iron-based superconductor, *Nat. Phys.* **15**(11), 1181 (2019).

[77] M. Chen, X. Chen, H. Yang, Z. Du, X. Zhu, E. Wang and H.-H. Wen, Discrete energy levels of Caroli-de Gennes-Matricon states in quantum limit in $FeTe_{0.55}Se_{0.45}$, *Nat. Commun.* **9**(1), 970 (2018).

[78] E. J. König and P. Coleman, Crystalline-symmetry-protected helical majorana modes in the iron pnictides, *Phys. Rev. Lett.* **122**, 207001 (May, 2019).

[79] S. Qin, L. Hu, C. Le, J. Zeng, F.-C. Zhang, C. Fang and J. Hu, Quasi-1d topological nodal vortex line phase in doped superconducting 3D *Dirac* semimetals, *Phys. Rev. Lett.* **123**, 027003 (2019).

[80] T. Kawakami and M. Sato, Topological crystalline superconductivity in dirac semimetal phase of iron-based superconductors, *Phys. Rev. B.* **100**, 094520 (2019).

[81] S. Qin, L. Hu, X. Wu, X. Dai, C. Fang, F.-C. Zhang and J. Hu, Topological vortex phase transitions in iron-based superconductors, *Sci. Bull.* **64**(17), 1207 (2019).

[82] W. A. Benalcazar, B. A. Bernevig and T. L. Hughes, Quantized electric multipole insulators, *Science* **357**(6346), 61 (2017).

[83] F. Schindler, A. M. Cook, M. G. Vergniory, Z. Wang, S. S. P. Parkin, B. A. Bernevig and T. Neupert, Higher-order topological insulators, *Sci. Adv.* **4**(6), eaat0346 (2018).

[84] C. W. Peterson, W. A. Benalcazar, T. L. Hughes and G. Bahl, A quantized microwave quadrupole insulator with topologically protected corner states, *Nature* **555**, 346 (2018).

[85] M. Serra-Garcia, V. Peri, R. Süsstrunk, O. R. Bilal, T. Larsen, L. G. Villanueva and S. D. Huber, Observation of a phononic quadrupole topological insulator, *Nature* **555**, 342 (2018).

[86] H. Xue, Y. Yang, F. Gao, Y. Chong and B. Zhang, Acoustic higher order topological insulator on a kagome lattice, *Nat. Mater.* **18**(2), 108 (2019).

[87] F. Schindler, Z. Wang, M. G. Vergniory, A. M. Cook, A. Murani, S. Sengupta, A. Y. Kasumov, R. Deblock, S. Jeon, I. Drozdov, H. Bouchiat, S. Guéron, A. Yazdani, B. A. Bernevig and T. Neupert, Higher-order topology in bismuth, *Nat. Phys.* **14**(9), 918 (2018).

[88] Z. Yan, F. Song and Z. Wang, Majorana corner modes in a high-temperature platform, *Phys. Rev. Lett.* **121**(9), 096803 (2018).

[89] Q. Wang, C.-C. Liu, Y.-M. Lu and F. Zhang, High-temperature majorana corner states, *Phys. Rev. Lett.* **121**(18), 186801 (2018).

[90] Y. Wang, M. Lin and T. L. Hughes, Weak-pairing higher order topological superconductors, *Phys. Rev. B.* **98**(16), 165144 (2018).

[91] X. Zhu, Tunable majorana corner states in a two-dimensional second-order topological superconductor induced by magnetic fields, *Phys. Rev. B.* **97**(20), 205134, (2018).

[92] X.-H. Pan, K.-J. Yang, L. Chen, G. Xu, C.-X. Liu and X. Liu, Lattice-symmetry-assisted second-order topological superconductors and majorana patterns, *Phys. Rev. Lett.* **123**, 156801 (2019).

[93] M. J. Gray, J. Freudenstein, S. Y. F. Zhao, R. O'Connor, S. Jenkins, N. Kumar, M. Hoek, A. Kopec, S. Huh, T. Taniguchi, K. Watanabe, R. Zhong, C. Kim, G. D. Gu and K. S. Burch, Evidence for helical hinge zero modes in an Fe-based superconductor, *Nano Lett.* **19**, 4890 (2019).

[94] L. Fu and C. L. Kane, Josephson current and noise at a superconductor/quantum-spin-hall-insulator/superconductor junction, *Phys. Rev. B.* **79**, 161408 (2009).

[95] Y. Zhang, J. J. Lee, R. G. Moore, W. Li, M. Yi, M. Hashimoto, D. H. Lu, T. P. Devereaux, D.-H. Lee and Z.-X. Shen, Superconducting gap anisotropy in monolayer FeSe thin film, *Phys. Rev. Lett.* **117**, 117001 (2016).

[96] Y. Sun, W. Zhang, Y. Xing, F. Li, Y. Zhao, Z. Xia, L. Wang, X. Ma, Q.-K. Xue, and J. Wang, High temperature superconducting FeSe films on $SrTiO_3$ substrates, *Sci. Rep.* **4**, 6040 (2014).

[97] F. Nabeshima, Y. Imai, A. Ichinose, I. Tsukada and A. Maeda, Growth and transport properties of FeSe/FeTe superlattice thin films, *Jpn J. Appl. Phys.* **56**(2), 020308 (2017).

[98] F. Li, H. Ding, C. Tang, J. Peng, Q. Zhang, W. Zhang, G. Zhou, D. Zhang, C.-L. Song, K. He, S. Ji, X. Chen, L. Gu, L. Wang, X.-C. Ma and Q.-K. Xue, Interface-enhanced high-temperature superconductivity in single-unit cell $FeTe_{1-x}Se_x$ films on $SrTiO_3$, *Phys. Rev. B.* **91**(22) 220503 (2015).

[99] M. B. Salamon, N. Cornell, M. Jaime, F. F. Balakirev, A. Zakhidov, J. Huang, and H. Wang, Upper critical field and Kondo effects in $Fe(Te_{0.9}Se_{0.1})$ thin films by pulsed field measurements, *Sci. Rep.* **6**, 21469 (2016).

[100] X. Wu, W. A. Benalcazar, Y. Li, R. Thomale, C.-X. Liu and J. Hu, Boundary-obstructed topological high-T_c superconductivity in iron pnictides, Preprint (2020); arXiv:2003.12204.

[101] C. Chen, K. Jiang, Y. Zhang, C. Liu, Y. Liu, Z. Wang and J. Wang, Atomic line defects and zero-energy end states in monolayer fe(te,se) high-temperature superconductors, *Nat. Phys.* (2020).

[102] Y. Zhang, K. Jiang, F. Zhang, J. Wang and Z. Wang, Atomic line defects in unconventional superconductors as a new route toward one dimensional topological superconductors, Preprint (2020) arXiv:2004.05860.

[103] X. Wu, J.-X. Yin, C.-X. Liu and J. Hu, Topological magnetic line defects in Fe(Te,Se) high-temperature superconductors, Preprint (2020); arXiv:2004.05848.

[104] H.-H. Sun, K.-W. Zhang, L.-H. Hu, C. Li, G.-Y. Wang, H.-Y. Ma, Z.-A. Xu, C.-L. Gao, D.-D. Guan, Y.-Y. Li, C. Liu, D. Qian, Y. Zhou, L. Fu, S.-C. Li, F.-C. Zhang and J.-F. Jia, Majorana zero mode detected with spin selective andreev reflection in the vortex of a topological superconductor, *Phys. Rev. Lett.* **116**, 257003 (2016).

[105] S. Zhu, L. Kong, L. Cao, H. Chen, M. Papaj, S. Du, Y. Xing, W. Liu, D. Wang, C. Shen, F. Yang, J. Schneeloch, R. Zhong, G. Gu, L. Fu, Y.-Y. Zhang, H. Ding and H.-J. Gao, Nearly quantized conductance plateau of vortex zero mode in an iron-based superconductor, *Science* **367**(6474), 189 (2020).

[106] H. Ren, F. Pientka, S. Hart, A. T. Pierce, M. Kosowsky, L. Lunczer, R. Schlereth, B. Scharf, E. M. Hankiewicz, L. W. Molenkamp, B. I. Halperin, and A. Yacoby, Topological superconductivity in a phase-controlled Josephson junction, *Nature* **569**(7754), 93 (2019).

[107] A. Fornieri, A. M. Whiticar, F. Setiawan, E. Portolés Marín, A. C. C. Drachmann, A. Keselman, S. Gronin, C. Thomas, T. Wang, R. Kallaher, G. C. Gardner, E. Berg, M. J. Manfra, A. Stern, C. M. Marcus and F. Nichele, Evidence of topological superconductivity in planar Josephson junctions, *Nature.* **569** (7754), 89 (2019).

[108] P. Dai, Antiferromagnetic order and spin dynamics in iron-based superconductors, *Rev. Mod. Phys.* **87**, 855 (2015).

[109] R. M. Fernandes, A. V. Chubukov and J. Schmalian, What drives nematic order in iron-based superconductors?, *Nat. Phys.* **10**, 97 (2014).

[110] X. Wu, Y. Liang, H. Fan and J. Hu, Nematic orders and nematicity-driven topological phase transition in FeSe, Preprint (2016); arXiv:1603.02055.

Chapter 4

Quaternion, Harmonic Oscillator, and High-Dimensional Topological States

Congjun Wu

Department of Physics, University of California,
San Diego, CA 92093, USA
wucj@physics.ucsd.edu

Quaternion, an extension of complex number, is the first discovered non-commutative division algebra by William Rowan Hamilton in 1843. In this chapter, we review the recent progress in building up the connection between the mathematical concept of quaternionic analyticity and the physics of high-dimensional topological states. Three- and four-dimensional harmonic oscillator wavefunctions are organized by the SU(2) Aharonov–Casher gauge potential to yield high-dimensional Landau levels possessing the full rotational symmetries and flat energy dispersions. The lowest Landau-level wavefunctions exhibit quaternionic analyticity, satisfying the *Cauchy–Riemann–Fueter* condition, which generalizes the two-dimensional complex analyticity to three and four dimensions. It is also the Euclidean version of the helical Dirac and the chiral Weyl equations. After dimensional reductions, these states become two- and three-dimensional topological states maintaining time-reversal symmetry but exhibiting broken parity. We speculate that quaternionic analyticity can provide a guiding principle for future researches on high-dimensional interacting topological states. Other progresses including high-dimensional Landau levels of Dirac fermions, their connections to high-energy physics, and high-dimensional Landau levels in the Landau-type gauges, are also reviewed. This research is also an important application of the mathematical subject of quaternion analysis to theoretical physics, and provides useful guidance for the experimental explorations on novel topological states of matter.

4.1 Introduction

I feel honored to contribute to this memorial volume for Professor Shoucheng Zhang. As one of his former Ph.D. students, I have been deeply influenced by his insights and tastes on physics along my research career. Shoucheng expressed that he liked our work on quaternion analyticity and high-dimensional Landau levels. Hence, I review the progress in this direction below.

Quaternions, also called Hamilton numbers, are the first noncommutative division algebra as a natural extension to complex numbers (see the quaternion plaque in Fig. 4.1). Imaginary quaternion units i, j and k are isomorphic to the anti-commutative SU(2) Pauli matrices $-i\sigma_{1,2,3}$. Hamilton used quaternions to represent three-dimensional (3D) and four-dimensional (4D) rotations, and performed the product of two rotations based on the quaternion multiplication. In fact, it is amazing that he was well ahead of his time — equivalently he was using the spin-$\frac{1}{2}$ fundamental representations of the SU(2) group, which was before quantum mechanics was discovered. Nevertheless, the development of quaternoinic analysis met significant difficulty since quaternions do not commute. An important progress was made by Fueter in 1935 as reviewed in [1] who defined the Cauchy–Riemann–Fueter condition for quaternionic analyticity. Amazingly again, this is essentially the Euclidean version of the Weyl equation proposed in 1929. Later on, there have been considerable efforts in constructing quantum mechanics and quantum field theory based on quaternions [2–4].

On the other hand, the past decade has witnessed a tremendous progress in the study of topological states of matter, in particular, time-reversal invariant topological insulators in two dimensions (2D) and 3D. Topological properties of their band structures are characterized by a $\mathbb{Z}_2$-index, which are stable against time-reversal invariant perturbations and weak interactions [5–15]. These studies are further developments of quantum anomalous Hall insulators characterized by the integer-valued Chern numbers [16, 17]. Later on, topological states of matter including both insulating and superconducting states have been classified into ten different classes in terms of their properties under the chiral, time-reversal, and particle–hole symmetries [18, 19]. These studies have mostly focused on lattice systems. The wavefunctions of the Bloch bands are complicated, and their energy spectra are dispersive, both of which are obstacles for the study of high-dimensional fractional topological states.

In contrast, the 2D quantum Hall states [20, 21] are early examples of topological states of matter studied in condensed matter physics. They arise from the Landau-level quantization due to the cyclotron motion of electrons in a magnetic field [22]. Their wavefunctions are simple and elegant, which are basically harmonic oscillator wavefunctions. They are reorganized to exhibit analytic properties by an external magnetic field.

Generally speaking, a 2D quantum mechanical wavefunction $\psi(x, y)$ is complex valued, but not necessarily complex analytic. We do not need the whole set of 2D harmonic oscillator wavefunctions, but would like to select a subset of them with nontrivial topological properties, then complex analyticity is a natural selection criterion. Indeed, the lowest Landau-level wavefunctions exhibit complex analyticity. Mathematically, it is imposed by the Cauchy–Riemann condition (see Eq. (4.4) in the text), and physically it is implemented by a magnetic field, which reflects the fact that the cyclotron motion is chiral. This fact greatly facilitated the construction of the Laughlin wavefunction in the study of fractional quantum Hall states [23].

How to generalize Landau levels to 3D and even higher dimensions is a challenging question. A pioneering work was done by Shoucheng and his former student Jiangping Hu in 2001 [24]. They constructed the Landau-level problem on the compact space of an S^4 sphere, which generalizes Haldane's formulation of the 2D Landau levels on an S^2 sphere. Haldane's construction is based on the first Hopf map [25], in which a particle is coupled to the vector potential from a $U(1)$ magnetic monopole. Zhang and Hu considered a particle lying on the S^4 sphere coupled to an SU(2) monopole gauge field, and employed the second Hopf map which maps a unit vector on an S^4 sphere to a normalized 4-component spinor. The Landau-level wavefunctions are expressed in terms of the four components of the spinor. Such a system is topologically nontrivial characterized by the second Chern number possessing time-reversal symmetry. This construction is very beautiful, however, it needs significantly advanced mathematical physics knowledge which may not be common for the general readers in the condensed matter physics, and atomic, molecular, and optical physics community.

We have constructed high-dimensional topological states (e.g., 3D and 4D) based on harmonic oscillator wavefunctions in flat spaces [26, 27]. They exhibit flat energy dispersions and nontrivial topological properties, hence, they are generalizations of the 2D Landau-level problem to high dimensions. Again we will select and reorganize a subset of wavefunctions in seeking for nontrivial topological properties. The strategy we employ is to use quaternion analyticity as the new selection criterion to replace the previous one of complex analyticity. Physically, it is imposed by spin–orbit coupling, which couples orbital angular momentum and spin together to form the helicity structure. In other words, the helicity generated by spin–orbit coupling plays the role of 2D chirality due to the magnetic field. Our proposed Hamiltonians can also be formulated in terms of spin-$\frac{1}{2}$ fermions coupled to an SU(2) gauge potential, or, an Aharonov–Casher potential. Gapless helical Dirac surface modes, or, chiral Weyl modes, appear on open boundaries manifesting the nontrivial topology of bulk states.

We have also constructed high-dimensional Landau levels of Dirac fermions [28], whose Hamiltonians can be interpreted in terms of complex quaternions. The zeroth Landau levels of Dirac fermions are a branch of half-fermion Jackiw–Rebbi modes [29], which are degenerate over all the 3D angular momentum quantum numbers. Unlike the usual parity anomaly and chiral anomaly in which massless Dirac fermions are minimally coupled to the background gauge fields, these Dirac Landau-level problems correspond to a nonminimal coupling between massless Dirac fermions and background fields. This problem lies at the interfaces among condensed matter physics, mathematical physics, and high-energy physics.

High-dimensional Landau levels can also be constructed in the Landau-type gauge, in which rotational symmetry is explicitly broken [30]. The helical, or, chiral plane-waves are reorganized by spatially dependent spin–orbit coupling to yield nontrivial topological properties. The 4D quantum Hall effect of the SU(2) Landau levels has also been studied in the Landau-type gauge, which exhibits the quantized nonlinear electromagnetic response as a spatially separated 3D chiral anomaly.

We speculate that quaternionic analyticity would act as a guiding principle for studying high-dimensional interacting topological states, which is a major challenging question. The high-dimensional Landau-level problems reviewed below provide an ideal platform for this research. This research is at the interface between mathematical and condensed matter physics, and has potential benefits to both fields.

This chapter is organized as follows. In Sec. 4.2, histories of complex number and quaternion, and the basic knowledge of complex analysis and quaternion analysis are reviewed. In Sec. 4.3, the 2D Landau-level problems are reviewed for both nonrelativistic particles and relativistic particles. The complex analyticity of the lowest Landau-level wavefunctions is presented. In Sec. 4.4, the constructions of high-dimensional Landau levels in 3D and 4D with explicit rotational symmetries are reviewed. The quaternionic analyticity of the lowest Landau-level wavefunctions, and the bulk–boundary correspondences in terms of the Euclidean and Minkowski versions of the Weyl equation are presented. In Sec. 4.5, we review the dimensional reductions from the 3D and 4D Landau-level problems to yield the 2D and 3D isotropic but parity-broken Landau levels, respectively. They can be constructed by combining a harmonic potential and a linear spin–orbit coupling. In Sec. 4.6, the high-dimensional Landau levels of Dirac fermions are constructed, which can be viewed as Dirac equations in phase spaces. They are related to gapless Dirac fermions nonminimally coupled to background fields. In Sec. 4.7, high-dimensional Landau levels in the anisotropic Landau-type gauge are reviewed. The 4D quantum Hall responses are derived as a spatially separated chiral anomaly. Conclusions and outlooks are presented in Sec. 4.8.

4.2 Histories of complex number and quaternion

4.2.1 *Complex number*

Complex number plays an essential role in mathematics and quantum physics. The invention of complex number was actually related to the history of solving the algebraic cubic equations, rather than solving the quadratic equation of $x^2 = -1$. If one lived in the 16th century, one could simply say that such an equation has no solution. But cubic equations are different. Consider a reduced cubic equation $x^3 + px + q = 0$, which can be solved by using radicals. Here is the Cardano formula,

$$x_1 = c_1 + c_2, \quad x_2 = c_1 e^{i\frac{2\pi}{3}} + c_2 e^{-i\frac{2\pi}{3}}, \quad x_3 = c_1 e^{-i\frac{2\pi}{3}} + c_2 e^{i\frac{2\pi}{3}}, \quad (4.1)$$

where

$$c_1 = \sqrt[3]{-\frac{q}{2} + \sqrt{\Delta}}, \quad c_2 = \sqrt[3]{-\frac{q}{2} - \sqrt{\Delta}}, \quad (4.2)$$

with the discriminant $\Delta = (\frac{q}{2})^2 + (\frac{p}{3})^3$. The key point of the expressions in Eq. (4.1) is that they involve complex numbers. For example, consider a cubic equation with real coefficients and three real roots $x_{1,2,3}$. It is purely a real problem: It starts with real coefficients and ends up with real solutions. Nevertheless, it can be proved by the Galois theory that there is no way to bypass i. Complex conjugate numbers appear in the intermediate steps, and finally they cancel to yield real solutions. As a concrete example, for the case that $p = -9$ and $q = 8$, complex numbers are unavoidable since $\sqrt{\Delta} = \sqrt{-11}$. The readers may check how to arrive at three real roots of $x_{1,2,3} = 1, -\frac{1}{2} \pm \frac{\sqrt{33}}{2}$.

Once the concept of complex number was accepted, it opened up an entire new field for both mathematics and physics. Early developments include the geometric interpretation of complex numbers in terms of the Gauss plane, the application of complex numbers for two-dimensional rotations, and the Euler formula

$$e^{i\theta} = \cos\theta + i\sin\theta. \quad (4.3)$$

The complex phase appears in the Euler formula, which is widely used in describing mechanical and electromagnetic waves in classic physics, and also quantum mechanical wavefunctions. Moreover, when a complex-valued function $f(x, y)$ satisfies the Cauchy–Riemann condition,

$$\frac{\partial f}{\partial x} + i\frac{\partial f}{\partial y} = 0, \tag{4.4}$$

it only depends on $z = x + iy$ but not on $\bar{z} = x - iy$. The Cauchy–Riemann condition sets up the foundation of complex analysis, giving rise to the Cauchy integral,

$$\frac{1}{2\pi i} \oint \frac{1}{z - z_0} dz f(z) = f(z_0). \tag{4.5}$$

For physicists, a practical use of complex analysis is to calculate loop integrals. Certainly, its importance is well beyond this. Complex analysis is the basic tool for many modern branches of mathematics. For example, it gives rise to the most elegant proof to *the fundamental theorem of algebra*: An algebraic equation $f(z) = 0$, i.e., $f(z)$ is an nth-order polynomial of z, has n complex roots. The proof is essentially to count the phase winding number of $1/f(z)$ as moving around a circle of radius $R \to +\infty$. On this circle, $1/f(z) \to z^{-n}$, then the winding number simply equals $-n$. On the other hand, the winding number is a topological invariant equal to the negative of the number of poles of $1/f(z)$. Hence, n equals the number of zeros of $f(z)$. Complex analysis is also the basic tool of number theory: The Riemann hypothesis, which aims at studying the distribution of prime numbers, is formulated as a complex analysis problem of the distributions of the zeros of the Riemann $\zeta(z)$-function.

Complex numbers actually are inessential in the entire scope of classical physics. It is well known that the complex number description for classic waves is only a convenience but not necessary. The first time that complex numbers are necessary is in quantum mechanics — the Schrödinger equation,

$$i\hbar\partial_t\psi = H\psi. \tag{4.6}$$

In contrast, classic wave equations only involve ∂_t^2, and i disappears since its square equals -1. In fact, Schrödinger attempted to eliminate i in his equation, but did not succeed. Hence, to a certain extent, i, or, the complex phase, is more important than $\hbar$ in quantum physics.

4.2.2 *Quaternion and quaternionic analyticity*

Since 2D rotations can be elegantly described by the multiplication of complex numbers, it is reasonable to expect that 3D rotations could also be described in a similar way by extending complex numbers to include the third dimension. Simply adding another imaginary unit j to construct $x + yi + zj$ does not work, since the product of two imaginary units $ij \neq i \neq j \neq \pm 1$.

Fig. 4.1.   The quaternion plaque on Brougham Bridge, Dublin. From wikipedia, https://en.wikipedia.org/wiki/History_of_quaternions.

It has to be a new imaginary unit defined as $k = ij$, and then the quaternion is constructed as

$$q = x + yi + zj + uk. \tag{4.7}$$

The quaternion algebra,

$$i^2 = j^2 = k^2 = ijk = -1, \tag{4.8}$$

was invented by Hamilton in 1843 when he passed the Brougham bridge in Dublin (see Fig. 4.1). He realized in a genius way that the product table of the imaginary units cannot be commutative. In fact, it can be derived based on Eq. (4.8) that i, j, and k anti-commute with one another, i.e.,

$$ij = -ji, \quad jk = -kj, \quad ki = -ik. \tag{4.9}$$

This is the first noncommutative division algebra discovered, and actually it was constructed before the invention of the concept of matrix. In modern mathematical language, quaternion imaginary units are isomorphic to the Pauli matrices $-i\sigma_1, -i\sigma_2, -i\sigma_3$.

Hamilton employed quaternions to describe the 3D rotations. Essentially he used the spin-$\frac{1}{2}$ spinor representation: Consider a 3D rotation R around the axis along the direction of $\hat{\Omega}$ and the rotation angle is γ. Define a unit imaginary quaternion,

$$\omega(\hat{\Omega}) = i \sin\theta \cos\phi + j \sin\theta \sin\phi + k \cos\theta, \tag{4.10}$$

where θ and ϕ are the polar and azimuthal angles of $\hat{\Omega}$. Then the unit quaternion associated with such a rotation is defined as

$$q = \cos\frac{\gamma}{2} + \omega(\hat{\Omega})\sin\frac{\gamma}{2}, \tag{4.11}$$

which is essentially an SU(2) matrix. A 3D vector $\vec{r}$ is mapped to an imaginary quaternion $r = xi + yj + zk$. After the rotation, $\vec{r}$ is transformed to $\vec{r}'$, and its quaternion form is

$$r' = qrq^{-1}. \tag{4.12}$$

This expression defines the homomorphism from SU(2) to SO(3). In fact, using quaternions to describe rotation is more efficient than using the 3D orthogonal matrix, hence, quaternions are widely used in computer graphics and aerospace engineering even today. If set $\vec{r} = \hat{z}$ in Eq. (4.12), and let q run over unit quaternions, which span the S^3 sphere, then a mapping from S^3 to S^2 is defined as

$$n = qkq^{-1}, \tag{4.13}$$

which is the first Hopf map.

Hamilton spent the last 20 years of his life to promote quaternion applications [8]. His ambition was to invent quaternion analysis which could be as powerful as complex analysis. Unfortunately, this was not successful because of the noncommutative nature of quaternions. Nevertheless, Fueter found the analogy to the Cauchy–Riemann condition for quaternion analysis [1,31]. Consider a quaternionically valued function $f(x, y, z, u)$: It is quaternionic analytic if it satisfies the following Cauchy–Riemann–Fueter condition,

$$\frac{\partial f}{\partial x} + i\frac{\partial f}{\partial y} + j\frac{\partial f}{\partial z} + k\frac{\partial f}{\partial u} = 0. \tag{4.14}$$

Equation (4.14) is the left-analyticity condition since imaginary units are multiplied from the left. A right-analyticity condition can also be similarly defined in which imaginary units are multiplied from the right. The left one is employed throughout this chapter for consistency. For a quaternionic analytic function, the analogy to the Cauchy integral is

$$\frac{1}{2\pi^2} \oiiint \frac{1}{|q - q_0|^2(q - q_0)} Dq f(q) = f(q_0), \tag{4.15}$$

where the integral is over a closed three-dimensional volume surrounding q_0. The measure of the volume element is

$$D(q) = dy \wedge dz \wedge du - i dx \wedge dz \wedge du + j dx \wedge dy \wedge du - k dx \wedge dy \wedge dz \tag{4.16}$$

and $K(q)$ is the four-dimensional Green's function,

$$K(q) = \frac{1}{q|q|^2} = \frac{x - yi - zj - uk}{(x^2 + y^2 + z^2 + u^2)^2}. \tag{4.17}$$

There have also been considerable efforts in formulating quantum mechanics and quantum field theory based on quaternions instead of complex numbers [2, 3]. Quaternions are also used to construct the Laughlin-like wavefunctions of the 2D fractional quantum Hall states [32].

As discussed in *"Selected Papers (1945–1980) of Chen Ning Yang with Commentary"* [4], C. N. Yang speculated that quaternion quantum theory would be a major revolution to physics, mostly based on the viewpoint of non-Abelian gauge theory. He wrote, *"... I continue to believe that the basic direction is right. There must be an explanation for the existence of* SU(2) *symmetry: Nature, we have repeatedly learned, does not do random things at the fundamental level. Furthermore, the explanation is most likely in quaternion algebra: its symmetry is exactly* SU(2). *Besides, the quaternion algebra is a beautiful structure. Yes, it is noncommutative. But we have already learned that nature chose noncommutative algebra as the language of quantum mechanics. How could she resist using the only other possible nice algebra as the language to start all the complex symmetries that she built into the universe?"*

4.3 Complex analyticity and two-dimensional Landau levels

In this section, I recapitulate the basic knowledge of the 2D Landau-level problem, including the Landau levels of both the nonrelativistic Schrödinger equation in Sec. 4.3.1 and the Dirac equation in Sec. 4.3.2. I explain the complex analyticity of the 2D lowest Landau-level wavefunctions.

4.3.1 *2D Landau levels for nonrelativistic electrons*

Why are the 2D Landau-level wavefunctions so interesting? The answer is their elegancy. The external magnetic field reorganizes the harmonic oscillator wavefunctions to yield analytic properties. To be concrete, the Hamiltonian for a 2D electron moving in an external magnetic field B reads,

$$H_{2D,sym} = \frac{(\vec{P} - \frac{q}{c}\vec{A})^2}{2M}. \tag{4.18}$$

In the symmetric gauge, i.e., $A_x = -\frac{1}{2}By$ and $A_y = \frac{1}{2}Bx$, the 2D rotational symmetry is explicit. The diamagnetic A^2-term corresponds to the harmonic potential, and the cross term becomes the orbital-Zeeman term. Then Eq. (4.18) can be reformulated as

$$H_{2D,sym} = \frac{P_x^2 + P_y^2}{2M} + \frac{1}{2} M\omega_0^2(x^2 + y^2) - \omega_0 L_z, \quad (4.19)$$

where ω_0 is half of the cyclotron frequency ω_c with $\omega_c = qB/(Mc)$ and $qB > 0$ is assumed. Equation (4.19) can be interpreted as the Hamiltonian of a rotating 2D harmonic potential, which is how the Landau-level physics is realized in cold atom systems.

Since the harmonic potential and orbital-Zeeman term commute, the Landau-level wavefunctions are just those of a 2D harmonic oscillator. In Fig. 4.2(a), the spectra of a 2D harmonic oscillator vs. the magnetic quantum number m are plotted, exhibiting a linear dependence on m as $E_{n_r,m} = \hbar\omega_0(2n_r + m + 1)$ where n_r is the radial quantum number. If we view this diagram horizontally, the degeneracies are finite and no nontrivial topology appears. But if they are viewed along the diagonal direction, they become Landau levels. This reorganization is due to the orbital-Zeeman term, which also disperses linearly $E_Z = -m\hbar\omega_0$. It cancels the same linear dispersion of a 2D harmonic oscillator, such that the Landau-level energies are flat. The states with $n_r = 0$ are the lowest Landau-level states, whose wavefunctions are given by

$$\psi_{LLL,m}(z) = z^m e^{-|z|^2/(4l_B^2)}, \quad (4.20)$$

where $m \geq 0$ and the magnetic length $l_B = \sqrt{\hbar c/(qB)}$.

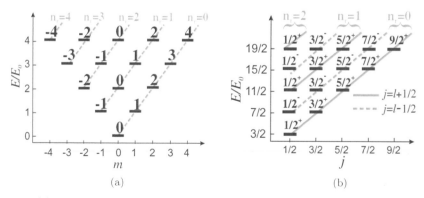

(a)   (b)

Fig. 4.2.   (a) The energy level diagram of a 2D harmonic oscillator vs. the magnetic quantum number m. The states along the tilted lines are reorganized into the 2D flat Landau levels. (b) The eigenstates of a 3D harmonic oscillator labeled by the total angular momentum $j_\pm = l \pm \frac{1}{2}$ Following the tilted solid (dashed) lines, these states are reorganized into the 3D Landau-level states with the positive (negative) helicity for $H_{3D,symm}^\pm$, respectively. From Ref. [27].

Now we impose complex analyticity, i.e., the Cauchy–Riemann condition, to select a subset of harmonic oscillator wavefunctions. Physically it is implemented by the magnetic field. It just means that the cyclotron motion is chiral. After suppressing the Gaussian factor, the lowest Landau-level wavefunction is simply as,

$$\psi_{LLL}(z) = f(z), \tag{4.21}$$

which has a one-to-one correspondence to a complex analytic function in the 2D plane. In fact, complex analyticity greatly facilitated the construction of the many-body Laughlin wavefunctions [23],

$$\psi_L(z_1, \ldots, z_n) = \prod_{i<j}(z_i - z_j)^3 e^{-\sum_i \frac{|z_i|^2}{4l_B^2}}, \tag{4.22}$$

which is actually an analytic function of several complex variables.

Along the edge of a 2D Landau-level system, the bulk flat states change to the 1D dispersive chiral edge modes. They satisfy the chiral wave equation [22],

$$\left(\frac{1}{v_f}\frac{\partial}{\partial t} - \frac{\partial}{\partial x}\right)\psi(x,t) = 0, \tag{4.23}$$

where v_f is the Fermi velocity.

4.3.2 2D Landau levels for Dirac fermions

This is a square-root problem of the Landau-level Hamiltonian of a Schrödinger fermion in Eq. (4.18). The Hamiltonian reads [33],

$$H_{2D}^D = l_0\omega\{(p_x - A_x)\sigma_x + (p_y - A_y)\sigma_y\}, \tag{4.24}$$

where $A_x = -\frac{1}{2}By$, $A_y = \frac{1}{2}Bx$, $l_0 = \sqrt{\frac{2\hbar c}{|qB|}}$, and $\omega = \frac{|qB|}{2mc}$. It can be recast in the form of

$$H_{2D}^D = \frac{\hbar\omega}{\sqrt{2}}\begin{bmatrix} 0 & a_y^\dagger + ia_x^\dagger \\ a_y - ia_x & 0 \end{bmatrix}, \tag{4.25}$$

where $a_i = \frac{1}{\sqrt{2}}(x_i/l_0 + ip_il_0/\hbar)$ $(i = x, y)$ are the phonon annihilation operators.

The square of Eq. (4.25) is reduced to the Landau-level Hamiltonian of a Schrödinger fermion with a supersymmetric structure as

$$(H_{2D}^D)^2 \Big/ \left(\frac{1}{2}\hbar\omega\right) = \begin{bmatrix} H_{2D,sym} - \frac{1}{2}\hbar\omega & 0 \\ 0 & H_{2D,sym} + \frac{1}{2}\hbar\omega \end{bmatrix}, \tag{4.26}$$

where $H_{2D,sym}$ is given in Eq. (4.19). The spectra of Eq. (4.25) are $E_{\pm n} = \pm\sqrt{n}\hbar\omega$ where n is the Landau-level index. The zeroth Landau-level states are singled out: Only the upper component of their wavefunctions is non-zero,

$$\Psi^D_{2D,LLL}(z) = \begin{pmatrix} \psi_{LLL}(z) \\ 0 \end{pmatrix}. \tag{4.27}$$

Here, $\psi_{LLL}(z)$ is the 2D lowest Landau-level wavefunctions of the Schrödinger equation, which is complex analytic. Other Landau levels with positive and negative energies distribute symmetrically around the zero energy.

Due to the particle–hole symmetry, each state of the zeroth Landau-level is a half-fermion Jackiw–Rebbi mode [29, 34]. When the chemical potential μ approaches $0^\pm$, the zeroth Landau-level is fully occupied, or, empty, respectively. The corresponding electromagnetic response is,

$$j_\mu = \pm\frac{1}{8\pi}\frac{q^2}{\hbar}\epsilon_{\mu\nu\lambda}F_{\nu\lambda}, \tag{4.28}$$

which is known as the 2D parity anomaly [33, 35–37]. The signs $\pm$ in Eq. (4.28) refer to $\mu = 0^\pm$, respectively. The two spatial components of Eq. (4.28) show the half-quantized quantum Hall conductance, and the temporal component is the half-quantized Streda formula [38].

4.4 3D Landau-level and quaternionic analyticity

We have seen the close connection between complex analyticity and the 2D topological states. In this section, we discuss how to construct high-dimensional topological states in flat spaces based on quaternionic analyticity.

4.4.1 *The 3D Landau-level Hamiltonian*

Our strategy is to construct the 3D Landau levels based on high-dimensional harmonic oscillator wavefunctions. Again we select a subset of them and reorganize them to exhibit nontrivial topological properties: *The selection criterion is quaternionic analyticity, and physically it is a consequence of spin–orbit coupling.* The physical picture of the 3D Landau-level wavefunctions in the symmetric-like gauge is intuitively presented in Fig. 4.3(a), which generalizes the fixed complex plane in the 2D Landau-level problem to a moving frame embedded in 3D. Define a frame with the orthogonal axes $\hat{e}_1$, $\hat{e}_2$, and $\hat{e}_3$, and the complex analytic wavefunctions are defined in

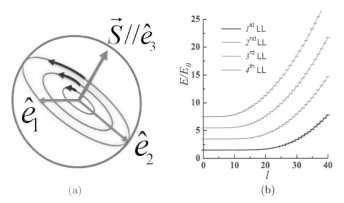

Fig. 4.3. (a) The coherent state picture for the 3D lowest Landau-level wavefunctions based on Eq. (4.31). $\hat{e}_1$–$\hat{e}_2$–$\hat{e}_3$ form an orthogonal triad. The lowest Landau-level wavefunction is complex analytic in the orbital plane $\hat{e}_1$–$\hat{e}_2$ with spin polarized along $\hat{e}_3$. (b) The surface spectra for the 3D Landau-level Hamiltonian equation (4.29). The open boundary condition is imposed for a ball with the radius $R_0/l_{so} = 8$. From Ref. [27].

the $\hat{e}_1$–$\hat{e}_2$ plane with spin polarized along the $\hat{e}_3$-direction. Certainly this frame can be rotated to an arbitrary configuration. The same strategy can be applied to any high dimensions.

Now we present the 3D Landau-level Hamiltonian as constructed in [27]. Consider coupling a spin-$\frac{1}{2}$ fermion to the 3D isotropic SU(2) Aharonov–Casher potential $\vec{A} = \frac{G}{2}\vec{\sigma} \times \vec{r}$ where G is the coupling constant and $\vec{\sigma}$'s are the Pauli matrices. The resultant Hamiltonian is

$$H_{3D,sym}^{\pm} = \frac{1}{2M}\left(\vec{P} - \frac{q}{c}\vec{A}(\vec{r})\right)^2 + V(\vec{r})$$

$$= \frac{P^2}{2M} + \frac{1}{2}M\omega_0^2 r^2 \mp \omega_0\vec{\sigma}\cdot\vec{L}, \qquad (4.29)$$

where $\pm$ refer to $G > 0(<0)$, respectively; $\omega_0 = \frac{1}{2}\omega_{so}$ and $\omega_{so} = |qG|/(Mc)$ is the analogy of the cyclotron frequency. $V(r) = -\frac{1}{2}M\omega_0^2 r^2$, nevertheless, the $\frac{1}{2M}(\frac{q}{c})^2 A^2(r)$ term in the kinetic energy contributes a quadratic scalar potential which equals $2|V(r)|$, hence, Eq. (4.29) is still bound from below. In contrast to the 2D case, $H_{3D,sym}^{\pm}$ preserve time-reversal symmetry. It can also be formulated as a 3D harmonic potential plus a spin–orbit coupling term. Again since these two terms commute, the 3D Landau-level wavefunctions are just the eigenstates of a 3D harmonic oscillator.

Consider the eigenstates of a 3D harmonic oscillator with an additional spin degeneracy $\uparrow$ and $\downarrow$. For later convenience, their eigenstates are organized into the eigenbases of the total angular momentum $j_{\pm} = l \pm \frac{1}{2}$, where $\pm$ represent the positive and negative helicities, respectively. The corresponding spectra are plotted in Fig. 4.2(b), showing a linear dispersion with respect to l as $E_{n_r, J_\pm = l \pm \frac{1}{2}, J_z} = \hbar\omega_0 \left(2n_r + l + \frac{3}{2}\right)$.

Again, if we view the spectra along the diagonal direction, the novel topology appears. The spin–orbit coupling term $\vec{\sigma} \cdot \vec{L}$ has two branches of eigenvalues, both of which disperse linearly as $l\hbar$ and $-(l+1)\hbar$ for the positive and negative helicity sectors, respectively. Combining the harmonic potential and spin–orbit coupling, we arrive at the flat Landau levels: For H_{3D}^+, the positive helicity states become dispersionless with respect to j_+, a main feature of Landau levels. Similarly, the negative helicity states become flat for H_{3D}^-. States in the 3D Landau levels exhibit the same helicity.

4.4.2   *The SU(2) group manifold for the lowest Landau-level wavefunctions*

Having understood why the spectra are flat, now we provide an intuitive picture for the lowest Landau-level wavefunctions with the positive helicity. If expressed in the orthonormal basis of (j_+, j_z), they are rather complicated,

$$\psi_{LLL, j_+ = l + \frac{1}{2}, j_z}(r, \hat{\Omega}) = r^l Y_{j_+ = l + \frac{1}{2}, j_z}(\hat{\Omega}) e^{-\frac{r^2}{4l_{so}^2}}, \qquad (4.30)$$

where $l_{so} = \sqrt{\hbar c / |qG|}$ is the analogy of the magnetic length and $Y_{j_+ = l + \frac{1}{2}, j_z}(\hat{\Omega})$ is the spin–orbit coupled spherical harmonic function with the positive helicity.

Instead, they become very intuitive in the coherent state representation. Let us start with the highest weight states with $j_+ = j_z$, whose wavefunctions are $\psi_{LLL, j_+ = j_z}(r, \hat{\Omega}) = (x + iy)^l \exp\{-\frac{r^2}{4l_{so}^2}\} \otimes | \uparrow \rangle$. Their spins are polarized along the z-direction and the orbital channel wavefunctions are complex analytic in the xy plane. We then perform a general SU(2) rotation such that the xyz-frame is rotated to the frame of $\hat{e}_1$–$\hat{e}_2$–$\hat{e}_3$. For a coordinate vector $\vec{r}$, its projection in the $\hat{e}_1$–$\hat{e}_2$-plane forms a complex variable $\vec{r} \cdot (\hat{e}_1 + i\hat{e}_2)$ based on in which plane we construct complex analytic functions.

Now it is clear why spin–orbit coupling is essential. Otherwise, if the orbital plane $\hat{e}_1$–$\hat{e}_2$ is flipped, then the complex variable changes to its conjugate, and the complex analyticity is lost. Nevertheless, the spin is polarized perpendicular along the $\hat{e}_3$, which also flips during the flipping of the orbital plane, such that the helicity remains invariant. In general, we can perform an arbitrary SU(2) rotation on the highest weight states and arrive at a set of coherent states forming the over-complete bases of the lowest Landau-level states as

$$\psi_{LLL,\hat{e}_{1,2,3},j_+}(r,\hat{\Omega}) = [(\hat{e}_1 + i\hat{e}_2)\cdot\vec{r}]^l e^{-\frac{r^2}{4l_{so}^2}} \otimes |\alpha_{\hat{e}_3}\rangle, \quad (l \geq 0), \quad (4.31)$$

where $(\hat{e}_3 \cdot \vec{\sigma})|\alpha_{\hat{e}_3}\rangle = |\alpha_{\hat{e}_3}\rangle$.

Now we can make comparisons among harmonic oscillator wavefunctions in different dimensions:

(1) In 1D, there are only real Hermite polynomials.
(2) In 2D, a subset of harmonic wavefunctions z^m (lowest Landau-level) are selected exhibiting the $U(1)$ structure.
(3) In 3D, define a moving frame $\hat{e}_1$–$\hat{e}_2$–$\hat{e}_3$, which is the same as the rigid-body configuration. The complex plane is $\hat{e}_1$–$\hat{e}_2$. In other words, the configuration space of the 3D lowest Landau-level states is that of a triad, or, the $SU(2)$ group manifold.

Since the $SU(2)$ group manifold is isomorphic to the space of unit quaternions, this motivates us to consider the analytic structure in terms of quaternions, which will be presented in Sec. 4.4.4.

4.4.3 *The off-centered solutions to the lowest Landau-level states*

Different from the 2D Landau-level Hamiltonian, which possesses the magnetic translation symmetry, the 3D Landau-level Hamiltonian equation (4.29) does not possess such a symmetry due to the non-Abelian nature of the $SU(2)$ gauge potential. Nevertheless, based on the coherent state representation described by Eq. (4.31), we can define magnetic translations within the $\hat{e}_1$–$\hat{e}_2$-plane, and organize the off-centered solutions in the 3D lowest Landau-level.

Consider all the coherent states in the $\hat{e}_1$–$\hat{e}_2$-plane described by Eq. (4.31). We define the magnetic translation for this set of states as

$$T_{\hat{e}_3}(\vec{\delta}) = \exp\left(-\vec{\delta}\cdot\vec{\nabla} + \frac{i}{4l_{so}^2}\vec{r}_{12}\cdot(\hat{e}_3\times\vec{\delta})\right), \quad (4.32)$$

where the translation vector $\vec{\delta}$ lies in the $\hat{e}_{1,2}$-plane and $\vec{r}_{12} = \vec{r} - \hat{e}_3(\vec{r}\cdot\hat{e}_3)$. Set $\hat{e}_1 = \hat{z}$, and the normal vector $\hat{e}_3$ lying in the xy-plane with an azimuthal angle ϕ', i.e., $\hat{e}_3(\phi') = \hat{x}\cos\phi' + \hat{y}\sin\phi'$, then $\alpha_{\hat{e}_3}(\phi') = \frac{1}{\sqrt{2}}(|\uparrow\rangle + e^{i\phi'}|\downarrow\rangle)$.

Consider the lowest Landau-level states localized at the origin,

$$\psi_{l=0,\hat{e}_3}(r, \hat{\Omega}) = e^{-\frac{r^2}{4l_{so}^2}} \otimes |\alpha_{\hat{e}_3}\rangle, \tag{4.33}$$

and translate it along $\hat{z}$ at the distance R. According to Eq. (4.32), we arrive at

$$\psi_{\phi',R}(\rho, \phi, z) = e^{i\frac{1}{2l_{so}^2}R\rho\sin(\phi-\phi')} e^{-|\vec{r}-R\hat{z}|^2/4l_{so}^2} \otimes \alpha_{\hat{e}_3}(\phi'), \tag{4.34}$$

where $\rho = \sqrt{x^2 + y^2}$ and ϕ is the azimuthal angular of $\vec{r}$ in the xy-plane.

We can restore the rotational symmetry around the $\hat{z}$-axis by performing the Fourier transform with respect to the angle ϕ', i.e., $\psi_{j_z=m+\frac{1}{2},R}(\rho, \phi, z) = \int_0^{2\pi} \frac{d\phi'}{2\pi} e^{im\phi'} \psi_{\phi',R}$. Then the resultant off-centered lowest Landau-level states are the eigenstates of j_z as

$$\psi_{j_z=m+\frac{1}{2},R}(\rho, \phi, z) = e^{\frac{-|\vec{r}-R\hat{z}|^2}{4l_{so}^2}} e^{im\phi} \{J_m(x)|\uparrow\rangle + J_{m+1}(x)e^{i\phi}|\downarrow\rangle\}, \tag{4.35}$$

where $x = R\rho/(2l_{so}^2)$. It describes a wavefunction with the shape of an ellipsoid, whose distribution in the xy-plane is within the distance of ml_{so}^2/R. The states $\psi_{\pm\frac{1}{2},R}$ have the narrowest waist sizes, and their aspect ratio scales as l_{so}/R as R goes large. On the other hand, for those states with $|m| < R/l_{so}$, they localize within the distance of l_{so} from the center located at $R\hat{z}$. As a result, the real space local density of states of the lowest Landau-level grows linearly with R.

4.4.4   *Quaternionic analyticity of the lowest Landau level wavefunctions*

In analogy to complex analyticity of the 2D lowest Landau-level states, we proved that the helicity structure of the 3D lowest Landau levels leads to quaternionic analyticity.

Just like two real numbers forming a complex number, a two-component complex spinor $\psi = (\psi_\uparrow, \psi_\downarrow)^T$ can be mapped to a quaternion by multiplying a j to the second component,

$$f = \psi_\uparrow + j\psi_\downarrow. \tag{4.36}$$

Then, the familiar symmetry transformations can be represented via multiplying quaternions. The time-reversal transformation $i\sigma_2\psi^*$ becomes $Tf = -fj$ satisfying $T^2 = -1$. The $U(1)$ phase $e^{i\theta} \to fe^{i\theta}$, and the $SU(2)$ rotation becomes

$$e^{i\frac{\phi}{2}\sigma_x}\psi \to e^{k\frac{\phi}{2}}f, \quad e^{i\frac{\phi}{2}\sigma_y}\psi \to e^{j\frac{\phi}{2}}f, \quad e^{i\frac{\phi}{2}\sigma_z}\psi \to e^{-i\frac{\phi}{2}}f. \qquad (4.37)$$

To apply the Cauchy–Reman–Fueter condition Eq. (4.14) to 3D, we simply suppress the fourth coordinate,

$$\frac{\partial f}{\partial x} + i\frac{\partial f}{\partial y} + j\frac{\partial f}{\partial z} = 0. \qquad (4.38)$$

We prove a remarkable property below that this condition (Eq. (4.38)) is rotationally invariant.

Lemma 4.1. *If a quaternionic wavefunction $f(x,y,z)$ is quaternionic analytic, i.e., it satisfies the Cauchy–Riemann–Fueter condition, then after an arbitrary $SU(2)$ rotation, the consequential wavefunction $f'(x,y,z)$ remains quaternionic analytic.*

Proof. Consider an arbitrary $SU(2)$ rotation $g(\alpha,\beta,\gamma) = e^{-i\frac{\alpha}{2}\sigma_z}e^{-i\frac{\beta}{2}\sigma_y}e^{-i\frac{\gamma}{2}\sigma_z}$, where α, β, γ are Euler angles. In the quaternion representation, it maps to $g = e^{i\frac{\alpha}{2}}e^{-j\frac{\beta}{2}}e^{i\frac{\gamma}{2}}$. After this rotation $f(x,y,z)$ transforms to

$$f'(x,y,z) = e^{i\frac{\alpha}{2}}e^{-j\frac{\beta}{2}}e^{i\frac{\gamma}{2}}f(x',y',z'), \qquad (4.39)$$

where (x',y',z') are the coordinates by applying g^{-1} on (x,y,z). It can be checked that

$$\left(\frac{\partial}{\partial x} + i\frac{\partial}{\partial y} + j\frac{\partial}{\partial z}\right)e^{i\frac{\alpha}{2}}e^{-j\frac{\beta}{2}}e^{i\frac{\gamma}{2}} = e^{i\frac{\alpha}{2}}e^{-j\frac{\beta}{2}}e^{i\frac{\gamma}{2}}\left(\frac{\partial}{\partial x'} + i\frac{\partial}{\partial y'} + j\frac{\partial}{\partial z'}\right).$$
$$(4.40)$$

Then we have

$$\left(\frac{\partial}{\partial x} + i\frac{\partial}{\partial y} + j\frac{\partial}{\partial z}\right)f'(x,y,z) = 0. \qquad (4.41)$$

Hence, the Cauchy–Riemann–Fueter condition is rotationally invariant. □

Based on this lemma, we prove the quaternionic analyticity of the 3D lowest Landau-level wavefunctions.

Theorem 4.1. *The 3D lowest Landau-level wavefunctions of $H^+_{3D,sym}$ in Eq. (4.29) have a one-to-one correspondence to the quaternionic analytic polynomials in 3D.*

Proof. We denote the quaternionic polynomials, which correspond to the orthonormal bases of the lowest Landau-level wavefunctions in Eq. (4.30), as f_{j_+,j_z}^{LLL} with $j_+ = l + \frac{1}{2}$, and $-j_+ \le j_z \le j_+$. The highest weight states $f_{j_+,j_+}^{LLL} = (x + iy)^l$ are complex analytic in the xy-plane, hence, they are obviously quaternionic analytic. Since all the coherent states can be obtained from the highest weight states via rotations, they are also quaternionic analytic. The coherent states form a set of overcomplex basis of the lowest Landau level wavefunctions, hence all the lowest Landau-level wavefunctions are quaternionic analytic.

Next we prove the completeness that f_{j_+,j_z}^{LLL}'s form the complete basis of the quaternionic analytic polynomials in 3D. By counting the degrees of freedom of the lth-order polynomials of x, y, z, and the number of the constraints from Eq. (4.38), we calculate the total number of the linearly independent lth-order quaternionic analytic polynomials as $C_{l+2}^2 - C_{l+1}^2 = l + 1$. On the other hand, any lowest Landau-level state in the sector of $j_+ = l + \frac{1}{2}$ can be represented as

$$f_l(x, y, z) = \sum_{m=0}^{l} f_{j_+=l+\frac{1}{2},j_z=m+\frac{1}{2}}^{LLL} q_m, \tag{4.42}$$

where q_m is a quaternion constant coefficient. Please note that q_{lm}'s are multiplied from right according to the convention equation (4.36). In Eq. (4.42), we have taken into account the fact $f_{j_+,-j_z}^{LLL} = -f_{j_+,-j_z}^{LLL} j$ due to the time-reversal transformation. Hence, the degrees of freedom in the lowest Landau-level with $j_+ = l + \frac{1}{2}$ are also $l+1$ in the quaternion sense. Hence, the lowest Landau-level wavefunctions are complete for the quaternionic analytic polynomials.   □

4.4.5   *Generalizations to 4D and above*

The above procedure can be straightforwardly generalized to four and even higher dimensions. To proceed, we need to employ the Clifford algebra Γ-matrices. Their ranks in different dimensions and concrete representations are presented in Appendix A. Then we employ the N-dimensional (ND) harmonic oscillator potential combined with spin–orbit coupling as

$$H^{ND,LL} = \frac{p_{ND}^2}{2m} + \frac{1}{2}m\omega_0^2 r_{ND}^2 - \omega_0 \sum_{1 \le i < j \le N} \Gamma_{ij} L_{ij}, \tag{4.43}$$

where $L_{ij} = r_i p_j - r_j p_i$. The spectra of Eq. (4.43) were studied in the context of the supersymmetric quantum mechanics [39]. However, its connection with Landau levels was not noticed there. The spin operators in N-dimensions are defined as $\frac{1}{2}\Gamma_{ij}$.

For the 4D case, the minimal representations for the Γ-matrices are still two-dimensional. They are defined as

$$\Gamma_{ij} = -\frac{i}{2}[\sigma_i, \sigma_j], \quad \Gamma_{i4} = \pm\sigma_i, \tag{4.44}$$

with $1 \leq i < j \leq 3$. The $\pm$ signs of Γ^{i4} correspond to two complex conjugate irreducible fundamental spinor representations of SO(4), and the $+$ sign will be taken below. The spectra of the positive helicity states are flat as $E_{+,n_r} = (2n_r + 2)\hbar\omega$. The coherent state picture for the 4D lowest Landau levels can be similarly constructed as follows: Again pick up two orthogonal axes $\hat{e}$ and $\hat{f}$ to form a 2D complex plane, and define complex analytic functions therein as,

$$(x_a\hat{e}_a + ix_a\hat{f}_a)^l e^{-\frac{r^2}{4l_{so}^2}} \otimes |\alpha_{\hat{e},\hat{f}}\rangle, \tag{4.45}$$

where $|\alpha_{\hat{e},\hat{f}}\rangle$ is the eigenstate of $\Gamma^{\hat{e},\hat{f}} = \hat{e}_a\hat{f}_b\Gamma^{ab}$ satisfying

$$\Gamma^{\hat{e},\hat{f}}|\alpha_{\hat{e},\hat{f}}\rangle = |\alpha_{\hat{e},\hat{f}}\rangle. \tag{4.46}$$

Hence, its spin is locked with its orbital angular momentum in the $\hat{e}$–$\hat{f}$-plane.

Following the similar methods in Sec. 4.4.4, we can prove that the 4D lowest Landau-level wavefunctions for Eq. (4.43) satisfy the 4D Cauchy–Riemann–Fueter condition (4.14), and thus are quaternionic analytic functions. Again it can be proved that they form the complete basis for the quaternionic left-analytic polynomials in 4D.

As for even higher dimensions, quaternions are not defined. Nevertheless, the picture of the complex analytic function defined in the moving frame still applies. If we still work in the spinor representation, we can express the lowest Landau-level wavefunctions as $\psi_{LLL}(x_i) = f_{LLL}(x_i)e^{-\frac{r^2}{2l_0^2}}$, where each component of the spinor f_{LLL} is a polynomial of r_i ($1 \leq i \leq N$). To work out the analytic properties of f_{LLL}, we factorize Eq. (4.43) as

$$H^{ND,LL} = \hbar\omega_0(\Gamma^i a_i^\dagger)(\Gamma^j a_j), \tag{4.47}$$

where a_i is the phonon operator in the ith-dimension defined as $a_i = \frac{1}{\sqrt{2}}\left(\frac{1}{l_0}r_i + i\frac{l_0}{\hbar}p_i\right)$, and $l_0 = \sqrt{\frac{\hbar}{m\omega_0}}$. Then $f_{LLL}(x_i)$ satisfies the following equation,

$$\Gamma^j \frac{\partial}{\partial x_j} f_{LLL}(x_i) = 0, \tag{4.48}$$

which can be viewed as the Euclidean version of the Weyl equation. When coming back to 3D and 4D, and following the mapping of Eq. (4.36), we arrive at quaternionic analyticity.

New let us construct the off-centered solutions to the lowest Landau-level states in 4D. We use $\vec{r}$ to denote a point in the subspace of x_1–x_2–x_3, and $\hat{\Omega}$ as an arbitrary unit vector in it. Set $\hat{e} = \hat{\Omega}$ and $\hat{f} = \hat{e}_4$ (the unit vector along the fourth-axis) in Eq. (4.45). $\alpha_{\hat{\Omega}\hat{e}_4}$ satisfies

$$(\sigma_{i4}\Omega_i)\alpha_{\hat{\Omega}\hat{e}_4} = (\vec{\sigma} \cdot \hat{\Omega})\alpha_{\hat{\Omega}\hat{e}_4} = \alpha_{\hat{\Omega}\hat{e}_4}, \qquad (4.49)$$

hence,

$$\alpha_{\hat{\Omega}\hat{e}_4} = \left(\cos\frac{\theta}{2}, \sin\frac{\theta}{2}e^{i\phi} \right)^T, \qquad (4.50)$$

where we use the gauge convention that the singularity is located at the south pole. Define the magnetic translation in the $\hat{\Omega}$–$\hat{e}_4$-plane,

$$T_{\hat{\Omega}x_4}(u_0\hat{x}_4) = \exp\left(-u_0\partial_{x_4} - \frac{i}{4l_{so}^2}(\vec{r} \cdot \hat{\Omega})u_0 \right), \qquad (4.51)$$

which translates along the $\hat{e}_4$-axis at the distance of u_0. Apply this translation to the state of $e^{-r^2/4l_{so}^2} \otimes \alpha_{\hat{\Omega}\hat{e}_4}$, we arrive at the off-center solution

$$\psi_{\Omega,u_0}(\vec{r}, x_4) = e^{-\frac{r^2+x_4^2}{4l_{so}^2}} e^{-i\frac{ru_0}{2l_{so}^2}} \otimes \alpha_{\hat{\Omega}\hat{e}_4}. \qquad (4.52)$$

Next, we perform the Fourier transform over the direction $\hat{\Omega}$,

$$\psi_{4D;j,j_z}(\vec{r}, x_4) = \int d\Omega Y_{l+\frac{1}{2},m+\frac{1}{2}}^{-\frac{1}{2}}(\hat{\Omega})\psi_{\Omega,w_0}(\vec{r}, x_4), \qquad (4.53)$$

where $j = l + \frac{1}{2}$ and $j_z = m + \frac{1}{2}$. Due to the Berry phase structure of $\alpha_{\hat{\Omega}\hat{e}_4}$ over $\hat{\Omega}$, the monopole spherical harmonic functions, $Y_{l+\frac{1}{2},m+\frac{1}{2}}^{-\frac{1}{2}}(\hat{\Omega})$, are used instead of the regular spherical harmonics. Then Eq. (4.53) possesses the 3D rotational symmetry around the new center $(0,0,0,w_0)$, and is characterized by the 3D angular momentum quantum numbers (j, j_z). The monopole harmonic function $Y_{jj_z}^q(\hat{\Omega})$ here is defined as

$$Y_{jj_z}^q(\hat{\Omega}) = \sqrt{\frac{2j+1}{4\pi}}e^{i(j_z+q)\phi}d_{j_z,-q}^l(\theta), \qquad (4.54)$$

where θ and ϕ are the polar and azimuthal angles of $\hat{\Omega}$, and $d_{j_z,-q}^l(\theta) = \langle jj_z|e^{-iJ_y\theta}|j-q\rangle$ is the standard Wigner rotation d-matrix. The gauge choice is consistent with that of Eq. (4.50).

4.4.6   *Boundary helical Dirac and Weyl modes*

The topological nature of the 3D Landau-level states is indicated clearly in the gapless surface spectra. Consider a ball of the radius $R_0 \gg l_{so}$ imposed by the open boundary condition. We have numerically solved the spectrum as shown in Fig. 4.3(b). Inside the bulk, the Landau-level spectrum is flat with respect to $j_+ = l + \frac{1}{2}$. As l increases to large values such that the classic orbital radiuses approach the boundary, the Landau levels become surface states and develop dispersive spectra.

We can derive the effective equation for the surface mode based on Eq. (4.29). Since r is fixed at the boundary, it becomes a rotor equation on the sphere. By linearizing the dispersion at the chemical potential μ, and replacing the angular momentum quantum number l by the operator $\vec{\sigma} \cdot \vec{L}$, we arrive at $H_{sf} = (v_f/R_0)\vec{\sigma} \cdot \vec{L} - \mu$ with v_f the Fermi velocity. This is the helical Dirac equation defined on the boundary sphere. When expanded in the local patch around the north pole $R_0 \hat{z}$, we arrive at

$$H_{sf} = \hbar v_f (\vec{k} \times \vec{\sigma}) \cdot \hat{z} - \mu. \tag{4.55}$$

The gapless surface states are robust against time-reversal invariant perturbations if odd numbers of helical Fermi surfaces exist according to the $\mathbb{Z}_2$ criterion [6,7]. Since each fully occupied Landau-level contributes one helical Dirac Fermi surface, the bulk is topologically nontrivial if odd numbers of Landau levels are occupied.

A similar procedure can be applied to the high-dimensional case by imposing the open boundary condition to Eq. (4.43). For example, around the north pole of $r_N = (0, \dots, R_0)$, the linearized low energy equation for the boundary modes is

$$H_{bd} = \hbar v_f \sum_{i=1}^{D-1} k_i \Gamma^{iN} - \mu. \tag{4.56}$$

On the boundary of the 4D sphere, it becomes the 3D Weyl equation that

$$H_{bd} = \hbar v_f \vec{k} \cdot \vec{\sigma} - \mu. \tag{4.57}$$

4.4.7   *Bulk–boundary correspondences*

We have already studied the bulk and boundary states of 2D, 3D and 4D lowest Landau-level states. They exhibit a series of interesting bulk–boundary correspondences as summarized in Table 4.1. In the 2D case, the bulk wavefunctions in the lowest Landau-level is complex analytic satisfying the Cauchy–Riemann condition. The 1D edge states satisfy the chiral wave

Table 4.1.   Bulk–boundary correspondences in the lowest Landau-level (LLL) states in 2D, 3D, and 4D.

	Bulk (Euclidean)	Boundary (Minkowski)
2D LLL	Complex analyticity	1D chiral wave
	$\partial_x f + i\partial_y f = 0$	$\partial_t \psi + \partial_x \psi = 0$
3D LLL	3D quaternionic analyticity	2D helical Dirac mode
	$\partial_x f + i\partial_y f + j\partial_z f = 0$	$\partial_t \psi + \sigma_2 \partial_x \psi - \sigma_1 \partial_y \psi = 0$
4D LLL	Quaternionic analyticity	3D Weyl mode
	$\partial_x f + i\partial_y f + j\partial_z f + k\partial_u f = 0$	$\partial_t \psi + \sigma_1 \partial_x \psi + \sigma_2 \partial_y \psi + \sigma_3 \partial_z \psi = 0$

equation (4.23). It is essentially the Weyl equation, which is of single-component in 1D. It can be viewed as the Minkowski version of the Cauchy–Riemann condition of Eq. (4.4). Or, conversely, the Cauchy–Riemann condition for the bulk wavefunctions can be viewed as the Euclidean version of the Weyl equation.

This correspondence goes in parallel in 3D and 4D lowest Landau-level wavefunctions. Their bulk wavefunctions satisfy the quaternionic analytic conditions, which can be viewed as the Euclidean version of the helical Dirac and Weyl equations, respectively.

4.4.8   *Many-body interacting wavefunctions*

It is natural to further investigate many-body interacting wavefunctions in the lowest Landau levels in 3D and 4D. As is well known that the complex analyticity of the 2D lowest Landau-level wavefunctions results in the elegant from of the 2D Laughlin wavefunction Eq. (4.22), which describes a 2D quantum liquid [22,23]. It is natural to further expect that the quaternionic analyticity of the 3D and 4D lowest Landau levels would work as a guidance in constructing high-dimensional SU(2) invariant quantum liquid. Nevertheless, the major difficulty is that quaternions do not commute. It remains challenging how to use quaternions to represent a many-body wavefunction with the spin degree of freedom.

Nevertheless, we present below the spin-polarized fractional many-body states in 3D and 4D Landau levels. In the 3D case, if the interaction is spin-independent, we expect spontaneous spin polarization at very low fillings due to the flatness of lowest Landau-level states in analogy to the 2D quantum Hall ferromagnetism [22, 40–43]. According to Eq. (4.31), fermions concentrate to the highest weight states in the orbital plane $\hat{e}_1$–$\hat{e}_2$ with spin polarized along $\hat{e}_3$, then it is reduced to a 2D quantum Hall-like problem on a membrane floating in the 3D space. Any 2D fractional quantum Hall-like state can be formed under suitable interaction pseudopotentials [25, 44, 45]. For example, the $\nu = \frac{1}{3}$ Laughlin-like state on this membrane is constructed as

$$\Psi_{\frac{1}{3}}(\vec{r}_1, \vec{r}_2, \ldots, \vec{r}_n)_{\sigma_1 \sigma_2 \cdots \sigma_n} = \prod_{i<j}[(\vec{r}_i - \vec{r}_j) \cdot (\hat{e}_1 + i\hat{e}_2)]^3$$

$$\otimes |\alpha_{\hat{e}_3}\rangle_{\sigma_1} |\alpha_{\hat{e}_3}\rangle_{\sigma_2} \cdots |\alpha_{\hat{e}_3}\rangle_{\sigma_n}, \qquad (4.58)$$

where $|\alpha_{\hat{e}_3}\rangle$ represents a polarized spin eigenstate along $\hat{e}_3$, and the Gaussian weight is suppressed for simplicity. Such a state breaks rotational symmetry and time-reversal symmetry spontaneously, thus it possesses low energy spin-wave modes. Due to the spin–orbit locked configuration in Eq. (4.31), spin fluctuations couple to the vibrations of the orbital motion plane, thus the metric of the orbital plane becomes dynamic. This is a natural connection to the work of geometrical description in fractional quantum Hall states [46–48].

Let us consider the 4D case, we assume that spin is polarized as the eigenstate $|\uparrow\rangle$ of $\Gamma^{12} = \Gamma^{34} = \sigma_3$. The corresponding spin-polarized lowest Landau-level wavefunctions are expressed as

$$\Psi^{4D}_{LLL,m,n} = (x + iy)^m (z + iu)^n \otimes |\uparrow\rangle, \qquad (4.59)$$

with $m, n \geq 0$. If all these spin-polarized lowest Landau-level states with $0 \leq m < N_m$ and $0 \leq n < N_n$ are filled, the many-body wavefunction is a Slater-determinant as

$$\Psi^{4D}(v_1, w_1; \ldots; v_N, w_N) = \det[v_i^\alpha w_i^\beta], \qquad (4.60)$$

where the coordinates of the ith particle form two pairs of complex numbers as $v_i = x_i + iy_i$ and $w_i = z_i + iu_i$; α, β and i satisfy $0 \leq \alpha < N_m, 0 \leq \beta < N_n$ and $1 \leq i \leq N = N_m N_n$. Such a state has a 4D uniform density as $\rho = \frac{1}{4\pi^4 l_G^2}$. A Laughlin-like wavefunction can be written down as $\Psi_k^{4D} = (\Psi^{4D})^k$ whose filling relative to ρ should be $1/k^2$. It would be interesting to further study its electromagnetic responses and fractional topological excitations based on Ψ_k^{4D}. Again such a state spontaneously breaks the rotational symmetry, and the coupled spin and orbital excitations would be interesting.

4.5   Dimensional reductions: 2D and 3D Landau levels with broken parity

In this section, we review another class of isotropic Landau-level-like states with time-reversal symmetry but broken parity in both 2D and 3D. The Hamiltonians are again the harmonic potentials plus spin–orbit couplings, but they are the couplings between spin and linear momentum, not orbital angular momentum [26, 49, 50]. They exhibit topological properties very similar to Landau levels.

An early study of these systems filled with bosons can be found in [51]. The spin–orbit coupled Bose–Einstein condensations (BECs) spontaneously break time-reversal symmetry, and exhibit the skyrmion-type spin textures coexisting with half-quantum vortices, which have been reviewed in [52]. Spin–orbit coupled BECs have become an active research direction of cold-atom physics, as extensively studied in [49, 53–57].

4.5.1   The 2D parity-broken Landau levels

We consider the Hamiltonian of the Rashba spin–orbit coupling combined with a 2D harmonic potential as

$$H_{2D,hm} = -\frac{\hbar^2 \nabla^2}{2M} + \frac{1}{2} M\omega^2 r^2 - \lambda(-i\hbar\nabla_x \sigma_y + i\hbar\nabla_y \sigma_x), \quad (4.61)$$

where λ is the spin–orbit coupling strength with the unit of velocity. Equation (4.61) possesses the $C_{v\infty}$-symmetry and time-reversal symmetry.

We fill the system with fermions and work on its topological properties. There are two different length scales: The trap length scale is defined as $l_T = \sqrt{\frac{\hbar}{M\omega}}$. If, without the trap, the single particle states $\psi_\pm(\vec{k})$ are eigenstates of the helicity operator $\vec{\sigma} \cdot (\vec{k} \times \hat{z})$ whose eigenvalues are ± 1, their spectra are $\epsilon_\pm(\vec{k}) = \hbar^2(k \mp k_0)^2/(2M)$, respectively. The lowest energy states are $\psi_+(\vec{k})$ located around a ring in momentum space with radius $k_0 = M\lambda/\hbar$. This introduces a spin–orbit length scale as $l_{so} = 1/k_0$. Then the ratio between these two length scales defines a dimensionless parameter $\alpha = l_T/l_{so}$, which describes the spin–orbit coupling strength relative to the harmonic potential.

In the case of strong spin–orbit coupling, i.e., $\alpha \gg 1$, a clear picture appears in momentum space. The low energy states are reorganized from the plane-wave states $\psi_+(\vec{k})$ with $k \approx k_0$. Since $\alpha \gg 1$, we can safely project out the negative helicity states $\psi_-(\vec{k})$ at high-energy, then the harmonic potential in the low energy sector becomes a Laplacian in momentum space subject to a Berry connection $\vec{A}_k$ as

$$V = \frac{M}{2}\omega^2 r^2 = \frac{M}{2}\omega^2(i\nabla_k - A_k)^2, \quad (4.62)$$

which drives the particle moving around the ring. It is well known that for the Rashba Hamiltonian, the Berry connection A_k gives rise to a π-flux at $\vec{k} = (0,0)$ but zero Berry curvature at $\vec{k} \neq 0$ [58]. The consequence is that the angular momentum eigenvalues become half-integers as $j_z = m + \frac{1}{2}$. The angular dispersion of the spectra can be estimated as $E_{agl}(j_z) = (j_z^2/2\alpha^2)\hbar\omega$, which is strongly suppressed by spin–orbit coupling. On the other hand, the radial energy quantization remains as usual $E_{rad}(n_r) = (n_r + \frac{1}{2})\hbar\omega$ up to a constant. Hence the total energy dispersion is

$$E_{n_r,j_z} \approx \left(n_r + \frac{1}{2} + \frac{j_z^2}{2\alpha^2}\right)\hbar\omega. \tag{4.63}$$

Similar results have also been obtained in [53–55]. Since $\alpha \gg 1$, the spectra are nearly flat with respect to j_z, we can treat n_r as a Landau-level index.

Next we define the edge modes of such systems, and their stability problem is quite different from that of the chiral edge modes of 2D magnetic Landau-level systems. In the regime that $\alpha \gg 1$, the spin–orbit length l_{so} is much shorter than l_T, such that l_T is viewed as the cutoff of the sample size. States with $|j_z| < \alpha$ are viewed as bulk states which localize within the region of $r < l_T$. For states with $|j_z| \sim \alpha$, their energies touch the bottom of the next higher Landau-level, and thus they should be considered as edge states. Due to time-reversal symmetry, each filled Landau-level of Eq. (4.61) gives rise to a branch of edge modes of Kramers' doublets $\psi_{n_r,\pm j_z}$. In other words, these edge modes are helical rather than chiral. Similarly to the Z_2 criterion in [6, 7], which was defined for Bloch wave states, in our case the following mixing term, $H_{mx} = \psi_{2D,n_r,j_z}^\dagger \psi_{2D,n_r,-j_z} + h.c.$, is forbidden by time-reversal symmetry. Consequently, the topological index for this system is Z_2.

4.5.2 *Dimensional reduction from 3D Landau levels*

In fact, we construct a Hamiltonian closely related to Eq. (4.61) such that its ground state is solvable exhibiting exactly flat dispersion. It is a consequence of the dimensional reduction based on the 3D Landau-level Hamiltonian equation (4.29). We cut a 2D off-centered plane perpendicular to the z-axis with the interception $z = z_0$. In this off-centered plane, inversion symmetry is broken, and Eq. (4.29) is reduced to

$$H_{2D,re} = H_{2D,hm} - \omega L_z \sigma_z. \tag{4.64}$$

The first term is just Eq. (4.61) by identifying $\lambda = \omega z_0$ and the frequency of the second term is the same as that of the harmonic trap. If $z_0 = 0$, the Rashba spin–orbit coupling vanishes, and Eq. (4.64) becomes the 2D quantum spin-Hall Hamiltonian, which is a double copy of Eq. (4.19). At $z_0 \neq 0$, σ_z is no longer conserved due to spin–orbit coupling.

In Sec. 4.4.3, we derived the off-centered ellipsoid type wavefunction. After setting $z = z_0$ in Eq. (4.35), we arrive at the following 2D wavefunction,

$$\psi_{2D,j_z}(r,\phi) = e^{-\frac{r^2}{4l_{so}^2}}\{e^{im\phi}J_m(k_0r)|\uparrow\rangle + e^{i(m+1)\phi}J_{m+1}(k_0r)|\downarrow\rangle\}, \quad (4.65)$$

where $J_m(k_0r)$ is the mth order Bessel functions. It is straightforward to prove that the simple reduction indeed gives rise to the solutions of the lowest Landau level to Eq. (4.64), since the partial derivative along the z-direction of the solution in Eq. (4.35) equals zero at $z = z_0$. We also prove that the energy dispersion is exactly flat as,

$$H_{2D,re}\psi_{2D,j_z} = \left(1 - \frac{\alpha^2}{2}\right)\hbar\omega\psi_{2D,j_z}. \quad (4.66)$$

The above two Hamiltonians equations (4.64) and (4.61) are nearly the same except the $L_z\sigma_z$ term, whose effect relies on the distance from the origin. Consider the lowest Landau-level solutions at $\alpha \gg 1$. The decay length of the Gaussian factor in Eq. (4.65) is l_T. Nevertheless, the Bessel functions peak around $k_0r_0 \approx m$, i.e., $r_0 \approx \frac{m}{\alpha}l_T$. Hence for states with $j_z < \alpha$, their wavefunctions already decay before reach l_T. Then the $L_z\sigma_z$-term compared to the Rashba one is a small perturbation at the order of $\omega r_0/\lambda = r_0/z_0 \ll 1$. In this regime, these two Hamiltonians are equivalent. In contrast, in the opposite limit that $j_z \gg \alpha^2$, the Bessel functions are cut off by the Gaussian factor, and only their initial power-law parts participate. The classic orbit radii are just $r_0 \approx \sqrt{m}l_T$, then the physics of Eq. (4.64) is controlled by the $L_s\sigma_z$-term as in the quantum spin-Hall systems. For the intermediate region that $\alpha < j_z < \alpha^2$, the physics is a crossover between the above two limits.

The many-body physics based on the above spin–orbit coupled Landau levels in Eq. (4.65) would be very interesting. Fractional topological states would be expected which are both rotationally and time-reversal invariant. However, σ_z is not a good quantum number and parity is also broken, hence, these states should be very different from a double copy of the fractional Laughlin states with spin-up and spin-down particles. The nature of topological excitations and properties of edge modes will be deferred to a future study.

4.5.3   The 3D parity-broken Landau levels

We have also considered the problem of a 3D harmonic potential plus a Weyl-type spin–orbit coupling, whose Hamiltonian is defined as [26],

$$H_{3D,hm} = -\frac{\hbar^2\nabla^2}{2M} + \frac{1}{2}M\omega^2r^2 - \lambda(-i\hbar\vec{\nabla}\cdot\vec{\sigma}). \quad (4.67)$$

The analysis can be performed in parallel to the 2D case. In the absence of spin–orbit coupling, the low energy states of Eq. (4.67) in momentum space form a spin–orbit sphere. The harmonic potential further quantizes the energy spectra as

$$E_{n_r,j,j_z} \approx \left(n_r + \frac{1}{2} + \frac{j(j+1)}{2\alpha^2} \right) \hbar\omega, \tag{4.68}$$

where n_r is the Landau-level index and j is the quantum number of the total angular momentum. Again j takes half-integer values because the Berry phase on the low-energy sphere exhibits a unit monopole structure.

Now we perform the dimensional reduction from the Hamiltonian equation (4.43) in the 4D case to 3D. We cut a 3D off-centered hyper-plane perpendicular to the fourth axis with the interception $x_4 = u_0$. Within this 3D hyper-plane of $(x_1, x_2, x_3, x_4 = u_0)$, Eq. (4.43) is reduced to

$$H_{3D,re} = H_{3D,hm} - \omega \vec{L} \cdot \vec{\sigma}, \tag{4.69}$$

where the first term is just Eq. (4.67) with the spin–orbit coupling strength set by $\lambda = \omega u_0$. Again, based on the center-shifted wavefunction in the lowest Landau level equation (4.53), and by setting $x_4 = u_0$, we arrive at the following wavefunction

$$\psi_{3D,JJ_z}(\vec{r}) = e^{-\frac{r^2}{4l_{so}^2}} \{ j_l(k_0 r) Y_{+,J,J_z}(\Omega_r) + i j_{l+1}(k_0 r) Y_{-,J,J_z}(\Omega_r) \}, \tag{4.70}$$

where $k_0 = u_0/l_T^2 = m\lambda/\hbar$; j_l is the lth order spherical Bessel function. $Y_{\pm,j,l,j_z}$'s are the spin–orbit coupled spherical harmonics defined as

$$Y_{+,j,l,j_z}(\Omega) = \left(\sqrt{\frac{l+m+1}{2l+1}} Y_{lm}, \sqrt{\frac{l-m}{2l+1}} Y_{l,m+1} \right)^T$$

with the positive eigenvalue of $l\hbar$ for $\vec{\sigma} \cdot \vec{L}$, and

$$Y_{-,j,l,j_z}(\Omega) = \left(-\sqrt{\frac{l-m}{2l+1}} Y_{lm}, \sqrt{\frac{l+m+1}{2l+1}} Y_{l,m+1} \right)^T$$

with the negative eigenvalue of $-(l+1)\hbar$ for $\vec{\sigma}\cdot\vec{L}$. It is straightforward to check that $\psi_{3D,j,j_z}(\vec{r})$ in Eq. (4.70) is the ground-state wavefunction satisfying

$$H_{3D,re}\psi_{3D,j,j_z}(\vec{r}) = \left(\frac{3}{2} - \frac{\alpha^2}{2} \right) \hbar\omega\psi_{3D,j,j_z}(\vec{r}). \tag{4.71}$$

4.6   High-dimensional Landau levels of Dirac fermions

In this section, we review the progress on the study of 3D Landau levels of relativistic Dirac fermions [28]. This is a square-root problem of the 3D Landau-level problem based on the Schrödinger equation reviewed in Sec. 4.4. This can also be viewed as Landau levels of complex quaternions.

4.6.1   *3D Landau levels for Dirac fermions*

In Eq. (4.25), two sets of phonon creation and annihilation operators $(a_x, a_y; a_x^\dagger, a_y^\dagger)$ are combined with the real and imaginary units to construct the Landau-level Hamiltonian for 2D Dirac fermions. Since in 3D there exist three sets of phonon creation and annihilation operators, complex numbers are insufficient.

The new strategy is to employ the Pauli matrices $\vec{\sigma}$ such that

$$H_{3D}^D = v\left\{\alpha_i p_i + \gamma_i i\hbar \frac{r_i}{l_0^2}\right\} = \frac{\hbar\omega}{\sqrt{2}} \begin{bmatrix} 0 & i\sigma_i a_i^\dagger \\ -i\sigma_i a_i & 0 \end{bmatrix}, \qquad (4.72)$$

where the repeated index i runs over x, y and z; $v = \frac{1}{2}l_0\omega$. The convention of γ-matrices is

$$\beta = \gamma_0 = \tau_3 \otimes I, \quad \alpha_i = \tau_1 \otimes \sigma_i, \quad \gamma_i = \beta\alpha_i = i\tau_2 \otimes \sigma_i. \qquad (4.73)$$

Equation (4.72) contains the complex combination of momenta and coordinates, thus it can be viewed as the generalized Dirac equation defined in the phase space. Apparently, Eq. (4.72) is rotationally invariant. It is also time-reversal invariant under the definition $T = \gamma_2\gamma_3 K$ where K is the complex conjugation, and $T^2 = -1$. Since $\beta H_{3D}^D \beta = -H_{3D}^D$, H_{3D}^D possesses the particle–hole symmetry and its spectra are symmetric with respect to the zero energy.

Similar to the 2D case, $(H_{3D}^D)^2$ has a supersymmetric structure. The square of Eq. (4.72) is block-diagonal, and two blocks are just the nonrelativistic 3D Landau-level Hamiltonians in Eq. (4.29),

$$\frac{(H_{3D}^D)^2}{\frac{1}{2}\hbar\omega} = \begin{bmatrix} H_{3D,sym}^+ - \frac{3}{2}\hbar\omega & 0 \\ 0 & H_{3D,sym}^- + \frac{3}{2}\hbar\omega \end{bmatrix}, \qquad (4.74)$$

where the mass M in $H_{3D,sym}^\pm$ is defined through the relation $l_0 = \sqrt{\hbar/(M\omega)}$. Based on Eq. (4.74), the energy eigenvalues of Eq. (4.72) are $E_{\pm n_r, j, j_z} = \pm\hbar\omega\sqrt{n_r}$, corresponding to positive and negative square roots of the nonrelativistic dispersion, respectively. The Landau-level wavefunctions

of the 3D Dirac electrons are expressed in terms of the nonrelativistic ones
of Eq. (4.29) as

$$\Psi_{\pm n_r,j,j_z}(\vec{r}) = \frac{1}{\sqrt{2}} \begin{pmatrix} \psi_{n_r,j_+,l,j_z}(\vec{r}) \\ \pm i\psi_{n_r-1,j_-,l+1,j_z}(\vec{r}) \end{pmatrix}. \tag{4.75}$$

Please note that the upper and lower two components possess different
values of orbital angular momenta. They exhibit opposite helicities of $j_\pm$,
respectively. The zeroth Landau-level ($n_r = 0$) states are special: There is
only one branch, and only the first two components of the wavefunctions are
nonzero as

$$\Psi_{n_r=0,j,j_z}(\vec{r}) = \begin{bmatrix} \Psi_{LLL,j_+,j_z}(\vec{r}) \\ 0 \end{bmatrix}, \tag{4.76}$$

where Ψ_{LLL,j_+,j_z}'s are the lowest Landau level solutions to the nonrelativistic
Hamiltonian Eq. (4.29).

Again the nontrivial topology of the 3D Dirac Landau problem manifests
in the gapless surface modes. Consider a spherical boundary with a large
radius R. The Hamiltonian takes the form of Eq. (4.72) inside the sphere, and
changes to the usual massive Dirac Hamiltonian $H_D = \alpha_i P_i + \beta\Delta$ outside.
We take the limit of $|\Delta| \to \infty$. Loosely speaking, this is a square-root version
of the open boundary problem of the 3D nonrelativistic case in Sec. 4.4.6.
Since square-roots can be taken as positive and negative, each branch of the
surface modes in the nonrelativistic Schrödinger case corresponds to a pair
of relativistic surface branches. These two branches disperse upward and
downward as increasing the angular momentum j, respectively. However,
the zeroth Landau-level branch is singled out. We can only take either
the positive or negative square root for its surface excitations. Hence, the
surface spectra connected to the bulk zeroth Landau-level disperse upward
or downward depending on the sign of the vacuum mass.

4.6.2 *Nonminimal Pauli coupling and anomaly*

Due to the particle–hole symmetry of Eq. (4.72), the 3D zeroth Landau-
level states are half-fermion modes in the same way as those in the 2D
Dirac case. Moreover, in the 3D case, the degeneracy is over the 3D angular
momentum numbers (j_+, j_z), thus the degeneracy is much higher than that of
2D. According to whether the chemical potential μ approaches 0^+ or 0^-, each
state in the zeroth lowest Landau level contributes a positive, or, negative
half fermion number, respectively. The Lagrangian of the 3D massless Dirac

Landau level problem is,

$$L = \bar{\psi}\{\gamma_0 i\hbar\partial_t - iv\gamma_i\hbar\partial_i\}\psi - v\hbar\bar{\psi}i\gamma_0\gamma_i\psi F^{0i}(r), \tag{4.77}$$

where $F^{0i} = x_i/l_0^2$. In all the dimensions higher than 2, $i\gamma_0\gamma_i$'s are a different set from γ_i's, thus Eq. (4.77) is an example of nonminimal coupling of the Pauli type. More precisely, it is a coupling between the electric field and the electric dipole moment. In the 2D case, the Lagrangian has the same form as Eq. (4.77), however, since $\gamma_{0,1,2}$ are just the usual Pauli matrices, it is reduced to the minimal coupling to the $U(1)$ gauge field.

Equation (4.77) is a problem of massless Dirac fermions coupled to a background field via a nonminimal Pauli coupling at 3D and above. The Fermion density is pumped by the background field from vacuum. This is similar to parity anomaly, and indeed it is reduced to parity anomaly in 2D. However, the standard parity anomaly only exists in even spatial dimensions [33, 35–37]. By contrast, the Landau-level problems of massless Dirac fermions can be constructed in any high spatial dimensions. Obviously, they are not chiral anomalies defined in odd spatial dimensions, either. It would be interesting to further study the nature of such kind of "anomaly".

In fact, Eq. (4.72) is just one possible representation for Landau levels of 3D massless Dirac fermions. A general 3D Dirac Landau-level Hamiltonian with a mass term can be defined as

$$H_{3D}^D(\hat{e}_1, \hat{e}_2, \hat{e}_3) = v[(\vec{\tau} \cdot \hat{e}_1) \otimes \sigma_i P_i + \hbar/l_0^2(\vec{\tau} \cdot \hat{e}_2)$$
$$\otimes \sigma_i r_i] + mv^2(\vec{\tau} \cdot \hat{e}_3) \otimes I, \tag{4.78}$$

where $\tau_{1,2,3}$ are Pauli matrices acting in the particle–hole channel, and $\hat{e}_{1,2,3}$ form an orthogonal triad in the 3D space. Equation (4.72) corresponds to the case of $\hat{e}_1 = \hat{x}$ and $\hat{e}_2 = \hat{y}$, and $m = 0$. The parameter space of $H_{3D}^D(\hat{e}_1, \hat{e}_2, \hat{e}_3)$ is the triad configuration space of SO(3).

Consider that the configuration of the triad $\hat{e}_{1,2,3}$ is spatially dependent. The first term in Eq. (4.78) should be symmetrized as $\frac{1}{2}\vec{\tau} \cdot [(\hat{e}_1(r)P_i + P_i\hat{e}_1(r)] \otimes \sigma_i$. The spatial distribution of the triad of $\hat{e}_{1,2,3}(\vec{r})$ can be in a topologically nontrivial configuration. If the triad is only allowed to rotate around a fixed axis, its configuration space is $U(1)$ which can form a vortex line type defect. There should be a Callan–Harvey type effect of the fermion zero modes confined around the vortex line [59]. In general, we can also have a 3D skyrmion type defect of the triad configuration. These novel defect problems and the associated zero energy fermionic excitations will be deferred to later studies.

4.6.3 Landau levels for Dirac fermions in four dimensions and above

The Landau-level Hamiltonian for Dirac fermions can be generalized to arbitrary N-dimensions (ND) by replacing the Pauli matrices in Eq. (4.72) with the Clifford algebra Γ-matrices in ND. We use the representation of the Γ-matrices as presented in Appendix A.

In odd dimensions $D = 2k + 1$, we use the kth rank Γ-matrices to construct the $D = 2k + 1$-dimensional Dirac Landau-level Hamiltonian,

$$H_{2k+1}^D = \frac{\hbar\omega_0}{2} \begin{pmatrix} 0 & i\Gamma_i^{(k)} a_i^\dagger \\ -i\Gamma_i^{(k)} a_i & 0 \end{pmatrix}, \qquad (4.79)$$

where $\Gamma_i^{(k)}$ is $2^k \times 2^k$-dimensional matrix, and $1 \le i \le 2k+1$. Again, $(H_{2k+1}^D)^2$ are reduced to a supersymmetric version of the $2k + 1$-dimensional Landau-level Hamiltonian for Schödinger fermions in Eq. (4.43). All other properties are parallel to the 3D case explained before.

For even dimensions $D = 2k$, we still take Eq. (4.79) by simply removing the terms of the $(2k + 1)$th dimension and keeping the terms from the first to the $(2k)$th dimension. Nevertheless, such a construction is reducible. In the representation presented in Appendix A, Eq. (4.79) after eliminating the $\Gamma_{2k+1}^{(k)}$ term can be factorized into a pair of Hamiltonians

$$H_{2k}^{\pm,D} = \frac{\hbar\omega_0}{2} \begin{pmatrix} 0 & \pm a_{2k}^\dagger + i\sum_{i=1}^k \Gamma_i^{(k-1)} a_i^\dagger \\ \pm a_k - i\sum_{i=1}^k \Gamma_i^{(k-1)} a_i & 0 \end{pmatrix}, \qquad (4.80)$$

where $\pm$ correspond to the pair of fundamental and anti-fundamental spinor representations in even dimensions.

For example, in four dimensions, we have

$$H_{4D}^{\pm,D} = \frac{\hbar\omega}{\sqrt{2}} \begin{bmatrix} 0 & \pm a_4^\dagger + i\sigma_i a_i^\dagger \\ \pm a_4 - i\sigma_i a_i & 0 \end{bmatrix}. \qquad (4.81)$$

Since three quaternionic imaginary units i, j, and k can be mapped to Pauli matrices $-i\sigma_1, -i\sigma_2$, and $-i\sigma_3$, respectively, and the annihilation and creation operators are essentially complex. $\pm a_4 - i\sigma_i a_i$ can be viewed as complex quaternions. Hence, Eq. (4.81) is a complex quaternionic generalization of the 2D Dirac Landau-level Hamiltonian equation (4.25).

4.7 High-dimensional Landau levels in the Landau-like gauge

We have discussed the construction of Landau levels in high dimensions for both Schrödinger and Dirac fermions in the symmetric-like gauge. In these problems, the rotational symmetry is explicitly maintained. Below we review the construction of Landau levels in the Landau-like gauge by reorganizing plane-waves to exhibit nontrivial topological properties [30]. It still preserves the flat spectra but not the rotational symmetry.

4.7.1 *Spatially separated 1D chiral modes: 2D Landau levels*

We recapitulate the Landau levels in the Landau gauge. By setting $A_x = By$ and $A_y = 0$ in the Hamiltonian equation (4.18), we arrive at

$$H_{2D,L} = \frac{P_y^2}{2M} + \frac{\left(P_x - \frac{e}{c}A_x\right)^2}{2M} = \frac{P_y^2}{2M} + \frac{1}{2}M\omega^2(y - l_B^2 P_x)^2, \quad (4.82)$$

with $l_B = \sqrt{\frac{\hbar}{M\omega}}$. The Landau-level wavefunctions are a product of a plane wave along the x-direction and a 1D harmonic oscillator wavefunction in the y-direction,

$$\psi_n(x, y) = e^{ik_x}\phi_n(y - y_0(k)), \quad (4.83)$$

where ϕ_n is the nth harmonic oscillator eigenstate with the characteristic length l_B, and its equilibrium position is determined by the momentum k_x, $y_0(k_x) = l_B^2 k_x$.

Hence, the Landau-level states with positive and negative values of k_x are shifted oppositely along the y-direction, and become spatially separated. If imposing the open boundary condition along the y-axis, chiral edge modes appear. The 2D quantum Hall effect is just the spatially separated 1D chiral anomaly in which the chiral current becomes the transverse charge current. After the projection to the lowest Landau level, we identify $y = l_B^2 k_x$, hence, the two spatial coordinates x and y become noncommutative as [60]

$$[x, y]_{LLL} = il_B^2. \quad (4.84)$$

In other words, the xy-plane is equivalent to the 2D phase space of a 1D system $(x; k_x)$ after the lowest Landau-level projection.

4.7.2 *Spatially separated 2D helical modes: 3D Landau levels*

The above picture can be generalized to the 3D Landau-level states: We keep the plane-wave modes with the good momentum numbers (k_x, k_y) and shift them along the z-axis. Spin–orbit coupling is introduced to generate the helical structure to these plane-waves, and the shifting direction is determined by the sign of helicity. To be concrete, the 3D Landau-level Hamiltonian in the Landau-like gauge is constructed as follows [30],

$$
\begin{aligned}
H_{3D,L}^{\pm} &= \frac{\vec{P}^2}{2M} + \frac{1}{2}M\omega_{so}^2 z^2 \mp \omega_{so}z(P_x\sigma_y - P_y\sigma_x) \\
&= \frac{P_z^2}{2M} + \frac{1}{2}M\omega_{so}^2 \left[z \mp \frac{1}{\hbar}l_{so}^2(P_x\sigma_y - P_y\sigma_x)\right]^2,
\end{aligned}
\tag{4.85}
$$

where $l_{so} = \sqrt{\hbar/(M\omega_{so})}$.

The key of Eq. (4.85) is the z-dependent Rashba spin–orbit coupling, such that it can be decomposed into a set of 1D harmonic oscillators along the z-axis coupled to 2D helical plane-waves. Define the helicity operator $\hat{\Sigma}_{2d}(\hat{k}_{2d}) = \hat{k}_x\sigma_y - \hat{k}_y\sigma_x$ where $\hat{k}$ is the unit vector along the direction of $\vec{k}$. $\chi_\Sigma(\hat{k}_{2d})$ is the eigenstate of $\hat{\Sigma}$ and $\Sigma = \pm 1$ is the eigenvalue. Then the 3D Landau-level wavefunctions are expressed as

$$
\Psi_{n,\vec{k}_{2d},\Sigma}(\vec{r}) = e^{i\vec{k}_{2d}\cdot\vec{r}_{2d}}\phi_n[z - z_0(k_{2d},\Sigma)] \otimes \chi_\Sigma(\hat{k}_{2d}),
\tag{4.86}
$$

where $\vec{k}_{2d} = (k_x, k_y)$, $\vec{r}_{2d} = (x, y)$, and $k_{2d} = (k_x^2 + k_y^2)^{\frac{1}{2}}$. The energy spectra of Eq. (4.86) is flat as $E_n = (n + \frac{1}{2})\hbar\omega_{so}$. The center of the oscillator wavefunction in Eq. (4.86) is shifted to $z_0 = l_{so}^2 k_{2d}\Sigma$.

The 3D Landau-level wavefunctions of Eq. (4.86) are spatially separated 2D helical plane-waves along the z-axis. As shown in Fig. 4.4(a), for states with opposite helicity eigenvalues, their central positions are shifted in opposite directions. If open boundaries are imposed perpendicular to the z-axis, each Landau level contributes a branch of gapless helical Dirac modes. For the system described by $H_{3D,L}^+$, the surface Hamiltonian is

$$
H_{bd} = \pm v_f(\vec{p} \times \vec{\sigma}) \cdot \hat{z} - \mu,
\tag{4.87}
$$

where $\pm$ apply to the upper and lower boundaries, respectively.

Unlike the 2D case in which the symmetric and Landau gauges are equivalent, the Hamiltonian in the symmetric-like gauge equation (4.29) and that in the Landau-like gauge equation (4.85) are *not* gauge equivalent. The Landau-like gauge explicitly breaks the 3D rotational symmetry while the symmetric-like gauge preserves it. Physical quantities calculated based on Eq. (4.85), such as density of states, are not 3D rotationally symmetric as those based on Eq. (4.29). Nevertheless, these two Hamiltonians belong to the same topological class.

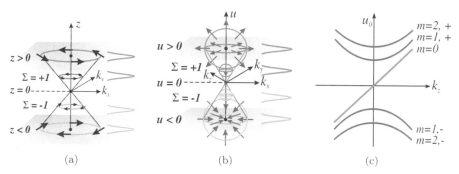

Fig. 4.4.   (a) 3D Landau-level wavefunctions as spatially separated 2D helical Dirac modes localized along the z-axis. (b) 4D Landau-level wavefunctions as spatially separated 3D Weyl modes localized along the u-axis. Note that 2D plane-wave modes with opposite helicities and the 3D ones with opposite chiralities are located at opposite sides of $z = 0$ and $u = 0$ planes, respectively. (c) The central positions $u_0(m, k_z, \nu)$ of the 4D Landau levels in the presence of the magnetic field $\vec{B} = B\hat{z}$. The branch of $m = 0$ runs across the entire u-axis, which gives rise to the quantized charge transport along the u-axis in the presence of $\vec{E} \parallel \vec{B}$ as indicated in Eq. (4.43). From Ref. [30].

4.7.3   *Spatially separated 3D Weyl modes: 4D Landau levels*

Again we can easily generalize the above procedure to any dimensions. For example, in four dimensions, we need to use the 3D helicity operator $\hat{\Sigma}_{3d} = \hat{P}_{3d} \cdot \vec{\sigma}$, whose eigenstates are denoted as χ_Σ with the eigenvalues $\Sigma = \pm 1$. Then the 4D Landau-level Hamiltonian is defined as [30]

$$
\begin{aligned}
H_{LL}^{4d,\pm} &= \frac{P_u^2 + \vec{P}_{3d}^2}{2M} + \frac{1}{2}M\omega^2 u^2 \mp \omega u \vec{P}_{3d} \cdot \vec{\sigma} \\
&= \frac{P_u^2}{2M} + \frac{1}{2}M\omega_{so}^2 \left(u \mp \frac{1}{\hbar} l_{so}^2 \vec{P}_{3d} \cdot \vec{\sigma} \right)^2,
\end{aligned}
\tag{4.88}
$$

where u and P_u are the coordinate and momentum in the fourth dimension, respectively, and $\vec{P}_{3d}$ is defined in the xyz-space. Inside each Landau level, the spectra are flat with respect to $\vec{k}_{3d}$ and Σ. Similarly to the 3D case, the 4D LL spectra and wavefunctions are solved by reducing Eq. (4.88) into a set of 1D harmonic oscillators along the u-axis as

$$
\Psi_{n,\vec{k}_{3d},\Sigma}(\vec{r}, u) = e^{i\vec{k}_{3d} \cdot \vec{r}} \phi_n[u - u_0(k_{3d}, \Sigma)] \otimes \chi_\Sigma(\vec{k}_{3d}).
\tag{4.89}
$$

The central positions $u_0(k_{3d}, \Sigma) = \Sigma l_{so}^2 k_{3d}$. This realizes the spatial separation of the 3D Weyl fermion modes with the opposite chiralities as shown in Fig. 4.4(b). With an open boundary imposed along the u-direction, the 3D chiral Weyl fermion modes appear on the boundary

$$
H_{bd} = \pm v_f(\vec{k}_{3D} \cdot \vec{\sigma}) - \mu.
\tag{4.90}
$$

4.7.4 *Phase space picture of high-dimensional Landau levels*

For the 2D case described by Eq. (4.82), the xy-plane is equivalent to the 2D phase space of a 1D system $(x; k_x)$ after the lowest Landau-level projection. The discrete step of k_x is $\Delta k_x = 2\pi/L_x$, and the momentum cutoff of the bulk state is determined by L_y as $k_{bk} = L_y/(2l_B^2)$. Since $|k_x| < k_{bk}$, the number of states $N_{2D,LL}$ scales with $L_x L_y$ as the usual 2D systems, but the crucial difference is that enlarging L_y does not change Δk_x but instead increases k_{bk}.

Similarly, the 3D Landau-level states (Eq. (4.85)) can be viewed as states in the 4D phase space $(xy; k_x k_y)$. The z-axis plays the double role of k_x and k_y. After the lowest Landau-level projection, z is equivalent to $z = l_{so}^2(p_x\sigma_y - p_y\sigma_x)/\hbar$, and thus

$$[x, z]_{LLL} = il_{so}^2\sigma_y, \quad [y, z]_{LLL} = -il_{so}^2\sigma_x, \quad [x, y]_{LLL} = 0. \quad (4.91)$$

The momentum cutoff of the bulk state is determined as $(k_x^2 + k_y^2)^{\frac{1}{2}} < k_{bk} = \hbar L_z/(2l_{so}^2)$, thus the total number of states N scales as $L_x L_y L_z^2$. As a result, the 3D local density of states linearly diverges as $\rho_{3D}(z) \propto |z|/l_{so}^4$ as $|z| \to \infty$. Similar divergence also occurs in the symmetric-like gauge as $\rho_{3D}(r) \propto r/l_{so}^4$. Now this seeming pathological result can be understood as the consequence of squeezing states of 4D phase space $(xy; k_x k_y)$ into the 3D real space (xyz). In other words, the correct thermodynamic limit should be taken according to the volume of 4D phase space. This reasoning is easily extended to the 4D LL systems (Eq. (4.88)), which can be understood as a 6D phase space of $(xyz; k_x k_y k_z)$.

4.7.5 *Charge pumping and the 4D quantum Hall effects*

The above 4D Landau-level states presented in Sec. 4.7.3 exhibit nonlinear electromagnetic response [13, 24, 61, 62] as the 4D quantum Hall effect. We apply the electromagnetic fields as

$$\vec{E} = E\hat{z}, \quad \vec{B} = B\hat{z}, \quad (4.92)$$

to the 4D Landau-level Hamiltonian equation (4.88) by minimally coupling fermions to the $U(1)$ vector potential,

$$A_{em,x} = 0, \quad A_{em,y} = Bx, \quad A_{em,z} = -cEt. \quad (4.93)$$

The $\vec{B}$-field further quantizes the chiral plane-wave modes inside the nth 4D spin–orbit Landau-level states into a series of 2D magnetic Landau-level states in the xy-plane as labeled by the magnetic Landau-level index m.

For the case of $m = 0$, the eigen-wavefunctions are spin polarized as

$$\Psi_{n,m=0}(k_y, k_z) = e^{ik_y y + ik_z z}\phi_n(u - u_0(k_z, m = 0))\varphi_{m=0}(x - x_0(k_y)) \otimes |\uparrow\rangle,$$
(4.94)

where ϕ_n is the nth order harmonic oscillator wavefunction with the spin–orbit length scale l_{so}, and φ_0 is the zeroth-order harmonic oscillator wavefunction with the magnetic length scale l_B. The central positions of the u-directional and x-directional oscillators are

$$x_0(k_y) = l_B^2 k_y, \quad u_0(k_z, m = 0) = l_{so}^2 k_z,$$
(4.95)

respectively. The key point is that $u_0(k_z, m = 0)$ runs across the entire u-axis. In contrast, wavefunctions $\Psi_{n,m}$ with $m \geq 1$ also exhibit harmonic oscillator wavefunctions along the u-axis. However, their central positions at $m \geq 1$ are,

$$u_0(k_z) = \pm l_{so}^2 \sqrt{k_z^2 + \frac{2m}{l_B^2}},$$
(4.96)

which only lie in half of the u-axis as shown in Fig. 4.4(c).

Since k_z increases with time in the presence of E_z, $u_0(m, k_z(t))$ moves along the u-axis. Only the $m = 0$ branch of the magnetic Landau-level states contribute to the charge pumping since their centers go across the entire u-axis, which results in an electric current along the u-direction. Since $k_z(t) = k_z(0) - \frac{eE}{\hbar}t$, during the time interval Δt, the number of electrons passing the cross-section at a fixed u is

$$\Delta N = \frac{L_x L_y}{2\pi l_B^2}\frac{eE_z \Delta t}{2\pi\hbar/L_z} = \frac{e^2}{4\pi^2\hbar^2 c}\vec{E} \cdot \vec{B}V\Delta t,$$
(4.97)

where V is the 3D cross-volume. Then the current density is calculated as

$$j_u = n_{occ}\frac{e\Delta N}{V\Delta t} = n_{occ}\alpha\frac{e}{4\pi^2\hbar}\vec{E} \cdot \vec{B},$$
(4.98)

where α is the fine-structure constant, and n_{occ} is the occupation number of the 4D spin–orbit Landau levels.

Equation (4.98) is in agreement with results from the effective field theory [13] as the 4D generalization of the quantum Hall effect. If we impose the open boundary condition perpendicular to the u-direction, the above charge pump process corresponds to the chiral anomalies of Weyl fermions with opposite chiralities on two opposite 3D boundaries, respectively. Since they are spatially separated, the chiral current corresponds to the electric current along the u-direction.

4.8 Conclusions and outlooks

I have reviewed a general framework for constructing Landau levels in high dimensions based on harmonic oscillator wavefunctions. By imposing spin–orbit coupling, their spectra are reorganized to exhibit flat dispersions and nontrivial topological properties. In particular, the lowest Landau-level wavefunctions in 3D and 4D in the quaternion representation satisfy the Cauchy–Riemann–Fueter condition, which is the generalization of complex analyticity to high dimensions. The boundary excitations are the 2D helical Dirac surface modes, or, the 3D chiral Weyl modes. There is a beautiful bulk–boundary correspondence that the Cauchy–Riemann–Fueter condition and the helical Dirac (chiral Weyl) equation are the Euclidean and Minkowski representations of the same analyticity condition, respectively. By dimensional reductions, we constructed a class of Landau levels in 2D and 3D which are time-reversal invariant but parity breaking. The Landau-level problem for Dirac fermions is a square-root problem of the nonrelativistic one, corresponding to complex quaternions. The zeroth-Landau-level states are a flat band of half-fermion Jackiw–Rebbi zero modes. It is at the interface between condensed matter and high-energy physics, related to a new type of anomaly. Unlike parity anomaly and chiral anomaly studied in field theory in which Dirac fermions are coupled to gauge fields through the minimal coupling, here Dirac fermions are coupled to background fields in a nonminimal way.

I speculate that high-dimensional Landau levels could provide a platform for exploring interacting topological states in high dimensions — due to the band flatness, and also the quaternionic analyticity of lowest Landau-level wavefunctions. It would stimulate the developments of various theoretical and numerical methods. This would be an important direction in both condensed matter physics and mathematical physics for studying high-dimensional topological states with both nonrelativistic and relativistic fermions. This research also provides interesting applications of quaternion analysis to theoretical physics.

Appendix A: Brief review on Clifford algebra

In this appendix, we review how to construct anti-commutative Γ-matrices. The familiar group is just the 2×2 Pauli matrices, i.e., rank-1. The rank-k

Γ-matrices can be defined recursively based on the rank-$(k-1)$ ones. At each level, there are $2k+1$ anti-commutative matrices, and their dimensions are $2^k \times 2^k$. In this chapter, we use the following representation:

$$\Gamma_i^{(k)} = \begin{bmatrix} 0 & \Gamma_a^{(k-1)} \\ \Gamma_a^{(k-1)} & 0 \end{bmatrix}, \quad \Gamma_{2k}^{(k)} = \begin{bmatrix} 0 & -iI \\ iI & 0 \end{bmatrix}, \quad \Gamma_{2k+1}^{(k)} = \begin{bmatrix} I & 0 \\ 0 & -I \end{bmatrix},$$

(A1)

where $i = 1, \ldots, 2k - 1$.

In $D = 2k + 1$-dimensional space, the $SO(2k+1)$ fundamental spinor is 2^k-dimensional. The generators are constructed $S_{ij} = \frac{1}{2}\Gamma_{ij}^{(k)}$ where

$$\Gamma_{ij}^{(k)} = -\frac{i}{2}[\Gamma_i^{(k)}, \Gamma_j^{(k)}].$$

(A2)

In the $D = 2k$-dimensional space, there are two irreducible fundamental spinor representations for the $SO(2k)$ group, both of which are with 2^{k-1}-dimensional. Their generators are denoted as S_{ij} and S'_{ij}, respectively, which can be constructed based on both rank-$(k-1)$ $\Gamma_i^{(k-1)}$ and $\Gamma_{ij}^{(k-1)}$-matrices. For the first $2k-1$ dimensions, the generators share the same form as that of the $SO(2k-1)$ group,

$$S_{ij} = S'_{ij} = \frac{1}{2}\Gamma_{ij}^{(k-1)}, \quad (1 \le i < j \le 2k - 1).$$

(A3)

Other generators $S_{i,2k}$ and $S'_{i,2k}$ differ by a sign — they are represented by the $\Gamma_i^{(k-1)}$ matrices,

$$S_{i,2k} = S'_{i,2k} = \pm\frac{1}{2}\Gamma_i^{(k-1)}, \quad (1 \le i \le 2k - 1).$$

(A4)

Acknowledgments

I thank Yi Li for collaborations on this set of works on high-dimensional topological states and for bringing in interesting concepts including the quaternionic analyticity. I also thank J.E. Hirsch for stimulating discussions, and S. C. Zhang, T. L. Ho, E. H. Fradkin, S. Das Sarma, F. D. M. Haldane, and C. N. Yang for their warm encouragements and appreciations. This work is partly supported by AFOSR FA9550-14-1-0168 and NSFC under the Grants No. 11729402.

References

[1] A. Sudbery, Quaternionic analysis, *Math. Proc. Cambridge Philos. Soc.* **85**(2), 199–224 (1979).

[2] S. L. Adler, *Quaternionic Quantum Mechanics and Quantum Fields.* vol. 88 (Oxford University Press, 1995).

[3] D. Finkelstein, Foundations of quaternion quantum mechanics, *J. Math. Phys.* **3**, 207 (1962).

[4] C. N. Yang, *Selected Papers (1945–1980) with Commentary* (World Scientific, 2005).

[5] B. A. Bernevig and S. C. Zhang, Quantum spin Hall effect, *Phys. Rev. Lett.* **96**(10), 106802 (2006).

[6] C. L. Kane and E. J. Mele, $\mathbb{Z}_2$ Topological order and the quantum spin Hall effect, *Phys. Rev. Lett.* **95**(14), 146802, (2005).

[7] C. L. Kane and E. J. Mele, Quantum spin Hall effect in graphene, *Phys. Rev. Lett.* **95**(22), 226801 (2005).

[8] L. Fu and C. L. Kane, Topological insulators with inversion symmetry, *Phys. Rev. B* **76**(4), 045302 (2007).

[9] L. Fu, C. L. Kane and E. J. Mele, Topological insulators in three dimensions, *Phys. Rev. Lett.* **98**(10), 106803 (2007).

[10] J. E. Moore and L. Balents, Topological invariants of time-reversal-invariant band structures, *Phys. Rev. B* **75**(12), 121306 (2007).

[11] B. A. Bernevig, T. L. Hughes, and S. C. Zhang, Quantum spin Hall effect and topological phase transition in HgTe quantum wells, *Science* **314**(5806), 1757 (2006).

[12] C. Wu, B. Bernevig, and S.-C. Zhang, Helical liquid and the edge of quantum spin Hall systems, *Phys. Rev. Lett.* **96**(10), 106401, (2006).

[13] X. L. Qi, T. L. Hughes, and S. C. Zhang, Topological field theory of time-reversal invariant insulators, *Phys. Rev. B* **78**(19), 195424 (2008).

[14] R. Roy, Topological phases and the quantum spin Hall effect in three dimensions, *Phys. Rev. B* **79**(19), 195322 (2009).

[15] R. Roy, Characterization of three-dimensional topological insulators by two-dimensional invariants, *New J. Phys.* **12**, 065009 (2010).

[16] D. J. Thouless, M. Kohmoto, M. P. Nightingale, and M. den Nijs, Quantized Hall conductance in a two-dimensional periodic potential, *Phys. Rev. Lett.* **49**(6), 405 (1982).

[17] F. D. M. Haldane, Model for a Quantum Hall effect without Landau levels: Condensed-matter realization of the "parity anomaly", *Phys. Rev. Lett.* **61**(18), 2015–2018 (1988).

[18] A. Kitaev, Periodic table for topological insulators and superconductors, *Amer. Inst. Phys. Conf. Ser.* **1134**, 22–30 (2009).

[19] A. Schnyder, S. Ryu, A. Furusaki, and A. Ludwig, Classification of topological insulators and superconductors in three spatial dimensions, *Phys. Rev. B* **78**(9), 195125 (2008).

[20] K. Klitzing, G. Dorda, and M. Pepper, New method for high-accuracy determination of the fine-structure constant based on quantized Hall resistance, *Phys. Rev. Lett.* **45**(6), 494–497 (1980).

[21] D. C. Tsui, H. L. Stormer, and A. C. Gossard, Two-dimensional magnetotransport in the extreme quantum limit, *Phys. Rev. Lett.* **48**, 1559 (1982); doi: 10.1103/PhysRevLett.48.1559; URL http://link.aps.org/doi/10.1103/PhysRevLett.48.1559.

[22] S. Girvin, The quantum hall effect: Novel excitations and broken symmetries, *Aspects topologiques de la physique en basse dimension. Topological aspects of low dimensional systems.* pp. 53–175 (1999).

[23] R. Laughlin, Anomalous quantum Hall effect: An incompressible quantum fluid with fractionally charged excitations, *Phys. Rev. Lett.* **50**(18), 1395–1398 (1983).

[24] S. C. Zhang and J. P. Hu, A four-dimensional generalization of the quantum Hall effect, *Science* **294**(5543), 823 (2001).

[25] F. D. M. Haldane, Fractional quantization of the hall effect: A hierarchy of incompressible quantum fluid states, *Phys. Rev. Lett.* **51**(7), 605–608 (1983).

[26] Y. Li, X. Zhou, and C. Wu, Two- and three-dimensional topological insulators with isotropic and parity-breaking Landau levels, *Phys. Rev. B* **85**, 125122 (2012).

[27] Y. Li and C. Wu, High-dimensional topological insulators with quaternionic analytic landau levels, *Phys. Rev. Lett.* **110**, 216802 (2013); doi: 10.1103/PhysRevLett. 110.216802; URL https://link.aps.org/doi/10.1103/PhysRevLett.110.216802.

[28] Y. Li, K. Intriligator, Y. Yu and C. Wu, Isotropic Landau levels of Dirac fermions in high dimensions, *Phys. Rev. B* **85**, 085132 (2012).

[29] R. Jackiw and C. Rebbi, Solitons with fermion number $\frac{1}{2}$, *Phys. Rev. D.* **13**, 3398–3409 (1976); doi: 10.1103/PhysRevD.13.3398; URL http://link.aps.org/doi/10.1103/PhysRevD.13.3398.

[30] Y. Li, S.-C. Zhang and C. Wu, Topological insulators with su(2) landau levels, *Phys. Rev. Lett.* **111**, 186803 (2013); doi: 10.1103/PhysRevLett.111.186803; URL https://link.aps.org/doi/10.1103/PhysRevLett.111.186803.

[31] I. Frenkel and M. Libine, Quaternionic analysis, representation theory and physics, *Adv. Math.* **218**(6), 1806–1877 (2008); doi: https://doi.org/10.1016/j.aim.2008.03.021; URL http://www.sciencedirect.com/science/article/pii/S0001870808000935.

[32] A. V. Balatsky, Quaternion generalization of the laughlin state and the three dimensional fractional QHE, Preprint (1992); arXiv:cond-mat/9205006.

[33] G. W. Semenoff, Condensed-matter simulation of a three-dimensional anomaly, *Phys. Rev. Lett.* **53**(26), 2449–2452 (1984).

[34] A. J. Heeger, S. Kivelson, J. R. Schrieffer and W. P. Su, Solitons in conducting polymers, *Rev. Mod. Phys.* **60**, 781–850 (1988); doi: 10.1103/RevModPhys.60.781; URL http://link.aps.org/doi/10.1103/RevModPhys.60.781.

[35] A. N. Redlich, Gauge noninvariance and parity nonconservation of three-dimensional fermions, *Phys. Rev. Lett.* **52**, 18–21 (1984); doi: 10.1103/PhysRevLett.52.18; URL http://link.aps.org/doi/10.1103/PhysRevLett.52.18.

[36] A. N. Redlich, Parity violation and gauge noninvariance of the effective gauge field action in three dimensions, *Phys. Rev. D* **29**, 2366–2374 (1984); doi: 10.1103/PhysRevD.29.2366; URL http://link.aps.org/doi/10.1103/PhysRevD.29.2366.

[37] A. J. Niemi and G. W. Semenoff, Fermion number fractionization in quantum field theory, *Phys. Rep.* **135**(3), 99–193 (1986).

[38] P. Streda, Quantised hall effect in a two-dimensional periodic potential, *J. Phys. C: Solid State Phys.* **15**(36), L1299 (1982); URL http://stacks.iop.org/0022-3719/15/i=36/a=006.

[39] B. K. Bagchi, *Supersymmetry in Quantum and Classical Mechanics* (Chapman & Hall/CRC, 2001).

[40] D. H. Lee and C. L. Kane, Boson-vortex-skyrmion duality, spin-singlet fractional quantum hall effect, and spin-1/2 anyon superconductivity, *Phys. Rev. Lett.* **64**, 1313–1317 (1990); doi: 10.1103/PhysRevLett.64.1313; URL http://link.aps.org/doi/10.1103/PhysRevLett.64.1313.

[41] S. L. Sondhi, A. Karlhede, S. A. Kivelson and E. H. Rezayi, Skyrmions and the crossover from the integer to fractional quantum hall effect at small zeeman energies, *Phys. Rev. B* **47**(24), 16419 (1993).

[42] H. A. Fertig, L. Brey, R. Côté and A. H. MacDonald, Charged spin-texture excitations and the hartree-fock approximation in the quantum hall effect, *Phys. Rev. B* **50**, 11018–11021 (1994); doi: 10.1103/PhysRevB.50.11018; URL http://link.aps.org/doi/10.1103/PhysRevB.50.11018.

[43] N. Read and S. Sachdev, Continuum quantum ferromagnets at finite temperature and the quantum hall effect, *Phys. Rev. Lett.* **75**, 3509–3512 (1995); doi: 10.1103/PhysRevLett.75.3509; URL http://link.aps.org/doi/10.1103/PhysRevLett.75.3509.

[44] F. D. M. Haldane and E. H. Rezayi, Finite-size studies of the incompressible state of the fractionally quantized hall effect and its excitations, *Phys. Rev. Lett.* **54**, 237–240 (1985); doi: 10.1103/PhysRevLett.54.237; URL http://link.aps.org/doi/10.1103/PhysRevLett.54.237.

[45] R. E. Prange and S. M. Girvin, Eds., *The Quantum Hall Effect*, 2nd edn. (Springer-Verlag, New York, 1990).

[46] D. Haldane, Geometrical description of the fractional quantum Hall effect, *Phys. Rev. Lett.* **107**, 116801 (2011).

[47] T. Can, M. Laskin, and P. Wiegmann, Fractional quantum Hall effect in a curved space: Gravitational anomaly and electromagnetic response, *Phys. Rev. Lett.* **113**(4), 046803 (2014); doi: 10.1103/PhysRevLett.113.046803.

[48] S. Klevtsov, X. Ma, G. Marinescu and P. Wiegmann, Quantum Hall effect and quillen metric, *Commun. Math. Phys.* **349**(3), 819–855 (2017); doi: 10.1007/s00220-016-2789-2.

[49] C. J. Wu, I. Mondragon-Shem and Z. Xiang-Fa, Unconventional Bose–Einstein condensations from spin–orbit coupling, *Chinese Phys. Lett.* **28**(9), 097102 (2011); URL http://stacks.iop.org/0256-307X/28/i=9/a=097102.

[50] Y. Li, X. Zhou and C. Wu, Three-dimensional quaternionic condensations, hopf invariants, and skyrmion lattices with synthetic spin–orbit coupling, *Phys. Rev. A* **93**, 033628 (2016); doi: 10.1103/PhysRevA.93.033628; URL https://link.aps.org/doi/10.1103/PhysRevA.93.033628.

[51] C. Wu and I. Mondragon-Shem, Exciton condensation with spontaneous time-reversal symmetry breaking, Preprint (2008); arXiv:0809.3532v1.

[52] X. Zhou, Y. Li, Z. Cai and C. Wu, Unconventional states of bosons with the synthetic spin–orbit coupling, *J. Phys. B Atomic Molecular Phys.* **46**(13), 134001 (2013); doi: 10.1088/0953-4075/46/13/134001.

[53] H. Hu, B. Ramachandhran, H. Pu and X. J. Liu, Spin–orbit coupled weakly interacting bose-einstein condensates in harmonic traps, *Phys. Rev. Lett.* **108**, 010402 (2012); doi: 10.1103/PhysRevLett.108.010402; URL http://link.aps.org/doi/10.1103/PhysRevLett.108.010402.

[54] S. Sinha, R. Nath and L. Santos, Trapped two-dimensional condensates with synthetic spin–orbit coupling, *Phys. Rev. Lett.* **107**, 270401 (2011); doi: 10.1103/PhysRevLett.107.270401; URL http://link.aps.org/doi/10.1103/PhysRevLett.107.270401.

[55] S. K. Ghosh, J. P. Vyasanakere and V. B. Shenoy, Trapped fermions in a synthetic non-abelian gauge field, *Phys. Rev. A* **84**, 053629 (2011); doi: 10.1103/PhysRevA.84.053629; URL http://link.aps.org/doi/10.1103/PhysRevA.84.053629.

[56] Z. Wang, X.-L. Qi and S.-C. Zhang, Equivalent topological invariants of topological insulators, *New J. Phys.* **12**, 065007 (2010).

[57] T.-L. Ho and S. Zhang, Bose–Einstein condensates with spin–orbit interaction, *Phys. Rev. Lett.* **107**, 150403 (2011); doi: 10.1103/PhysRevLett.107.150403; URL http://link.aps.org/doi/10.1103/PhysRevLett.107.150403.

[58] D. Xiao, M. Chang and Q. Niu, Berry phase effects on electronic properties, *Rev. Mod. Phys.* **82**, 1959–2007 (2010).

[59] C. G. Callan and J. A. Harvey, Anomalies and fermion zero modes on strings and domain walls, *Nuc. Phys. B* **250**, 427–436 (1985); doi: 10.1016/0550-3213(85)90489-4; URL http://www.sciencedirect.com/science/article/pii/0550321385904894.

[60] D. H. Lee and J. M. Leinaas, Mott insulators without symmetry breaking, *Phys. Rev. Lett.* **92**, 096401 (2004); doi: 10.1103/PhysRevLett.92.096401; URL http://link.aps.org/doi/10.1103/PhysRevLett.92.096401.

[61] P. Werner, (4+1)-dimensional quantum hall effect & applications to cosmology, Preprint (2012); arXiv:1207.4954.

[62] J. Fröhlich and B. Pedrini, New applications of the chiral anomaly, Preprint (2000); arXiv:hep-th/0002195.

Chapter 5

Right and Left in Quantum Dynamics of Solids

Naoto Nagaosa*,† and Takahiro Morimoto*

*Department of Applied Physics, The University of Tokyo,
Hongo, Tokyo, 113-8656, Japan
†RIKEN Center for Emergent Matter Sciences (CEMS),
Wako, Saitama, 351-0198, Japan

Chirality in quantum dynamics of solids is discussed in a pedagogical fashion. In sharp contrast to the chirality in static structures of objects, the chirality in dynamics including currents offers much richer physics; especially, the role of time-reversal symmetry $\mathcal{T}$ in addition to the inversion symmetry $\mathcal{P}$, and the geometrical Berry phase are stressed. Recent studies on various transport and optical phenomena are reviewed from a unified point of view.

5.1 Introduction

Chirality, i.e., right-handedness and left-handedness, is a key issue penetrating though the whole sciences including physics, chemistry, and biology [1]. For example, when the four atoms attached to a carbon atom are different, the right-handed and left-handed molecules are distinct. Also in crystal, right-handed and left-handed structures show different physical properties especially their nonreciprocal effects.

When one considers the currents, i.e., the flow of particles, the situation is less trivial. For example, the Newtonian equation is invariant with respect to the time-reversal symmetry $\mathcal{T}$, the motion where the particles are moving in the right direction and that in the left direction are equivalent even when the potential $V(x)$ is asymmetric, i.e., $V(x) \neq V(-x)$. However, once the friction or dissipation is there, the solutions of the equation of motion are different between the right and left directions.

In quantum mechanics, one can formulate the transmission/reflection problem of an asymmetric potential $V(x)$. The scattering matrix S describes this problem, and the unitary nature of S, i.e., $S^\dagger S = 1$, indicates that the transition/reflection probability is identical for the incident waves from right and left. Therefore, the time-reversal symmetry in both the microscopic

103

dynamics ($\mathcal{T}$) and the macroscopic (ir)reversibility plays an essential role in addition to the spatial inversion symmetry $\mathcal{P}$.

In solids, $\mathcal{T}$ relates the two states $(\boldsymbol{k}, \sigma)$ and $(-\boldsymbol{k}, \bar{\sigma})$ with $\bar{\sigma}$ being the opposite spin to σ. Therefore, the energy dispersion of Bloch wavefunction satisfies $\varepsilon_\sigma(\boldsymbol{k}) = \varepsilon_{\bar{\sigma}}(-\boldsymbol{k})$. Therefore, in the conventional treatment of the electron transport, i.e., Boltzmann equation approach, the directional asymmetry between the right and left is missing as long as the spin index is summed, i.e., charge transport, while the spin current can show the nonreciprocal transport. Therefore, the nonreciprocal charge transport in solids becomes an intriguing issue since it contains physics beyond this conventional theory.

In this chapter, we review theoretical aspects of nonreciprocal responses in solids, mainly focusing on the nonlinear responses. A review article has been published by one of the authors including both experiments and theories [2], and the present paper includes some recent advances.

5.2   Onsager's reciprocal theorem and nonreciprocal linear responses

As an introduction to the nonlinear nonreciprocal responses, we briefly discuss the Onsager's theorem and linear nonreciprocal responses [3, 4]. Onsager derived the relation of the linear response functions based on the time-reversal symmetry $\mathcal{T}$ of the microscopic dynamics given as

$$\kappa_{AB}(\omega, \mathcal{B}) = \varepsilon_A \varepsilon_B \kappa_{BA}(\omega, -\mathcal{B}), \tag{5.1}$$

where $\mathcal{B}$ is the magnetic field, and $\kappa_{AB}(\omega)$ is the response function of the observable A to the field h_B conjugate to the physical quantity B, i.e.,

$$\langle A \rangle(\omega, \mathcal{B}) = \kappa_{AB}(\omega, \mathcal{B}) h_B(\omega). \tag{5.2}$$

Here, $\varepsilon_A = \pm 1$ is specifies the even $(+1)$ or odd (-1) nature of A with respect to the time-reversal. When the translational symmetry is there, the wavevector $\boldsymbol{q}$ is well defined and Eq. (5.1) can be written as

$$\kappa_{AB}(\boldsymbol{q}, \omega, \mathcal{B}) = \varepsilon_A \varepsilon_B \kappa_{BA}(-\boldsymbol{q}, \omega, -\mathcal{B}). \tag{5.3}$$

Equation (5.3) gives the basis to discuss the various nonreciprocal optical responses such as the directional dichroism, where $\boldsymbol{q}$ is the wavevector of light. The dielectric function $\varepsilon_{\alpha\beta}(\boldsymbol{q}, \omega, \mathcal{B})$ $(\alpha, \beta = x, y, z)$ can contain the term

$$\eta \delta_{\alpha\beta} \boldsymbol{q} \cdot \mathcal{B}, \tag{5.4}$$

which describes the different transmission or reflection of light between q and $-q$ under a magnetic field [5].

5.3   Magnetochiral anisotropy of nonlinear conduction

Broken inversion symmetry $\mathcal{P}$ often leads to the ferroelectricity, i.e., spontaneous electric polarization, in insulators. In metals, on the other hand, the consequence of broken $\mathcal{P}$ is much less trivial, since the electric field inside the material is screened. Therefore, it should be sought in the transport properties. Rikken *et al.* was the first to observe the nonreciprocal transport in noncentrosymmetric materials under an external magnetic field $\mathcal{B}$ [6–9] The resistivity R is expressed in terms of the formula

$$R = R_0(1 + \beta \mathcal{B}^2 + \gamma \mathcal{B} I), \tag{5.5}$$

where R_0 is the resistivity at zero magnetic field, I is the current, β is the coefficient of the magnetoresistance, and the last term represent the directional nonlinear resistivity induced by $\mathcal{B}$. This nonlinear nonreciprocal transport is called magnetochiral anisotropy (MCA). Equation (5.5) is derived by the heuristic argument by replacing the wavevector q in Eq. (5.4) by the current I. Note, however, that Eq. (5.4) is valid for the linear response, and Eq. (5.5) is for the nonlinear response. The generalization of the Onsager's theorem should be considered seriously, and we will see in Sec. 5.5.3 that the broken $\mathcal{T}$ is not necessarily required for the nonreciprocal responses.

5.3.1   *Magnetochiral anisotropy in polar semiconductors*

Equation (5.5) gives an excellent description of the nonreciprocal transport in various materials, but the effect is usually very small, i.e., γ is typically $10^{-3} - 10^{-1} \mathrm{T}^{-1} \mathrm{A}^{-1}$ [6–9]. The reason for this smallness is that both the spin-orbit interaction λ and the magnetic energy $\mu_B \mathcal{B}$ are needed to realize the MCA, and the energy denominator for these perturbations is the kinetic energy of electrons, which is typically the Fermi energy ε_F. Usually λ and $\mu_B \mathcal{B}$ are much smaller than ε_F, and hence the effect is very weak. One possible way to enhance MCA is to reduce ε_F, i.e., low carrier density system or to enhance λ. A polar semiconductor BiTeBr satisfies both of these conditions [10]. The crystal structure of this system is shown in Fig. 5.1(a), where the polarization occurs along the c-axis, which results in the Rashba spin–orbit interaction and hence the spin-splitting of the bands. Under an external magnetic field perpendicular to the c-axis, e.g., along b-axis, the band dispersion becomes asymmetric between $+a$ and $-a$

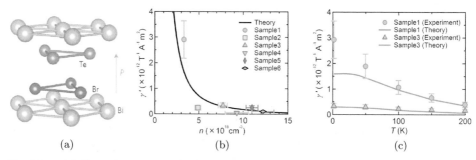

Fig. 5.1.   (a) Crystal structure of polar semiconductor BiTeBr. The polarization direction is along c-axis. (b) The low temperature ($T = 2K$) value of $\gamma' = \gamma A$ (A: area of the cross-section of the sample) as a function of carrier density. Solid curve is the theoretical result based on the Boltzmann transport equation expended to the second order in the electric field. There is no fitting parameter. (c) Temperature dependence of γ'. Reproduced from Ref. [10].

directions. An analysis in terms of the Boltzmann theory has been done for MCA, where γ is independent of the relaxation time τ similarly to the Hall coefficient. Therefore, γ can be regarded as the intrinsic quantity to the band structure. Figures 5.1(b) and 5.1(c) show the comparison between the theoretical calculation without the fitting parameter (since the band structure is already known from the first-principles calculation and angle-resolved photo-emission spectroscopy), and the experimental results. One can see that both the carrier number dependence (b) and temperature dependence (c) show good agreement between theory and experiment.

5.3.2   *Magnetochiral anisotropy in Weyl semimetals*

Weyl semimetals with broken spatial inversion symmetry also provides an interesting platform for the MCA due to chiral anomaly [11]. Realization of Weyl semimetals requires either broken time-reversal symmetry $\mathcal{T}$ or inversion symmetry $\mathcal{P}$, and the famous Weyl semimetal TaAs is an example with broken $\mathcal{P}$. The MCA in Weyl semimetals arises from the chiral anomaly as follows. In the presence of the external electric and magnetic fields, the chiral anomaly in Weyl fermions induces charge imbalance between Weyl and anti-Weyl fermions as $Q_5 \simeq \tau \boldsymbol{E} \cdot \boldsymbol{B}$ with the relaxation time τ in the steady state. This leads to the change of the Fermi surface, and can modify the linear conductivity as $\Delta\sigma \propto \boldsymbol{E} \cdot \boldsymbol{B}$ as schematically illustrated in Fig. 5.2. This means that the chiral anomaly gives rise to the current response

$$\boldsymbol{J} \propto (\boldsymbol{E} \cdot \boldsymbol{B})\boldsymbol{E}, \tag{5.6}$$

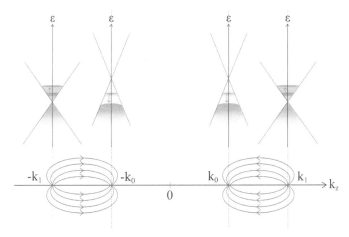

Fig. 5.2. Schematic picture of magnetochiral anisotropy of Weyl fermions. Applying an electric field $\boldsymbol{E}$ (parallel to $\boldsymbol{B}$) transfers electrons from the Weyl node to the anti-Weyl node due to chiral anomaly, which induces nonlinear current response. Reproduced from Ref. [11].

which is nothing but the MCA. MCA in Weyl semimetals can show a significant enhancement when the chemical potential approaches the Weyl points (where the energy denominator $\epsilon_F \to 0$), since the change of the Fermi surface can become large in this case. We estimate the MCA for TaAs from this mechanism as $\gamma \simeq 0.3\ T^{-1}A^{-1}$, for a typical sample of a cross-section $A = 0.1\text{mm}^2$ [11].

5.3.3 *Magnetochiral anisotropy in superconductors*

As the large energy denominator ε_F is the reason why the MCA is a small effect, one may look for the possibility to replace ε_F by a smaller energy scale to enhance MCA. Superconductivity is a possible way where the energy denominator can be the superconducting gap Δ. In 2D superconductors, the amplitude of the gap develops at the mean field transition temperature T_0, and its phase is locked at the Kosterlitz–Thouless transition temperature T_{KT} when the penetration depth λ is much longer than the sample size [12]. Slightly above T_0, the superconducting fluctuation gives rise to the additional conductivity called para-conductivity. This contribution can be calculated in terms of the time-dependent Ginzburg–Landau theory [13]. For a noncentrosymmetric superconductor, the Ginzburg–Landau-free energy contains the third-order derivative terms, which results in the nonreciprocal nonlinear para-conductivity. This possibility has been explored in transition metal dichalcogenide MoS_2 [14]. The band structure of this material shows the trigonal warping as described by

$$H_{k\sigma,\tau} = \frac{\hbar^2 k^2}{2m} + \alpha\tau_z k_x(k_x^2 - 3k_y^2) - \Delta_z\sigma_z - \Delta_{SO}\sigma_z\tau_z, \qquad (5.7)$$

where σ's are Pauli matrices for spin while τ's for orbital. α-term describes the trigonal warping, Δ_z comes from the Zeeman coupling due to external magnetic field, and Δ_{SO} is the Ising spin splitting due to the spin–orbit interaction. Assuming the on-site pairing, the Ginzburg–Landau–free energy is given by

$$F = \int d^2 r \Psi^* \left[a + \frac{p^2}{4m} + \frac{\Lambda\mathcal{B}}{\hbar^3}(p_x^3 - 3p_x p_y^2) \right] \Psi + \int d^2 r \frac{b}{2} |\Psi|^4, \qquad (5.8)$$

where

$$\Lambda = \frac{93\zeta(5)}{28\zeta(3)} \frac{g\mu_B \Delta_{SO}\alpha}{\pi(k_B T_0)^2}. \qquad (5.9)$$

With this Ginzburg–Landau–free energy, the additional current due to the superconducting fluctuation is given by

$$\mathbf{j} = \frac{e^2}{16\hbar\epsilon}\mathbf{E} - \frac{\pi e^3 m\Lambda\mathcal{B}}{64\hbar^3 k_B T_0 \epsilon^2}\mathbf{F}(\mathbf{E}) \qquad (5.10)$$

with $\epsilon = (T - T_0)/T_0$ and $\mathbf{F}(\mathbf{E}) = (E_x^2 - E_y^2, 2E_x E_y)$. Translating this result into the MCA, the γ-value, γ_S, due to the para-conductivity is enhanced from that in the normal state γ_N as

$$\frac{\gamma_S}{\gamma_N} \sim \left(\frac{\varepsilon_F}{k_B T_0}\right)^3, \qquad (5.11)$$

which indicates a gigantic enhancement of MCA due to the superconducting fluctuation. This can be interpreted as the replacement of the energy denominator ε_F by the superconducting gap $\Delta \sim k_B T_0$. More importantly, the quantum nature of the electrons tends to suppress MCA while the dynamics of the superconducting order parameter is classical as a consequence of the general principle of spontaneous symmetry breaking. Experimentally, it has been observed that γ is negligible in the normal state and increases as the temperature approaches to $T_0 \cong 9K$, and $\gamma(T = 9K) \cong 1000\text{T}^{-1}\text{A}^{-1}$, while the theoretical estimate is $\gamma(T_0) \cong 400\text{T}^{-1}\text{A}^{-1}$ [14]. This semi-quantitative agreement indicates that the superconducting fluctuation causes the giant MCA. However, the analysis above does not describe the γ which continues to increase below T_0. In the temperature region $T_{KT} < T < T_0$, the resistivity comes from the motion of vortices, i.e., the defects of the phase of the order parameter. Under an external magnetic field perpendicular to the plane, the Kosterlitz–Thouless transition is suppressed and T_{KT} becomes zero. On the other hand, when the MCA is induced by the in-plane magnetic field, the Kosterlitz–Thouless transition remains and one expects the critical behavior of γ near T_{KT} as $\gamma \sim (T - T_{KT})^{-3/2}$ [15]. This prediction has been recently confirmed experimentally for superconductivity in $Bi_2Te_3/FeTe$ interface [16]. More systematic study on the MCA in 2D superconductors has been also developed, and readers are referred to [15] for more details.

5.4 Ratchet

Ratchet is a typical system where the right and left symmetry is broken by the asymmetric periodic potential. At thermal equilibrium, the fluctuation does not drive any one-direction motion, while the noise in nonequilibrium state causes it. Namely, it is needed to supply the energy to keep the one-directional motion which is consumed as the dissipation. Once the external force F is applied, on the other hand, the velocity v as a function of F can show the asymmetry between positive and negative sign of F. This problem is relevant to the Josephson junction systems where the periodic potential $U(x)$ for the position of the flux x can be designed to be asymmetric [17]. The velocity v of the vortex is directly related to the voltage drop V, while F corresponds to the current. At high temperature, the quantum mechanical tunneling between the neighboring minima of the potential can be neglected, and the transition rate is determined by the activation factor $e^{-E_B/(k_BT)}$ with the potential barrier E_B. At low temperature, the quantum mechanics becomes relevant, and the wave nature of x enters. Namely, the quantum-classical crossover manifests itself in the nonreciprocal transport. We will discuss this issue in the following two subsections.

5.4.1 *Classical ratchet*

In the high-temperature region, the motion of the coordinate x is treated classically by the following stochastic equation of motion:

$$\eta \frac{dx}{dt} = -\frac{\partial U}{\partial x} + F - \sqrt{2\eta T} \xi(t), \qquad (5.12)$$

where F is the external force, U is the periodic potential with period L, i.e., $U(x+L) = U(x)$ [18]. The last term is the Langevin force by Gaussian white noise, and η is the viscous friction coefficient, both of which are related by the fluctuation–dissipation theorem. Corresponding Fokker–Planck equation for the distribution function $p(x, t)$ reads

$$\frac{\partial p}{\partial t} = \frac{1}{\eta} \frac{\partial}{\partial x} \left[\left(\frac{\partial U}{\partial x} - F \right) p + T \frac{\partial p}{\partial x} \right]. \tag{5.13}$$

The steady solution to Eq. (5.13) has been already obtained, and the velocity is expressed as [18]

$$v = \frac{L}{\beta \eta} \frac{1 - e^{-\beta F L}}{\int_0^L dy \, I_0(y) e^{-\beta F y}}, \tag{5.14}$$

where

$$I_0(y) = \int_{x_0}^{x_0 + L} dx \, e^{\beta [U(x) - U(x - y)]} \tag{5.15}$$

with β being the inverse temperature. When the velocity v is expanded with respect to F, we obtain

$$v = \mu_1 F + \mu_2 F^2 + O\left(F^3\right), \tag{5.16}$$

$$\mu_1 = \frac{L}{\beta \eta} \frac{\beta L}{\int_0^L dy \, I_0(y)}, \tag{5.17}$$

$$\mu_2 = \frac{L}{\beta \eta} \frac{\beta^2 L \int_0^L dy \left(y - \frac{L}{2}\right) I_0(y)}{\left[\int_0^L dy \, I_0(y)\right]^2}, \tag{5.18}$$

where μ_2 describing the nonreciprocal transport is nonzero for asymmetric potentials. Note that the system does not break the time-reversal symmetry $\mathcal{T}$ microscopically, while the dissipation η represents the irreversibility, i.e., the macroscopic time-reversal symmetry breaking. These two sources of the asymmetry between t and $-t$ play crucial role in the nonlinear and nonreciprocal phenomena. Assuming the potential

$$U(x) = U_0 \frac{x}{L} \quad \text{mod } L, \tag{5.19}$$

we obtain the asymptotic form of μ_1 and μ_2 in the limit of large $\beta U_0 \gg 1$ as

$$\mu_1 = \frac{\beta^2 U_0^2}{\eta} e^{-\beta U_0}, \tag{5.20}$$

$$\mu_2 = \frac{L \beta^3 U_0^2}{2\eta} e^{-\beta U_0}, \tag{5.21}$$

and the ratio $\mu_2/\mu_1 = L\beta/2$ is insensitive to the strength of the potential U_0 or the friction coefficient η [15]. Note here that when one starts from the band theory and Boltzmann transport theory, one cannot obtain the second-order mobility μ_2 since the band dispersion is symmetric between $+$ and $-$ directions. Therefore, the quantum Brownian motion in the periodic potential should be considered carefully, as discussed in the next subsection.

5.4.2 *Schmid transition and scaling theory of quantum Ratchet*

Quantum Brownian motion of a particle in the periodic potential with dissipation has been studied intensively over a long term [19]. The quantum dissipation is introduced by the coupling to harmonic bosons as Caldeira–Leggett proposed [20, 21]. The mobility of this model has been studied in influence integral formalism [22], and it has been revealed by employing the renormalization group analysis that a quantum phase transition occurs at the critical value of the dimensionless friction α; the ground state is the extended Bloch wave for $\alpha < \alpha_c = 1$, while it becomes the localized one for $\alpha > \alpha_c = 1$ [23–30]. As for the linear mobility μ_1, it approaches to a finite value $\mu_1 \propto 1/\alpha$ when $\alpha < 1$, while μ_1 vanishes as $\mu_1 \sim T^{2(\alpha-1)}$ when $\alpha > 1$ as $T \to 0$. Then the issue is how the nonreciprocal nonlinear mobility μ_2 behaves as a function of temperature and α for the asymmetric periodic potential

$$V(x) = V_1 \cos\left(2\pi\frac{x}{a}\right) + V_2 \sin\left(4\pi\frac{x}{a}\right). \tag{5.22}$$

Theoretically, the instanton approach [32–34] and perturbation theory [35] have been employed for this problem, but the global understanding of μ_2 has not yet been obtained. Experimentally, the quantum ratchet effects has been studied in.[17, 36, 37].

We employed the perturbative expansion with respect to V_1 and V_2, and the third-order terms in μ_2 gives the leading order nonreciprocal response [39]. Figure 5.3(a) shows the numerically obtained μ_2 as a function of temperature for various values of α, and one can see that it shows the nonmonotonous T-dependence corresponding to the quantum-classical crossover. Namely, μ_2 decreases as $T \to 0$ when the quantum nature of the particle becomes relevant. In this low-temperature limit, μ_2 scales as $\mu_2 \sim T^{6/\alpha-4}$, When $\alpha > 1$, the perturbation theory breaks down at low temperature, where one should treat the tunneling amplitude between the neighboring potential minima as the perturbation. Generalizing these considerations, we propose the following scaling law for the velocity as

$$v = \frac{F}{\eta} - T^{2/\alpha-1}g_o^<(F/T) - T^{6/\alpha-2}g_e^<(F/T) \tag{5.23}$$

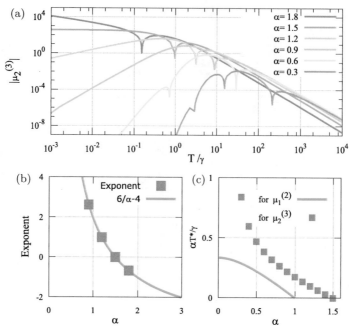

Fig. 5.3.   (a) Temperature dependence of the second-order mobility μ_2 evaluated perturbatively for the asymmetric potential $V(x)$ given in Eq. (5.22) for various values of α. There are two temperature regions with different power law dependence, i.e., $\sim T^{6/\alpha-4}$ at low temperature and $\sim T^{-11/4}$ at high temperature. (b) The power law exponent of μ_2 as a function of α showing the asymptotic behavior $\mu_2 \sim T^{6/\alpha-4}$. (c) The crossover temperature T^* defined as the peak position in panel (a). Reproduced from Ref. [39].

for $\alpha < 1$, while

$$v = t(F)^2 T^{2\alpha-1} g_o^>(F/T), \qquad (5.24)$$

with $t(F) = t_0 + cF$ for $\alpha > 1$, and $g_o^>(F/T)$ is the odd function of its argument, i.e., it contains only the odd-order term in the Taylor expansion.

5.5   Electron correlation and nonreciprocal transport

In contrast to MCA that requires broken $\mathcal{T}$ symmetry, nonreciprocal transport can be achieved with preserving microscopic $\mathcal{T}$ once dissipation effect (macroscopic $\mathcal{T}$ breaking) is incorporated as we have seen in rachet motions. In this section, focusing on nonreciprocal transport in (noncentrosymmetric) crystals, we first extend the Onsager's relation to nonlinear responses to see the necessary ingredients for nonreciprocal transport, and then consider nonreciprocal current from electron correlation.

5.5.1 *Generalized Onsager's theorem*

We consider extension of Onsager's theorem to nonlinear current responses, focusing on the second-order current response,

$$J_i(\omega_1 + \omega_2) = \sigma_{ijj}(\omega_1, \omega_2)E_j(\omega_1)E_j(\omega_2). \tag{5.25}$$

For systems of noninteracting electrons, computing the nonlinear conductivity $\sigma_{ijj}(i\omega_{n_1}, i\omega_{n_2})$ in the imaginary time Green's function formalism leads to the relationship

$$\sigma_{ijj}(i\omega_{n_1}, i\omega_{n_2}) = -\sigma_{ijj}(-i\omega_{n_2}, -i\omega_{n_1}), \tag{5.26}$$

under time-reversal symmetry, which gives the nonlinear generalization of Onsager's relation [40]. At the first glance, the above expression seems to suggest that the nonlinear conductivity $\sigma_{ijj}(\omega_1, \omega_2)$ vanishes in the dc limit ($\omega_1 \to 0$ and $\omega_2 \to 0$), and no nonreciprocal current response is allowed in the noninteracting system. However, there is a subtlety in the analytic continuation to real frequencies. Namely, there exists a branch cut at the real axis in the complex frequency plane due to the dissipation effect, and the analytic continuation of $i\omega_n \to 0$ for the two quantities, $\sigma_{ijj}(i\omega_1, i\omega_2)$ and $\sigma_{ijj}(-i\omega_2, -i\omega_1)$, are generally different under dissipation. This situation is similar to the case of the linear conductivity $\sigma_{xx}(\omega)$ where the limit of $\omega \to +0i$ gives a dissipative current response which is proportional to the relaxation time τ. Therefore, the nonreciprocal current response requires dissipation and should be proportional to the relaxation time τ.

5.5.2 *Keldysh formalism of nonlinear transport and NO-GO theorem*

We can also show that dc nonreciprocal current response does not appear in the system of noninteracting electrons without dissipation, by using the gauge invariant formulation of Keldysh Green's functions [41, 42]. In this formulation, we consider expansion of the Green's function G with respect to the applied electric field E as

$$G(\omega, k) = G_0(\omega, k) + \frac{E}{2}G_E(\omega, k) + \frac{E^2}{8}G_{E^2}(\omega, k) + O(E^3). \tag{5.27}$$

The second-order current response which gives nonreciprocal transport can be computed with the expansion of G as

$$J_{E^2} = -i \int d\omega dk \, \mathrm{tr}[v(k)G_{E^2}^<(\omega, k)]. \tag{5.28}$$

By using the explicit expression for G_{E^2}, we can show that the nonreciprocal current J_{E^2} identically vanishes under the $\mathcal{T}$ symmetry [40]. Furthermore, this argument for a clean system can be generalized to systems with static disorder potential (which does not cause relaxation) by introducing the phase twist at the boundary of the disordered system. This procedure is basically the same as the discussion that is used to define the Chern number in quantum Hall systems with disorder potential [43]. Thus the elastic scatterings for noninteracting electrons does not lead to nonreciprocal current response without dissipation. We note that the situation becomes different once the dissipation effect is included and nonreciprocal current can emerge, as we discuss in Sec. 5.7 in terms of Landau Zener tunneling.

5.5.3   *Electron correlation and nonreciprocal transport*

So far we have seen that nonreciprocal current response requires some dissipation from the generalized Onsager's relation and that noninteracting electrons with elastic scattering neither show nonreciprocal current. In contrast to these situations, nonreciprocal current does appear once we incorporate electron–electron interactions under the dissipation effect.

Let us consider a two-band model and suppose that the valence band is partially filled. When the external electric field E is applied, the electrons in the valence band are shifted toward the direction of E in the momentum space as illustrated in Fig. 5.4. This causes the change of Fermi momenta $\simeq eE\tau$ for the right and the left movers. (We note that the appearance of the relaxation time τ signals that the dissipation effect is involved.) In the presence of electron–electron interaction, this change of distribution in the k space modifies the effective band structure through the change of the Hartree term. In particular, the modification of effective band structure depends on the direction of the E in the case of the inversion broken systems. This means that the Fermi velocity changes in a different manner depending on the direction of E as $v_F(E) = v_{F,0} + bE$, where the conductivity acquires

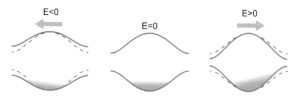

Fig. 5.4.   Schematic picture of nonreciprocal current from electron interaction. An external electric field shifts the distribution of electrons in the momentum space, which modifies the effective band structure through electron interaction. The change of the band structure differs depending on the direction of $\boldsymbol{E}$, and the asymmetric change of the group velocity with respect to $\boldsymbol{E}$ induces nonreciprocal current response. Reproduced from Ref. [40].

E linear correction as $\sigma(E) = \sigma_0 + cE$ (b, c: some constants). This E linear correction in the conductivity leads to the current response,

$$J = \sigma(E)E = \sigma_0 E + cE^2, \tag{5.29}$$

where the second term $\propto E^2$ is the nonreciprocal current response.

We can quantify this E-dependent modification of the band structure by studying the self-energy proportional to E within the gauge invariant Keldysh Green's function formalism. This allows us to estimate the nonreciprocity ratio γ (the ratio of the nonlinear current to the original current) as

$$\gamma \equiv \frac{\delta J}{J} \simeq \frac{U}{E_g} \frac{eEa}{W}, \tag{5.30}$$

where U is the strength of the electron–electron interaction, E_{g,k_F} is the energy gap, W is the band width and a is the lattice constant. As an example, let us estimate γ from this mechanism for the molecular conductor TTF-CA which is a strongly correlated insulator. The typical order of electric field that can be applied is $E \simeq 10^5 \text{V/m}$ [44, 45]. Since the lattice constant is $a \simeq 1\text{nm}$, the electric voltage in the unit cell becomes $eEa \simeq 10^{-4}$ eV, and the band width is given by $W \simeq 0.2$ eV. Thus the nonreciprocity ratio can be $\gamma \simeq 10^{-3}$ which may be comparable with that in magnetochiral anisotropy in Bi helix [6].

5.6 Shift current

So far we considered nonlinear current response with applied dc electric field. In this section, we review on another nonlinear current responses with applied *ac* electric field, called shift current.

5.6.1 *Berry connection and intracell coordinates*

Before explaining shift current response, let us briefly review on the Berry connection of Bloch electrons. We consider the Bloch wavefunction of the momentum k and the band index n that is given by $\phi_{nk}(x) = e^{ikx}u_{nk}(x)$ where $u_{nk}(x)$ is the periodic part, i.e., $u_{nk}(x+L) = u_{nk}(x)$ with L being the translational vector. In the momentum space picture, the position operator is given by the k-derivative of the wavefunction, as

$$r = i\frac{\partial}{\partial k}. \tag{5.31}$$

This allows us to calculate the expectation value of the position operator $\langle r \rangle$ for the wave packet made of the Bloch wavefunction

$$\Psi_B(r) = \int \frac{dk}{\sqrt{2\pi}} c(k) e^{ikr} u_{nk}(r), \tag{5.32}$$

with a coefficient $c(k)$. We obtain the expectation value of the position operator as

$$\langle r \rangle = \int dr \, \Psi_B^*(r) r \Psi_B(r)$$

$$= \int dk c^*(k) \left[i \frac{\partial}{\partial k} + i \langle u_{nk} | \frac{\partial}{\partial k} | u_{kn} \rangle \right] c(k). \tag{5.33}$$

Therefore, one can define the position operator by the expression in the parenthesis at the last line of Eq. (5.33), where the (minus of the) second term,

$$a_n(k) = -i \langle u_{nk} | \nabla_k u_{nk} \rangle, \tag{5.34}$$

is called Berry connection. Berry connection $a_n(k)$ is regarded as an intracell coordinate of the Bloch electron, i.e., the center position of the Bloch wavefunction in the unit cell [46, 47].

5.6.2 *Floquet formalism*

The Floquet formalism provides a concise description of periodically driven systems in terms of static band picture. It is based on the Floquet theorem which is a time-direction analog of Bloch's theorem. Instead of solving the time-dependent Schrödinger equation explicitly, Floquet formalism enables to study dynamics of the nonequilibrium system with the effectively static Floquet Hamiltonian

$$(H_F)_{mn} = \frac{1}{T} \int_0^T dt e^{i(m-n)\Omega t} H_0(t) - \delta_{mn} m \hbar \Omega. \tag{5.35}$$

The Floquet Hamiltonian is spanned over the extended Hilbert space with an extra integer labeled by m, n (Floquet index), and the eigenstates of the Floquet Hamiltonian encode the information of the dynamics of the system which is equivalent to solving the time-dependent Schrödinger equation.

The virtue of the Floquet formalism is that one can study nonequilibrium/nonlinear phenomena with effectively static Hamiltonian H_F. In the case of the nonlinear optical response that we are interested in, external electric field gives a time-periodic perturbation and can be treated by the Floquet formalism. In the Floquet picture of the nonlinear optical response, the nonequilibrium steady state that realizes under optical transition (from the valence band to the conduction band) is described by the anticrossings of a valence band dressed with one photon and a conduction band dressed with no photon, as schematically shown in Fig. 5.5(a). The anticrossings of

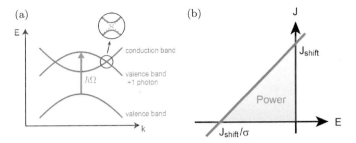

Fig. 5.5.   (a) Floquet two band model. Under the drive by the monochromatic light, energy bands evolve into Floquet bands which describe Bloch states dressed with photons. When two Floquet bands cross, they show an anticrossing, which gives a concise description of the nonequilibrium steady state. (b) $I - V$ characteristics of shift current photovoltaics. We consider current density $J(E)$ as a function of the external dc electric field E. $J(E = 0)$ corresponds to the shift current, and the slope of $J(E)$ determines the efficiency of the energy conversion. Reproduced from Refs. [48, 52].

those two Floquet bands are captured by the Floquet Hamiltonian [48]

$$H_F = \begin{pmatrix} \epsilon_1 + \Omega & -iA^* v_{12} \\ iA v_{21} & \epsilon_2 \end{pmatrix} \equiv d_0 + \boldsymbol{d} \cdot \boldsymbol{\sigma}, \qquad (5.36)$$

where subscripts 1 and 2 refer to the valence and the conduction bands, respectively, $A = E/\Omega$ and $v = \partial H_0(A = 0)/\partial k$. Correspondingly, the current operator $\tilde{v}$ in the Floquet description is given by

$$\tilde{v} = \begin{pmatrix} v_{11} & -iA^*(\partial_k v)_{12} \\ iA(\partial_k v)_{21} & v_{22} \end{pmatrix} \equiv b_0 + \boldsymbol{b} \cdot \boldsymbol{\sigma}. \qquad (5.37)$$

With this Floquet formalism, combined with Keldysh formalism that incorporates the dissipation effect, we are able to study the current responses in the nonequilibrium steady states under light irradiation.

5.6.3   *Shift current in noncentrosymmetric materials*

Shift current is a dc current response caused by photoirradiation in noncentrosymmetric crystals and is described by the nonlinear conductivity $\sigma^{(2)}(\omega)$ as [49, 50]

$$J_{\text{shift}} = \sigma^{(2)}(\omega)|E(\omega)|^2. \qquad (5.38)$$

Shift current responses require broken inversion symmetry $\mathcal{P}$, since the current is odd under $\mathcal{P}$ while $|E|^2$ is even under $\mathcal{P}$. Thus shift current is generally expected in noncentrosymmetric crystals. An interesting feature is that the shift current does not require time-reversal breaking in contrast to MCA. Namely, shift current does not arise from the group velocity of the photocarriers (which should be canceled out due to the $\mathcal{T}$ constraint on the band structure $\epsilon(\boldsymbol{k}) = \epsilon(-\boldsymbol{k})$), but rather it arises from the quantum mechanical nature of the Bloch wavefunctions.

Applying the Floquet formalism to the dc current response under the light irradiation leads to the expression,

$$J = -i\text{Tr}(\tilde{v}G^<) = \int dk \frac{\frac{\Gamma}{2}(-d_x b_y + d_y b_x)}{d^2 + \frac{\Gamma^2}{4}}, \tag{5.39}$$

where Tr denotes a trace for the two by two matrix and integration over ω and $\boldsymbol{k}$. Some computation with respect to the matrix element of $\partial_k v$ leads to the expression,

$$-d_x b_y + d_y b_x = A^2 |v_{12}|^2 R_k, \tag{5.40}$$

where R_k is a shift vector given by

$$R_k = \left[\frac{\partial}{\partial k} \text{Im}(\log v_{12}) + a_1 - a_2 \right]. \tag{5.41}$$

Thus we end up with the expression for the shift current as

$$J_{\text{shift}} = \frac{2\pi e^3}{\hbar^2 \omega^2} |E(\omega)|^2 \int [dk] |v_{12}|^2 R_k \delta(\omega_{21} - \omega), \tag{5.42}$$

where we recovered e and $\hbar$, and $[dk] = dk/(2\pi)^d$ for d-dimensional systems [48].

An intuitive picture of shift current is polarization current from the photoexcited electron–hole pairs. Photocreated electron–hole pairs have nonzero polarization $\simeq R_k \simeq a_1 - a_2$ from the difference in the intracell coordinates between the valence and conduction bands. Light irradiation constantly creates electron–hole pairs, resulting in polarization that increases linearly in time. Since the time derivative of the polarization gives current, this process leads to dc current generation, which can generally happen in noncentrosymmetric crystals where Berry connection a_n becomes finite.

5.6.4   *Shift current by excitons in insulators*

Shift current also appears in systems of interacting electrons when the inversion symmetry is broken. As an example of interaction effect on semi-conductors, let us consider the case of exciton formation due to the attractive interaction between photoexcited electrons and holes. We can study the excitonic processes in the nonlinear optical effect within the Floquet formalism, where we treat the attractive interaction between photoexcited electron and holes with the mean field approximation [51]. The Floquet treatment shows that the nonvanishing shift current appears at the exciton resonance for noncentrosymmetric crystals. It is interesting that the dc current flows with exciton creation even though there does not exist free carriers (i.e., the excitons are bound states of photoexcited electron hole pair). The current arises from nonzero polarization of the exciton. In noncentrosymmetric crystals the excitons have nonzero polarization since the intracell coordinates for the conduction and valence states generally differ. Similarly to the noninteracting case, when excitons are constantly created by light irradiation, the polarization from the excitons increases in time and this constant increase of polarization gives rise to the shift current.

5.6.5 *I-V characteristics and shot noise of shift current*

When one considers application of shift current for solar cell actions, an important quantity is the I–V characteristics, i.e., how much current I flows under a certain applied voltage V (Fig. 5.5(b)). In the case of shift current photovoltaics, nonzero shift current J_{shift} appears at $V = 0$ under light irradiation. If we apply the external voltage (dc electric field) in the opposite direction to the shift current, photoexcited free carriers should flow in the opposite direction and starts to cancel shift current. If we assume this excess current induced by the dc electric field E_{dc} obeys Ohm's law (i.e., the excess current is linearly proportional to E_{dc}), we can write the I–V characteristic as

$$J(E_{\text{dc}}) = J_{\text{shift}} + \sigma_E E_{\text{dc}}, \tag{5.43}$$

which is schematically illustrated in Fig. 5.5(b). The shift current is completely canceled at $E_{\text{dc}} = -J_{\text{shift}}/\sigma_E$, and the area encircled by the I–V characteristics and the two axes gives the total power that can be generated from the shift current photovoltaics. Thus the slope σ_E determines the efficiency of the power conversion; smaller σ_E is better for the power conversion. By using the Floquet formalism, the slope σ_E can be obtained as [52]

$$\sigma_E = \frac{4\pi e^4}{\hbar^3 \omega^2} |E(\omega)|^2 \tau^2 \int [dk] |v_{12}|^2 (v_{11} - v_{22}) R' \delta(\omega_{21} - \omega), \tag{5.44}$$

where $[dk] \equiv dk/(2\pi)^d$ with the dimension d, $R' = \text{Re}[(\partial_k v)_{12}/v_{12}]$, and $\omega_{21} = (\epsilon_2 - \epsilon_1)/\hbar$. This expression indicates that the slope of the I–V characteristics strongly depends on the relaxation time τ while the shift current J_{shift} does not show such dependence on τ. Also, the slope is proportional to the group velocity difference $v_{11} - v_{22}$ between the two bands participating in the optical transition, and especially it vanishes when two bands are parallel, for example, in Landau levels.

The Floquet formalism also allows us to compute the current noise of the shift current photovoltaics, which is important when we consider its application to the photodetectors [52]. The noise (the zero frequency component of the current autocorrelation) is expressed as

$$S = \frac{e^4}{\hbar^2\omega^2}E^2\tau \int [dk]|v_{11} - v_{22}||v_{12}|^2\delta(\omega_{21} - \omega). \qquad (5.45)$$

This shows that the noise S also depends on the relaxation time τ and the group velocity difference $v_{11} - v_{22}$. Again, the noise is significantly suppressed in the flat band systems such as the Landau levels. Furthermore, the noise does not have a part corresponding to the shot noise which is proportional to current J_{shift}. This reflect the nature of shift current which is different from the usual (dissipative) current response and is more like a polarization current. The absence of the shot noise part indicates that the shift current photovoltaics can provide efficient photodetectors.

5.6.6   *First-principles calculation of shift current*

There have been intensive efforts in studying shift current responses using first-principles calculations. Early first-principle studies are done for typical ferroelectric materials $BaTiO_3$ [53] and bismuth ferrite [54]. In particular, the calculation for $BaTiO_3$ shows the sign change of current depending on the frequency and light polarization which reproduces experimental photocurrent direction. This agreement established that the photocurrent in the ferroelectrics $BaTiO_3$ originates from the shift current mechanism. First principle calculations of shift current also reproduces experimental results of photovoltaic effects in other ferroelectric materials such as SbSI [55] and $Sn_2P_2S_6$ [56]. Also, there have been efforts to predict and design photovoltaic responses for low-dimensional materials from the view point of the first-principles calculations, including 1D ferroelectric polymer films [57] and 2D monochalcogenides GeS [58]. Another interesting shift current photovoltaics is Weyl semimetals, where the photovoltaic effect of inversion broken Weyl semimetal TaAs has been studied using the first principles calculations [59].

5.7   Discussion and conclusions

As described above, the nonreciprocal responses of quantum materials are a highly nontrivial issue in condensed matter physics. In addition to the spatial inversion symmetry $\mathcal{P}$, the time-reversal symmetry $\mathcal{T}$ plays an essential role.

It is noted here that there are two types of the asymmetry between t and $-t$; one is the broken $\mathcal{T}$ in the microscopic Hamiltonian, and the other is the macroscopic irreversibility. The latter corresponds to the dissipation, which introduces the difference between right and left directions even at the classical equation of motion. When the quantum mechanical coherence is introduced, i.e., the Bloch wavefunction is formed, and the nonreciprocal nature is suppressed at zero temperature even with the dissipation. Therefore, the quantum nature tends to suppress the nonreciprocal responses, and the nonmonotonous temperature dependence is the signature of the quantum-classical crossover. Superconductivity is the limit of long-range quantum coherence, and hence appears to be against the nonreciprocal responses. However, here, another fundamental principle of physics, i.e., the spontaneous symmetry breaking enters where the order parameter behaves as a classical object even when its origin is quantum mechanical. This is basically the reason why γ is enhanced by the superconducting fluctuation.

There is also an essential difference between the noninteracting single-particle system and many-body system. The electron-electron interaction acts as the feedback mechanism to change the band structure modified by the external electric field, which is the origin of the rectification effect by pn-junction. Even for the bulk crystal with translational symmetry, this feedback effect is finite but very small, and it is the reason for the negligible nonreciprocal transport in polar metals. Also the dc and ac responses are quite different as seen in the comparison between the NO–GO theorem discussed in Sec. 5.5.2 and the shift current. The dc current induced by the ac electric field can be nonzero even without the broken $\mathcal{T}$ for the shift current as it involves energy absorption from photons which is a dissipative process. The geometry of the Bloch wavefunction is relevant to this dc current via the shift vector, i.e., the Berry connection. This shift vector is also relevant to the nonreciprocal Landau–Zener tunneling in the band insulator [60], which leads to nonreciprocal current response once one incorporates dissipation effect [61]. An interesting theoretical issue is to study the crossover from the Landau–Zener tunneling and shift current in the plane of the strength of the electric field E and the frequency Ω of the electric field. This crossover is called Keldysh crossover, and the issue is how the nonreciprocal responses behave in this plane.

Nonlinear responses in nonequilibrium states are the subject of broad interest in physics, chemistry and biology. "Direction" is always the crucial issue when dynamical processes are considered, and hopefully one can control it by external knob. Chirality is considered as the property of static structures in most of the cases as in the molecules, while the dynamical chirality is considered in the higher dimensions including space and time. Therefore, the physics of space–time as relativistic quantum field theory addresses is naturally relevant to the problem of chirality as evidenced in the dynamics of condensed matters. We expect more and more rich results will be explored in this field.

Acknowledgments

We would like to dedicate this article to late Prof. Shoucheng Zhang, who had been the constant source of original ideas and physical intuition. His legacy will continue to inspire and encourage us for long. The authors thank M. Ezawa, C. Felser, J. Fujioka, K. Hamamoto, S. Hoshino, T. Ideue, H. Ishizuka, Y. Iwasa, Y. Kaneko, M. Kawasaki, S. Kitamura, S. Koshikawa, R. Nakai, M. Nakamura, M. Sotome, N. Ogawa, M. Ogino T. Park, Y. Saito, S. Shimizu, K.S. Takahashi, Y. Tokura, J. van den Brink, R. Wakatsuki, Binghai Yan K. Yasuda, H. Yasuda, R. Yoshimi, Y. Zhang, for collaborations and useful discussion. This work was supported by JSPS KAKENHI Grant (No. 18H03676), and by JST CREST Grant No. JPMJCR1874, Japan.

References

[1] M. Gardner, *"The New Ambidextrous Universe: Symmetry and Asymmetry from Mirror Reflections to Superstrings"*, 3rd revised edn. (Dover, 2005).

[2] Y. Tokura and N. Nagaosa, *Nat. Comm.* **9**, 3740 (2018).

[3] L. Onsager, *Phys. Rev.* **37**, 405 (1931).

[4] R. Kubo, *J. Phys. Soc. Jpn.* **12**, 570 (1957).

[5] G. L. J. A. Rikken and E. Raupach, *Nature* **390**, 493 (1997).

[6] G. L. J. A. Rikken, J. Folling and P. Wyder, *Phys. Rev. Lett.* **87**, 236602 (2001).

[7] V. Krstić, S. Roth, M. Burghard, K. Kern and G. L. J. A. Rikken, *J. Chem. Phys.* **117**, 11315 (2002).

[8] G. L. J. A. Rikken and P. Wyder, *Phys. Rev. Lett.* **94**, 016601 (2005).

[9] F. Pop, P. Auban-Senzier, E. Canadell, G. L. J. A. Rikken and N. Avarvari, *Nat. Commun.* **5**, 3757 (2014).

[10] T. Ideue *et al.*, *Nat. Phys.* **13**, 578 (2017).

[11] T. Morimoto and N. Nagaosa, *Phys. Rev. Lett.* **117**, 146603 (2016).

[12] B. I. Halperin and D. R. Nelson, *J. Low Temp. Phys.* **36**, 599 (1979).

[13] M. Tinkham, *Introduction to Superconductivity* (McGraw-Hill, New York, 1996).

[14] R. Wakatsuki *et al.*, *Sci. Adv.* **3**, e1602390 (2017).

[15] S. Hoshino, R. Wakatsuki, K. Hamamoto and N. Nagaosa, *Phys. Rev. B* **98**, 054510 (2018).

[16] K. Yasuda *et al.*, *Nat. Commun.* **10**, 2734 (2019).

[17] J. B. Majer, J. Peguiron, M. Grifoni, M. Tusveld and J. E. Mooij, *Phys. Rev. Lett.* **90**, 056802 (2003).

[18] P. Reimann, C. Van den Broeck, H. Linke, P. Hänggi, J. M. Rubi and A. Përez-Madrid, *Phys. Rev. Lett.* **87**, 010602 (2001).

[19] U. Weiss, *Quantum Dissipative Systems* (World Scientific, 2012).

[20] A. O. Caldeira and A. J. Leggett, *Ann. Phys.* (N.Y.) **149**, 374 (1983).
[21] A. J. Leggett, S. Chakravarty, A. T. Dorsey, Matthew P. A. Fisher, A. Garg and W. Zwerger, *Rev. Mod. Phys.* **59**, 1 (1987).
[22] R. P. Feynman and F. L. Vernon, *Ann. Phys.* (N.Y.) **24**, 118 (1963).
[23] A. Schmid, *Phys. Rev. Lett.* **51**, 1506 (1983).
[24] F. Guinea, V. Hakim and A. Muramatsu, *Phys. Rev. Lett.* **54**, 263 (1985).
[25] F. Guinea, *Phys. Rev. B* **32**, 7518 (1985).
[26] M. P. A. Fisher and W. Zwerger, *Phys. Rev. B* **32**, 6190 (1985).
[27] W. Zwerger, *Phys. Rev. B* **35**, 4737 (1987).
[28] C. L. Kane and M. P. A. Fisher, *Phys. Rev. Lett.* **68**, 1220 (1992).
[29] C.L. Kane and M. P. A. Fisher, *Phys. Rev. B* **46**, 15233 (1992).
[30] A. Furusaki and N. Nagaosa, *Phys. Rev. B* **47**, 4631 (1993).
[31] P. Jung, J. G. Kissner and P. Hänggi, *Phys. Rev. Lett.* **76**, 3436 (1996).
[32] P. Reimann, M. Grifoni and P. Hänggi, *Phys. Rev. Lett.* **79**, 10 (1997).
[33] S. Yukawa, M. Kikuchi, G. Tatara and H. Matsukawa, *J. Phys. Soc. Jpn.* **66**, 2953 (1997).
[34] G. Tatara, M. Kikuchi, S. Yukawa and H. Matsukawa, *J. Phys. Soc. Jpn.* **67**, 1090 (1998).
[35] S. Scheidl and V. M. Vinokur, *Phys. Rev. B* **65**, 195305 (2002).
[36] H. Linke, T. E. Humphrey, A. Löfgren, A. O. Sushkov, R. Newbury, R. P. Taylor and P. Omling, *Science* **286**, 2314 (1999).
[37] R. Menditto, H. Sickinger, M. Weides, H. Kohlstedt, D. Koelle, R. Kleiner and E. Goldobin, *Phys. Rev. E* **94**, 042202 (2016).
[38] J. L. Mateos, *Phys. Rev. Lett.* **84**, 258 (1999).
[39] K. Hamamoto, T. Park, H. Ishizuka and N. Nagaosa, *Phys. Rev. B* **99**, 064307 (2019).
[40] T. Morimoto and N. Nagaosa, *Sci. Rep.* **8**, 2973 (2018).
[41] S. Onoda, N. Sugimoto and N. Nagaosa, *Prog. Theoret. Phys.* **116**, 61 (2006).
[42] N. Sugimoto, S. Onoda and N. Nagaosa, *Phys. Rev. B* **78**, 155104 (2008).
[43] Q. Niu, D. J. Thouless and Y.-S. Wu, *Phys. Rev. B* **31**, 3372 (1985).
[44] Y. Tokura, H. Okamoto, T. Koda, T. Mitani and G. Saito, *Phys. Rev. B* **38**, 2215 (1988).
[45] T. Mitani, Y. Kaneko, S. Tanuma, Y. Tokura, T. Koda and G. Saito, *Phys. Rev. B* **35**, 427 (1987).
[46] R. Resta, *Rev. Mod. Phys.* **66**, 899 (1994).
[47] N. Nagaosa and T. Morimoto, *Adv. Mat.* **29**, 1603345 (2017).
[48] T. Morimoto and N. Nagaosa, *Sci. Adv.* **2**, e1501524 (2016).
[49] R. von Baltz, W. Kraut, *Phys. Rev. B* **23**, 5590 (1981).
[50] J. E. Sipe, A. I. Shkrebtii, *Phys. Rev. B* **61**, 5337 (2000).
[51] T Morimoto, N Nagaosa. *Phys. Rev. B* **94**, 035117 (2016).
[52] T. Morimoto, M. Nakamura, M. Kawasaki and N. Nagaosa, *Phys. Rev. Lett.* **121**, 267401 (2018).
[53] S. M. Young and A. M. Rappe, *Phys. Rev. Lett.* **109**, 116601 (2012).
[54] S. M. Young, F. Zheng and A. M. Rappe, *Phys. Rev. Lett.* **109**, 236601 (2012).
[55] M. Sotome, M. Nakamura, J. Fujioka, M. Ogino, Y. Kaneko, T. Morimoto, Y. Zhang, M. Kawasaki, N. Nagaosa, Y. Tokura and N. Ogawa, *Proc. Nat. Acad. Sci.* **116**, 1929 (2019).
[56] M. Sotome, M. Nakamura, J. Fujioka, M. Ogino, Y. Kaneko, T. Morimoto, Y. Zhang, M. Kawasaki, N. Nagaosa, Y. Tokura and N. Ogawa, *Appl. Phys. Lett.* **114**, 151101 (2019).
[57] A. M. Cook, B. M. Fregoso, F. De Juan, S. Coh and J. E. Moore, *Nature commun.* **8**, 14176 (2017).

[58] T. Rangel, B. M. Fregoso, B. S. Mendoza, T. Morimoto, J. E. Moore and J. B. Neaton, *Phys. Rev. Lett.* **119**, 067402 (2017).

[59] Y. Zhang, H. Ishizuka, J. van den Brink, C. Felser, B. Yan and N. Nagaosa, *Phys. Rev. B* **97**, 241118 (2018).

[60] S. Kitamura, N. Nagaosa and T. Morimoto, *Commun. Phys.* **3**, 63 (2020).

[61] S. Kitamura, N. Nagaosa and T. Morimoto, *Phys. Rev. B* **102**, 245141 (2020).

Chapter 6

Spintronics Meets Topology

Shuichi Murakami

Department of Physics, Tokyo Institute of Technology,
Ookayama, Meguro-ku, Tokyo 152-8551, Japan
TIES, Tokyo Institute of Technology, Ookayama,
Meguro-ku, Tokyo 152-8551, Japan

Since the discovery of topological insulators, topological phases have been the subject of intensive theoretical and experimental studies in condensed matter. In this article, I show how this discovery has been achieved through an encounter between spintronics and topology. Through this encounter, importance of gauge field in condensed matter has been reacknowledged. I also explain several examples to show how this encounter has led to births of new fields in condensed matter.

6.1 Introduction

Topological materials have been studied intensively, since the discoveries of topological insulators. It is surprising that topological materials are abundant in nature and even well-known materials turn out to be topological, as has been revealed only recently. In retrospect, the proposals of the intrinsic spin Hall effect [1, 2] have led us to the discoveries of topological insulators [3–5], which triggered a breakthrough in material science. In the twentieth century, the only known topological systems realized in condensed matter are quantum Hall systems and topological systems were regarded as special cases in condensed matter, e.g., the quantum Hall systems which can be seen only in high magnetic fields and specially designed systems to mimic some models having special topological characteristics.

Through the study of the quantum Hall effect, it has been recognized that Hall effects are closely related to topological phases. Meanwhile, since the usual Hall effect requires breaking of the time-reversal symmetry, the topological phase in the quantum Hall effect, i.e., the Chern insulator, also requires breaking of the time-reversal symmetry. In this sense, the spin Hall effect is the first example of the Hall effects which can be realized with time-reversal symmetry. Then, it is a natural step to consider an insulating analog of the spin Hall effect, that is, topological insulators. In this chapter, we review the studies on spin Hall effect and topological insulators in this context, with some reviews on our works and related works.

6.2 Intrinsic spin Hall effect

Through the study of the Hall effect, it has been realized that the gauge field in k space affects electron motions. An intuitive way to understand it is to use the following semiclassical equation of motion [6]. We consider the Bloch states in the nth band in the crystal. Out of the Bloch wavefunctions, we make a wavepacket, which is localized both in k-space and in real-space. Under an electric field $\boldsymbol{E}$, its position in these spaces evolves as

$$\dot{\boldsymbol{x}} = \frac{1}{\hbar}\frac{\partial E_n(\boldsymbol{k})}{\partial \boldsymbol{k}} - \dot{\boldsymbol{k}} \times \boldsymbol{B}_n(\boldsymbol{k}), \tag{6.1}$$

$$\hbar\dot{\boldsymbol{k}} = -e\boldsymbol{E}, \tag{6.2}$$

where $E_n(\boldsymbol{k})$ is the energy eigenvalue of the nth band, $-e$ is an electron charge, $\boldsymbol{x}$ and $\boldsymbol{k}$ are the real-space and k-space positions of the wavepacket, respectively and $\hbar = h/(2\pi)$ is the Planck constant. We introduced the Berry curvature $\boldsymbol{B}_n(\boldsymbol{k})$ for the nth band [7], defined by

$$\boldsymbol{B}_n(\boldsymbol{k}) = i\left\langle \frac{\partial u_{n,\boldsymbol{k}}}{\partial \boldsymbol{k}} \middle| \times \middle| \frac{\partial u_{n,\boldsymbol{k}}}{\partial \boldsymbol{k}} \right\rangle, \tag{6.3}$$

where $u_{n,\boldsymbol{k}}$ is the periodic part of the Bloch wavefunction $\psi_{n,\boldsymbol{k}}(\boldsymbol{x}) = u_{n,\boldsymbol{k}}(\boldsymbol{x})e^{i\boldsymbol{k}\cdot\boldsymbol{x}}$. In these semiclassical equations of motion, the second term on the right-hand side of Eq. (6.1), $-\dot{\boldsymbol{k}} \times \boldsymbol{B}_n(\boldsymbol{k})$ ($\propto \boldsymbol{E} \times \boldsymbol{B}_n(\boldsymbol{k})$), is called anomalous velocity. This term is caused by the Berry curvature and leads to the velocity perpendicular to the electric field, which naturally gives rise to the Hall effect.

The types of the Hall effect expected in each system depend on symmetries of the system. At each $\boldsymbol{k}$, the Berry curvature $\boldsymbol{B}_n(\boldsymbol{k})$ is generally nonzero. Nonetheless, in nonmagnetic systems, the sum of all the currents due to this anomalous velocity is zero due to the time-reversal symmetry, meaning an absence of Hall effect as is naturally expected. In a ferromagnet, the total transverse current due to the anomalous velocity is nonzero and it is the anomalous Hall effect shown in Fig. 6.1(a). In particular, if we apply this theory to magnetic insulators, it leads to quantization of Hall conductivity and it is called the quantum anomalous Hall effect. Namely, in a two-dimensional (2D) insulator, the semiclassical equations of motion

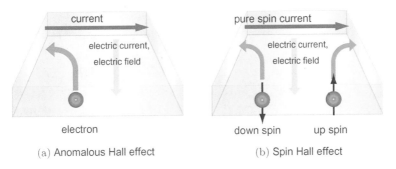

(a) Anomalous Hall effect   (b) Spin Hall effect

Fig. 6.1.   Schematic figures of (a) anomalous Hall effect and (b) spin Hall effect.

lead to

$$\sigma_{xy} = -\frac{e^2}{h}\nu, \quad \nu = \sum_{n:\text{occ.}} \int_{\text{BZ}} \frac{d^2 k}{2\pi} B_{n,z}(\boldsymbol{k}). \tag{6.4}$$

Here, ν is called the Chern number, defined as an integral of the Berry curvature over the Brillouin zone, summed over the occupied bands n. Since ν is proven to be always an integer, the quantization of the Hall conductivity as an integer multiple of e^2/h results [8,9] and this Chern number represents the number of chiral edge states going along the edge of the system at the Fermi energy.

Next, let us apply the semiclassical equations of motion (Eqs. (6.1) and (6.2)) to a nonmagnetic metal with the spin–orbit coupling (SOC). In a two-dimensional system along the xy-plane as an example, the Berry curvature is along the z-axis. In nonmagnetic systems with the SOC, the Berry curvature satisfies $B_z^{\uparrow}(\boldsymbol{k}) = B_z^{\downarrow}(-\boldsymbol{k})$, where $B_z^{\alpha}(\boldsymbol{k})$ ($\alpha = \uparrow, \downarrow$) is the Berry curvature in a α-spin sector. Therefore, the transverse currents in the up-spin and down-spin sectors are the opposite, meaning that the transverse current is zero, but the transverse spin curernt is nonzero as shown in Fig. 6.1(b). This is called spin Hall effect [1,2].

Through the studies on these Hall effects, it is now established that the Berry curvature in k-space affects the electron motion. In Ref. 1, in collaboration with Shoucheng Zhang and Naoto Nagaosa, the Luttinger model describing the valence bands of cubic semiconductors with the SOC was used to show the spin Hall effect due to the Berry curvature. In fact, in the paper [10] by Jiangping Hu and Shoucheng Zhang, the SU(2) gauge-field structure is studied and it is shown that an SU(2) spin current is induced by the electric field, which is a four-dimensional generalization of the quantum Hall effect. This SU(2) gauge field can be induced by the SOC. This idea led to the proposal of the spin Hall effect in [1].

The SOC is required in the spin Hall effect [1, 2], because without the spin–orbit coupling the motions of electrons cannot depend on spin directions. In the above semiclassical equations of motion, the Berry curvature becomes dependent on spin in the presence of the SOC, leading to the spin Hall effect. Thus the spin Hall effect becomes significant when the SOC is large. In spintronics, via this spin Hall effect, one can easily convert between the charge current and spin current with a simple setup using a heavy metal and the spin Hall effect has been studied intensively. Now the spin Hall effect has been measured in a broad range of nonmagnetic metals and semiconductors and in the field of spintronics it is widely used as a probe of spin currents in solids.

We note that in a strict sense, the spin current itself is not well-defined as an observable. A current should be defined so as to satisfy the continuity equation and therefore, spin current, j_j^i, representing the spin along the j-direction flowing along the i-direction, should satisfy the continuity equation $\frac{\partial}{\partial x_i} j_j^i + \frac{\partial}{\partial t} S_j = 0$ for the spin S_j along the j-direction. This equation, however, requires the conservation of the spin $\frac{d}{dt} \int dV S_j = 0$, which is satisfied when the Hamiltonian commutes with the spin S_j. In reality, in the presence of the SOC, the spins are not conserved in general and therefore the spin current is not well-defined. Nonetheless, one can introduce the spin current as a phenomenological quantity to discuss spin transport phenomena to compare the theory and experiment [11].

6.3   Topological insulators

This relationship in Sec. 6.2 between the (intrinsic) anomalous Hall effect and the topological phase in the quantum anomalous Hall effect, i.e., the Chern insulator, is naturally extended to the spin Hall effect as pioneered in Refs. [3–5]. It leads to the notion of the quantum spin Hall effect, i.e., the proposal of two-dimensional topological insulators. The simplest case of a two-dimensional topological insulator is a superposition of two quantum Hall systems for the up-spin and the down-spin, with opposite Chern numbers $\nu_\uparrow = 1$, $\nu_\downarrow = -1$. This superposition naturally leads to helical edge states, carrying a pure spin current. Topologically nontrivial nature of the topological insulators is characterized in terms of Z_2 topology [4, 12]. In the above realization of the topological insulator, the spin S^z is assumed to be a conserved quantity. Nonetheless, even when this spin conservation is violated, as is always the case in real materials, Z_2 topology survive, as long as the time-reversal symmetry is preserved. In other words, although there is a problem in defining spin current as a physical observable, it does not affect the Z_2 topological classification.

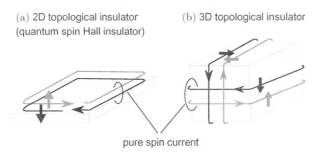

(a) 2D topological insulator
(quantum spin Hall insulator)

(b) 3D topological insulator

pure spin current

Fig. 6.2.   Schematic figures of (a) the two-dimensional topological insulator and (b) the three-dimensional topological insulator.

In Fig 6.2, we show schematic figures of the two-dimensional (2D) and three-dimensional (3D) topological insulators. Due to the bulk–boundary correspondence of the Z_2 topology, the topological insulators have helical boundary states on the edges in two dimensions and on the surfaces in three dimensions. These helical states carry pure spin current, consisting of two states with opposite spins flowing in opposite directions, although a precise and quantitative definition of "pure spin current" is ambiguous.

This topological insulator phase is characterized by the Z_2 topological invariant [4, 12] and it is defined in terms of inversion parities when the system has inversion symmetry [13]. Under various symmetries [14] including crystallographic symmetries [15], various types of topologically insulating phases appear and several methods have been established to study the list of topological phases under the given symmetry [16–18]. From the methods in Refs 16, 17 one can obtain topological invariants expressed in terms of irreducible representations (irreps) at high-symmetry points in the k-space. On the other hand, the K-theory approach [18] gives information on topological invariants expressed in terms of k-space integrals. Thus by combining their results one can reach a full understanding of topological invariants under various symmetries. For example, systems with glide symmetry are classified in terms of the Z_2 topological invariant and one can see directly that by adding inversion symmetry this glide-Z_2 topological invariant becomes equal to a half of the Z_4 topological invariant defined from inversion parities [19, 20], which is used to characterize higher-order topological insulators.

6.4   Topological semimetals

Another beautiful manifestation of topology in electronic band structure in crystals is topological semimetals. It is well known that in the band structure in crystals, band degeneracy usually appears only at high-symmetry points or on high-symmetry lines, because of the higher-dimensional irreducible representation of the space group (or magnetic space group) considered. Nonetheless, interestingly, another type of degeneracy may appear due to topology. Such kind of degeneracy cannot be explained solely by representation theory, but can be explained in terms of topological protection. When such kind of topological degeneracy appears at or near the Fermi energy and when the bands are gapped except for these bands forming the topological degeneracy, the material is called topological semimetal or topological metal. The most remarkable example of topological semimetals is a Weyl semimetal. The Weyl semimetal (WSM) is a three-dimensional system with a nondegenerate Dirac cone in the band dispersion close to the Fermi energy.

The idea of the Weyl semimetal [21–23] is easily understood by using a simple model. Although the following discussion is based on a simple model, it applies also to real materials as well, because it is governed by topology. We consider a problem of when two bands are degenerate in a band structure in a three-dimensional crystal, without assuming any crystallographic symmetry except for translation symmetries. Because only two bands are involved, one can simply consider a two-band model dependent on the Bloch wavevector $k = (k_x, k_y, k_z)$,

$$H(k) = a_0(k)\sigma_0 + \sum_{j=x,y,z} a_j(k)\sigma_j \qquad (6.5)$$

where $a_j(k)$ $(j = 0, x, y, z)$ are functions of k, σ_0 is a 2×2 identity matrix and σ_j $(j = x, y, z)$ are the Pauli matrices. Let us consider what condition is required for degeneracy of the two energy eigenvalues. It is given by three conditions $a_j(k) = 0$ $(j = x, y, z)$. Since there are three variables k_x, k_y and k_z, these conditions can have some solutions in general. Such solutions are given by intersections of the three surfaces in k-space described by $a_j(k) = 0$ and at these points the two energy eigenvalues are exactly degenerate. We note that this degeneracy does not stem from any crystallographic symmetry. Around such a point the dispersion is shown to be linear, forming a Dirac cone without spin degeneracy and the point of degeneracy is called Weyl node (Fig. 6.3(a)). Moreover, the Berry curvature has a source or a sink at this point (Fig. 6.3(b)), with quantized strength. In other words around these points the Berry curvature has a δ-function divergence at these points: $\frac{1}{2\pi}\nabla_k \cdot B(k) = q\delta(k - k_0)$ $(k \sim k_0)$ with $q = \pm 1$. This means that the Berry curvature has a monopole with a quantized monopole charge $q = \pm 1$. Since the monopole charge is shown to be always quantized, this degeneracy is topological.

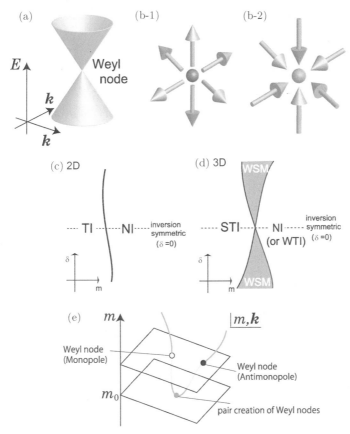

Fig. 6.3. Weyl semimetal. (a) Schematic figure of a dispersion around a Weyl node. (b) Weyl nodes are classified into (b-1) a monopole and (b-2) an antimonopole, which are a source and a sink of the Berry curvature, respectively. (c) and (d) Universal phase diagram between a topological insulator (TI) and a normal insulator (NI) phases in (c) 2D and (d) 3D. m is a parameter driving the phase transition and δ represents the degree of inversion-symmetry breaking. (c) In 2D, the NI–TI phase transition occurs directly, with the gap closing only on the phase transition line in the phase diagram. (d) In 3D, when the inversion symmetry is broken ($\delta \neq 0$), the Weyl semimetal (WSM) phase should intervene between the NI and the TI phases. (e) Gap closing by changing a parameter m. The gap closes at $m = m_0$ and a monopole–antimopole pair is created. The gap remains closed along the curve in the $(\boldsymbol{k}, m)$ space.

Thus, the degeneracy at Weyl points is topologically protected in three dimensions. This monopole charge is a topological invariant associated with this Weyl node. In other words, the surface integral of the Berry curvature over a surface surrounding the Weyl node (times $\frac{1}{2\pi}$) is always quantized to be $+1$ or -1, which corresponds to a monopole and an antimonopole, respectively. It can be considered as a Chern number associated with this Weyl node.

Because of this quantization of the monopole charge, it cannot appear or disappear, by a perturbative change of the system. The only chance for the Weyl nodes to appear or to disappear is via pair creation and annhilation of a monopole and an antimonopole [22]. To describe it, we consider how the system evolves by changing a system parameter m. By introducing a real parameter m, we can write

$$H(\boldsymbol{k}, m) = a_0(\boldsymbol{k}, m)\sigma_0 + \sum_{j=x,y,z} a_j(\boldsymbol{k}, m)\sigma_j. \tag{6.6}$$

Here we assume that when $m < m_0$ the system is gapped, which means that the eigenvalues are nondegenerate and at $m = m_0$ the gap closes. Similar to the previous discussion, the gap closes when three conditions $a_j(\boldsymbol{k}, m) = 0$ $(j = x, y, z)$ are simultaneously satisfied. In the four-dimensional $(\boldsymbol{k}, m)$ space, these three conditions determine a curve, along which the eigenvalues are degenerate (Fig. 6.3(e)). This curve does not have endpoints, since the Berry phase around this curve is quantized to be π. Therefore, as we increase m across m_0, the number of points in $\boldsymbol{k}$ in the given value of m is one at $m = m_0$ and becomes two for $m > m_0$. This splitting of degenerate points in k space is shown to be a pair creation of Weyl nodes into a monopole and an antimonopole [22].

This pair creation of Weyl nodes via gap closing indeed occurs in many cases between two insulating phases with different topological invariants. For example, in many materials one can have a transition between a Z_2 topological insulator (TI) phase and a normal insulator (NI) phase and it is shown that when the inversion symmetry is broken, there should be a Weyl semimetal phase in between these two insulating phases [22, 24]. Its exception is systems with inversion symmetry. In spinful systems with time-reversal and inversion symmetries, all the states are doubly degenerate by Kramers theorem and the condition for band gap closing changes. In this case, the band gap can close only at time-reversal invariant momenta, when the parities of the valence band and conduction band are opposite and this band gap closing leads to band inversion, i.e., an NI–TI topological phase transition. As a result, in three dimensions, the NI–TI phase diagram is universally of the form shown in Fig. 6.3(d). This conclusion can be directly confirmed by using a simple model [25]. On the other hand, in two dimensions, the NI–TI phase diagram is shown in Fig. 6.3(c), where the NI and TI phases are always next to each other without an intermediate WSM phase. There are many material realization for this appearance of a Weyl semimetal phase between two insulating phases with different values of topological invariants. For example, in $Pb_{1-x}Sn_xTe$, where the inversion

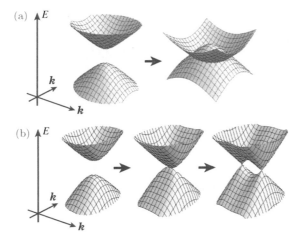

Fig. 6.4. Gap closing in inversion-asymmetric spinful systems. After the gap closing, the system becomes either (a) a nodal-line semimetal, protected by mirror symmetry, or (b) a Weyl semimetal. The details of the resulting topological semimetal, e.g., the number and the positions of the Weyl nodes, depend on the gap-closing k-point and the irreducible representations of the valence and the conduction bands.

symmetry is broken, external pressure induces an NI-WSM-TCI phase transition as experimentally observed in Ref. 26. The experimental data on its intermediate WSM phase is in good agreement with the *ab initio* calculation.

In fact, one can show a stronger conclusion. Suppose we consider a spinful semiconductor without inversion symmetry and tune some system parameter to close the gap. In this case, an evolution of the band structure across the closing of the band gap is determined by symmetry of the system and the k-point of the gap closing. By considering all possibilities for space groups and k-points for the gap closing, we can show that after the gap closing, the gap will not open immediately after the gap closing, but it becomes either a nodal-line semimetal or a Weyl semimetal (Fig. 6.4). One can find many materials showing this behavior, such as tellurium under pressure [24, 27].

6.5 Material realization of topological phases

Once we derive a formula for some topological invariant, one can search for materials showing nontrivial topological phases from various viewpoints. One can search for topological materials based on the physical picture for the topological phase, as we show in some examples in this section. Alternatively, if the formula is expressed in terms of irreps at high-symmetry points, one can search for topological materials over materials databases to obtain a number of candidate materials.

6.5.1   *Topological insulators*

The proposals of the topological insulators have brought about a renewed viewpoint onto material science. As the Z_2 topological insulator is driven by the SOC, its candidate should have a strong SOC, involving heavy elements. The first realistic proposals of the topological insulators were made for bismuth (Bi) bilayer [28] and the CdTe/HgTe/CdTe quantum well [29], both proposed to be two-dimensional topological insulators. In the latter proposal by Andrei Bernevig, Taylor Hughes and Shoucheng Zhang on the quantum well, a change of the width of the quantum well is proposed to lead to a phase transition between the topological and trivial insulator phases. Moreover, the Fermi energy can be changed by the gate voltage, allowing us to measure the dependence of the transport properties on the Fermi energy, as experimentally measured in [30] in order to distinguish between the TI and the NI. On the other hand, in the former proposal, the Bi (111) bilayer [28], being the bulk-like structure, is in fact relatively unstable and it easily transits into black phosphorus structure. Therefore, it took years to observe an evidence of the 2D TI in bismuth bliayer. One is by measuring transport immediately after peeling off a bilayer out of the bulk crystal [31] and the other is by measuring the states on the bilayer terrace on the surface of the bulk bismuth crystal [32], mimicking the edges of the bilayer film. In fact, bismuth itself is a semimetal as a three-dimensional bulk material, having a very small hole pocket and electron pockets. It means that it is very close to an insulator and indeed by making it into a bilayer film, it becomes an insulator. Moreover, its small gap is due to the large spin–orbit coupling, meaning that it is considered as a promising candidate for a TI and indeed it is the case [28]. On the other hand, in the three-dimensional crystal of bismuth, the system is a semimetal and the Z_2 topological invariant is trivial. Nonetheless, by doping antimony, the material $Bi_{1-x}Sb_x$ becomes a 3D TI for $0.07 < x < 0.22$, which was proposed as a first example of the 3D TI [33]. Now, many materials are known as topological insulators, including the most famous and well-studied topological insulators Bi_2Se_3 and Bi_2Te_3. It is interesting that they are well-known materials, but their topological properties have not been noticed before. Many other topological insulator materials have been identified, with the aid of materials informatics approach [34–36].

6.5.2   *Topological insulators with π Zak phase*

In addition to the Z_2 topological insulators, various topological insulating phases are known. One of the topological insulating phases commonly found in real materials is the topological insulator phase with a Zak phase π. In one-dimensional crystalline insulators, the Zak phase is defined as

$$\theta = i \sum_n^{\text{occ.}} \int_0^{2\pi/a} dk \, \langle u_{n,k} | \frac{\partial}{\partial k} | u_{n,k} \rangle, \tag{6.7}$$

where k is the wavenumber, a is the lattice constant, $u_{n,k}$ is the periodic part of the Bloch wavefunction of the nth eigenstate and the summation is over the occupied states. This Zak phase is defined modulo 2π and it is quantized as 0 or π in inversion-symmetric systems. When $\theta = \pi$, there exist two degenerated edge states and they are in the middle of the gap when the chiral symmetry anti-commutes with the Hamiltonian. In the modern theory of polarization [37–39], the Zak phase is related with the electric polarization P via

$$P = -e \frac{\theta}{2\pi} \ (\text{mod } e). \tag{6.8}$$

This equation indicates that the polarization is $e/2$ (mod e) in a system with the π Zak phase.

This notion of Zak phase can be naturally extended to three dimensions. In this case, one can define the Zak phase similarly to Eq. (6.7) by choosing one crystal plane, which will be related to a surface orientation. Then the Zak phase is defined as an integral along the surface normal $k_\perp$ in three dimensions. The resulting Zak phase depends on the wavevector along the surface $\boldsymbol{k}_\parallel$ as expressed by

$$\theta(\boldsymbol{k}_\parallel) = i \sum_n^{\text{occ.}} \int_0^{b_\perp} dk_\perp \, \langle u_{n,\boldsymbol{k}} | \frac{\partial}{\partial k_\perp} | u_{n,\boldsymbol{k}} \rangle \ (\text{mod } 2\pi), \tag{6.9}$$

where $b_\perp$ is the width of the 3D Brillouin zone along the direction perpendicular to the surface. In systems with both inversion and time-reversal symmetries, θ is quantized as 0 or π (mod 2π). Since it is quantized, it is constant when $\boldsymbol{k}_\parallel$ is continuously changed, as long as the gap remains open. On the other hand, in such systems with both inversion and time-reversal symmetries, nodal lines can appear between the valence and conduction bands and it is characterized by the π Berry phase around them, i.e., the Berry phase along a loop enclosing the nodal line is π. When $\boldsymbol{k}_\parallel$ changes across the nodal line, the Zak phase changes between π and 0, because of the π Berry phase around the nodal lines (see Fig. 6.5(a)). Thus, the projection of the nodal line onto the surface Brillouin zone is a loop dividing the area with the Zak phase $\theta = 0$ and that with $\theta = \pi$. Because the Berry phase is quantized, this nodal line is topological.

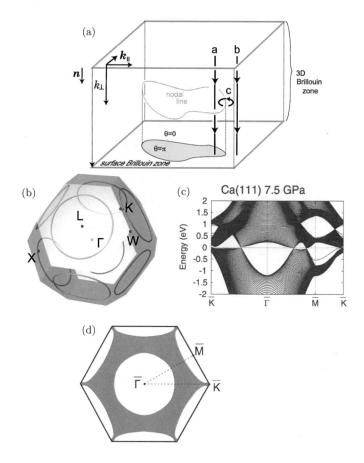

Fig. 6.5. Calcium (Ca) as a nodal-line semimetal. (a) Schematic figure for a nodal-line semimetal. The Zak phase is defined along a straight path (shown as "a" and "b") perpendicular to the surface. The values along the paths "a" and "b" are different by π because the nodal line lies in between. The Berry phase along the path "c" around the nodal line is π. (b) Nodal lines in calcium. (c) Band structure of a (111) slab of calcium. The drumhead surface states are shown in red. (d) Zak phase in the 2D Brillouin zone along the (111) plane. In the while and the purple regions, the Zak phase is 0 and π, respectively.

Such a behavior can be seen in various materials. One simple example is alkaline earth metals, calcium (Ca) and strontium (Sr) [40]. In these alkaline earth metals in the face-centered-cubic structure, valence and conduction bands are degenerate along nodal lines in k-space. At higher pressure like 7.5GPa, Ca is a nodal-line semimetal with nodal lines almost at the Fermi energy as shown in Fig. 6.5(b). It means that the Zak phase along the (111) direction changes between 0 and π in the 2D Brillouin zone along the (111) plane (Fig. 6.5(d)). As a result, in the region with the π Zak phase, there appear topological midgap surface states, in the form of drumhead surface states (Fig. 6.5(c)). These drumhead surface states appear within the region surrounded by the projection of the nodal lines onto the (111) plane. Other well-known examples are (111) surfaces of silicon (Si) and diamond [38,41]. In these covalent crystals, per surface unit cell on the (111) surface, one covalent bond is cut and it leads to one dangling bond, giving rise to the π Zak phase in the whole surface Brillouin zone, if no surface reconstruction occurs. Then this π Zak phase leads to midgap surface states, as is confirmed by *ab initio* calculation [41]. Moreover, even after surface reconstructions, this nontrivial π Zak phase survives when the surface reconstruction multiplies the surface unit cell odd times (e.g., 7×7 as typically seen in Si (111) surfaces) [41].

Some electride materials can be good candidates for various topological phases. Electrides are ionic crystals with electrons serving as anions and because of the low work functions, they can easily have band inversion in their band structure, leading to various topological phases [42]. For example, Y_2C is a two-dimensional electride, which means that it is a layered material with interstitial electron layers between atomic layers along the (111) plane. This material has nodal lines (Fig. 6.6(a)) with π Berry phase, which is protected by a combination of time-reversal and inversion symmetries.

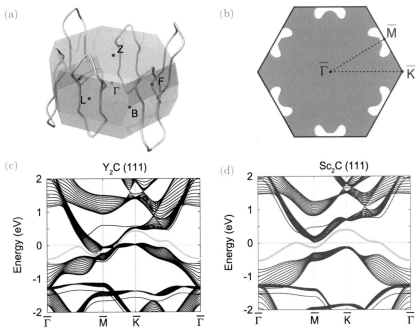

Fig. 6.6. Topological electrides. (a) Nodal lines in Y_2C. (b) Zak phase in the 2D Brillouin zone along (111) direction in Y_2C. In the white and the green regions, the Zak phase is 0 and π, respectively. (c) Band structure of a (111) slab of Y_2C. (d) Band structure of a (111) slab of Sc_2C.

Therefore, in the 2D Brillouin zone along the (111) plane, some region has π Zak phase while others have 0 Zak phase, separated by projections of the nodal lines (Fig. 6.6(b)). As a result, on the (111) surface, within the π Zak phase region, there exist in-gap surface states (Fig. 6.6(c)), which is analogous to the in-gap edge states in the Su–Schrieffer–Heeger (SSH) model with π Zak phase. On the other hand, as compared with Y_2C, the nodal lines shrink and disappear in Sc_2C. Hence, Sc_2C is an insulator and the Zak phase in the 2D Brillouin zone along the (111) plane is π everywhere and topological surface states exist over the whole 2D Brillouin zone (Fig. 6.6(d)) [42].

6.5.3 Higher-order topological insulators

As we explained thus far, one can find various types of topologically insulating and semimetallic phases in real materials. In particular, we introduce here an example of higher-order topological insulator. In contrast to topological insulators where the topological boundary states reside in $(D-1)$ dimensions, where D is the dimension of the system, in higher-order topological insulators, the topological boundary states reside in the $(D-d)$-dimensional boundaries, where $d \geq 2$ [43–47]. For example, in Ref. [47], the authors proposed a higher-order topological insulator protected by C_n rotational symmetry. In these cases, a fractional charge is trapped at the corners of the crystal preserving the C_n symmetry and the amount of the charge is quantized topologically [48].

We show an example of apatite $A_6B_4(SiO_4)_6$ in Fig. 6.7 as a C_6-protected higher-order topological insulator. The apatite $A_6B_4(SiO_4)_6$ is an electride and its three-dimensional crystal has one-dimensional hollow spaces, forming a triangular lattice in the xy-plane. In the La-apatite, the bulk is gapped and these hollow spaces support electronic states near the Fermi energy, offering one-dimensional transport channels and therefore, it is called a one-dimensional electride. Suppose this material forms a crystal with a shape of a hexagonal prism, with its hexagonal shape shown in Fig. 6.7(a). Here, the one-dimensional hollow spaces come to the corners of the hexagonal crystal. This gives rise to a filling at the corner, different from that in bulk. Indeed, per lattice constant along the z-direction, the bulk filling is $\nu_{bulk} = 2$ and the edge filling is $\nu_{edge} = 1$, whereas that at the corner is $\nu_{corner} = 2/3$. Thus only the corner is metallic and is 2/3-filled. Indeed, the first-principle calculation in Fig. 6.7(b) justifies this argument. An important point is that this corner charge is topologically determined, as described in [48]. In short, this quantized topological corner charge is determined from the Wyckoff positions of the Wannier centers of the electronic states and from the building

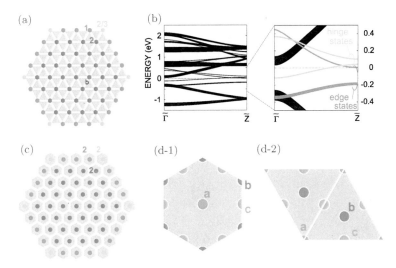

Fig. 6.7. Apatite as higher-order topological insulator protected by C_6 symmetry. (a) Schematic figure of an apatite crystal with a shape of a hexagonal cylinder. It consists of triangular blocks and the 1D electron channels are represented by solid circles. (b) Band structure of the La-apatite with hinges. (c) Crystal consisting of hexagonal blocks, to be compared with (a). (d-1) and (d-2) Comparison of the Wyckoff positions between (d-1) hexagonal blocks and (d-2) triangular blocks.

blocks constituting the crystal. In the present case, the charges are at the Wyckoff position $1a$. On the other hand, the building block of the apatite has a shape of a regular triangle as in Fig. 6.7(a) and it is different from a hexagonal building block shown in Fig. 6.7(c), assumed in [47]. The Wyckoff position $1a$ is in the middle of the hexagonal building block (Fig. 6.7(e)) but at the boundary of the triangular building block (Fig. 6.7(f)). Therefore, in the apatite with triangular blocks, there appears a quantized fractional corner charge, while with the hexagonal block, a fractional corner charge does not appear with electrons at $1a$ Wyckoff positions. This suggests a significant role of the shape of the building block of the crystal in determining the value of the corner charge.

6.6 Non-hermitian systems

So far we discussed hermitian systems and study topology in condensed materials. When we turn to nonhermitian systems, theoretical frameworks change drastically. While theories on topological phases in nonhermitian systems have been developed, there remain a number of unsolved problems. For example, the bulk–edge correspondence, representing the correspondence between bulk topological invariants and topological boundary modes, seems

to be violated in nonhermitian systems. Moreover, even the "bulk band structure", calculated from the Bloch wavefunctions with a real wavevector $\boldsymbol{k}$,
do not correctly reproduce the spectrum for a long open system. This has
been resolved by constructing nonBloch band theory [49, 50]. For example,
in a one-dimensional system, the wavenumber k takes complex values within
the nonBloch band theory for nonhermitian systems, in order to describe
a long open chain [49, 50]. It is in strong contrast with hermitian systems,
where the wavenumber is always real.

Thus, before studying the bulk–edge correspondence in nonhermitian
systems, one needs to find how to describe the bands in a long open chain. In
hermitian systems, it is known that the spectrum of an open chain and that
of a periodic chain are asymptotically the same in the limit of a large system
size. Therefore, one can use real k, used for the periodic chain, to describe also
a long open chain. Nonetheless, it is not the case in nonhermitian crystals and
a method to calculate the asymptotic spectrum of a long open chain is needed
(Fig. 6.8(a)). The nonBloch band theory [49, 50] gives a solution to the question. In this theory, one can determine a generalized Brillouin zone (GBZ),

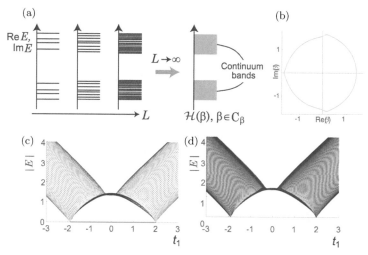

Fig. 6.8.   Bloch band theory for nonhermitian crystals. (a) Schematic figure of the
spectrum for a nonhermitian open chain with length L. The nonBloch band theory
describes its $L \to \infty$ limit, where the states form continuum bands. (b) Example of
the generalized Brillouin zone for $\beta \equiv e^{ik}$, where k is the complex Bloch wavevector
for the continuum bands. The model is the nonhermitian SSH model (6.13) with $t_1 =$
$0.3, t_2 = 1.1, t_3 = 1/5, \gamma_1 = 0$ and $\gamma_2 = -4/3$. (c) Spectrum for the extended SSH
model in a finite open chain, with the parameters $t_2 = 1.4$, $t_3 = 1/5, \gamma_1 = 5/3$ and
$\gamma_2 = 1/3$. It has topological edge states in the range $-1.905 < t_1 < 1.984$. (d) Spectrum
constructed from the nonBloch band theory, for the same parameters as (c). Within the
range $-1.905 < t_1 < 1.984$, the winding number is $w = 1$.

defined as a set of $\beta \equiv e^{ik}$ with k being the wavenumber, which describes continuum bands in a long open chain. It becomes a closed loop around the origin, but it is not necessarily a unit circle, unlike the Hermitian cases where the GBZ is always the unit circle. To determine the GBZ [50], we begin with the Bloch eigenvalue equation

$$\det[H(\beta) - E] = 0, \tag{6.10}$$

where E is the energy eigenvalue, $H(\beta)$ is the Bloch Hamiltonian matrix expressed in terms of $\beta = e^{ik}$. In 1D tight-binding models, this eigenvalue equation is an algebraic equation with an even degree $2M$ in general, where M is an integer. Let $\beta = \beta_1, \ldots, \beta_{2M}$ be the solutions of Eq. (6.10), in order of their absolute values: $|\beta_1| \leq |\beta_2| \leq \cdots \leq |\beta_{2M}|$. Then, we showed that the condition to determine the GBZ is given by $|\beta_M| = |\beta_{M+1}|$ [50]. From this condition, one can determine the GBZ as a trajectory of β_M and β_{M+1} for a given model. As an example, we consider a nonhermitian extension of the SSH model, whose Bloch Hamiltonian is given by a 2×2 matrix

$$\mathcal{H}(\beta) = \begin{pmatrix} 0 & R_+(\beta) \\ R_-(\beta) & 0 \end{pmatrix}, \tag{6.11}$$

$$R_+(\beta) = \left(t_2 - \frac{\gamma_2}{2}\right)\beta^{-1} + \left(t_1 + \frac{\gamma_1}{2}\right) + t_3\beta, \tag{6.12}$$

$$R_-(\beta) = t_3\beta^{-1} + \left(t_1 - \frac{\gamma_1}{2}\right) + \left(t_2 + \frac{\gamma_2}{2}\right)\beta, \tag{6.13}$$

where t_1, t_2, t_3, γ_1 and γ_2 are constants. In Fig. 6.8(c), we show an example of the GBZ for this model in [50]. Remarkably, the GBZ can have cusps, which never emerge in Hermitian systems. Moreover, it is not a unit circle, i.e., $|\beta| \neq 1$ in general, which represents nonhermitian skin effect. Once the GBZ is obtained, one can calculate the continuum bands by using Eq. (6.10). It is expected to be the same as the spectrum for a long open chain and indeed they are the same as we see in Figs. 6.8(d) and 6.8(e). The only difference between the continuum band (Fig. 6.8(e)) and the spectrum for a long open chain (Fig. 6.8(d)) is the existence of the in-gap states in the latter. This can be perfectly explained in terms of a nontrivial value of the winding number, $w = 1$, in the range $-1.905 < t_1 < 1.984$, where the winding number w is a topological invariant defined in terms of the phase winding of $R_\pm(\beta)$ as β goes along the GBZ.

These unique properties in nonhermitian systems give rise to quantitatively different physics as contrasted with hermitian cases. For example, in nonhermitian one-dimensional systems with both time-reversal and sublattice symmetries, which is the case for the nonhermitian SSH model

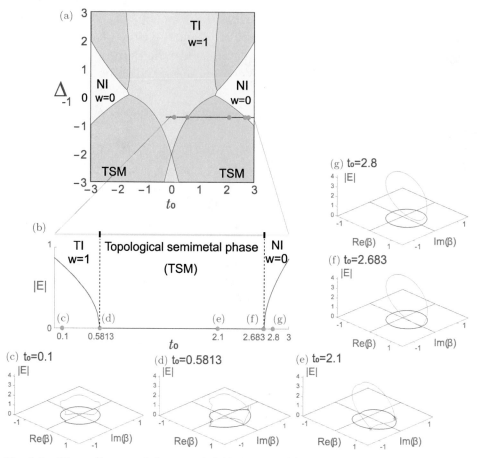

Fig. 6.9.   Phase diagram of the extended SSH model. (a) Phase diagram by changing two parameters t_0 and Δ_{-1} of the model. TI, NI and TSM stand for the topological insulator with $w = 1$, the normal insulator with $w = 0$ and the topological semimetal phases, respectively. (b) Energy spectrum along the line shown in (a). Because of the sublattice symmetry, the two energy eigenvalues appear in a pair, E and $-E$ and for illustration we show their absolute value only. The gap remains closed in the TSM phase, which never occurs in the hermitian case. (c)–(g) Band structures as a function of the wavenumber along the GBZ expressed in terms of $\beta \equiv e^{ik}$. As the system parameter t_0 changes, the GBZ is deformed so that the gap is always closed somewehere on the GBZ.

Eq. (6.11), a topological semimetal phase is stabilized by a novel mechanism [51]. In Fig. 6.9, we show an example of the evolution of the band structure for the nonhermitian SSH model, with $R_{\pm}(\beta) = (1.2 \pm 0.3)\beta + t_0 + (0.5 \mp \Delta_{-1})\beta^{-1}$, where t_0 and Δ_{-1} are constants. It shows three phases, the TI phase with the winding number $w = 1$, the NI phase with the winding number $w = 0$ and a topological semimetal (TSM) phase. The phase diagram is in Fig. 6.9(a), with its evolution of the band gap is in

Fig. 6.9(b). The band structures are shown as a function of the wavenumber along the GBZ expressed in terms of $\beta \equiv e^{ik}$ in Figs. 6.9(c)–(g). In this case, as the system parameter t_0 is changed along the black line in Fig. 6.9(a), the GBZ is also changed so that the system always remains gapless when $0.5813 < t_0 < 2.683$. When t_0 is within this range, the gap-closing points move along the GBZ. As we increase the value of t_0, at $t_0 = 0.5813$ (Fig. 6.9(d)), the gap-closing points are generated at the cusps of the GBZ and at $t_0 = 2.683$ (Fig. 6.9(f)), the gap-closing points are annihilated in pair. Thus this evolution of bands, stabilizing the TSM phase over a finite range of t_0, is unique to nonhermitian systems, arising from the characteristic behavior of the GBZ.

6.7 Summary

In the past two decades, an important role of the gauge field and emergent topology in condensed materials has brought about renewed interest even in well-known materials, and in various fields in condensed matter. It turned out that many known materials are topological, and it opened up enormous possibilities for further experimental and theoretical investigations.

Acknowledgments

I would like to express my sincere gratitude to Prof. Shoucheng Zhang for fruitful collaborations, valuable advice and his guidance to the beauty of gauge field and topology in condensed matter physics. I also would like to thank my collaborators, Prof. Naoto Nagaosa, Dr. Motoaki Hirayama, Dr. Takashi Miyake, Dr. Shoji Ishibashi, Dr. Ryo Okugawa, Dr. Shunichi Kuga, Dr. Ken Shiozaki, Dr. Heejae Kim, Mr. Yusuke Aihara, Prof. Hideo Hosono, Prof. Satoru Matsuishi, Mr. Ryo Takahashi, Mr. Kazuki Yokomizo and many others for the collaborations in the works introduced in this paper. This work was supported by Japan Society for the Promotion of Science (JSPS) KAKENHI Grant Nos. JP18H03678 and JP20H04633.

References

[1] S. Murakami, N. Nagaosa and S.-C. Zhang, *Science* **301**, 1348 (2003).
[2] J. Sinova, D. Culcer, Q. Niu, N. A. Sinitsyn, T. Jungwirth and A. H. MacDonald, Universal intrinsic spin Hall effect, *Phys. Rev. Lett.* **92**, 126603 (2004).
[3] B. A. Bernevig and S.-C. Zhang, *Phys. Rev. Lett.* **96**, 106802 (2006).
[4] C. L. Kane and E. J. Mele, *Phys. Rev. Lett.* **95**, 146802 (2005).
[5] C. L. Kane and E. J. Mele, *Phys. Rev. Lett.* **95**, 226801 (2005).
[6] G. Sundaram and Q. Niu, *Phys. Rev. B.* **59**, 14915 (1999).
[7] M. V. Berry, *Proc. Roy. Soc. London Ser A.* **392** 45, (1984).

[8] D. J. Thouless, M. Kohmoto, M. P. Nightingale and M. den Nijs, *Phys. Rev. Lett.* **49**, 405 (1982).

[9] F. D. M. Haldane, Model for a quantum hall effect without landau levels: Condensed-matter realization of the "parity anomaly", *Phys. Rev. Lett.* **61**, 2015 (1988).

[10] S.-C. Zhang and J. Hu, *Science* **294**, 823 (2001).

[11] M. Obstbaum, M. Decker, A. K. Greitner, M. Haertinger, T. N. G. Meier, M. Kronseder, K. Chadova, S. Wimmer, D. Ködderitzsch, H. Ebert and C. H. Back, *Phys. Rev. Lett.* **117**, 167204 (2016).

[12] L. Fu and C. L. Kane, *Phys. Rev. B.* **74**, 195312 (Nov, 2006).

[13] L. Fu and C. L. Kane, *Phys. Rev. B.* **76**, 045302 (2007).

[14] S. Ryu, A. P. Schnyder, A. Furusaki and A. W. Ludwig, *New J. Phys.* **12**, 065010 (2010).

[15] L. Fu, Topological crystalline insulators, *Phys. Rev. Lett.* **106**, 106802 (2011).

[16] B. Bradlyn, L. Elcoro, J. Cano, M. Vergniory, Z. Wang, C. Felser, M. Aroyo and B. A. Bernevig, *Nature* **547**, 298 (2017).

[17] H. C. Po, A. Vishwanath and H. Watanabe, *Nat. Commun.* **8** (1), 50 (2017).

[18] K. Shiozaki, M. Sato and K. Gomi, arXiv:1802.06694.

[19] H. Kim, K. Shiozaki and S. Murakami, *Phys. Rev. B.* **100**, 165202 (2019).

[20] H. Kim and S. Murakami, arXiv:2006.03607.

[21] G. E. Volovik, *The Universe in a Helium Droplet* (Oxford University Press, 2003).

[22] S. Murakami, *New J. Phys.* **9**, 356 (2007).

[23] X. Wan, A. M. Turner, A. Vishwanath and S. Y. Savrasov, *Phys. Rev. B.* **83**, 205101 (2011).

[24] S. Murakami, M. Hirayama, R. Okugawa and T. Miyake, *Sci. Adv.* **3** (5), (2017).

[25] S. Murakami and S.-i. Kuga, *Phys. Rev. B.* **78**, 165313 (2008).

[26] T. Liang, S. Kushwaha, J. Kim, Q. Gibson, J. Lin, N. Kioussis, R. J. Cava and N. P. Ong, *Science Adv.* **3** (2017).

[27] M. Hirayama, R. Okugawa, S. Ishibashi, S. Murakami and T. Miyake, *Phys. Rev. Lett.* **114**, 206401 (2015).

[28] S. Murakami, *Phys. Rev. Lett.* **97**, 236805 (2006).

[29] B. A. Bernevig, T. L. Hughes and S.-C. Zhang, *Science* **314**, 1757 (2006).

[30] M. König, S. Wiedmann, C. Brüne, A. Roth, H. Buhmann, L. W. Molenkamp, X.-L. Qi and S.-C. Zhang, *Science* **318**, 766 (2007).

[31] C. Sabater, D. Gosálbez-Martínez, J. Fernández-Rossier, J. G. Rodrigo, C. Untiedt and J. J. Palacios, *Phys. Rev. Lett.* **110**, 176802 (2013).

[32] I. Drozdov, A. Alexandradinata, S. Jeon, S. Nadj-Perge, H. Ji, R. J. Cava, B. A. Bernevig and A. Yazdani, *Nature Phys.* **10**, 664 (2014).

[33] L. Fu, C. L. Kane and E. J. Mele, *Phys. Rev. Lett.* **98** (10), 106803 (2007).

[34] T. Zhang, Y. Jiang, Z. Song, H. Huang, Y. He, Z. Fang, H. Weng and C. Fang, *Nature* **566**, 475 (2019).

[35] M. Vergniory, L. Elcoro, C. Felser, N. Regnault, B. A. Bernevig and Z. Wang, *Nature* **566**, 480 (2019).

[36] F. Tang, H. Po, A. Vishwanath and X. Wan, *Nature* **566**, 486 (2019).

[37] R. D. King-Smith and D. Vanderbilt, *Phys. Rev. B.* **47**, 1651 (1993).

[38] D. Vanderbilt and R. D. King-Smith, *Phys. Rev. B.* **48**, 4442 (1993).

[39] R. Resta, *Ferroelectrics.* **136**, 51 (1992).

[40] M. Hirayama, R. Okugawa, T. Miyake and S. Murakami, *Nat. Commun.* **8**, 14022 (2017).

[41] Y. Aihara, M. Hirayama and S. Murakami, *Phys. Rev. Research* **2**, 033224 (2020).

[42] M. Hirayama, S. Matsuishi, H. Hosono and S. Murakami, *Phys. Rev. X.* **8**, 031067 (2018).

[43] J. Langbehn, Y. Peng, L. Trifunovic, F. von Oppen and P. W. Brouwer, *Phys. Rev. Lett.* **119**, 246401 (2017).

[44] Z. Song, Z. Fang and C. Fang, *Phys. Rev. Lett.* **119**, 246402 (2017).

[45] W. A. Benalcazar, B. A. Bernevig and T. L. Hughes, *Science* **357**, 61 (2017).

[46] W. A. Benalcazar, B. A. Bernevig and T. L. Hughes, *Phys. Rev. B.* **96**, 245115 (2017).

[47] W. A. Benalcazar, T. Li and T. L. Hughes, *Phys. Rev. B.* **99**, 245151 (2019).

[48] M. Hirayama, R. Takahashi, S. Matsuishi, H. Hosono and S. Murakami, *Phys. Rev. Research* **2**, 043131 (2020).

[49] S. Yao and Z. Wang, Edge states and topological invariants of nonhermitian systems, *Phys. Rev. Lett.* **121**, 086803 (2018).

[50] K. Yokomizo and S. Murakami, Non-bloch band theory of nonhermitian systems, *Phys. Rev. Lett.* **123**, 066404 (2019).

[51] K. Yokomizo and S. Murakami, *Phys. Rev. Research* **2**, 043045 (2020).

Chapter 7

Thermalization and Its Absence within Krylov Subspaces of a Constrained Hamiltonian

Sanjay Moudgalya[*,†], Abhinav Prem[‡], Rahul Nandkishore[§],
Nicolas Regnault[*,†] and B. Andrei Bernevig[*]

*Department of Physics, Princeton University, NJ 08544, USA
†Laboratoire de Physique de l'Ecole normale supérieure
ENS, Université PSL, CNRS, Sorbonne Université
Université Paris-Diderot, Sorbonne Paris Cité, Paris, France
‡Princeton Center for Theoretical Science
Princeton University, NJ 08544, USA
§Department of Physics and Center for Theory of Quantum Matter
University of Colorado, Boulder, CO 80309, USA

We study the quantum dynamics of a simple translation invariant, center-of-mass (CoM) preserving model of interacting fermions in one dimension (1D), which arises in multiple experimentally realizable contexts. We show that this model naturally displays the phenomenology associated with fractonic systems, wherein single charges can only move by emitting dipoles. This allows us to demonstrate the rich Krylov fractured structure of this model, whose Hilbert space shatters into exponentially many dynamically disconnected subspaces. Focusing on *exponentially large* Krylov subspaces, we show that these can be either integrable or nonintegrable, thereby establishing the notion of *Krylov-restricted* thermalization. We analytically find a *tower* of integrable Krylov subspaces of this Hamiltonian, all of which map onto spin-1/2 XX models of various system sizes. We also discuss the physics of the nonintegrable subspaces, where we show evidence for weak Eigenstate Thermalization Hypothesis (ETH) restricted to each nonintegrable Krylov subspace. Further, we show that constraints in some of the thermal Krylov subspaces cause the long-time expectation values of local operators to deviate from behavior typically expected from translation invariant systems. Finally, we show using a Schrieffer–Wolff transformation that such models naturally appear as effective Hamiltonians in the large electric field limit of the interacting Wannier–Stark problem and comment on connections of our work with the phenomenon of Bloch many-body localization.

7.1 Introduction

Rapid advances in the coherent control and manipulation of cold atoms have enabled experiments to study the nonequilibrium dynamics of closed quantum many-body systems [1–6]. Consequently, the question of how (and whether) an arbitrary quantum state achieves thermal equilibrium while evolving under unitary dynamics has moved to the forefront of contemporary research. An important theoretical development along these lines is the Eigenstate Thermalization Hypothesis (ETH) [7–10], which, in its strong form, states that, as far as expectation values of local observables are concerned, *all* eigenstates of an ergodic system display thermal behavior [11–13]. Although lacking a formal proof, it is widely held that generic interacting systems obey the strong version of ETH, as evinced by several numerical studies [9,11,14,15]. Notable exceptions are integrable models, which possess extensively many conserved quantities and many-body localized (MBL) systems [16–18], where the *emergence* of extensively many local integrals of motion prohibits the system from exploring all allowed configurations in Hilbert space [19,20]. MBL systems thus evade ergodicity even at high energy densities and are able to retain a memory of their initial conditions in local observables for arbitrarily long times, leading to rich new physics which has been extensively studied numerically (see Refs. [21–23] for a review). An important open question is whether similar phenomena, e.g., violation of ETH or memory of initial conditions at long times, can occur in translation invariant nonintegrable systems [24–33].

This has generated much interest in identifying nonintegrable models which violate strong ETH but obey *weak* ETH, where the latter consists of a measure zero set of nonthermal eigenstates and is sufficient for preventing complete thermalization of the system [34, 35]. One recent line of attack has been to identify *exact* excited eigenstates [36, 38] in the middle of the spectrum of nonintegrable Hamiltonians that could shed significant light on ETH and its violation, given that the dynamics of a quantum system is governed by the properties of the full many-body spectrum and not only its low-lying features. There has been promising progress in this direction — Refs. [37,38] identified and analyzed an infinite tower of exact eigenstates of the celebrated 1D Affleck–Kennedy–Lieb–Tasaki (AKLT) models [39, 40], where some states of the tower are present in the bulk of the energy spectrum and are nonthermal, thus representing a novel type of strong ETH violation. Moreover, Refs. [41–44] have recently found similar exact ETH-violating eigenstates in a variety of models. In addition, Ref. [45] proposed a general construction of "embedding" ETH-violating eigenstates into a thermal spectrum, which has also been applied to construct systems with topological eigenstates in the middle of the spectrum [46].

Concurrently, an experiment on a 1D chain of Rydberg atoms observed persistent revivals upon quenching the system from certain initial conditions, while other initial conditions led to the system thermalizing rapidly [47]. This striking dependence on initial conditions was numerically demonstrated to be caused by a vanishing number of nonthermal states that coexist with an otherwise thermal spectrum [48–52], dubbed "quantum many-body scars". Several explanations for the origin of quantum scars have been proposed: analog to single-particle scarring [48,52–54], proximity to integrability [55], existence of approximate quasiparticle towers of states [41,56,57], confinement [58–60] and an emergent SU(2) symmetry [51]. Furthermore, recent works have constructed several models that show similar characteristics [49,61–63], studied the stability of the scars to perturbations [64] and found quantum scars in Floquet settings [66,67,80]. These discoveries thus reveal new possibilities for quantum dynamics which may occur between the extremes of thermalization and the complete breaking of ergodicity.

Although it remains largely unclear what the general desiderata are for the presence of scar states, systems with constrained dynamics, such as kinetically constrained models [68–70] and the PXP model [71–73], offer a promising platform for exploring ergodicity breaking. *Fractonic* systems, whose defining feature is the presence of excitations with restricted mobility, are natural candidates displaying constrained dynamics (see Ref. [74] for a review). Indeed, alongside their novel ground-state features, 3D gapped fracton models have garnered attention also for their slow quantum dynamics in the absence of spatial disorder [75–77]. As first observed in Ref. [78], the conservation of higher (e.g., dipole or angular) moments in $U(1)$ symmetric systems places stringent constraints on the mobility of excitations, rendering isolated charges completely immobile. This insight has allowed the characteristic physics of fractons to be realized away from their initial conception in exactly solvable 3D lattice models, potentially even in 1D[a] (see, e.g., Refs. [79,80]). This perspective was recently taken in Ref. [81], where random unitary dynamics in 1D with conserved dipole moment were shown to localize for reasons beyond usual locator-expansion techniques.

In this chapter, we investigate the quantum dynamics of a translation invariant, nonintegrable 1D fermionic chain with conserved center-of-mass (CoM). Rather than imposing constraints by hand, we show that the CoM conserving model we study has a natural origin in two distinct physical settings: in the thin-torus limit of the fractional quantum Hall effect and

---

[a]For our purposes, fractonic behavior refers to the strict immobility of isolated charges and (possibly) restricted mobility of bound charges (e.g., dipoles). It remains an open question whether the nontrivial topological features, such as a sub-extensive ground-state degeneracy, associated with 3D gapped fracton models are possible in spatial dimension less than three.

in the strong electric field limit of the interacting Wannier–Stark problem, a regime accessible to current cold-atom experiments[b] [82]. Focusing on systems close to half-filling, we define composite degrees of freedom in terms of which CoM conservation maps onto dipole moment conservation, revealing the underlying fractonic nature of the model.

Once we resolve the Hamiltonian into its disparate symmetry sectors, we find that the Hilbert space further shatters into exponentially many dynamically disconnected sectors or *Krylov subspaces*, which have previously been studied under various settings [62, 83–86]. This shattering is a consequence of charge and center-of-mass conservation and, as discussed in Refs. [85, 86], the presence of exponentially many small (finite size in the thermodynamic limit) closed Krylov subspaces can lead to effectively localized dynamics. Here, we instead focus on a new phenomenon within exponentially *large* Krylov subspaces, which are of infinite size in the thermodynamic limit and unveil a rich structure within these sectors, leading to new notions of *Krylov-restricted* integrability and thermalization.

Specifically, we find that several such large Krylov subspaces are *integrable*, thereby establishing the phenomenon of emergent integrability and further breaking of ergodicity *within* closed Krylov sectors. Meanwhile, other large sectors remain nonintegrable. To bring this distinction into focus, we propose that a modified version of ETH applies to Krylov fractured systems, wherein conventional diagnostics of nonintegrability, such as level statistics, are defined with respect to a symmetry sector *and a Krylov subspace*. Using our modified definition, we conclude that the problem "thermalizes" within each nonintegrable Krylov subspace, in that the long-time behavior of a state belonging to a particular Krylov subspace coincides with the Gibbs ensemble *restricted* to that subspace. Remarkably, we find that this restricted thermalization within some of the Krylov subspaces leads to the "infinite temperature" state within the Krylov sectors showing atypical behavior, in that the late-time charge density deviates from that expected from unconstrained translation invariant systems. Violations of this modified or "Krylov-restricted ETH" require either integrability, conventional "disorder induced" many-body localization, or existence of further symmetries within the Krylov subspace. Armed with this understanding, we also revisit the problem of interacting Wannier-Stark localization [87, 88], which we argue requires the ideas introduced in this paper for a more complete understanding.

---

[b]For example, tilting an optical lattice subjects the trapped ultracold atoms to a linear field.

This paper is organized as follows: We introduce the pair-hopping model, also studied in Ref. 62, in Sec. 7.2 and show that it conserves center-of-mass. We then briefly discuss its origins in the thin torus limit of the fractional quantum Hall effect (FQHE) and in the limit of strong electric field in the interacting Wannier–Stark problem. In Sec. 7.3, we introduce a convenient formalism to study this model at half-filling and show that it exhibits fractonic phenomenology. In Sec. 7.4, we discuss the notion of Krylov fracture, i.e., the phenomenon where systems exhibit several closed subspaces that are dynamically disconnected with respect to product states. We show examples of integrable and nonintegrable dynamically disconnected Krylov subspaces in Secs. 7.5 and 7.6, respectively. The integrable subspaces we study exactly map onto XX models of various sizes and the nonintegrable subspaces show features that are typically not expected in nonintegrable models, which we discuss in Sec. 7.7. Finally, we make connections to Bloch MBL in Sec. 7.8 and conclude in Sec. 7.9. Various details are relegated to appendices.

7.2 Model and its symmetries

The "pair-hopping model" we study is a one-dimensional chain of interacting spinless fermions with translation and inversion symmetry, with the Hamiltonian [62, 89]

$$H = \sum_{j=1}^{L_b} H_j = \sum_{j=1}^{L_b} \left(c_j^\dagger c_{j+3}^\dagger c_{j+2} c_{j+1} + h.c. \right), \tag{7.1}$$

where $L_b = L - 3$ for open boundary conditions (OBC), $L_b = L$ for periodic boundary conditions (PBC) and the subscripts are defined modulo L for PBC. Note that we have set the overall energy scale equal to one for convenience. Each term H_j of Eq. (7.1) vanishes on all spin configurations on sites j to $j+3$ except for

$$\begin{aligned} \overset{j \quad\quad\quad j+3}{H_j|0\ 1\ 1\ 0\rangle} &= \overset{j \quad\quad\quad j+3}{|1\ 0\ 0\ 1\rangle}, \\ \overset{j \quad\quad\quad j+3}{H_j|1\ 0\ 0\ 1\rangle} &= \overset{j \quad\quad\quad j+3}{|0\ 1\ 1\ 0\rangle}, \end{aligned} \tag{7.2}$$

where $|a\ b\ c\ d\rangle$ represents the occupation of sites j to $j+3$. In the rest of the paper, we will use the following shorthand notation

$$|1\ 0\ 0\ 1\rangle \leftrightarrow |0\ 1\ 1\ 0\rangle \tag{7.3}$$

to represent Eq. (7.2), i.e., the action of individual terms of the Hamiltonian equation (7.1). This pair-hopping model preserves the center-of-mass position, i.e., the center-of-mass position operator [89]

$$
\widehat{C} \equiv \begin{cases} \sum_{j=1}^{L} j\hat{n}_j & \text{if } OBC, \\ \exp\left(\dfrac{2\pi i}{L}\sum_{j=1}^{L} j\hat{n}_j\right) & \text{if } PBC, \end{cases} \tag{7.4}
$$

where the number operator $\hat{n}_j \equiv c_j^\dagger c_j$ commutes with the Hamiltonian of Eq. (7.2). Hamiltonians with such conservation laws, including the model given by Eq. (7.1), were first discussed in Ref. [89] in the quest to build featureless Mott insulators.

As emphasized by Ref. [89], the spectra of center-of-mass preserving Hamiltonians have some unusual features. For example, at a filling $\nu = p/q$ (with p and q coprime), the full spectrum is q-fold degenerate, which stems from the fact that the center-of-mass position operator $\widehat{C}$ and the translation operator $\widehat{T}$ do not commute. More precisely, consider a 1D chain of length L with periodic boundary conditions. As shown in Ref. [89],

$$
\widehat{C}\widehat{T} = e^{2\pi i\nu}\widehat{T}\widehat{C}, \tag{7.5}
$$

where ν is the filling fraction $\nu = p/q$. This results in a q-fold degeneracy of the spectrum with PBC.

The pair-hopping model (7.1), with even system size $L = 2N$ and with PBC, has an additional symmetry: sublattice particle number conservation. That is, the operators

$$
\hat{n}_e = \sum_{j=1}^{N}\hat{n}_{2j}, \hat{n}_o = \sum_{j=1}^{N-1}\hat{n}_{2j+1}, \tag{7.6}
$$

both commute with Eq. (7.1). This can be seen by writing the action of the terms of the pair-hopping Hamiltonian as

$$
\overset{e\ o\ e\ o}{|1\ 0\ 0\ 1\rangle} \leftrightarrow \overset{e\ o\ e\ o}{|0\ 1\ 1\ 0\rangle}, \overset{o\ e\ o\ e}{|1\ 0\ 0\ 1\rangle} \leftrightarrow \overset{o\ e\ o\ e}{|0\ 1\ 1\ 0\rangle}, \tag{7.7}
$$

where the superscripts o and e label the parity of the sites. The actions of Eq. (7.7) conserve the particle number on the odd and even sites separately. Sublattice number conservation of Eq. (7.6) trivially implies the conservation of total particle number ($n_e + n_o$). Note that the sublattice number conservation is a special property of the truncated Hamiltonian equation (7.1) and does not hold in general for center-of-mass preserving Hamiltonians. For example, the extended pair-hopping Hamiltonian $\sum_j \left(c_j^\dagger c_{j+3}^\dagger c_{j+2} c_{j+1} + c_j^\dagger c_{j+4}^\dagger c_{j+3} c_{j+1} + \text{h.c.}\right)$ preserves the center-of-mass position but does not conserve the sublattice particle number.

7.2.1 *Experimental relevance*

An especially appealing feature of center-of-mass preserving terms, including the pair-hopping term (7.1), is their natural appearance in multiple experimentally relevant systems. The first setting in which such models appear is in the quantum Hall effect, when translation invariant interactions are projected onto a single Landau level [62, 90, 91]. We refer the reader to Ref. [62] for a derivation, but summarize the general idea here: one works in the Landau gauge, such that the single particle orbitals in a Landau level can be written as eigenstates of the magnetic translation operators in the $\hat{y}$-direction, in which case the position in the $\hat{x}$-direction is the momentum quantum number in the $\hat{y}$-direction. The matrix elements of a translation invariant interaction between the single particle orbitals are hence momentum conserving in the $\hat{y}$-direction, which translates to center-of-mass conservation in the $\hat{x}$-direction of the effective one-dimensional model [90]. A general interaction operator projected to a Landau level of an $L_x \times L_y$ quantum Hall system has the form

$$H = \sum_{j=1}^{N_\Phi} \sum_{k,m} V_{km} \left(c_j^\dagger c_{j+k+m}^\dagger c_{j+k} c_{j+m} + \text{h.c.}\right), \qquad (7.8)$$

where $N_\Phi = L_x L_y / (2\pi)$ is the number of flux quanta and $V_{km} \sim \exp\left(-2\pi^2 \left(k^2 + m^2\right)/L_y^2\right)$ with the magnetic length set to unity. Thus, in the "thin-torus" limit ($L_y \to 0$), one of the dominant terms is the pair-hopping Hamiltonian equation (7.1). We note that such Hamiltonians also appear in the thin torus limit of the pseudopotential Hamitonians for several Fractional Quantum Hall states [62, 92–95].

A second origin of such center-of-mass preserving models is in the well-known Wannier–Stark problem [96]: spinless fermions hopping on a finite one-dimensional lattice, subject to an electric field. While localization at the single-particle level has been long established [97], an interacting version of the problem has recently been studied and found to display behavior associated with MBL systems at strong fields [87, 88]; this phenomenon goes under the name Bloch (or Stark) MBL. In Sec. 7.8, we show that the dynamics of the Bloch MBL model in the limit of an infinitely strong electric field is governed by an effective center-of-mass preserving Hamiltonian, with the lowest order "hopping" term given precisely by equation (7.1).

Specifically, the resulting Hamiltonian is again of the form equation (7.8), with N_Φ replaced by the system size.[c] This mapping hence allows us to present a new perspective on the phenomenon of Bloch MBL (see Sec. 7.8), in addition to providing a natural experimental setting, accessible to current cold-atom experiments, for realizing the model studied here.

7.3 Hamiltonian at 1/2 filling

We now proceed to study the spectrum of the pair hopping Hamiltonian equation (7.1). In this work, we will be focusing on systems at, or close to, half filling and will restrict ourselves to even system sizes $L = 2N$. For the study of this Hamiltonian at other filling factors, see Refs. [62, 98].

7.3.1 *Composite degrees of freedom*

To study this model and to elucidate its relation to the physics of fractons, we define composite degrees of freedom formed by grouping neighboring sites of the original model. Assuming an even number of sites, we group sites $2j - 1$, $2j$ of the original lattice into a new site j so as to form a new chain with $N = L/2$ sites. We define new degrees of freedom for these composite sites as follows:

$$|\uparrow\rangle \equiv |0\ 1\rangle, |\downarrow\rangle \equiv |1\ 0\rangle,$$
$$|+\rangle \equiv |1\ 1\rangle, |-\rangle \equiv |0\ 0\rangle. \tag{7.9}$$

The choice of grouping is unambiguously defined for OBC and we stick to it for most of this paper. Writing the action of the Hamiltonian equation (7.2) in terms of these composite degrees of freedom, we find

$$\left|\ \boxed{01}\ \boxed{10}\ \right\rangle \leftrightarrow \left|\ \boxed{10}\ \boxed{01}\ \right\rangle$$
$$\Longleftrightarrow |\uparrow\downarrow\rangle \leftrightarrow |\downarrow\uparrow\rangle, \tag{7.10}$$

$$\left|\ \boxed{10}\ \boxed{11}\ \boxed{00}\ \right\rangle \leftrightarrow \left|\ \boxed{11}\ \boxed{00}\ \boxed{10}\ \right\rangle$$
$$\Longleftrightarrow |\downarrow + -\rangle \leftrightarrow |+ - \downarrow\rangle, \tag{7.11}$$

$$\left|\ \boxed{00}\ \boxed{11}\ \boxed{01}\ \right\rangle \leftrightarrow \left|\ \boxed{01}\ \boxed{00}\ \boxed{11}\ \right\rangle$$
$$\Longleftrightarrow |- + \uparrow\rangle \leftrightarrow |\uparrow - +\rangle, \tag{7.12}$$

---

[c]Note that for both the FQHE and the Bloch MBL case, the dominant center-of-mass conserving terms are nearest-neighbor ($\hat{n}_j \hat{n}_{j+1}$) and next nearest-neighbor electrostatic terms ($\hat{n}_j \hat{n}_{j+2}$), but the lowest order "hopping" is the pair-hopping Hamiltonian of equation (7.1).

$$\left| \boxed{10} \boxed{11} \boxed{01} \right\rangle \leftrightarrow \left| \boxed{11} \boxed{00} \boxed{11} \right\rangle$$

$$\Longleftrightarrow \; |\downarrow + \uparrow\rangle \leftrightarrow |+ - +\rangle , \tag{7.13}$$

$$\left| \boxed{01} \boxed{00} \boxed{10} \right\rangle \leftrightarrow \left| \boxed{00} \boxed{11} \boxed{00} \right\rangle$$

$$\Longleftrightarrow \; |\uparrow - \downarrow\rangle \leftrightarrow |- + -\rangle , \tag{7.14}$$

where $\boxed{\cdots}$ represents a grouping of some sites $2j - 1$ and $2j$ and $|a\rangle \leftrightarrow |b\rangle$ represents the action of a single term of the Hamiltonian on $|a\rangle$ resulting in $|b\rangle$ and vice versa (see Eqs. (7.2) and (7.3)). For reasons that will become clear forthwith, we set the nomenclature of the composite degrees of freedom as follows:

$$|+\rangle , |-\rangle : Fractons$$

$$|+-\rangle , |-+\rangle : Dipoles$$

$$|\uparrow\rangle , |\downarrow\rangle : Spins$$

Here, Eqs. (7.11)–(7.14) resemble the rules restricting the mobility of fractons and are similar to those discussed in Ref. [81] (see Ref. [74] for a review on fractons).

In particular, Eqs. (7.11) and (7.12) represent the free propagation of dipoles when separated by spins and Eqs. (7.13) and (7.14) encode the characteristic movement of a fracton through the emission or absorption of a dipole, i.e., dipole assisted hopping. However, in contrast to usual fracton phenomenology, here the movement of fractons is also sensitive to the background spin configuration. For example, the fracton in the configuration $|\cdots \downarrow + \uparrow \cdots\rangle$ can move by emitting a dipole (see equation (7.13)) while that in the configuration $|\cdots \uparrow + \downarrow \cdots\rangle$ cannot. In our convention, the fractons $|+\rangle$ and $|-\rangle$ have spin 0 and charges $+1$ and -1, respectively, while the spins $|\uparrow\rangle$ and $|\downarrow\rangle$ have charge 0 and spins $+1$ and -1, respectively. Thus, the unit cell charge and spin operators in terms of the original fermionic degrees of freedom read

$$\widehat{Q}_j \equiv \hat{n}_{2j-1} + \hat{n}_{2j} - 1, \; \widehat{S}_j^z \equiv -\hat{n}_{2j-1} + \hat{n}_{2j}, \tag{7.15}$$

where j is the unit cell index and $2j - 1$, $2j$ are the site indices of the original configuration. We represent the total number of $+$, $-$, $\uparrow$ and $\downarrow$ by N_+, N_-, $N_\uparrow$, $N_\downarrow$, respectively. Thus, the *total charge* is $N_+ - N_-$ and the *total spin* is $N_\uparrow - N_\downarrow$.

7.3.2 Symmetries in terms of the composite degrees

We now study the symmetries of the Hamiltonian whose terms act on the composite degrees of freedom through Eqs. (7.10)–(7.14). As discussed in Sec. 7.3, the pair-hopping model (7.1) has several symmetries: sublattice charge conservation, center-of-mass conservation, inversion and translation (for PBC). Using Eqs. (7.10)–(7.14), we now interpret these symmetries in terms of the composite degrees of freedom defined in equation (7.9).

The model in terms of the composite degrees of freedom conserves the total spin and the total charge, as is evident from Eqs. (7.10)–(7.14). In other words, $N_\uparrow - N_\downarrow$ and $N_+ - N_-$ are separately conserved. Indeed, using the definitions of spin and charge in equation (7.15), the total spin operator $\widehat{S}^z$ and total charge operator $\widehat{Q}$ can be expressed in terms of the operators in the original Hilbert space as follows:

$$\widehat{Q} \equiv \sum_{j=1}^{N} \widehat{Q}_j = \hat{n}_e + \hat{n}_o - N, \quad \widehat{S}^z \equiv \sum_{j=1}^{N} \widehat{S}_j^z = \hat{n}_o - \hat{n}_e, \qquad (7.16)$$

where $\hat{n}_e$ and $\hat{n}_o$ are the sublattice particle numbers defined in Eq. (7.6). Thus, the conservation of total charge and total spin in the fracton model is a direct consequence of the sublattice number conservation of the pair-hopping model.

Moreover, the fractonic behavior inherent in the rules specified by Eqs. (7.10)–(7.14) suggests that the *dipole moment* of the composite degrees of freedom is a conserved quantity [78]. This operator is defined similarly to the center-of-mass operator Eq. (7.4) as:

$$\widehat{D} \equiv \begin{cases} \sum_{j=1}^{N} j\widehat{Q}_j & \text{if } OBC, \\ \exp\left(i\frac{2\pi}{N}\sum_{j=1}^{N} j\widehat{Q}_j\right) & \text{if } PBC. \end{cases} \qquad (7.17)$$

To explicitly show that $\widehat{D}$ is in fact a conserved quantity of the composite fractonic model, we observe that

$$\sum_{j=1}^{N} j\widehat{Q}_j = \sum_{j=1}^{N} j\left(\hat{n}_{2j-1} + \hat{n}_{2j} - 1\right)$$

$$= \sum_{j=1}^{N} \frac{(2j-1)\hat{n}_{2j-1} + 2j\hat{n}_{2j}}{2} + \sum_{j=1}^{N} \frac{\hat{n}_{2j-1}}{2} - \sum_{j=1}^{N} j$$

$$= \frac{1}{2}\sum_{j=1}^{L} j\hat{n}_j + \frac{\hat{n}_o}{2} - \frac{N(N+1)}{2}. \qquad (7.18)$$

Then, using Eqs. (7.4), (7.6) and (7.18), in terms of the original operators in the pair-hopping model, the operator $\widehat{D}$ can be expressed as

$$
\widehat{D} = \begin{cases} \dfrac{1}{2}\widehat{C} + \dfrac{1}{2}\hat{n}_o - \dfrac{N(N+1)}{2} & \text{if } OBC, \\[3mm] \widehat{C}^{\frac{1}{2}} e^{i\frac{\pi}{L}\hat{n}_o} e^{-i\frac{\pi N(N+1)}{L}} & \text{if } PBC. \end{cases} \tag{7.19}
$$

Since $\widehat{C}$ and $\hat{n}_o$ are conserved operators of the pair-hopping Hamiltonian, as discussed in Sec. 7.2, it follows from Eq. (7.19) that $\widehat{D}$ is conserved in the composite model. To complete our discussion, we note that the composite model also preserves inversion as well as translation symmetry (with PBC), neither of which commute with $\widehat{D}$. Details of the symmetries are relegated to Appendix A.

7.4 Krylov fracture

We now study the dynamics of H and show that it exhibits exponentially many dynamically disconnected subspaces. More precisely, we construct *Krylov subspaces* of the form

$$
\mathcal{K}\left(H, |\psi_0\rangle\right) \equiv \operatorname{span}\{|\psi_0\rangle, H|\psi_0\rangle, H^2|\psi_0\rangle, \ldots\} \tag{7.20}
$$

that are by definition closed under the action of the Hamiltonian H. While $|\psi_0\rangle$ in equation (7.20) can in principle be an arbitrary state, we are interested in the dynamics of initial product states, which are more easily accessible to experiments. Hence, we focus on Krylov subspaces generated by product states $|\psi_0\rangle$, which we dub *root states* of the Krylov subspace $\mathcal{K}\left(H, |\psi_0\rangle\right)$. For a generic nonintegrable Hamiltonian H without any symmetries, one expects that $\mathcal{K}\left(H, |\psi_0\rangle\right)$ for *any* initial product state $|\psi_0\rangle$ is the *full* Hilbert space of the system. For a nonintegrable Hamiltonian with some symmetry and with $|\psi_0\rangle$ an eigenstate of the symmetry, one typically expects that $\mathcal{K}\left(H, |\psi_0\rangle\right)$ spans *all* states with the same symmetry quantum number as $|\psi_0\rangle$.

Surprisingly, however, we show that the pair-hopping Hamiltonian (7.1) exhibits *Krylov fracture* i.e., even after resolving the charge and center-of-mass symmetries, we find generically that $\mathcal{K}\left(H, |\psi_0\rangle\right)$ does *not* span all states with the same symmetry quantum numbers as $|\psi_0\rangle$. Thus, the full Hilbert space of the system $\mathcal{H}$ is of the form

$$
\mathcal{H} = \bigoplus_{\mathrm{s}} \mathcal{H}^{(\mathrm{s})}, \quad \mathcal{H}^{(\mathrm{s})} = \bigoplus_{i=1}^{K^{(\mathrm{s})}} \mathcal{K}(H, |\psi_i^{(\mathrm{s})}\rangle), \tag{7.21}
$$

where s labels the distinct symmetry quantum numbers, such as charge and center-of-mass, $K^{(s)}$ denotes the number of disjoint Krylov subspaces generated from product states with the same symmetry quantum numbers and $|\psi_i^{(s)}\rangle$ are the root states generating the Krylov subspaces. Note that the root states in Eq. (7.21) are chosen such that they generate distinct disconnected Krylov subspaces, since the same subspace can be generated by different root states. Stated symbolically,

$$\mathcal{K}(H, |\psi_i^{(s)}\rangle) \cap \mathcal{K}(H, |\psi_{i'}^{(s')}\rangle) = \delta_{s,s'}\delta_{i,i'}\mathcal{K}(H, |\psi_i^{(s)}\rangle). \qquad (7.22)$$

Fracture of the form (7.21), where the total number of Krylov subspaces $K^{(s)}$ is exponentially large in the system size, was recently shown to *always* exist in Hamiltonians and random-circuit-models with center-of-mass conservation [85, 86] (alternatively referred to as "dipole moment" conservation). While the presence of these symmetries guarantees fracture, one can distinguish between "strong" and "weak" fracture [85,86], depending respectively on whether or not the ratio of the largest Krylov subspace to the Hilbert space within a given global symmetry sector vanishes in the thermodynamic limit. Strong (respectively, weak) fracture is associated with the violation of weak (respectively, strong) ETH with respect to the full Hilbert space. The pair-hopping model (7.1) (which is equivalent to the Hamiltonian H_4 in Ref. [85] with spin-1/2) numerically appears to exhibit *strong* fracture within several symmetry sectors. However, the addition of longer-range CoM preserving terms numerically appears to cause the Hilbert space to fracture only weakly [85], with the fracture disappearing with the addition of infinite-range CoM preserving terms, even if the interaction strength decays exponentially with range [99].

By definition, distinct Krylov subspaces are *dynamically disconnected*, i.e., no state initialized completely within one of the Krylov subspaces can evolve out to a different Krylov subspace. Indeed, exponentially many of these Krylov subspaces are one-dimensional *static* configurations — product states that are eigenstates of H. For instance, the Hamiltonian vanishes on any product state that does not contain the patterns "$\ldots 0110 \ldots$" or "$\ldots 1001 \ldots$", since those are the only configurations on which terms of H act nontrivially (see Eq. (7.2)). The charge-density-wave (CDW) state

$$|1111000011110000 \ldots \ldots 1111000011110000\rangle$$

is one example of a static configuration that is an eigenstate. In terms of the composite degrees of freedom, we can equivalently consider configurations with only $+$, $-$ and no spins, such as

$$|\cdots + + - - + + - - \cdots\rangle,$$

with a pattern that alternates between $+$ and $-$ with "domain walls" that are at least two sites apart. According to Eqs. (7.10)–(7.14), all terms of the Hamiltonian vanish on these configurations: since there are *exponentially* many such patterns, there are equally many one-dimensional Krylov subspaces. We can also construct small Krylov subspaces by embedding finite nontrivial blocks, on which the Hamiltonian acts nontrivially, into the static configurations, thereby leading to exponentially many Krylov subspaces of every size [85, 86]. For example, the following configurations $|\psi_\pm\rangle$

$$|\psi_\pm\rangle = \frac{1}{\sqrt{2}} (|+ + - - \cdots + + - - \uparrow\downarrow + + - - \cdots + + - -\rangle$$

$$\pm |+ + - - \cdots + + - - \downarrow\uparrow + + - - \cdots + + - -\rangle) \quad (7.23)$$

are composed of one nontrivial block $\uparrow\downarrow$ sandwiched within a frozen configuration and they thus have energies $E_\pm = \pm 1$. Exponentially many configurations with energies $E = \pm 1$ can be constructed by changing the frozen configuration around the nontrivial block.

The presence of exponentially many static states (within each symmetry sector) in the Hilbert space leaves an imprint on the dynamical behavior of such systems. Specifically, time-evolution starting from randomly chosen product states looks highly nongeneric from the perspective of the full Hilbert space. For example, in the absence of Krylov fracture one typically expects that the bipartite entanglement entropy evolves to the Page value [100], the average bipartite entanglement entropy of states in the Hilbert space. For a system of Hilbert space dimension $D[L] = 2^L$, the Page value is $\log D[L/2] \approx L/2 \log 2$. However, in the presence of Krylov fracture, we expect that the late-time bipartite entanglement entropy of product states $|\psi_0\rangle$ is smaller and typically $\sim \log D_\mathcal{K}[L/2]$, where $D_\mathcal{K}[L]$ is the dimension of the Krylov subspace $\mathcal{K}(H, |\psi_0\rangle)$ for a system size L. The phenomenon of Krylov fracture can thus be regarded as a breaking of ergodicity with respect to the full Hilbert space, resulting in (at the very least) violation of strong ETH.

However, what remains unclear is whether, for systems exhibiting Krylov fracture, thermalization occurs *within* each of the Krylov subspaces. Of course, thermalization or ETH-violation are only well-posed concepts for *large* Krylov subspaces $\mathcal{K}$ (with dimension $\mathcal{D}_\mathcal{K}[L] \to \infty$ as $L \to \infty$)[d] and

---

[d]Note that the dimension of the Krylov subspace $D_\mathcal{K}[L]$ could in principle scale polynomially with L; however, we are not aware of any such example in the pair-hopping model (7.1).

do not have a clear meaning when the Krylov subspace has a finite dimension in the thermodynamic limit, as is the case for the exponentially many static configurations discussed above. Indeed, there exist exponentially large Krylov subspaces of the Hamiltonian equation (7.1) at filling $\nu = p/(2p+1)$ for which Krylov-restricted thermalization appears to hold for most initial states, as recently demonstrated by some of the present authors [62]. There, we demonstrated the existence of Krylov subspaces with Wigner–Dyson level statistics, despite such Krylov subspaces hosting *quantum scars*, i.e., evenly spaced towers of anomalous states in the spectrum that lead to revivals in the fidelity of time evolution from particular initial states. Those Krylov subspaces are examples of ones that violated Krylov–restricted *strong* ETH, although Krylov-restricted *weak* ETH is satisfied. However, it has not yet been established if Krylov-restricted *weak* ETH is *necessarily* satisfied for large-dimensional Krylov subspaces, or if there are examples of *semi-integrable* systems with both integrable and nonintegrable Krylov subspaces, opening the door to further violations of ergodicity within Krylov sectors.

Thus, in what follows, we will focus on high-dimensional *irreducible* Krylov subspaces $\mathcal{K}(H, |\psi\rangle)$, defined as those with exponentially large dimension $\mathcal{D}_\mathcal{K}[L] \sim \alpha^L$ as $L \to \infty$ ($\alpha > 1$) and which satisfy

$$\mathcal{K}(H, |\psi\rangle) \neq \mathcal{K}(H, |\psi_1\rangle) \oplus \mathcal{K}(H, |\psi_2\rangle) \tag{7.24}$$

for any product states $|\psi_1\rangle$ and $|\psi_2\rangle$, after resolving charge and center-of-mass symmetries. Remarkably, we find several examples of both integrable and nonintegrable subspaces in the model (7.1), demonstrating the rich dynamical structure inherent in systems with fractured Hilbert spaces. Studying the dynamics of root states that generate large irreducible Krylov subspaces thus allows us to establish that integrability or nonintegrability of a system is correctly defined only *within* each Krylov subspace.

7.5   Integrable subspaces

In this section, we illustrate several *integrable* irreducible Krylov subspaces with exponentially large dimension present in the pair-hopping model (7.1).

7.5.1   *Spin subspace*

The simplest example of a large integrable Krylov subspace can be generated by a root state $|\psi_0\rangle$ (see equation (7.20)) which is any product state of only spin degrees of freedom: $\uparrow$ and $\downarrow$ as defined in equation (7.9). From equation (7.10), we find that the Hamiltonian restricted to this subspace can be written as a nearest-neighbor Hamiltonian with actions:

$$|\uparrow\uparrow\rangle \to 0, |\downarrow\downarrow\rangle \to 0, |\uparrow\downarrow\rangle \leftrightarrow |\downarrow\uparrow\rangle, \tag{7.25}$$

where $|a\rangle \to 0$ and $|a\rangle \leftrightarrow |b\rangle$ represent the action of a single term of the Hamiltonian. Thus, starting from a root state with $N_\uparrow$ spin $\uparrow$'s (and hence $(N - N_\uparrow)$ spin $\downarrow$'s), such as

$$|\uparrow\downarrow\uparrow\uparrow\downarrow\rangle, \quad (N, N_\uparrow) = (5, 3),$$

the action of the Hamiltonian only rearranges the spins.

In particular, note that: (i) The number of $\uparrow$'s and $\downarrow$'s in the root state $N_\uparrow$ and $N - N_\uparrow$, respectively, are preserved upon the action of the Hamiltonian, (ii) no fractons (i.e., $+$'s or $-$'s) are created and (iii) *all* product configurations with N spins and a fixed value of $N_\uparrow$ are part of the Krylov subspace $\mathcal{K}(H, |\psi_0\rangle)$ associated with the root state $|\psi_0\rangle$. Furthermore, since the Hamiltonian restricted to this subspace only interchanges the spins (see Eq. (7.25)), it maps *exactly* onto that of the spin-1/2 XX model:

$$H_{XX}[N] \equiv \sum_{j=1}^{N} \left(\sigma_j^+ \sigma_{j+1}^- + \sigma_j^- \sigma_{j+1}^+ \right), \tag{7.26}$$

where $\{\sigma_j^+\}$ and $\{\sigma_j^-\}$ are onsite Pauli matrices. This mapping was first noted in earlier works on half-filled Landau levels [90,91,101] and is formally illustrated in Appendix B. As is well known, the Hamiltonian equation (7.26) can be solved using a Jordan-Wigner transformation [102], upon which it maps onto a noninteracting problem. We numerically observe that the full ground state of the Hamiltonian equation (7.1) belongs this Krylov subspace with $(N, N_\uparrow) = \left(N, \lfloor \frac{N}{2} \rfloor \right)$. We refer to Appendix C for a complete discussion of the structure of the eigenstates within this Krylov subspace.

An important note regarding symmetries: each Krylov subspace generated from a root state with only spins and with a fixed $N_\uparrow$ (dubbed the spin Krylov subspace) only generates one symmetry sector of the XX model with a fixed S_z. All symmetry sectors of the XX model can be generated by starting from root states with different $N_\uparrow$, so that the full spectrum of the XX model of N sites is embedded within the spectrum of the pair-hopping Hamiltonian H (7.1), both for OBC and PBC.

With respect to the symmetries of H, these Krylov subspaces lie within the sector $(Q, D, S^z) = (0, 0, 2N_\uparrow - N)$, where Q, D and S^z are the total charge, dipole moment and spin respectively, discussed in Sec. 7.3.2. However, these are *not* the only states within that (Q, D, S^z) symmetry sector, providing evidence for the Krylov fracture in the pair-hopping Hamiltonian H. For example, the product state

$$|* \cdots * + - - + * \cdots *\rangle, \tag{7.27}$$

where $* = \uparrow, \downarrow$ and with $(N_\uparrow - 1)$ $\uparrow$'s (and hence $(N - N_\uparrow - 1)$ $\downarrow$'s) lies within the symmetry sector $(Q, D, S^z) = (0, 0, 2N_\uparrow - N)$ but outside the spin Krylov subspace constructed above.

7.5.2 Single-dipole subspace

Restricting our attention to OBC, we now demonstrate the existence of another set of integrable Krylov subspaces $\mathcal{K}(H, |\psi_0\rangle)$, which are generated from root states containing only a single dipole. Such root states are of the form

$$|\psi_0\rangle = |* \cdots * + - * \cdots *\rangle, |\psi_0\rangle = |* \cdots * - + * \cdots *\rangle, \tag{7.28}$$

where $* = \uparrow, \downarrow$. The action of the Hamiltonian equation (7.1) on configurations of the form (7.28) is given by

$$|\downarrow + -\rangle \leftrightarrow |+ - \downarrow\rangle, |\uparrow - +\rangle \leftrightarrow |- + \uparrow\rangle,$$

$$|\uparrow + -\rangle \to 0, |+ - \uparrow\rangle \to 0,$$

$$|\downarrow - +\rangle \to 0, |- + \downarrow\rangle \to 0. \tag{7.29}$$

Since dipole moment is conserved, the dipole does not "disintegrate" under the action of the Hamiltonian equation (7.29), i.e., the dipole does not separate into its constituent $+$ and $-$ fractons. As it turns out, Krylov subspaces generated by root states of the form (7.28) with N sites are *isomorphic* to Hilbert spaces of $(N - 1)$ spin-1/2's, with the effective Hamiltonians within these Krylov subspaces given by XX models of $(N - 1)$ sites. In the following, we focus on the Krylov subspace corresponding to a $+-$ dipole. As we discuss later, the generalization to $-+$ dipoles follows similarly (Table 7.1).

   To show this, we first observe that as a consequence of equation (7.29), a dipole $+-$ in the root state can *never* cross an $\uparrow$ spin to its left or to its right. In other words, the dipole $+-$ can only hop left (right) if there is a $\downarrow$ spin immediately to its left (right). Hence, all product states in the Krylov subspace generated by a root state $|\psi_0\rangle$ with one dipole $+-$ preserve the number of $\uparrow$ spins to the left and right of the dipole separately. Denoting these conserved quantities by $N_\uparrow^{(1)}$ and $N_\uparrow^{(2)}$ respectively, we see that product

Table 7.1. Table of integrable Krylov subspaces (by no means an exhaustive list) of the pair-hopping model for system size $L = 2N$, with OBC at half-filling.

Krylov Subspace	Root Configuration	Quantum Numbers	Restricted Hamiltonian
Spin	$\lvert\uparrow\downarrow\cdots\downarrow\uparrow\rangle$	$N_\uparrow$	$H_{XX}[N]$
Single $+-$ dipole	$\lvert\uparrow\cdots\downarrow+-\uparrow\cdots\downarrow\rangle$	$N_\uparrow^{(1)}, N_\uparrow^{(2)}$	$H_{XX}[N-1]$
Two separated $+-$ dipoles	$\lvert\uparrow\cdots\downarrow+-\uparrow\cdots\downarrow+-\uparrow\cdots\downarrow\rangle$	$N_\uparrow^{(1)}, N_\uparrow^{(2)} \geq 1, N_\uparrow^{(3)}$	$H_{XX}[N-2]$
Two adjacent $+-$ dipoles	$\lvert\uparrow\cdots\downarrow+-+-\uparrow\cdots\downarrow\rangle$	$N_\uparrow^{(1)}, N_\uparrow^{(2)} = 0, N_\uparrow^{(3)}$	$H_{XX}[N-1]$
X separated $+-$ dipoles	$\lvert\uparrow\cdots\downarrow+-\cdots+-\cdots\downarrow+-\uparrow\cdots\downarrow\rangle$	$N_\uparrow^{(1)}, \{N_\uparrow^{(2)}, \ldots, N_\uparrow^{(X-1)}\} \geq 1, N_\uparrow^{(X)}$	$H_{XX}[N-(X-1)]$
X adjacent $+-$ dipoles	$\lvert\uparrow\cdots\downarrow+-+-\cdots+-\uparrow\cdots\downarrow\rangle$	$N_\uparrow^{(1)}, \{N_\uparrow^{(2)}, \ldots, N_\uparrow^{(X-1)}\} = 0, N_\uparrow^{(X)}$	$H_{XX}[N-1]$

For each type of Krylov subspace, we provide the root configuration generating it, the associated quantum numbers and the Hamiltonian restricted to that subspace. Dipole subspaces for the oppositely oriented $-+$ dipoles can be constructed analogously (see main text for discussion).

states in the Krylov subspace $\mathcal{K}(H, |\psi_0\rangle)$ always have the form

$$|* \cdots * + - * \cdots *\rangle, \tag{7.30}$$

$$\underbrace{}_{N_\uparrow^{(1)}} \underbrace{}_{N_\uparrow^{(2)}}$$

where $* = \uparrow, \downarrow$. This Krylov subspace can thus be uniquely labeled by the tuple $(N, N_\uparrow^{(1)}, N_\uparrow^{(2)})$. For example, the Krylov subspace $\mathcal{K}(H, |\psi_0\rangle)$ generated by the configuration $|\psi_0\rangle = |\uparrow\downarrow + - \uparrow\downarrow\rangle$ with OBC consists of the following basis states:

$$|\uparrow\downarrow + - \uparrow\downarrow\rangle, |\downarrow\uparrow + - \uparrow\downarrow\rangle, |\uparrow\downarrow + - \downarrow\uparrow\rangle, |\downarrow\uparrow + - \downarrow\uparrow\rangle$$

$$|\uparrow + - \downarrow\uparrow\downarrow\rangle, |\uparrow + - \uparrow\downarrow\downarrow\rangle, |\uparrow + - \downarrow\downarrow\uparrow\rangle$$

$$|\uparrow\downarrow\downarrow + - \uparrow\rangle, |\downarrow\uparrow\downarrow + - \uparrow\rangle, |\downarrow\downarrow\uparrow + - \uparrow\rangle. \tag{7.31}$$

Note that all the states in $\mathcal{K}(H, |\psi_0\rangle)$ are labeled by $(N, N_\uparrow^{(1)}, N_\uparrow^{(2)}) = (6, 1, 1)$. In order to map configurations of the form (7.30) onto an effective spin-1/2 Hilbert space, note that the rules of equation (7.29) are *identical* to those of equation (7.10) when the dipole $+-$ is replaced by an $\uparrow$ spin. This observation allows us to establish two crucial results on the single-dipole Krylov subspace $\mathcal{K}(H, |\psi_0\rangle)$.

Firstly, product states in the single-dipole Krylov subspace consisting of a $+-$ dipole can be *uniquely* mapped onto product states of $(N-1)$ spin-1/2's with $(N_\uparrow^{(1)} + N_\uparrow^{(2)} + 1)$ $\uparrow$'s by replacing the $+-$ dipole with an $\uparrow$. For example, the following condition holds:

$$\underset{\text{(A)}}{|\uparrow\uparrow\downarrow + - \uparrow\downarrow\uparrow\uparrow\uparrow\rangle} \iff \underset{\text{(B)}}{|\uparrow\uparrow\downarrow\uparrow\uparrow\downarrow\uparrow\uparrow\uparrow\rangle}, \tag{7.32}$$

where configuration (A) in the Krylov subspace with $(N, N_\uparrow^{(1)}, N_\uparrow^{(2)}) = (10, 2, 4)$ maps onto the configuration (B) in the spin subspace with $(N, N_\uparrow) = (9, 6)$ by replacing the $+-$ dipole with an $\uparrow$. The inverse mapping from the spin-1/2 Hilbert space of $(N-1)$ sites and $(N_\uparrow^{(1)} + N_\uparrow^{(2)} + 1)$ $\uparrow$'s to the single-dipole Krylov subspace $(N, N_\uparrow^{(1)}, N_\uparrow^{(2)})$ proceeds by identifying one $\uparrow$ to be the $+-$ dipole such that the resulting configuration has the correct $N_\uparrow^{(1)}$ and $N_\uparrow^{(2)}$. For instance in equation (7.32), given $(N, N_\uparrow^{(1)}, N_\uparrow^{(2)}) = (10, 2, 4)$, the mapping from (B) to (A) is possible only if the third $\uparrow$ in the configuration (B) is replaced by a $+-$ dipole.

The mapping for the single $-+$ dipole subspace follows analogously, with $\uparrow$ replaced by $\downarrow$, i.e., by identifying $-+$'s with $\downarrow$'s instead. In that case, the quantities $N_\downarrow^{(1)}$ and $N_\downarrow^{(2)}$, defined as

$$|* \cdots * - + * \cdots *\rangle, \tag{7.33}$$

$$\underbrace{\qquad}_{N_\downarrow^{(1)}} \underbrace{\qquad}_{N_\downarrow^{(2)}}$$

are preserved within the Krylov subspace. Thus, the single-dipole Krylov subspace with OBC and a fixed $(N, N_\uparrow^{(1)}, N_\uparrow^{(2)})$ (respectively, $(N, N_\downarrow^{(1)}, N_\downarrow^{(2)})$) is *isomorphic* to the Hilbert space of $(N-1)$ spin-1/2's with $(N_\uparrow^{(1)} + N_\uparrow^{(2)} + 1)$ $\uparrow$'s (respectively, $(N_\downarrow^{(1)} + N_\downarrow^{(2)} + 1)$ $\downarrow$'s). Secondly, since equation (7.29) is identical to equation (7.25) when the dipole $+-$ (respectively, $-+$) is replaced with an $\uparrow$ (respectively, $\downarrow$), the effective Hamiltonian within each such Krylov subspace is the XX model of $(N-1)$ sites with OBC.[e] In particular, the spectrum of H in equation (7.26) restricted to the single Krylov subspace labelled by $(N, N_\uparrow^{(1)}, N_\uparrow^{(2)})$ (respectively, $(N, N_\downarrow^{(1)}, N_\downarrow^{(2)})$) is precisely the spectrum of the quantum number sector $S_z = (2(N_\uparrow^{(1)} + N_\uparrow^{(2)}) + 3 - N)$ (respectively, $S_z = -(2(N_\downarrow^{(1)} + N_\downarrow^{(2)}) + 3 - N))$ of the XX model.

Note that with PBC this Krylov subspace is no longer isomorphic to the spin-1/2 Hilbert space of the XX model, since the inverse mapping from the spin-1/2 Hilbert space to the dipole subspace is not unique. Thus, the effective Hamiltonian within this Krylov subspace cannot map exactly onto the XX model of equation (7.26) with PBC and it remains unclear whether or not the resulting Hamiltonian is integrable for any finite system size.

7.5.3 *Multidipole subspaces*

We now consider Krylov subspaces generated by root configurations containing multiple identically oriented dipoles. All such subspaces turn out to be integrable and governed by effective XX Hamiltonians of various sizes. As with a single dipole discussed in the previous section, spins and dipoles interact according to equation (7.29). A crucial property of these rules, which we will make use of throughout this section, is that the $+-$ (respectively, $-+$) dipole cannot cross any $\uparrow$ (respectively, $\downarrow$) spins under the action of the Hamiltonian H.

We first illustrate the case where the root state contains two $+-$ dipoles before discussing the general setting. Since the dipoles $+-$ cannot cross $\uparrow$'s,

---

[e]Once an $\uparrow$ spin is identified, note that the action of the XX Hamiltonian also preserves $N_\uparrow^{(1)}$ and $N_\uparrow^{(2)}$, the number of $\uparrow$ spins to the left and to the right of the identified $\uparrow$ spin respectively.

the Krylov subspace generated from a root state with two identically oriented dipoles preserves three quantities of the root state: $(N_\uparrow^{(1)}, N_\uparrow^{(2)}, N_\uparrow^{(3)})$, depicted schematically by the following configurations:

$$|\ast \cdots \ast + - \ast \cdots \ast + - \ast \cdots \ast\rangle, \qquad (7.34)$$

$$\underbrace{\qquad}_{N_\uparrow^{(1)}} \quad \underbrace{\qquad}_{N_\uparrow^{(2)}} \quad \underbrace{\qquad}_{N_\uparrow^{(3)}}$$

where $\ast = \uparrow, \downarrow$. That is, for a Krylov subspace generated by root states with two $+-$ dipoles, the number of $\uparrow$ spins to the left of the left dipole, in between the two dipoles and to the right of the right dipole are each separately conserved. Thus, the quantities $(N, N_\uparrow^{(1)}, N_\uparrow^{(2)}, N_\uparrow^{(3)})$ uniquely label the Krylov subspace.

We now restrict our discussion to the Krylov subspace containing two $+-$ dipoles, with the generalization to the two $-+$ dipole subspace being straightforward. Provided $N_\uparrow^{(2)} \geq 1$ in the root state $|\psi_0\rangle$, the two dipoles are always separated by an $\uparrow$ spin and can never be adjacent to each other; the action of the Hamiltonian is therefore entirely specified by equation (7.29). Product states in the Krylov subspace can be mapped onto configurations of $(N - 2)$ spin-1/2's with $(N_\uparrow^{(1)} + N_\uparrow^{(2)} + N_\uparrow^{(3)} + 2)$ $\uparrow$'s by replacing the $+-$ dipoles by $\uparrow$'s. For example,

$$\underset{(A)}{|\uparrow\downarrow\uparrow + - \uparrow\downarrow\uparrow + - \uparrow\downarrow\rangle} \iff \underset{(B)}{|\uparrow\downarrow\uparrow\uparrow\uparrow\downarrow\uparrow\uparrow\uparrow\downarrow\rangle}, \qquad (7.35)$$

where the configuration (A) in the two-dipole Krylov subspace labeled by $(N, N_\uparrow^{(1)}, N_\uparrow^{(2)}, N_\uparrow^{(3)}) = (12, 2, 2, 1)$, maps onto configuration (B).

Similar to the single-dipole case, the inverse mapping is unique once $(N, N_\uparrow^{(1)}, N_\uparrow^{(2)}, N_\uparrow^{(3)})$ are specified. This inverse mapping proceeds by identifying two of the $\uparrow$'s to be $+-$ dipoles such that the resulting configuration has the required values of $N_\uparrow^{(1)}$, $N_\uparrow^{(2)}$ and $N_\uparrow^{(3)}$. For example, given that $(N, N_\uparrow^{(1)}, N_\uparrow^{(2)}, N_\uparrow^{(3)}) = (12, 2, 2, 1)$, the two-dipole configuration (A) in equation (7.35) is the unique two-dipole configuration corresponding to spin configuration (B).

The mapping for the two-dipole subspace with $-+$ dipoles follows analogously, with $\uparrow$ replaced by $\downarrow$, i.e., by identifying $-+$'s with $\downarrow$'s instead. The action of the Hamiltonian is completely specified by equation (7.29) when the dipoles are not allowed to be adjacent each other; as discussed in Sec. 7.5.2, equation (7.29) is identical to equation (7.25) when the $+-$ (respectively, $-+$) dipole is identified with $\uparrow$ (respectively, $\downarrow$) spin. Thus, the Hamiltonian restricted to the two $+-$ (respectively, $-+$) dipole Krylov subspace is identical to the XX model of $(N - 2)$ sites within the

$S_z = (2(N_\uparrow^{(1)} + N_\uparrow^{(2)} + N_\uparrow^{(3)}) + 6 - N)$ (respectively, $S_z = -(2(N_\downarrow^{(1)} + N_\downarrow^{(2)} + N_\downarrow^{(3)}) + 6 - N))$ sector.

We emphasize that the two-dipole Krylov subspace of N is isomorphic to the spin-1/2 Hilbert space of $(N - 2)$ sites *only* when the two $+-$ (respectively, $-+$) dipoles have at least one $\uparrow$ (respectively, $\downarrow$) spin between them, i.e., only if $N_\uparrow^{(2)} \geq 1$ (respectively, $N_\downarrow^{(2)} \geq 1$). When the two dipoles are adjacent to each other, using Eqs. (7.13) and (7.14) we find that the action of the Hamiltonian H reads

$$|+--+-\rangle \leftrightarrow |\downarrow + \uparrow -\rangle, |+--+-\rangle \leftrightarrow |+\uparrow - \downarrow\rangle,$$

$$|-+-+\rangle \leftrightarrow |\uparrow - \downarrow +\rangle, |-+--+\rangle \leftrightarrow |-\downarrow + \uparrow\rangle. \tag{7.36}$$

As a consequence, the action of the Hamiltonian on root states of the form $|\cdots + - + - \cdots\rangle$ result in the "disintegration" of dipoles, resulting in configurations of the form:

$$|\cdots \downarrow + \uparrow - \cdots\rangle, \quad |\cdots + \uparrow - \downarrow \cdots\rangle,$$

which cannot be mapped onto a configuration of $(N - 2)$ spin-1/2's through the map described earlier in this section. Nevertheless, we find that such Krylov subspaces *does* map onto the XX model, albeit one with $(N - 1)$ spin-1/2's; we discuss this mapping in Appendix D.

The preceding discussion straightforwardly generalizes to three or more dipoles. For a Krylov subspace generated by a root state containing n identically oriented dipoles, with OBC the system can be partitioned into $(n + 1)$ segments separated by the dipoles. We introduce the quantities $N_\uparrow^{(1)}, N_\uparrow^{(2)}, \ldots, N_\uparrow^{(n+1)}$, where $N_\uparrow^{(j)}$ (respectively, $N_\downarrow^{(j)}$) represents the number of $\uparrow$ (respectively, $\downarrow$) spins in the jth segment of the chain in the root state:

$$|\cdots + \overset{1}{-} \cdots + \overset{2}{-} \cdots + \overset{n-1}{-} \cdots + \overset{n}{-} \cdots\rangle, \tag{7.37}$$
$$\underbrace{\qquad}_{N_\uparrow^{(1)}} \underbrace{\qquad}_{N_\uparrow^{(2)}} \qquad \underbrace{\qquad}_{N_\uparrow^{(n)}} \underbrace{\qquad}_{N_\uparrow^{(n+1)}}$$

with the superscripts $1, 2, \ldots, n$ indexing the dipoles. Since a $+-$ dipole is not allowed to cross an $\uparrow$ spin under the action of the Hamiltonian, the quantities $\{N_\uparrow^{(j)} \geq 1\}$ are invariant under the dynamics, i.e., these quantities are identical for all product states within the Krylov subspace generated by the root state of the form (7.37). As with two dipoles, this is true provided no dipoles are adjacent in the root state, which corresponds to the constraint $N_\uparrow^{(j)} \geq 1$ for any j.

In this case ($N^{(j)} \neq 0 \forall j$), the n dipole Krylov subspace exactly maps onto a spin-1/2 Hilbert space with $(N - n)$ sites and $(n + \sum_{j=1}^{n+1} N_{\uparrow}^{(j)})$ ↑'s by identifying each $+-$ dipole with an ↑ spin. For example,

$$|{\uparrow\downarrow} +- {\downarrow\uparrow\uparrow} +- {\downarrow\uparrow} +- {\uparrow\uparrow}\rangle \iff |{\uparrow\downarrow\uparrow\downarrow\uparrow\uparrow\uparrow\downarrow\uparrow\uparrow\uparrow}\rangle, \qquad (7.38)$$
$$\underset{(A)}{} \qquad\qquad\qquad\qquad \underset{(B)}{}$$

where $n = 3$ and where the three-dipole configuration (A) with $(N_{\uparrow}^{(1)}, N_{\uparrow}^{(2)}, N_{\uparrow}^{(3)}, N_{\uparrow}^{(4)}) = (1, 2, 1, 2)$ maps onto the spin configuration (B). This mapping onto the spin-1/2 Hilbert space is invertible provided the tuple $(N_{\uparrow}^{(1)}, N_{\uparrow}^{(2)}, \ldots, N_{\uparrow}^{(n+1)})$ is known and it proceeds by identifying n ↑ spins in each product configuration with $+-$ dipoles such that the resulting configuration has the requisite $(N_{\uparrow}^{(1)}, N_{\uparrow}^{(2)}, \ldots, N_{\uparrow}^{(n+1)})$ values. For example, given $(N_{\uparrow}^{(1)}, N_{\uparrow}^{(2)}, N_{\uparrow}^{(3)}, N_{\uparrow}^{(4)}) = (1, 2, 1, 2)$, configuration (B) in equation (7.38) uniquely maps onto (A) by identifying the appropriate ↑ spins with $+-$ dipoles.

The mapping with $-+$ dipoles proceeds in a similar way by replacing the $-+$ dipole by ↓. The quantities $\{N_{\downarrow}^{(j)}\}$ are thus preserved within the Krylov subspaces, where

$$|\cdots \underset{N_{\downarrow}^{(1)}}{\underbrace{- + \cdots}} \overset{1}{} \underset{N_{\downarrow}^{(2)}}{\underbrace{- + \cdots}} \overset{2}{} \underset{N_{\downarrow}^{(n)}}{\underbrace{- + \cdots}} \overset{n-1}{} \underset{N_{\downarrow}^{(n+1)}}{\underbrace{- + \cdots}} \overset{n}{} \rangle. \qquad (7.39)$$

Since the Hamiltonian equation (7.29) is identical to equation (7.10) upon the identification of dipoles with spins, the Hamiltonian restricted to the Krylov subspace for the n $+-$ (respectively, $-+$) dipole case is the XX model with $(N - n)$ sites within the quantum number sector $S_z = (3n + \sum_{j=1}^{n+1} N_{\uparrow}^{(j)} - N)$ (respectively, $S_z = -(3n + \sum_{j=1}^{n+1} N_{\downarrow}^{(j)} - N)$).

When $N_{\uparrow}^{(j)} = 0$ or $N_{\downarrow}^{(j)} = 0$ for some j in the root state (7.37), the mapping prescribed above fails because the action of the Hamiltonian causes the adjacent dipoles to disintegrate, as shown in equation (7.36). Nevertheless, as we show in Appendix D, we find that the Krylov subspace remains integrable even if some dipoles in the root state are adjacent. Specifically, we find that the Hamiltonian restricted to a Krylov subspace with *only* n $+-$ (respectively, $-+$) dipoles is the XX model of $(N - n + X)$ sites, where X is the number of segments j containing no spins, such that $N_{\uparrow}^{(j)} = 0$ (respectively, $N_{\downarrow}^{(j)} = 0$). For example, the effective Hamiltonians restricted to the Krylov subspaces generated by the root states

$$|*\cdots*+-+-+-*\cdots*\rangle$$

and

$$|*\cdots*+-+-*\cdots*+-*\cdots*\rangle,$$

where $* = \uparrow, \downarrow$ are the XX models acting on $(N-1)$ and $(N-2)$ spin-1/2's respectively.

As was the case for a single dipole, the mapping onto XX models does not work with PBC. However, it is not clear if the effective Hamiltonian restricted to this sector with PBC is solvable for a finite system size, although integrability of this sector should be restored in the thermodynamic limit and the energy spectrum should display Poisson-level statistics for a large enough system size. Finally, we note that upon the addition of electrostatic terms or disorder (discussed in Appendix E), the spin subspace described in Sec. 7.5.1 maps onto the XXZ model or disordered XX model and thus remains integrable. However, the dipole subspaces are no longer integrable and they show all the signs of usual nonintegrability, including GOE-level statistics [103].

7.5.4   *Systematic construction of integrable subspaces*

Having illustrated the existence of several integrable Krylov subspaces of the pair-hopping model (7.1), we briefly discuss a general prescription for constructing additional irreducible integrable subspaces by using the integrable subspaces of Secs. 7.5.1–7.5.3 as building blocks. As also emphasized in Refs. [85, 86], one can introduce *blockades*, i.e., regions of the chain on which terms of the Hamiltonian vanish. For example, consider the following root state with a configuration of the form:

$$|*\cdots*++\cdots++*\cdots*\rangle, \tag{7.40}$$

where $* = \uparrow, \downarrow$, with $N_+ \geq 2$ and $N_- = 0$. Following the rules (7.10)–(7.14), the Hamiltonian can act nontrivially only on sites contained within regions A and B of the root state equation (7.40).

Due to this, all basis states of the Krylov subspace generated from the root state equation (7.40) retain the same schematic form, with $++\cdots++$ ($N_+ \geq 2$) acting as a blockade that spatially disconnects two parts of the Krylov subspace.

Thus, one can show that the effective Hamiltonian restricted to such blockaded Krylov subspaces is simply given by the sum of two independent XX models acting on distinct degrees of freedom lying in regions A and B. Note that blockades can also be constructed using exponentially many other "static" patterns [85, 86], such as $-\,-\,\cdots\,-\,-,\,+\,+\,-\,-\,\cdots\,+\,+\,-\,-,$ or $+\,+\,\uparrow\,\cdots\,\uparrow\,+\,+$, which in turn lead to exponentially many integrable subspaces.

Similarly, we can also introduce blockades for the dipole Krylov subspaces considered in Secs. 7.5.2–7.5.3, as long as the dipoles do not interact with the blockade. For example, consider the root configuration of the form of equation (7.40) where region A is a root configuration for an integrable subspace with one or more $-+$ dipoles and region B is a root configuration for an integrable subspace with $+-$ dipoles:

$$|*\cdots *-+*\cdots *++\cdots ++*\cdots *+-*\cdots *\rangle, \qquad (7.41)$$

$$\underbrace{}_{A} \qquad \underbrace{}_{B}$$

where $* = \uparrow, \downarrow$. Upon successive applications of the Hamiltonian on the root state of equation (7.41), the dipoles in regions A and B do not interact with the string of $+$'s in between the regions. Thus, the string of $+$'s acts as a blockade and the Krylov subspaces generated by such root configurations are integrable, since the restricted Hamiltonian is a sum of XX models on regions A and B. While we have only illustrated the simplest cases where blockades are introduced between regions A and B, each of which are integrable regions that do not interact with the blockade, we can of course generalize by introducing n blockades separating $n+1$ regions, each of which contain the integrable subspaces that do not interact with the neighboring blockades. In such a case, the Hamiltonian restricted to the Krylov subspace is a sum of $n+1$ independent XX models.

A detailed study delineating *all* integrable subspaces of the pair-hopping model (7.1) is beyond the scope of this work. Nevertheless, the above examples suffice to illustrate the existence of *exponentially many* integrable Krylov subspaces, clearly establishing the possibility of emergent, Krylov-restricted integrability in systems exhibiting Krylov fracture.

7.6   Nonintegrable subspaces and Krylov-restricted ETH

Given that large swaths of the spectrum of the pair-hopping Hamiltonian are solvable, it is natural to ask whether this model is *completely* integrable. The standard diagnostic for probing nonintegrability of some Hamiltonian is the appearance of random matrix behavior *within* a sector resolved by symmetries of that Hamiltonian. For example, the energylevel statistics [21, 103] and the matrix elements of local operators in the energy eigenbasis (according to ETH) [8] are expected to follow random matrix behavior for nonintegrable systems.

Generally, in unconstrained models, symmetry sectors are themselves examples of well-defined dynamically disconnected Krylov subspaces. In other words, a root-state which is an eigenstate of the symmetry typically generates a Krylov subspace which spans all states within that symmetry sector. However, for systems exhibiting Krylov fracture, there exist several dynamically disconnected Krylov subspaces *within* each symmetry sector. As was also emphasized by Refs. [85, 86], resolving eigenstates by symmetries alone may hence be insufficient for identifying ergodicity, given the possibility of Krylov fracture.

Thus, we pose the crucial question that motivates the title of the paper: Whether symmetries are only a subset of the more general phenomena of Krylov fracture and if ergodicity or its absence should correspondingly be defined within dynamically disconnected irreducible Krylov subspaces. In the previous section, we encountered examples of Krylov subspaces within symmetry sectors which display the characteristic trademarks of integrable systems, e.g., Poisson-level statistics. Now, we wish to ask whether Krylov subspaces that are *not* integrable exhibit conventional diagonostics of ergodic systems, such as Wigner–Dyson-level statistics and ETH [8]. Of course, random matrix theory is a statement about "large" matrices, i.e., in the limit that the size of the matrix goes to infinity; consequently, the question of thermalization within Krylov subspaces is only well-posed for "large" Krylov subspaces, whose size tends to infinity in the thermodynamic limit. Thus, we explore some simple nonintegrable Krylov subspaces of the pair-hopping model (7.1) and, in the process, establish the notion of Krylov-restricted ETH.

Indeed, there exist Krylov subspaces of the pair-hopping Hamiltonian which are not integrable. Consider for instance the Krylov subspace generated by the root state containing both $+-$ *and* $-+$ dipoles:

$$|\psi_0\rangle = |* \cdots * - + + - * \cdots *\rangle, \qquad (7.42)$$

where $* = \uparrow, \downarrow$. Since the dipoles are of opposite orientation, the mapping of the $+-$ and $-+$ dipoles to $\uparrow$ and $\downarrow$ spins would only be justified if $|-++-\rangle \leftrightarrow |+--+\rangle$ under the action of the Hamiltonian, which is strictly prohibited by the rules given in Eqs. (7.10)–(7.14). As a result, the Hamiltonian restricted to this Krylov subspace does not need to map onto an integrable model. Another example is the Krylov subspace generated by

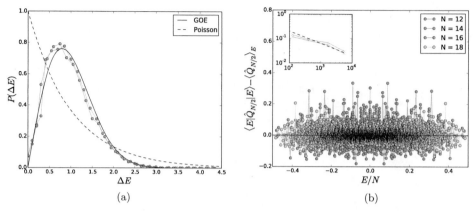

Fig. 7.1.   (a) Level statistics within the Krylov subspace $\mathcal{K}(H, |\psi_0\rangle)$ generated by various root states $|\psi_0\rangle$ with OBC, where H is the pair-hopping Hamiltonian equation (7.1). Red: $|\psi_0\rangle = |\downarrow\uparrow\uparrow\uparrow\downarrow\downarrow - + + - \downarrow\downarrow\uparrow\uparrow\uparrow\downarrow\rangle$ (Krylov subspace dimension $\mathcal{D_K} = 18849$), Blue: $|\psi_0\rangle = |\uparrow\downarrow\uparrow\downarrow\uparrow + \uparrow\uparrow\uparrow\uparrow\downarrow\downarrow - \downarrow\uparrow\downarrow\uparrow\downarrow\rangle$ (Krylov subspace dimension $\mathcal{D_K} = 21660$). The configurations of spins in the root configurations have been chosen to ensure that the Krylov subspace does not have any symmetries. The standard $\langle r \rangle$ parameter [104] in these subspaces is 0.5331 and 0.5276 respectively, close to the GOE value of 0.53. (b) Evidence for the Eigenstate Thermalization Hypothesis (ETH) in the nonintegrable Krylov subspace $\mathcal{K}(H, |\psi_0\rangle)$ generated by the root state shown in equation (7.45), which for $N = 18$ reads $|\psi_0\rangle = |\uparrow\downarrow\uparrow\downarrow\uparrow\downarrow\uparrow - + + - \downarrow\uparrow\downarrow\uparrow\downarrow\uparrow\downarrow\rangle$. In order to break the symmetries within this Krylov subspace, the couplings $\{J_j\}$ of the terms of the pair-hopping Hamiltonian (see equation (E10)) are chosen from a uniform distribution $[1 - W, 1 + W]$, with $W = 0.1$. This disorder preserves the Krylov fractured structure of the Hilbert space. Main: The difference between $\langle E| \widehat{Q}_{N/2} |E\rangle$, the expectation value of the charge operator in an eigenstate at energy E and $\langle \widehat{Q}_{N/2}\rangle_E$, the thermal expectation value at that energy determined by averaging $\langle E| \widehat{Q}_{N/2} |E\rangle$ in an energy window of $\Delta E = 0.05$ [105]. Inset: The standard deviations of that difference as a function of the Hilbert dimension $\mathcal{D_K}$ scales $\sim 1/\sqrt{\mathcal{D_K}}$ (dotted line) for two operators: the charge operator $\widehat{Q}_{N/2}$ (blue) and the spin operator $\widehat{S}^z_{N/2}$ (red), consistent with ETH within the Krylov subspace $\mathcal{K}(H, |\psi_0\rangle)$.

the root state containing two separated fractons:

$$|\psi_0\rangle = |* \cdots * + * \cdots * - * \cdots *\rangle, \tag{7.43}$$

where $* = \uparrow, \downarrow$ and the $* \cdots *$ in between the $+$ and $-$ contains both $\uparrow$ and $\downarrow$ spins. The latter condition is required to ensure that $|\psi_0\rangle$ does not belong to any of the integrable multidipole Krylov subspaces discussed in Appendix D.

We have numerically studied the behavior of Krylov subspaces generated by root states such as those given in Eqs. (7.42) and (7.43). As shown in Fig. 7.1(a), we find that eigenstates of the Hamiltonian within these Krylov subspaces $\mathcal{K}(H, |\psi_0\rangle)$ exhibit GOE-level statistics, providing evidence for the nonintegrability of the Krylov subspace. We further **conjecture** that such nonintegrable Krylov subspaces satisfy the Eigenstate Thermalization

Hypothesis (ETH) [7–11]. ETH states that the matrix elements of local operators in the energy eigenstates of a nonintegrable model take the form [11]

$$\langle E_m | \widehat{O} | E_n \rangle = \bar{O}(E) \delta_{m,n} + R_{m,n} e^{-S(E)/2} f_O(E, \omega), \qquad (7.44)$$

where $\widehat{O}$ is a local operator that is invariant under the symmetries of the Hamiltonian, $|E_m\rangle$ and $|E_n\rangle$ are the energy eigenstates with energies E_m and E_n with the same symmetry quantum numbers, $E = (E_m + E_n)/2$, $\omega = E_m - E_n$, $R_{m,n}$ is a random variable with zero mean and unit variance, $\bar{O}(E)$ is a smooth function of E and represents the thermal expectation value of $\widehat{O}$ at energy E,[f] $f_O(E, \omega)$ is a smooth function of E and ω which do not scale with the system size [11] and $S(E)$ is the thermodynamic entropy at energy E. In equation (7.44), since $S(E) \sim \log \mathcal{D}$ for states in the middle of the spectrum, where $\mathcal{D}$ is the Hilbert space dimension, the standard deviation of expectation values of operators in the eigenstates is expected to scale as $\sim 1/\sqrt{\mathcal{D}}$ for eigenstates in the middle of the spectrum [105].

Here, we want to test whether equation (7.44) holds within a nonintegrable Krylov subspace. We focus on the Krylov subspace with the root states (with OBC):

$$|\psi_0\rangle = \begin{cases} |\uparrow\downarrow \cdots \uparrow\downarrow - + +- \uparrow\downarrow \cdots \uparrow\downarrow\rangle N = 4p, \\ |\uparrow\downarrow \cdots \uparrow\downarrow\uparrow - + +- \downarrow\uparrow\downarrow \cdots \uparrow\downarrow\rangle N = 4p + 2, \end{cases} \qquad (7.45)$$

with two dipoles $-+$ and $+-$ placed at the center of the chain. Furthermore, to probe the validity of Eq. (7.44), we need to choose an operator $\widehat{O}$ that preserves the Krylov subspaces. Hence we choose the charge operator on the $(N/2)$th site $\widehat{O} = \widehat{Q}_{N/2}$, which is diagonal in the basis of product states. Since the Krylov subspace $\mathcal{K}(H, |\psi_0\rangle)$ has symmetries (e.g., inversion symmetry), we add disorder to the couplings of the pair-hopping Hamiltonian (see Eq. (E10)), which does not affect the structure of the Krylov subspaces of the Hamiltonian and focus on testing the ergodicity within the Krylov subspace. To probe the validity of Eq. (7.44) within nonintegrable Krylov subspaces, in Fig. 7.1(b) we plot the quantity $\left(\langle E | \widehat{O} | E \rangle - \bar{O}(E) \right)$, as a function of E, where $|E\rangle$ is the eigenstate with energy E. The inset show the variance of the difference as a function of the Krylov subspace dimension $\mathcal{D}_\mathcal{K}$.

---

[f]The thermal value here is determined by averaging the eigenstate expectation values $\langle E | \widehat{O} | E \rangle$ over a small energy window ΔE, where $|E\rangle$ is an eigenstate with energy E [105].

Two observations in Fig. 7.1(b) suggest the validity of ETH within the Krylov subspace. Firstly, the quantity $\langle E|\widehat{O}|E\rangle - \bar{O}(E)$ is centered about 0, which shows that eigenstate expectation values approach the thermal expectation value. Secondly, the standard deviation of the difference (shown in the inset) scales as $\sim 1/\sqrt{\mathcal{D}_\mathcal{K}}$, the dimension of the Krylov subspace. Hence these observations provide evidence for "diagonal ETH" *within* nonintegrable Krylov subspaces, supporting the existence of Krylov-restricted ETH in systems exhibiting Krylov fracture.

7.7 Quasilocalization from thermalization

Based on the results of the previous section, which established the phenomenon of Krylov-restricted ETH, we expect that the long-time behavior of typical states within a particular nonintegrable (respectively, integrable) Krylov subspace coincides with the Gibbs ensemble (respectively, generalized Gibbs ensemble) restricted to that subspace. Such Krylov-restricted thermalization can lead to surprising behavior within some Krylov subspaces. For example, in the following we show that the *thermal* expectation value of charge density on the chain within a particular Krylov subspace is spatially nonuniform for any finite system size.

To illustrate this behavior, we consider the dynamics of a single fracton immersed in a spin background, i.e., we study the Krylov subspace generated by the root state with PBC:

$$|\psi_0\rangle = |* \cdots * + * \cdots *\rangle, \tag{7.46}$$

where $* = \uparrow, \downarrow$ such that $N_\uparrow = N_\downarrow$. The configuration $|\psi_0\rangle$ thus belongs to the quantum number sector $Q = 1, D = \exp(i\pi(N+1)/N), S^z = 0$, where N is the length of the chain. Since we impose PBC here, all configurations of $*$'s in the root state generate the same Krylov subspace as the spins can rearrange among themselves under the action of the Hamiltonian (see Eq. (7.10)). Hence, in the following, we only explicitly describe the action of the Hamiltonian on the fractons, given that all possible spin configurations (with $N_\uparrow = N_\downarrow$) are generated within this subspace. An explicit example of the complete list of product configurations in the Krylov subspace generated by $|\psi_0\rangle = |\uparrow\uparrow\uparrow + \downarrow\downarrow\downarrow\rangle$ for $N = 7$ is given in Appendix F. There are two possibilities for how the state $|\psi_0\rangle$ of Eq. (7.46) evolves under one application of the Hamiltonian H: either the spins can rearrange amongst themselves or the fracton moves by emitting a dipole, according to Eq. (7.13). Since we are only focusing on the fracton, in the latter case, the new basis state reads

$$|\psi_1\rangle = |* \cdots * + - + * \cdots *\rangle, \tag{7.47}$$

where $* = \uparrow, \downarrow$. Upon further actions of the Hamiltonian, the emitted $+-$ or $-+$ dipole in equation (7.47) can propagate in the spin background to the left or to the right, leaving behind a free $+$ fracton and resulting in one of the following two configurations:

$$|\psi_2\rangle = \begin{cases} |* \cdots * + - \downarrow \cdots \downarrow + * \cdots *\rangle, \\ |* \cdots * + \uparrow \cdots \uparrow - + * \cdots *\rangle, \end{cases} \tag{7.48}$$

where $* = \uparrow, \downarrow$ such that $N_\uparrow = N_\downarrow$. With either a string of $\downarrow$'s or $\uparrow$'s (upper and lower situation in Eq. (7.48), respectively), further actions of the Hamiltonian enable the isolated fracton in Eq. (7.48) to move through the emission of an additional dipole, which can then propagate in the spin background. This results in configurations of the form:

$$|\psi_3\rangle = \begin{cases} |* \cdots * + - + - \downarrow \cdots \downarrow + * \cdots *\rangle, \\ |* \cdots * + \uparrow \cdots \uparrow - + - + * \cdots *\rangle, \end{cases} \tag{7.49}$$

where $* = \uparrow, \downarrow$ such that $N_\uparrow = N_\downarrow$. Once configurations of the form Eq. (7.49) are generated, a fracton can absorb a dipole when acted upon by the Hamiltonian, as allowed by Eq. (7.14). The resulting configurations are of the form:

$$|\psi_4\rangle = \begin{cases} |* \cdots * + \uparrow - \downarrow \cdots \downarrow + * \cdots *\rangle, \\ |* \cdots * + \uparrow \cdots \uparrow - \downarrow + * \cdots *\rangle, \end{cases} \tag{7.50}$$

where $* = \uparrow, \downarrow$ such that $N_\uparrow = N_\downarrow$. Following the above discussion, one can show that the repeated emission and absorption of multiple dipoles generates product states within the Krylov subspace that are necessarily of the form:

$$|\cdots * + \uparrow \cdots \uparrow - \downarrow \cdots \downarrow + \uparrow \cdots \uparrow - \downarrow \cdots \downarrow + * \cdots\rangle, \tag{7.51}$$

i.e., with strings of only $\uparrow$'s or $\downarrow$'s between consecutive fractons. Given the symmetries of the Hamiltonian, only strings of the form equation (7.51), that have the same (Q, D, S^z) quantum numbers as the root state $|\psi_0\rangle$, are allowed in the Krylov subspace. Hence, this subspace is characterized by the presence of an emergent string-order (equivalently, it is nonlocally constrained).

To illustrate the novel features of this Krylov subspace, we compare the time evolution of the charge density on the middle site (the site on which the fracton resides initially) with that on a different site, which initially hosts

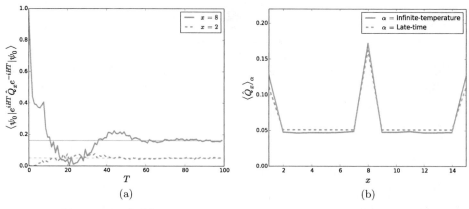

Fig. 7.2.   (Color online) (a) Time-evolution of the expectation value of on-site charge operators for the middle site and a site away from the middle under the pair hopping Hamiltonian with PBC, starting from an initial state of the form $|\psi_0\rangle = |* \cdots * + * \cdots *\rangle$ where $* = \uparrow, \downarrow$ for $N = 15$ with total spin $S^z = 0$. The horizontal lines show the infinite-temperature expectation values of the same charge operators. Data averaged over 10 configurations of the $*$'s such that $S^z = 0$. (b) Late-time charge profile on sites of the chain matches the infinite temperature value *within* the Krylov subspace $\mathcal{K}(H, |\psi_0\rangle)$. They both show a peak on the middle site, providing an example of *quasilocalization from thermalization*.

a spin. The results are shown in Fig. 7.2(a), which compares the charge density at the middle site (in blue) to that at a different site (in green) as a function of time. Irrespective of the spin configuration in the initial state, we consistently find that the middle site exhibits a higher charge density as compared to any other site. Moreover, as shown in Fig. 7.2(b), we find that this late-time charge density *matches* that predicted by ETH, assuming the initial state lies in the middle of the spectrum of the Krylov subspace. The charge density at an inverse temperature β *restricted* to the Krylov subspace $\mathcal{K}$ is then given by

$$\langle \widehat{Q}_{\text{mid}} \rangle_\beta = \frac{\text{Tr}\left(\widehat{Q}_{\text{mid}} e^{-\beta \mathcal{H}_{|\mathcal{K}}}\right)}{\text{Tr}\left(e^{-\beta \mathcal{H}_{|\mathcal{K}}}\right)}, \tag{7.52}$$

where $\mathcal{H}_{|\mathcal{K}}$ is the restriction of the Hamiltonian H to the Krylov subspace $\mathcal{K}$ and $\widehat{Q}_{\text{mid}}$ is the charge operator of the middle site, using the established convention: spins are charge neutral, whereas $+$ and $-$ fractons have charges $+1$ and -1, respectively.

Assuming infinite temperature ($\beta = 0$) in equation (7.52), we obtain

$$\langle \widehat{Q}_{\text{mid}} \rangle_\beta = \frac{\text{Tr}\left(\widehat{Q}_{\text{mid}}\right)}{\text{Tr}\left(\mathbb{1}_{|\mathcal{K}}\right)} \equiv \frac{\mathcal{Q}_N}{\mathcal{D}_N} = \frac{3}{N}, \tag{7.53}$$

where $\mathbb{1}|_{\mathcal{K}}$ is the identity restricted to the Krylov subspace $\mathcal{K}$ and thus $\mathrm{Tr}\,(\mathbb{1}|_{\mathcal{K}}) = \mathcal{D}_N$, the Hilbert space dimension of $\mathcal{K}(H, |\psi_0\rangle)$ of the chain of N sites. We provide analytical and numerical arguments for the result of $3/N$ in Eq. (7.53) in Appendix F (see Eq. (F8)). On the other hand, the latetime expectation value of the charge density on any other site in the middle of the chain is $1/N$. We dub this phenomenon as *quasilocalization* of the fracton, since it is localized for any finite system size although the localization vanishes in the thermodynamic $(N \to \infty)$ limit. We emphasize that unlike usual mechanisms for localization, which rely on the existence of localized eigenstates [21,85,86], the phenomenon here is *quasilocalization from thermalization*, which is a consequence of ergodicity, albeit ergodicity *within* a constrained Krylov subspace.

7.8 Connections with Bloch MBL

Having established some consequences of Krylov fracture, we now discuss the relationship between our model and the Bloch (or Stark) MBL problem [87, 88]. The latter is an interacting extension of the well-known single particle Wannier–Stark localization [97], with the Hamiltonian given by

$$H_{\mathrm{Bloch}} = t \sum_{j=1}^{L-1} \left(c_j^\dagger c_{j+1} + h.c.\right) + E \sum_{j=1}^{L} j \hat{n}_j$$

$$+ V_0 \sum_{j=1}^{L} w_j \hat{n}_j + V_1 \sum_{j=1}^{L-1} \hat{n}_j \hat{n}_{j+1}, \tag{7.54}$$

where $\hat{n}_j = c_j^\dagger c_j$ is the fermionic number operator, t is the hopping strength, w_j is an on-site disorder (w_j random) or curvature ($w_j \sim j^2$) whose strength is set by V_0 and V_1 is the nearest-neighbor repulsion strength. Here, the model is defined on a chain with L sites and with open boundary conditions.

Observe that the term $\sum_j j \hat{n}_j$, representing the uniform electric field, is *precisely* the center-of-mass operator $\widehat{C}$ for OBC, defined in equation (7.4). As detailed in Appendix G, we can perform a Schrieffer–Wolff transformation [106] perturbatively at large E/t for an infinite chain to derive the effective CoM preserving Hamiltonian (see equation (G33)):

$$H_{\mathrm{eff}} = V_0 \sum_j \widetilde{w}_j \hat{n}_j + \widetilde{V}_1 \sum_j \hat{n}_j \hat{n}_{j+1} + \widetilde{V}_2 \sum_j \hat{n}_j \hat{n}_{j+2}$$

$$- \frac{t^2 V_1}{E^2} \sum_j \left(c_j^\dagger c_{j+3}^\dagger c_{j+2} c_{j+1} + h.c.\right) + \mathcal{O}\left(\frac{t^3}{E^3}\right), \tag{7.55}$$

where $\widetilde{w}_j$, $\widetilde{V}_1$, $\widetilde{V}_2$ are defined in equation (G34). $\widetilde{w}_j$ and $\widetilde{V}_1$ are the disorder and nearest-neighbor interaction strengths, respectively "renormalized" by corrections of $\mathcal{O}\left(t^2/E^2\right)$ and $\widetilde{V}_2$ is the effective next-nearest neighbor interaction of $\mathcal{O}\left(t^2/E^2\right)$. Hence, the leading order hopping term in the effective Hamiltonian governing the Wannier–Stark model is the pair-hopping term studied in this paper, given by Eq. (7.1). Longer range center-of-mass preserving terms, including n-body terms for $n > 2$ appear at higher orders in perturbation theory and are therefore suppressed by higher powers of t/E; we thus expect their strength to drop off exponentially with range as $\sim t^n/E^n$, for terms which have support over $\sim n$ sites.

Given this mapping, we now comment briefly on the phenomenon of Bloch MBL, as discussed in Refs. [87, 88]. We begin by noting that the electric field in itself is *not* sufficient to give MBL, since while the electric field "switches off" single particle hopping, it leaves in place the correlated center-of-mass preserving hopping processes discussed above. As we have discussed in the preceding sections, eigenstates of such processes are by no means guaranteed to be localized. Thus, different physics must underlie the numerical observation of MBL in the Bloch MBL problem.

Strictly in the $E/t \to \infty$ limit, the effective Hamiltonian consists only of the nearest-neighbor electrostatic term $V_1 \sum_j \hat{n}_j \hat{n}_{j+1}$ and the onsite potential term $V_0 \sum_j w_j \hat{n}_j$. When $w_j = 0$, i.e., without disorder or curvature, the eigenstates are clearly not localized since the spectrum of $V_1 \sum_j \hat{n}_j \hat{n}_{j+1}$ is highly degenerate. However, that degeneracy is lifted by small disorder or curvature; thus, when w_j is random or $w_j \sim j^2$, all the eigenstates of Eq. (7.55) have low entanglement. This is consistent with the fact that Refs. [87, 88] do not observe MBL without curvature or disorder, respectively.

Moving away from the $E/t \to \infty$ limit, we obtain the effective Hamiltonian of equation (7.55) for large but finite E/t, which exhibits Krylov fracture. The fracture is said to be "strong" [85, 86] if the dimension of the largest Krylov subspace is a vanishing fraction of the full Hilbert space dimension in the thermodynamic limit. This leads to the nonthermalization of generic initial product states with respect to the entire Hilbert space [85, 86], for example the entanglement entropy does not saturate to the maximum value allowed by the full Hilbert space. For a "minimal" center-of-mass preserving Hamiltonian, such as the pair-hopping model of Eq. (7.1), obtained by retaining only the leading order hopping terms in the effective Hamiltonian, strong fracture indeed occurs.[g] A simple example of such nonthermalization is the CDW state $|0101 \cdots 01\rangle$ used as a diagnostic of localization in Ref. [87]. This state forms a one-dimensional Krylov subspace under the pair-hopping Hamiltonian of equation (7.1): it maps onto the state $|\uparrow\uparrow \cdots \uparrow\uparrow\rangle$ under the mapping defined in Sec. 7.3. Clearly, once initialized with this state, the system will forever retain memory of its initial condition under time evolution with the minimal pair-hopping Hamiltonian.

However, we note that the effective Hamiltonian H_{eff} of Eq. (7.55) is a good approximation to the Bloch MBL Hamiltonian H_{Bloch}, given by Eq. (7.54), only for large values of E/t. To test the effectiveness of H_{eff}, we study the quantity

$$\mathcal{O}(T) = \sum_{|\phi_n\rangle \in \mathcal{K}(H_{\text{eff}}, |\psi_0\rangle)} |\langle \phi_n | e^{-iH_{\text{Bloch}}T} |\psi_0\rangle|^2, \tag{7.56}$$

which is the weight of the state $e^{-iH_{\text{Bloch}}T} |\psi_0\rangle$ within the Krylov subspace $\mathcal{K}(H_{\text{eff}}, |\psi_0\rangle)$.[h] We expect H_{eff} to correctly capture the dynamics of H_{Bloch} only for values of E/t when

$$\langle \mathcal{O} \rangle_T \equiv \lim_{\tau \to \infty} \frac{1}{\tau} \int d\tau \, \mathcal{O}(\tau) \approx 1. \tag{7.57}$$

In Fig. 7.3, we show the behavior of $\langle \mathcal{O} \rangle_T$ for the initial state $|\psi_0\rangle = |\uparrow\downarrow\uparrow\downarrow \cdots\rangle = |01100110 \cdots\rangle$.

Thus, we find that H_{eff} is a good approximation for H_{Bloch} only for $E/t \gtrsim 50$ when $V_0, V_1 \sim \mathcal{O}(1)$ and for system sizes up to $L = 14$. In Fig. 7.3, we also find that for a fixed value of E/t, H_{eff} becomes a worse approximation for H_{Bloch} with increasing system size. Thus, it is not clear whether Krylov fracture of the pair-hopping model of Eq. (7.1) plays a significant role in the observations of Refs. [87,88], which focus on the regimes where $E/t \sim \mathcal{O}(10)$.

---

[g]The pair-hopping Hamiltonian equation (7.1) is equivalent to a $S = 1/2$ spin Hamiltonian for which evidence of strong fracture was found in Ref. [85]. We have also verified numerically up to $L = 24$ that the size of the largest Krylov subspace $\sim 2^L$ while the Hilbert space dimension $\sim 4^L$, consistent with strong fracture.

[h]Note that since H_{eff} of equation (7.55) and H of equation (7.1) only differ by diagonal terms, $\mathcal{K}(H_{\text{eff}}, |\psi_0\rangle) = \mathcal{K}(H, |\psi_0\rangle)$.

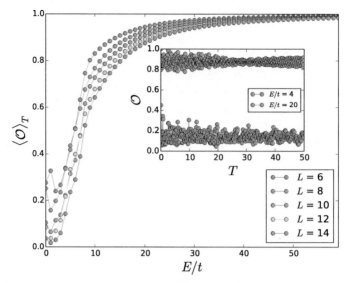

Fig. 7.3. Inset: Weight of the state $e^{-iH_{\text{Bloch}}T}|\psi_0\rangle$ within the Krylov subspace $\mathcal{K}(H_{\text{eff}}, |\psi_0\rangle)$, captured by the quantity $\mathcal{O}(T)$ (defined in equation (7.56)) for two values of the electric field E. Main: Time-average of $\mathcal{O}(T)$, denoted by $\langle\mathcal{O}\rangle_T$ as a function of E/t. $\mathcal{O}$ is close to 1 for larger values of E, justifying that the pair-hopping Hamiltonian H is a good approximation for H_{Bloch}. Data is shown for $V_0/t = 0$, $V_1/t = 1$ and $|\psi_0\rangle = |\uparrow\downarrow\uparrow\downarrow\cdots\rangle = |01100110\cdots\rangle$.

To conclude this section, we speculate on two mechanisms that give rise to localized eigenstates at the smaller values of E/t with disorder, which could provide a partial explanation for the Bloch MBL phenomenon in Refs. [87, 88]: (i) At smaller values of E/t, terms at higher order in perturbation theory cannot be neglected in the effective Hamiltonian. However, since terms generated at all orders in perturbation theory are necessarily center-of-mass preserving, the hopping term of the effective Hamiltonian at any finite order exhibits exponentially many frozen eigenstates [85, 86]. The addition of disorder breaks the exponentially large degeneracy of these frozen states under the effective Hamiltonian, which results in exponentially many product eigenstates of the effective Hamiltonian at any finite order. (ii) When disorder is added in equation (7.54) (as is done in Ref. [88]), then this can give rise to conventional "disorder-induced" MBL [21] within Krylov subspaces of the effective Hamiltonian. This can happen even when the disorder is weak compared to the bare single particle hopping t, because the disorder may be strong compared to the largest hopping term: from equation (7.55), we see that the hopping term is of $\mathcal{O}(t^2 V_1/E^2)$, while the disorder is an $\mathcal{O}(V_0)$ term, which suggests the possibility of conventional MBL in the effective Hamiltonian.

7.9   Conclusions and open questions

In this chapter, we have studied a simple translation invariant model which conserves both charge and center-of-mass and which provides a natural platform for realizing the physics of fractonic systems. Specifically, we find that the pair-hopping model (7.1) exhibits the phenomenon of Krylov fracture, wherein various regions of Hilbert space are dynamically disconnected even if they belong to the same global symmetry sectors. In addition to exponentially many product eigenstates, whose effect on quantum dynamics was studied in Refs. [85, 86], the pair-hopping model also hosts several *large* closed Krylov subspaces with dimensions that grow exponentially in the system size at half-filling.

We find that exponentially many of such large Krylov subspaces admit a mapping onto spin-1/2 XX models of various sizes and hence, constitute examples of *integrable* Krylov subspaces. However, not all large Krylov subspaces show signs of integrability; instead, the model also possesses exponentially many nonintegrable subspaces, many of which show level-repulsion and behavior consistent with ETH. Moreover, some of these Krylov subspaces are highly constrained, which leads to atypical dynamical behavior even within a *thermal* Krylov subspace, an effect we dub "quasilocalization due to thermalization". By this, we specifically mean that the late-time expectation values of local operators within such subspaces deviate from the expected behavior in generic translation invariant systems. Finally, since the pair-hopping model appears as the leading order hopping term in the strong-field limit of the interacting Wannier–Stark problem, we make contact between our work and Bloch MBL. Besides shedding new light on Bloch MBL, our work hence also provides an experimentally relevant setting for studying the dynamics of center-of-mass preserving systems.

Our results, which illustrate the rich structure that can arise as a consequence of Krylov fracture, harbor several implications for the dynamics of isolated quantum systems. Firstly, in the presence of Krylov fracture, we have demonstrated that notions of ergodicity and its violation are well-defined once restricted to large Krylov subspaces. Moreover, we showed that usual diagnostics, such as energy-level statistics, accurately capture whether such Krylov subspaces are integrable or not. These results thus suggest that a modified version of ETH, restricted to large Krylov subspaces, holds for systems with fractured Hilbert spaces.

Secondly, our results provide a clear example of a "semi-integrable" model, i.e., one where integrable as well as nonintegrable exponentially large Krylov subspaces coexist [83, 84]. When viewed from the perspective of the entire Hilbert space (within a particular symmetry sector), the integrable Krylov subspaces are examples of quantum many-body scars, since they are ETH-violating states embedded within the entire many-body spectrum. Unlike the exponentially many static configurations (one-dimensional Krylov subspaces) which necessarily exist for any center-of-mass (dipole moment) conserving Hamiltonian [85,86], these integrable subspaces have an exponentially large dimension, which can lead to nontrivial dynamics in an otherwise nonintegrable model. For the cognoscenti, we note that such subspaces are qualitatively distinct from subspaces generated from states containing a blockaded region [107]. Thus, the existence of such integrable Krylov subspaces of dimension much smaller than that of the full Hilbert space, even if only approximately closed, might be related to quantum many-body scars which by now have been observed in several constrained systems, including the PXP model [48,61].

Additionally, even large nonintegrable subspaces show ergodicity breaking with respect to the entire Hilbert space [85] and instead obey ETH only once restricted to the Krylov subspace, resulting in highly nongeneral *thermal* expectation values of local operators within such Krylov subspaces. Note also that we have only focused on Krylov subspaces generated by root states that are product states (see equation (7.21)), but one could also study closed Krylov subspaces generated by other low-entanglement states; whether this leads to further fracturing within the Krylov subspaces of the pair-hopping model is a question for future work.

On a different note, we described the emergent fractonic behavior of composite degrees of freedom in a simple model, one which can be realized by subjecting fermions hopping on a chain to a strong electric field. It would be interesting to study whether similar emergent behavior appears in higher dimensions. For instance, one can impose the conservation of *quadrupole moment* in two-dimensions, which could be arranged, e.g., by adding strong field-gradients. Such a system could allow one to study the relation, if any, between the dynamics of fracton models [75–77] and Krylov fracture.

Note added: During the completion of this work, there appeared Refs. [108, 109] which also discuss connections between center-of-mass preserving models and the Bloch MBL phenomenon and Ref. [110] which discusses labeling the Krylov subspaces of a related model by nonlocal symmetries. Our results agree wherever there is overlap.

Appendix A: Symmetries of the pair-hopping Hamiltonian in terms of composite degrees of freedom

In this appendix, we discuss some of the symmetries of the pair-hopping Hamiltonian equation (7.1) in terms of the composite degrees of freedom defined in equation (7.9). Similarly to the center-of-mass operator equation (7.4), for PBC the dipole moment operator $\widehat{D}$ in equation (7.17) does not commute with translation by one unit cell (which corresponds to translation by two sites in the original degrees of freedom). To see this, note that under translation $j \to j + 1$,

$$\sum_{j=1}^{N} j\widehat{Q}_j \mapsto \sum_{j=1}^{N} j\widehat{Q}_j + \widehat{Q}, \tag{A1}$$

with $\widehat{Q}$ the total charge operator. This operator obeys nontrivial commutation relations with translations along the chain, since

$$\widehat{T}\widehat{D}\widehat{T}^{-1} = \widehat{D} \exp\left(\frac{2\pi i}{L}\widehat{Q}\right) = \widehat{D} \exp\left(2\pi i \frac{p}{q}\right), \tag{A2}$$

where $\widehat{T}$ is the operator for translation by one unit cell (two sites of the original system) and we are focusing on states with a fixed charge Q such that $Q/N = p/q$. Thus,

$$\left[\widehat{T}^q, \widehat{D}\right] = 0, \tag{A3}$$

within the charge Q sector.

Further, as discussed in Sec. 7.2, the pair-hopping Hamiltonian H is inversion symmetric (i.e., under the exchange of sites j and $L - j + 1$). After grouping sites using Eq. (7.9), the inversion symmetry of H also flips the composite spin degrees of freedom $|\uparrow\rangle \leftrightarrow |\downarrow\rangle$ in addition to interchanging the sites j and $N - j + 1$. For example, when $L = 10$ ($N = 5$), under inversion about the center bond in the third unit cell, the configuration

$$\left| \boxed{01}\,\boxed{10}\,\boxed{11}\,\boxed{01}\,\boxed{00} \right\rangle \to \left| \boxed{00}\,\boxed{10}\,\boxed{11}\,\boxed{01}\,\boxed{10} \right\rangle.$$

In terms of composite degrees of freedom, this corresponds to the transformation

$$|\uparrow\ \downarrow + \uparrow\ -\rangle \to |-\ \downarrow + \uparrow\ \downarrow\rangle,$$

which is the usual inversion about the center site followed by a spin flip. However, inversion also does not commute with translation symmetry (with PBC). Under inversion, a momentum eigenstate with momentum k goes to a state with momentum $-k$. Similarly, under inversion symmetry, note that

$$\sum_{j=1}^{N} j\widehat{Q}_j \to \sum_{j=1}^{N} (N + 1 - j)\,\widehat{Q}_j = (N + 1)\,\widehat{Q} - \sum_{j=1}^{N} j\widehat{Q}_j, \tag{A4}$$

such that the dipole moment operator transforms as

$$\hat{D} \to \begin{cases} (N+1)\,\hat{Q} - \hat{D} & \text{if } OBC, \\ \exp\left(\dfrac{2\pi i Q}{N}\right)\hat{D}^{-1} & \text{if } PBC. \end{cases} \tag{A5}$$

Thus, for PBC, the inversion symmetry can be diagonalized only in sectors with dipole moment D that satisfies:

$$D^2 = \exp\left(\frac{2\pi i Q}{N}\right).$$

Appendix B: Formal mapping of the spin Krylov subspace to the XX model

In this appendix, we show the formal mapping from the spin Krylov subspace in the pair-hopping Hamiltonian equation (7.1) at half-filling to the XX model. We define spin-1/2 raising and lowering operators using the fermionic operators c_j and $c_j^\dagger$,

$$\sigma_j^+ \equiv c_{2j-1}^\dagger c_{2j},$$

$$\sigma_j^- \equiv c_{2j}^\dagger c_{2j-1}. \tag{B1}$$

Using equation (B1), we obtain

$$\{\sigma_j^+, \sigma_j^-\} = \{c_{2j-1}^\dagger c_{2j},\ c_{2j}^\dagger c_{2j-1}\}$$

$$= c_{2j-1}^\dagger c_{2j-1} c_{2j} c_{2j}^\dagger + c_{2j}^\dagger c_{2j} c_{2j-1} c_{2j-1}^\dagger$$

$$= \hat{n}_{2j-1}(1 - \hat{n}_{2j}) + \hat{n}_{2j}(1 - \hat{n}_{2j-1})$$

$$= \hat{n}_{2j-1} + \hat{n}_{2j} - 2\hat{n}_{2j-1}\hat{n}_{2j}. \tag{B2}$$

Since $n_{2j-1}, n_{2j} \in \{0,1\}$, σ_j^+ and σ_j^- are valid Pauli operators only within the subspace of configurations that satisfy

$$n_{2j-1} + n_{2j} = 1, \quad n_{2j-1} n_{2j} = 0 \tag{B3}$$

$$\implies \{\sigma_j^+, \sigma_j^-\} = 1. \tag{B4}$$

The conditions in equation (B4) are only satisfied if the composite degrees of freedom on unit cells j and $j+1$ are $|\!\uparrow\rangle$ or $|\!\downarrow\rangle$ (see equation (7.9)) and hence the mapping from fermions to effective spin degrees of freedom is restricted only to the spin Krylov subspace. First, we rewrite the pair-hopping Hamiltonian equation (7.1) (with PBC) as

$$H = \sum_{j=1}^{N} \left(c_{2j-1}^\dagger c_{2j+2}^\dagger c_{2j+1} c_{2j} + c_{2j}^\dagger c_{2j+3}^\dagger c_{2j+2} c_{2j+1} + h.c. \right)$$

$$= \sum_{j=1}^{N} \left(c_{2j-1}^\dagger c_{2j} c_{2j+2}^\dagger c_{2j+1} + c_{2j}^\dagger c_{2j+3}^\dagger c_{2j+2} c_{2j+1} + h.c. \right). \tag{B5}$$

Given the conditions in Eq. (B3), either $n_{2j+1} = 0$, $n_{2j+2} = 1$ or $n_{2j+1} = 1$, $n_{2j+2} = 0$ for every j for configurations within the spin Krylov subspace. Since the second term of Eq. (B5) contains $c_{2j+2} c_{2j+1}$, it and its Hermitian conjugate always vanish on states within the spin Krylov subspace. Thus, we obtain

$$H_{XX}[N] = \sum_{j=1}^{N} \left(\sigma_j^+ \sigma_{j+1}^- + h.c. \right)$$

$$= \frac{1}{2} \sum_{j=1}^{N} \left(\sigma_j^x \sigma_{j+1}^x + \sigma_j^y \sigma_{j+1}^y \right), \tag{B6}$$

which is the familiar XX model.

The XX model is solved via the Jordan–Wigner transformation [102], which proceeds by defining the operators

$$\sigma_j^+ = (-1)^{\sum_{l<j} d_l^\dagger d_l} d_j^\dagger,$$

$$\sigma_j^- = (-1)^{\sum_{l<j} d_l^\dagger d_l} d_j,$$

$$\sigma_j^z = 2 d_j^\dagger d_j - 1, \tag{B7}$$

where d_j's and $d_j^\dagger$'s are fermionic operators. Using Eq. (B7), the Hamiltonian $H_{XX}[N]$ is mapped onto a noninteracting fermionic hopping Hamiltonian:

$$H_d = \sum_{j=1}^{N} \left(d_j^\dagger d_{j+1} + h.c. \right). \tag{B8}$$

Thus, the many-body ground state is a Fermi sea of the d fermions, with the Fermi momentum $k_F = \pm \pi/2$:

$$|G\rangle = \prod_{k<k_F} d_k^\dagger |0\rangle, \tag{B9}$$

where the vacuum $|0\rangle$ is defined by

$$d_j \, |0\rangle = 0, \quad 1 \le j \le N. \tag{B10}$$

Appendix C: Energies of the integrable Krylov subspaces

We now discuss the energies of the various integrable Krylov subspaces discussed in Sec. 7.5, which map onto XX models of various sizes. The ground-state energies of $H_{XX}[N]$ with PBC (and approximately for OBC) can be written as (see Ref. [111])

$$E_N = \begin{cases} -2 \sin\left(\dfrac{p\pi}{2p+1}\right) \csc\left(\dfrac{\pi}{2p+1}\right) & \text{if } N = 2p+1, \\[4mm] -2 \csc\left(\dfrac{\pi}{2p}\right) & \text{if } N = 2p. \end{cases} \tag{C1}$$

As described in Sec. 7.5, starting with a root state with a single dipole in the spin background (with restrictions discussed in Sec. 7.5.3) results in a Krylov subspace for which the Hamiltonian maps onto an XX model with $(N-1)$ sites. Interestingly, this state is separated by a finite gap from the ground state of the full pair-hopping model, which we numerically observe to be in the spin subspace discussed in Sec. 7.5.1.

Since this finite gap corresponds to the insertion of a dipole, we associate it with the energy of creating a single dipole. Using equation (C1), this *dipole gap* in the thermodynamic limit (where the OBC and PBC spectra are the same) is

$$\Delta E_d = -2 \lim_{p \to \infty} \left(\sin\left(\frac{p\pi}{2p+1}\right) \csc\left(\frac{\pi}{2p+1}\right) - \csc\left(\frac{\pi}{2p}\right) \right)$$

$$= \frac{2}{\pi} \approx 0.64. \tag{C2}$$

The dipole gap also corresponds to the gap between the ground states of the single-dipole and two-dipole Krylov subspaces illustrated in Secs. 7.5.2 and 7.5.3, and more generally, between the ground states of the n dipole and the $(n+1)$ dipole Krylov subspaces illustrated in Sec. 7.5.3. Thus, the pair-hopping model (7.1) exhibits an equally spaced tower of integrable Krylov subspaces.

Appendix D: Mapping to the XX model when dipoles are adjacent to each other

Here, we study the dipole Krylov subspaces generated by root states containing adjacent, identically oriented dipoles. The discussion will focus on OBC throughout this appendix.

D.1   *Multidipole subspace*

We first consider the two-dipole Krylov subspace $\mathcal{K}(H, |\psi_0\rangle)$ generated by the root state

$$|\psi_0\rangle = |\underbrace{* \cdots *}_{N_\uparrow^{(1)}} + - + - \underbrace{* \cdots *}_{N_\uparrow^{(3)}}\rangle, \tag{D1}$$

where $* = \uparrow, \downarrow$ and where $N_\uparrow^{(1)}$ and $N_\uparrow^{(3)}$ represent the number of $\uparrow$'s to the left and right of the dipoles, respectively. Here $N_\uparrow^{(2)} = 0$, where $N_\uparrow^{(2)}$ is the number of $\uparrow$'s between the two dipoles (see equation (7.34)). Under the action of the Hamiltonian, which results in one of the dipoles moving via repeated applications of equation (7.11), we obtain product states of the form

$$|\psi_1\rangle = |\underbrace{* \cdots *}_{N_\uparrow^{(1)}} + - \underbrace{\downarrow \cdots \downarrow}_{d} + - \underbrace{* \cdots *}_{N_\uparrow^{(3)}}\rangle, \tag{D2}$$

within the Krylov subspace, where $* = \uparrow, \downarrow$, d is the number of $\downarrow$ spins in between the dipoles and the total number of $\downarrow$ spins is conserved.

Similarly, applying the Hamiltonian on $|\psi_0\rangle$ and using the rules (7.13) and (7.14), which results in fractons absorbing a dipole, product states of the form

$$|\psi_2\rangle = |\underbrace{* \cdots *}_{N_\uparrow^{(1)}} + \uparrow - \underbrace{* \cdots *}_{N_\uparrow^{(3)}}\rangle \tag{D3}$$

are generated within the Krylov subspace. Due to Eqs. (7.13) and (7.14), the $+$ and $-$ fractons in equation (D3) cannot move without the emission of a dipole (which results in equation (D1)), *all* product states in $\mathcal{K}(H, |\psi_0\rangle)$ are of the form specified by Eqs. (D1), (D2), or (D3).

We now map each product state of the form $|\psi_0\rangle$, $|\psi_1\rangle$, or $|\psi_2\rangle$ onto product states in a spin-1/2 Hilbert space of $(N - 1)$ sites. The mapping proceeds similarly to the multi-dipole case discussed in Sec. 7.5.3, i.e., by identifying the $+-$ dipoles as $\uparrow$'s. However, to account for the action of the Hamiltonian that results in product states of the form $|\psi_2\rangle$ in the Krylov subspace, we introduce a $\downarrow$ spin between the dipoles and map $|\psi_0\rangle$, $|\psi_1\rangle$ and

$|\psi_2\rangle$ according to

$$|* \cdots * + - + - * \cdots *\rangle \iff |* \cdots * \uparrow\downarrow\uparrow * \cdots *\rangle$$

$$\underbrace{\qquad}_{N_\uparrow^{(1)}} \qquad \underbrace{\qquad}_{N_\uparrow^{(3)}} \qquad \underbrace{\qquad}_{N_\uparrow^{(1)}} \qquad \underbrace{\qquad}_{N_\uparrow^{(3)}}$$

$$|* \cdots * + - \downarrow \cdots \downarrow + - * \cdots *\rangle \iff |* \cdots * \uparrow\downarrow \cdots \downarrow\uparrow * \cdots *\rangle$$

$$\underbrace{\qquad}_{N_\uparrow^{(1)}} \underbrace{\quad}_{d} \underbrace{\qquad}_{N_\uparrow^{(3)}} \qquad \underbrace{\qquad}_{N_\uparrow^{(1)}} \underbrace{\quad}_{d+1} \underbrace{\qquad}_{N_\uparrow^{(3)}}$$

$$|* \cdots * + \uparrow - * \cdots *\rangle \iff |* \cdots * \uparrow\uparrow * \cdots *\rangle, \qquad \text{(D4)}$$

$$\underbrace{\qquad}_{N_\uparrow^{(1)}} \qquad \underbrace{\qquad}_{N_\uparrow^{(3)}} \qquad \underbrace{\qquad}_{N_\uparrow^{(1)}} \qquad \underbrace{\qquad}_{N_\uparrow^{(3)}}$$

where the $*$'s remain in the same configurations as in the original configurations.

The reverse mapping from the spin-1/2 Hilbert space is unique provided the quantities $(N_\uparrow^{(1)}, N_\uparrow^{(3)})$ are fixed and it proceeds as follows. In the spin-1/2 configuration, we identify two $\uparrow$ spins such that there are $N_\uparrow^{(1)}$ and $N_\uparrow^{(3)}$ $\uparrow$ spins to the left and right of them, respectively. Since the spin-1/2 configuration has $(N_\uparrow^{(1)} + N_\uparrow^{(3)} + 2)$ $\uparrow$ spins, we are guaranteed that the two chosen $\uparrow$'s only have $\downarrow$'s between them. Depending on the number of $\downarrow$'s between the chosen $\uparrow$'s, we then use equation (D4) to obtain the corresponding configuration in the Krylov subspace $\mathcal{K}(|\psi_0\rangle, H)$.

Through the mapping equation (D4), the action of the Hamiltonian restricted to this Krylov subspace is equivalent to the XX model of $(N-1)$ sites. To see this, note that the action of the terms of the Hamiltonian on the states in equation (D4) can be of three kinds, which can be mapped onto the action on spin degrees of freedom through equation (D4):

$$|\cdots + - \downarrow \cdots\rangle \leftrightarrow |\cdots \downarrow + - \cdots\rangle$$
$$\iff |\cdots \uparrow\downarrow \cdots\rangle \leftrightarrow |\cdots \downarrow\uparrow \cdots\rangle, \qquad \text{(D5)}$$

$$|\cdots + - + - \cdots\rangle \leftrightarrow |\cdots \downarrow + \uparrow - \cdots\rangle$$
$$\iff |\cdots \uparrow\downarrow\uparrow \cdots\rangle \leftrightarrow |\cdots \downarrow\uparrow\uparrow \cdots\rangle, \qquad \text{(D6)}$$

$$|\cdots + - + - \cdots\rangle \leftrightarrow |\cdots + \uparrow - \downarrow \cdots\rangle$$
$$\iff |\cdots \uparrow\downarrow\uparrow \cdots\rangle \leftrightarrow |\cdots \uparrow\uparrow\downarrow \cdots\rangle, \qquad \text{(D7)}$$

which are precisely the actions of the XX model on spin-1/2's. The mapping for two adjacent $-+$ dipoles proceeds analogously, with $\downarrow$ and $\uparrow$ interchanged in Eqs. (D4) and (D5)–(D7).

The preceding discussion for the two-dipole subspace can be extended to the Krylov subspace generated by a root state containing n dipoles with no spins between them ($*$'s are $\uparrow$ or $\downarrow$):

$$|\psi_0\rangle = \left| * \cdots * \boxed{+-+- \cdots +-+-} * \cdots * \right\rangle, \qquad (D8)$$

$$\underbrace{}_{N_\uparrow^{(1)}} \quad \underbrace{\phantom{n \text{ dipoles}}}_{n \text{ dipoles}} \quad \underbrace{}_{N_\uparrow^{(n+1)}}$$

and $N_\uparrow^{(1)}$ and $N_\uparrow^{(n+1)}$ denote the number of $\uparrow$'s to the left and right of the string of dipoles, respectively. We map the root state Eq. (D8) to a spin configuration by identifying each $+-$ dipole by an $\uparrow$ and inserting a $\downarrow$ spin between the dipoles. Thus, we find that $|\psi_0\rangle$ maps onto a spin-1/2 configuration by replacing the n consecutive dipoles by a "Néel state" of $(2n-1)$ spins:

$$|\psi_0\rangle \iff \left| * \cdots * \boxed{\uparrow\downarrow\uparrow \cdots \uparrow\downarrow\uparrow} * \cdots * \right\rangle \qquad (D9)$$

$$\underbrace{}_{N_\uparrow^{(1)}} \quad \underbrace{\phantom{(2n-1) \text{ spins}}}_{(2n-1) \text{ spins}} \quad \underbrace{}_{N_\uparrow^{(n+1)}}$$

We do not attempt to rigorously prove the mapping for arbitrary n, but instead illustrate the mapping for the case when $n = 3$ and provide a conjecture for arbitrary n. When the dipoles interact among themselves according to Eqs. (7.13) and (7.14), the mappings read ($*$'s are $\uparrow$ or $\downarrow$):

$$|* \cdots * +-+-+- * \cdots *\rangle \iff |* \cdots * \uparrow\downarrow\uparrow\downarrow\uparrow * \cdots *\rangle$$

$$|* \cdots * \downarrow +\uparrow -+- * \cdots *\rangle \iff |* \cdots * \downarrow\uparrow\uparrow\downarrow\uparrow * \cdots *\rangle$$

$$|* \cdots * +\uparrow -\downarrow +- * \cdots *\rangle \iff |* \cdots * \uparrow\uparrow\downarrow\downarrow\uparrow * \cdots *\rangle$$

$$|* \cdots * +-\downarrow +\uparrow - * \cdots *\rangle \iff |* \cdots * \uparrow\downarrow\downarrow\uparrow\uparrow * \cdots *\rangle$$

$$|* \cdots * +-+\uparrow -\downarrow * \cdots *\rangle \iff |* \cdots * \uparrow\downarrow\uparrow\uparrow\downarrow * \cdots *\rangle$$

$$|* \cdots * \downarrow +\uparrow\uparrow -\downarrow * \cdots *\rangle \iff |* \cdots * \downarrow\uparrow\uparrow\uparrow\downarrow * \cdots *\rangle$$

$$|* \cdots * +\uparrow -+-\downarrow * \cdots *\rangle \iff |* \cdots * \uparrow\uparrow\downarrow\uparrow\downarrow * \cdots *\rangle$$

$$|* \cdots * +\uparrow\uparrow -\downarrow\downarrow * \cdots *\rangle \iff |* \cdots * \uparrow\uparrow\uparrow\downarrow\downarrow * \cdots *\rangle$$

$$|* \cdots * \downarrow +-+\uparrow - * \cdots *\rangle \iff |* \cdots * \downarrow\uparrow\downarrow\uparrow\uparrow * \cdots *\rangle$$

$$|* \cdots * \downarrow\downarrow +\uparrow\uparrow - * \cdots *\rangle \iff |* \cdots * \downarrow\downarrow\uparrow\uparrow\uparrow * \cdots *\rangle, \qquad (D10)$$

where the quantities $N_\uparrow^{(1)}$ and $N_\uparrow^{(4)}$ (shown in equation (D9)) are conserved in each of the above configurations. Apart from these, applying the Hamiltonian to configurations in Eq. (D10) where dipoles move according to Eq. (D5), we derive the following maps (∗'s are ↑ or ↓):

$$|*\cdots*+-\downarrow\cdots\downarrow+-\downarrow\cdots\downarrow+-*\cdots*\rangle \iff |*\cdots*\uparrow\downarrow\cdots\downarrow\uparrow\downarrow\cdots\downarrow\uparrow*\cdots*\rangle$$
$$\underbrace{\qquad}_{d'}\ \underbrace{\qquad}_{d}\qquad\qquad\qquad\underbrace{\qquad}_{d'+1}\ \underbrace{\qquad}_{d+1}$$

$$|*\cdots*+\uparrow-\downarrow\cdots\downarrow+-*\cdots*\rangle \iff |*\cdots*\uparrow\uparrow\downarrow\cdots\downarrow\uparrow*\cdots*\rangle$$
$$\underbrace{\qquad}_{d}\qquad\qquad\qquad\underbrace{\qquad}_{d+1}$$

$$|*\cdots*+-\downarrow\cdots\downarrow+\uparrow-*\cdots*\rangle \iff |*\cdots*\uparrow\downarrow\cdots\downarrow\uparrow\uparrow*\cdots*\rangle.$$
$$\underbrace{\qquad}_{d}\qquad\qquad\qquad\underbrace{\qquad}_{d+1}$$

$$\text{(D11)}$$

Note that in order to derive the mapping for a particular configuration within this Krylov subspace, one should start from the mapping of the root state in equation (D9) and follow the actions of the Hamiltonian in Eqs. (D5)–(D7). Note that the mappings of Eqs. (D8), (D10) and (D11) are only valid if there are no other dipoles or fractons other than the ones shown. We discuss the case of multiple dipole blocks in the next subsection.

D.2 *Systematic construction of integrable Krylov subspaces*

In the previous section, we conjectured that the Krylov subspace generated by a root state with n contiguous dipoles is integrable and maps onto a particular quantum number sector of an XX model with $(N-1)$ sites and we showed an example for $n = 3$. This mapping can be extended to Krylov subspaces generated by root states containing configurations with m blocks of contiguous $+-$ dipoles, with at least one ↑ separating the blocks.

We start by illustrating the case when $m = 2$. The root state with two blocks of dipoles reads

$$|\psi_0\rangle = |*\cdots*\overbrace{+-\cdots+-}^{n_1\ \text{dipoles}}*\cdots*\overbrace{+-\cdots+-}^{n_2\ \text{dipoles}}*\cdots*\rangle, \qquad \text{(D12)}$$
$$\underbrace{\qquad}_{N_\uparrow^{(1)}}\qquad\underbrace{\qquad}_{N_\uparrow^{(n_1+1)}}\qquad\underbrace{\qquad}_{N_\uparrow^{(n_1+n_2+2)}}$$

where $* = \uparrow, \downarrow$, $N_\uparrow^{(j)}$ denotes the number of ↑ spins in the jth segment of the chain, which is the part of the chain between the jth and $(j+1)$th $+-$ dipole. In Eq. (D12), $N_\uparrow^{(j)} = 0$ if $2 \leq j \leq n_1$ or $n_1 + 2 \leq j \leq n_1 + n_2 + 1$ and we are considering the case where $N_\uparrow^{(n_1+1)} \geq 1$. The mapping from $|\psi_0\rangle$ onto a configuration of spin-1/2's proceeds as follows. The two groups of n_1 and n_2 dipoles are mapped onto Néel states of $(2n_1 - 1)$ and $(2n_2 - 1)$ spins, respectively. Since the Hamiltonian acts on the state $|\psi_0\rangle$ in equation (D12) according to Eqs. (D5)–(D7), the Hamiltonian H restricted to this Krylov subspace is the XX model of size $(N-2)$. The full dictionary of mappings to the spin-1/2 Hilbert space can be derived by starting from the mapping for the root configuration and following the actions of the Hamiltonian in Eqs. (D5)–(D7).

This mapping directly generalizes to a root state with m dipole groups with the form

$$|\psi_0\rangle = |\ast \cdots \ast \overbrace{+-\cdots+-}^{n_1 \text{ dipoles}} \underset{N_\uparrow^{(1)}}{\underbrace{}} \diamond \cdots \cdots \diamond \overbrace{+-\cdots+-}^{n_m \text{ dipoles}} \underset{N_\uparrow^{(\sum\limits_{k=1}^m n_k + m)}}{\underbrace{}} \ast \cdots \ast\rangle, \qquad (D13)$$

where $\ast = \uparrow, \downarrow$ and $\diamond \cdots \diamond$ consists of spins and $(m-2)$ blocks of $+-\cdots+-$, with two adjacent blocks separated by at least one $\uparrow$. The mapping to the spin-1/2 chain proceeds by mapping each sequence of n_l adjacent dipoles in equation (D13) onto a Néel state of $(2n_l - 1)$ spins, as depicted in equation (D9). Since the Hamiltonian acts on this subspace according to Eqs. (D5)–(D7), the Hamiltonian restricted to this Krylov subspace is the XX model with $(N-m)$ sites. In general, the Krylov subspace consisting of n $+-$ (respectively, $-+$) dipoles with m values of j such that $N_\uparrow^{(j)} = 0$ (respectively, $N_\downarrow^{(j)} = 0$), precisely maps onto the XX model with $(N-n+m)$ sites.

Appendix E: Effect of electrostatic terms and disorder

E.1 *Electrostatic terms*

We now briefly discuss the effect of adding electrostatic terms to the analysis of the integrable Krylov subspaces discussed in Sec. 7.5. In particular, we consider two simple perturbations to the Hamiltonian

$$\delta H_1 = V_1 \sum_{j=1}^{L_b} \hat{n}_j \hat{n}_{j+1}, \, \delta H_2 = V_2 \sum_{j=1}^{L_b'} \hat{n}_j \hat{n}_{j+2}, \qquad (E1)$$

where $L_b = L-1$ (respectively, $L_b = L$) and $L_b' = L-2$ (respectively, $L_b' = L$) for OBC (respectively, PBC). The terms in equation (E1) are the simplest two electrostatic terms. In experimentally relevant settings, these terms typically have strengths greater than or comparable to that of the pair-hopping Hamiltonian H; see Eq. (7.8) and Sec. 7.2 for a discussion of their sizes.

The electrostatic terms are nearest-neighbor terms that are diagonal in the basis of product states of the composite degrees of freedom, i.e., of the spins and fractons defined in Eq. (7.9). Since composite degrees of freedom are formed by grouping pairs of neighboring sites, in terms of composite degrees of freedom the electrostatic Hamiltonians δH_1 and δH_2 in Eq. (E1) map onto nearest-neighbor Hamiltonians $\delta \mathcal{H}_1$ and $\delta \mathcal{H}_2$, which have the forms:

$$\delta \mathcal{H}_1 = \sum_{j=1}^{N_b} (\delta \mathcal{H}_1)_{j,j+1} + \sum_{j=1}^{N} (\delta \mathcal{H}_1)_j, \delta \mathcal{H}_2 = \sum_{j=1}^{N_b} (\delta \mathcal{H}_2)_{j,j+1}, \qquad \text{(E2)}$$

where $N_b = N$ (respectively, $N_b = N - 1$) for PBC (respectively, OBC) and $\{(\delta \mathcal{H}_\alpha)_{j,j+1}\}$ and $\{(\delta \mathcal{H}_\alpha)_j\}$ are nearest-neighbor and onsite terms, respectively. The actions of each of the nearest-neighbor terms follow directly by using Eq. (E1) and the definitions Eq. (7.9) and can be tabulated as:

Config.	$\delta\mathcal{H}_1$	$\delta\mathcal{H}_2$	Config.	$\delta\mathcal{H}_1$	$\delta\mathcal{H}_2$
$\lvert ++\rangle$	V_1	$2V_2$	$\lvert \downarrow +\rangle$	0	V_2
$\lvert + \uparrow\rangle$	0	V_2	$\lvert \downarrow\uparrow\rangle$	0	0
$\lvert + \downarrow\rangle$	V_1	V_2	$\lvert \downarrow\downarrow\rangle$	0	V_2
$\lvert +-\rangle$	0	0	$\lvert \downarrow -\rangle$	0	0
$\lvert \uparrow +\rangle$	V_1	V_2	$\lvert -+\rangle$	0	0
$\lvert \uparrow\uparrow\rangle$	0	V_2	$\lvert - \uparrow\rangle$	0	0
$\lvert \uparrow\downarrow\rangle$	V_1	0	$\lvert - \downarrow\rangle$	0	0
$\lvert \uparrow -\rangle$	0	0	$\lvert --\rangle$	0	0

$$\text{(E3)}$$

whereas the onsite terms read:

Config.	$\delta\mathcal{H}_1$	Config.	$\delta\mathcal{H}_1$
$\lvert +\rangle$	V_1	$\lvert \uparrow\rangle$	0
$\lvert -\rangle$	0	$\lvert \downarrow\rangle$	0

$$\text{(E4)}$$

Importantly, since these terms are diagonal in the product basis, they do not change the structure of Krylov subspaces generated from product states. In other words, the full Hilbert space is still expressed in the same form as Eq. (7.21) irrespective of whether H contains electrostatic terms or not.

Within the integrable spin subspace discussed in Sec. 7.5.1, the onsite terms always vanish according to Eq. (E4). Further, according to Eq. (E3), the actions of the nearest-neighbor terms of $\delta\mathcal{H}_1$ and $\delta\mathcal{H}_2$ read

$$
\begin{aligned}
(\delta\mathcal{H}_1 + \delta\mathcal{H}_2)_{j,j+1} \left|\downarrow\uparrow\right\rangle &= 0, \\
(\delta\mathcal{H}_1 + \delta\mathcal{H}_2)_{j,j+1} \left|\uparrow\downarrow\right\rangle &= V_1 \left|\uparrow\downarrow\right\rangle, \\
(\delta\mathcal{H}_1 + \delta\mathcal{H}_2)_{j,j+1} \left|\uparrow\uparrow\right\rangle &= V_2 \left|\uparrow\uparrow\right\rangle, \\
(\delta\mathcal{H}_1 + \delta\mathcal{H}_2)_{j,j+1} \left|\downarrow\downarrow\right\rangle &= V_2 \left|\downarrow\downarrow\right\rangle.
\end{aligned}
\tag{E5}
$$

The above actions of the electrostatic terms are succinctly encoded in the Hamiltonian $\delta\mathcal{H} = \delta\mathcal{H}_1 + \delta\mathcal{H}_2$

$$
\begin{aligned}
\delta\mathcal{H} &= \sum_{j=1}^{N_b} \left(\frac{V_2}{2} \left(1 + \sigma_j^z \sigma_{j+1}^z\right) + \frac{V_1}{4} \left(1 + \sigma_j^z\right)\left(1 - \sigma_{j+1}^z\right) \right) \\
&= \sum_{j=1}^{N_b} \left(\frac{V_1 + 2V_2}{4} + \frac{2V_2 - V_1}{4} \sigma_j^z \sigma_{j+1}^z \right) + \frac{V_1}{4} \sum_{j=1}^{N_b} (\sigma_j^z - \sigma_{j+1}^z),
\end{aligned}
\tag{E6}
$$

where the unit cell index j is defined modulo N for PBC. Thus, equation (E6) reduces to

$$
\delta\mathcal{H} = \sum_{j=1}^{N_b} \left(\frac{V_1 + 2V_2}{4} + \frac{2V_2 - V_1}{4} \sigma_j^z \sigma_{j+1}^z \right) +
\begin{cases}
\frac{V_1}{4} (\sigma_1^z - \sigma_N^z) & \text{if } OBC, \\
0 & \text{if } PBC.
\end{cases}
\tag{E7}
$$

Thus, the restriction of the total Hamiltonian—the pair-hopping Hamiltonian in addition to the electrostatic terms—to the spin Krylov subspace maps onto (for PBC and infinite chain for OBC)

$$
H_T = \sum_j \left(\frac{V_1 + 2V_2}{4} + \sigma_j^+ \sigma_{j+1}^- + \sigma_j^- \sigma_{j+1}^+ + \frac{2V_2 - V_1}{4} \sigma_j^z \sigma_{j+1}^z \right),
\tag{E8}
$$

which is the translation invariant **XXZ** model and is thus Bethe Ansatz integrable.

In contrast, the integrable dipole subspaces discussed in Secs. 7.5.2 and 7.5.3 become nonintegrable upon the addition of electrostatic terms. To see this, consider the action of the nearest-neighbor terms of $(\delta\mathcal{H}_1 + \delta\mathcal{H}_2)$

on the dipole, which are given by (using equation (E3))

$$(\delta\mathcal{H}_1 + \delta\mathcal{H}_2)\,|\uparrow +-\rangle = (V_1 + V_2)\,|\uparrow +-\rangle,$$

$$(\delta\mathcal{H}_1 + \delta\mathcal{H}_2)\,|\downarrow +-\rangle = V_2\,|\downarrow +-\rangle,$$

$$(\delta\mathcal{H}_1 + \delta\mathcal{H}_2)\,|+- \uparrow\rangle = 0,$$

$$(\delta\mathcal{H}_1 + \delta\mathcal{H}_2)\,|+- \downarrow\rangle = 0. \tag{E9}$$

When $V_1 \neq 0$, the actions encoded in equation (E9) break the symmetry between the configurations $|+ - *\rangle$ and $|* + -\rangle$, where $* = \uparrow, \downarrow$. Thus the dipole *cannot* be identified with an $\uparrow$, as is the case in the absence of electrostatic terms. We have verified that upon addition of electrostatic terms, the energy levels within any quantum number sector of the dipole subspace show GOE-level statistics. The same is true for Krylov subspaces with $-+$ dipoles, for which the action of the electrostatic terms follows from equation (E9) upon the application of inversion symmetry.

E.2   *Disorder*

Consider the disordered pair-hopping Hamiltonian,

$$H = \sum_{j=1}^{L_b} H_j = \sum_{j=1}^{L_b} J_j \left(c_j^\dagger c_{j+3}^\dagger c_{j+2} c_{j+1} + h.c. \right), \tag{E10}$$

where $L_b = L - 3$ (respectively, $L_b = L$) for OBC (respectively, PBC) and $\{J_j\}$ are the disordered couplings. Assuming $L = 2N$, we divide the Hamiltonian equation (E10) into two parts to preempt the mapping onto composite degrees of freedom, defined in equation (7.9):

$$H = \sum_{j=1}^{N_b^{(o)}} J_{2j-1} \left(c_{2j-1}^\dagger c_{2j+2}^\dagger c_{2j+1} c_{2j} + h.c. \right)$$

$$+ \sum_{j=1}^{N_b^{(e)}} J_{2j} \left(c_{2j}^\dagger c_{2j+3}^\dagger c_{2j+2} c_{2j+1} + h.c. \right), \tag{E11}$$

where $N_b^{(o)} = N-1$ (respectively, $N_b^{(o)} = N$) and $N_b^{(e)} = N-2$ (respectively, $N_b^{(e)} = N$) for OBC (respectively, PBC). Once the sites $(2j-1)$ and $2j$ are grouped into one unit cell, the actions of the Hamiltonian terms are as follows (see Eqs. (7.10)–(7.14)):

$$
\left|\ \boxed{\begin{smallmatrix}2j & 2j+1\\ 0\ 1\end{smallmatrix}}\ \boxed{1\ 0}\ \right\rangle \xleftrightarrow{J_{2j-1}} \left|\ \boxed{\begin{smallmatrix}2j & 2j+1\\ 1\ 0\end{smallmatrix}}\ \boxed{0\ 1}\ \right\rangle
$$

$$
\Longleftrightarrow |\uparrow\downarrow\rangle \xleftrightarrow{J_{2j-1}} |\downarrow\uparrow\rangle, \tag{E12}
$$

$$
\left|\ \boxed{\begin{smallmatrix}2j\\ 1\ 0\end{smallmatrix}}\ \boxed{\begin{smallmatrix}2j+3\\ 1\ 1\end{smallmatrix}}\ \boxed{0\ 0}\ \right\rangle \xleftrightarrow{J_{2j}} \left|\ \boxed{\begin{smallmatrix}2j\\ 1\ 1\end{smallmatrix}}\ \boxed{0\ 0}\ \boxed{\begin{smallmatrix}2j+3\\ 1\ 0\end{smallmatrix}}\ \right\rangle
$$

$$
\Longleftrightarrow |\downarrow +-\rangle \xleftrightarrow{J_{2j}} |+-\downarrow\rangle, \tag{E13}
$$

$$
\left|\ \boxed{\begin{smallmatrix}2j\\ 0\ 0\end{smallmatrix}}\ \boxed{\begin{smallmatrix}2j+3\\ 1\ 1\end{smallmatrix}}\ \boxed{0\ 1}\ \right\rangle \xleftrightarrow{J_{2j}} \left|\ \boxed{\begin{smallmatrix}2j\\ 0\ 1\end{smallmatrix}}\ \boxed{0\ 0}\ \boxed{\begin{smallmatrix}2j+3\\ 1\ 1\end{smallmatrix}}\ \right\rangle
$$

$$
\Longleftrightarrow |-+\uparrow\rangle \xleftrightarrow{J_{2j}} |\uparrow -+\rangle, \tag{E14}
$$

$$
\left|\ \boxed{\begin{smallmatrix}2j\\ 1\ 0\end{smallmatrix}}\ \boxed{\begin{smallmatrix}2j+3\\ 1\ 1\end{smallmatrix}}\ \boxed{0\ 1}\ \right\rangle \xleftrightarrow{J_{2j}} \left|\ \boxed{\begin{smallmatrix}2j\\ 1\ 1\end{smallmatrix}}\ \boxed{0\ 0}\ \boxed{\begin{smallmatrix}2j+3\\ 1\ 1\end{smallmatrix}}\ \right\rangle
$$

$$
\Longleftrightarrow |\downarrow +\uparrow\rangle \xleftrightarrow{J_{2j}} |+-+\rangle, \tag{E15}
$$

$$
\left|\ \boxed{\begin{smallmatrix}2j\\ 0\ 1\end{smallmatrix}}\ \boxed{\begin{smallmatrix}2j+3\\ 0\ 0\end{smallmatrix}}\ \boxed{1\ 0}\ \right\rangle \xleftrightarrow{J_{2j}} \left|\ \boxed{\begin{smallmatrix}2j\\ 0\ 0\end{smallmatrix}}\ \boxed{1\ 1}\ \boxed{\begin{smallmatrix}2j+3\\ 0\ 0\end{smallmatrix}}\ \right\rangle
$$

$$
\Longleftrightarrow |\uparrow -\downarrow\rangle \xleftrightarrow{J_{2j}} |-+-\rangle, \tag{E16}
$$

where $|a\rangle \xleftrightarrow{J} |b\rangle$ denotes that the action of a term of the Hamiltonian on the configuration $|a\rangle$ results in $|b\rangle$ with a coefficient J and vice versa.

Since the spin Krylov subspace discussed in Sec. 7.5.1 is only sensitive to the action of the Hamiltonian on the spin degrees of freedom, according to equation (E12) the Hamiltonian restricted to the Krylov subspace maps onto the disordered XX model:

$$
H = \sum_{j=1}^{N_b} J_{2j-1}\left(\sigma_j^+ \sigma_{j+1}^- + \sigma_j^- \sigma_{j+1}^+\right), \tag{E17}
$$

where $N_b = N-1$ (respectively, $N_b = N$) for OBC (respectively, PBC). Thus, we expect that the spin Krylov subspace exhibits Anderson localization [16] upon the addition of disorder.

We now analyze the effect of disorder on the single- or multi-dipole Krylov subspaces. Recall that the Hamiltonian restricted to the dipole Krylov subspaces in Secs. 7.5.2 and 7.5.3 maps onto the XX model by identifying the $+-$ (respectively, $-+$) dipole with an $\uparrow$ (respectively, $\downarrow$) and noting that Eq. (7.11) (respectively, Eq. (7.12)) is identical to Eq. (7.10) upon this identification. However, if $J_{2j-1} \neq J_{2j}$, Eq. (E13) (respectively, Eq. (E14)) is *no longer* identical to Eq. (E12) when $+-$ (respectively, $-+$) is identified with $\uparrow$ (respectively, $\downarrow$). Hence, the Hamiltonian restricted to dipole Krylov subspaces does not map onto the disordered XX model as one would naively expect.

Appendix F: Properties of the fracton Krylov subspace

Here, we discuss some properties of the fracton Krylov subspace discussed in Sec. 7.6. To understand the effects of this constrained Krylov subspace, we focus on odd system sizes N and on the root state consisting of a $+$ fracton on the center site $(N+1)/2$ along with an equal number of $\uparrow$ and $\downarrow$ spins enveloping it, as shown in Eq. (7.46). Such a configuration has charge $Q = 1$, spin $S^z = 0$ and dipole moment $D = \exp(i\pi(N+1)/N)$ (see Eq. (7.17) for the definition of dipole moment with PBC). We refer to the site containing the fracton in the root state as the *middle site*.

For purposes of illustration, we consider the root state $|\psi_0\rangle = |\uparrow\uparrow\uparrow + \downarrow\downarrow\downarrow\rangle$ with $N = 7$ and with PBC, which has charge $Q = 1$, spin $S = 0$ and dipole moment $D = e^{8\pi i/7}$. Using the actions of Eqs. (7.10)-(7.14), the product state configurations in $\mathcal{K}(H, |\psi_0\rangle)$ are then

$$|\uparrow\uparrow\uparrow + \downarrow\downarrow\downarrow\rangle,$$

$$|\downarrow\uparrow\uparrow + \downarrow\downarrow\uparrow\rangle,$$

$$|\uparrow\downarrow\uparrow + \downarrow\downarrow\uparrow\rangle, |\downarrow\downarrow\uparrow\uparrow + \downarrow\uparrow\downarrow\rangle,$$

$$|\uparrow\uparrow\downarrow + \downarrow\downarrow\uparrow\rangle, |\uparrow\downarrow\uparrow + \downarrow\uparrow\downarrow\rangle, |\downarrow\uparrow\uparrow + \uparrow\downarrow\downarrow\rangle$$

$$|\uparrow\uparrow\downarrow + \downarrow\uparrow\downarrow\rangle, |\uparrow\downarrow\uparrow + \uparrow\downarrow\downarrow\rangle, |\downarrow\downarrow\uparrow + \downarrow\uparrow\uparrow\rangle$$

$$|\uparrow\uparrow\downarrow + \uparrow\downarrow\downarrow\rangle, |\downarrow\uparrow\downarrow + \uparrow\downarrow\uparrow\rangle, |\downarrow\downarrow\uparrow + \uparrow\downarrow\uparrow\rangle, |\downarrow\uparrow\downarrow + \downarrow\uparrow\uparrow\rangle$$

$$|\uparrow\uparrow + - + \downarrow\downarrow\rangle, |\downarrow\uparrow + - + \downarrow\uparrow\rangle, |\uparrow\downarrow\downarrow + \downarrow\uparrow\uparrow\rangle, |\uparrow\downarrow\downarrow + \uparrow\downarrow\uparrow\rangle,$$

$$|\downarrow\downarrow\uparrow + \uparrow\uparrow\downarrow\rangle, |\downarrow\uparrow\downarrow + \uparrow\uparrow\downarrow\rangle$$

$$|\uparrow\downarrow + - + \downarrow\uparrow\rangle, |\downarrow\uparrow + - + \uparrow\downarrow\rangle, |\uparrow\downarrow\downarrow + \uparrow\uparrow\downarrow\rangle$$

$$|\uparrow + -\downarrow + \downarrow\uparrow\rangle, |\uparrow\downarrow + - + \uparrow\downarrow\rangle, |\downarrow\uparrow + \uparrow - + \downarrow\rangle, |\downarrow\downarrow\downarrow + \uparrow\uparrow\uparrow\rangle$$

$$|\downarrow\downarrow + - + \uparrow\uparrow\rangle, |\uparrow + -\downarrow + \uparrow\downarrow\rangle, |\uparrow\downarrow + \uparrow - + \downarrow\rangle$$

$$|\uparrow + - + - + \downarrow\rangle, |\downarrow\downarrow + \uparrow - + \uparrow\rangle, |\downarrow + - \downarrow + \uparrow\uparrow\rangle$$

$$|\downarrow + - + - + \uparrow\rangle, |+ - \downarrow\downarrow + \uparrow\uparrow\rangle, |\downarrow\downarrow + \uparrow\uparrow - +\rangle, |\uparrow + \uparrow - \downarrow + \downarrow\rangle$$

$$|+-\downarrow+-+\uparrow\rangle, |\downarrow+-+\uparrow-+\rangle, |\downarrow+\uparrow-\downarrow+\uparrow\rangle,$$

$$|+-+-\downarrow+\uparrow\rangle, |\downarrow+\uparrow-+-+\rangle, |+-\downarrow+\uparrow-+\rangle$$

$$|+\uparrow-\downarrow\downarrow+\uparrow\rangle, |+-+-+-+\rangle, |\downarrow+\uparrow\uparrow-\downarrow+\rangle$$

$$|+-+\uparrow-\downarrow+\rangle, |+\uparrow-\downarrow+-+\rangle, |+\uparrow-+-\downarrow+\rangle, |+\uparrow\uparrow-\downarrow\downarrow+\rangle,$$

$$\text{(F1)}$$

where configurations on the nth row are product configurations belonging to $\text{span}\{|\psi_0\rangle, H|\psi_0\rangle, \ldots, H^{n-1}|\psi_0\rangle\}$ but not to $\text{span}\{|\psi_0\rangle, H|\psi_0\rangle, \ldots, H^{n-2}|\psi_0\rangle\}$. That is, they are the new product configurations obtained on the $(n-1)$th action of the Hamiltonian H on $|\psi_0\rangle$.

As discussed in Sec. 7.6 (see Eq. (7.53)), in order to obtain the infinite temperature expectation value of the charge on the middle site *within* the Krylov subspace, we must compute the Hilbert space dimension $\mathcal{D}_N$ of the Krylov subspace as well as $\mathcal{Q}_N$, the difference between the number of product states with a + fracton and with a − fracton on the middle site. These quantities can be enumerated numerically for various system sizes and are tabulated in the following:

N	$\mathcal{D}_N$	$\mathcal{Q}_N$
3	3	1
5	12	1
7	50	14
9	210	15
11	882	56
13	3696	210
15	15444	792
17	64350	3003

$$\text{(F2)}$$

For example, the total number of configurations in equation (F1) ($N = 7$) is 50 and it can be explicitly verified that the total middle site charge summed over all configurations equals 14.

We now compute the infinite temperature expectation value of the middle site charge. We find that the Hilbert space dimension $\mathcal{D}_N$ for odd system sizes, tabulated in equation (F2) for $N \le 17$, corresponds to the integer sequence OEIS A092443 [112], which takes the standard form

$$\mathcal{D}_{N=2n+1} = \frac{n+2}{2}\binom{2n}{n}.$$

$$\text{(F3)}$$

Similarly, we find that $\mathcal{Q}_N$, tabulated in Eq. (F2) for $N \leq 17$, corresponds to the integer sequence OEIS A051924 [112], which has the closed form

$$\mathcal{Q}_{N=2n+1} = \frac{3n-2}{n}\binom{2(n-1)}{n-1}. \tag{F4}$$

Although we do not attempt to prove this rigorously here, we posit that Eqs. (F3) and (F4) accurately represent $\mathcal{D}_{N=2n+1}$ and $\mathcal{Q}_{N=2n+1}$ for all values of n. With these expressions in hand, we can then analytically obtain the infinite temperature charge density from the ratio $\mathcal{Q}_N/\mathcal{D}_N$.

To find the asymptotic behavior for large N, we use Stirling's approximation

$$n! \approx \sqrt{2\pi n}\left(\frac{n}{e}\right)^n. \tag{F5}$$

The asymptotic behavior of $\mathcal{D}_N$ of equation (F3) is then

$$\mathcal{D}_{N=2n+1} = \frac{n+2}{2}\frac{(2n)!}{(n!)^2} \approx \frac{n+2}{2\sqrt{\pi n}}\frac{\left(\frac{2n}{e}\right)^{2n}}{\left(\frac{n}{e}\right)^{2n}}$$

$$\sim \sqrt{\frac{n}{4\pi}}2^{2n} \sim \sqrt{\frac{N}{32\pi}}2^N, \tag{F6}$$

whereas $\mathcal{Q}_N$ asymptotes to

$$\mathcal{Q}_{N=2n+1} = \frac{3n-2}{n}\frac{(2(n-1))!}{((n-1)!)^2}$$

$$\approx \frac{(3n-2)}{n\sqrt{\pi(n-1)}}\frac{\left(\frac{2(n-1)}{e}\right)^{2(n-1)}}{\left(\frac{n-1}{e}\right)^{2(n-1)}}$$

$$\sim \frac{3}{\sqrt{16\pi n}}2^{2n} \sim \frac{3}{\sqrt{32\pi N}}2^N. \tag{F7}$$

Hence, for large N, the infinite temperature expectation value of the charge on the middle site is given by

$$\frac{Q_N}{\mathcal{D}_N} \sim \frac{3}{N}. \tag{F8}$$

Appendix G: Schrieffer–Wolff transformation for the Bloch MBL Hamiltonian

In this appendix, we explicitly derive the pair-hopping Hamiltonian equation (7.1) in the large E/t limit of the Bloch MBL Hamiltonian equation (7.54):

$$H_{\text{Bloch}} = t\sum_j (c_j^\dagger c_{j+1} + h.c.) + E\sum_j j\hat{n}_j + V_0\sum_j \hat{n}_j + V_1\sum_j w_j\hat{n}_j\hat{n}_{j+1},$$
(G1)

where we have omitted the limits on the sums since we consider a chain of infinite length. Furthermore, we treat t, V_0, V_1 perturbatively and hence rescale equation (7.54) by E, such that the Hamiltonian is recast as

$$H \equiv \frac{H_{\text{Bloch}}}{E} = \hat{C} + \lambda(\hat{T}_+ + \hat{T}_- + \hat{V}),$$
(G2)

where $\hat{C}$ is the CoM operator (for OBC)

$$\hat{C} = \sum_j jn_j,$$
(G3)

$\hat{T} = \hat{T}_+ + \hat{T}_-$ and $\hat{V} = \hat{V}_0 + \hat{V}_1$, with

$$\hat{T}_+ = \sum_j c_{j+1}^\dagger c_j, \quad \hat{T}_- = \sum_j c_j^\dagger c_{j+1} = \hat{T}_+^\dagger,$$

$$\hat{V}_0 = \alpha_0\sum_j w_j\hat{n}_j, \quad \hat{V}_1 = \alpha_1\sum_j \hat{n}_j\hat{n}_{j+1}.$$
(G4)

Here, the parameters are defined as

$$\lambda = \frac{t}{E}, \quad \alpha_\nu = \frac{V_\nu}{t},$$
(G5)

for $\nu \in \{0, 1\}$ and where we work in the regime where $\alpha_\sigma \sim \mathcal{O}(1)$.

As is clear from Eq. (G4), $\hat{T}_+$ and $\hat{T}_-$ correspond to hopping processes that increase and decrease energies by one unit with respect to the CoM term $\hat{C}$. That is,

$$\hat{C}|\mu\rangle = \mathcal{E}_\mu|\mu\rangle \implies \hat{C}(\hat{T}_\pm|\mu\rangle) = (\mathcal{E}_\mu \pm 1)\hat{T}_\pm|\mu\rangle.$$
(G6)

Following the standard Schrieffer–Wolff procedure [106], we divide the Hilbert space into "blocks", which are subspaces degenerate under the leading order term $\widehat{C}$. Terms of the Hamiltonian which only have nonvanishing matrix elements within the same block are called "block diagonal" whereas terms which only have nonvanishing matrix elements between different blocks are called "block off-diagonal". For the Hamiltonian H, the "block diagonal" and "block off-diagonal" parts H_d and H_{od} respectively read

$$H = \underbrace{\widehat{C} + \lambda\widehat{V}}_{H_d} + \underbrace{\lambda\widehat{T}}_{H_{od}}. \tag{G7}$$

Next, we wish to perturbatively find a unitary transformation such that the resultant Hamiltonian has no "block off-diagonal" parts:

$$H_{\text{eff}} = e^{\lambda S} H e^{-\lambda S}, \tag{G8}$$

where S is anti-Hermitian. Here, each block diagonal subspace of H_{eff} corresponds to a subspace degenerate under $\widehat{C}$—since $\widehat{C}$ is the center-of-mass operator with OBC (see equation (7.4)), different block diagonal parts of H_{eff} correspond to subspaces labeled by distinct center-of-mass quantum numbers. The effective Hamiltonian can be expressed as

$$H_{\text{eff}} = e^{\lambda S} H e^{-\lambda S} = \sum_{n=0}^{\infty} \lambda^n H_{\text{eff}}^{(n)}, \tag{G9}$$

where $H_{\text{eff}}^{(n)}$ is the effective Hamiltonian in nth-order perturbation theory. In what follows, we show that the pair-hopping term arises in $H_{\text{eff}}^{(3)}$, i.e., in the effective Hamiltonian restricted to one center-of-mass sector of the Bloch MBL Hamiltonian.

We now derive the expression for H_{eff} up to third order in perturbation theory. Expanding H_{eff}, defined in Eq. (G8), in powers of λ, we obtain

$$H_{\text{eff}} = H + \lambda\,[S, H] + \frac{\lambda^2}{2}\,[S, [S, H]] + \frac{\lambda^3}{6}\,[S, [S, [S, H]]] + \mathcal{O}\left(\lambda^4\right). \tag{G10}$$

We also expand S in powers of λ as

$$S = S_0 + \lambda S_1 + \lambda^2 S_2 + \mathcal{O}\left(\lambda^3\right). \tag{G11}$$

Using Eqs. (G7) and (G10), we obtain

$$H_{\text{eff}} = \widehat{C} + \lambda\{\widehat{V} + \widehat{T} + [S_0, \widehat{C}]\}$$

$$+ \lambda^2\left\{\frac{1}{2}[S_0, [S_0, \ \widehat{C}]] + [S_0, \ \widehat{V} + \widehat{T}] + [S_1, \widehat{C}]\right\}$$

$$+ \lambda^3\left\{\frac{1}{6}[S_0, [S_0, [S_0, \widehat{C}]]]\right.$$

$$+ \frac{1}{2}([S_1, [S_0, \widehat{C}]] + [S_0, [S_1, \widehat{C}]] + [S_0, [S_0, \widehat{V} + \widehat{T}]])$$

$$\left. + [S_1, \widehat{V} + \widehat{T}] + [S_2, \widehat{C}]\right\} + \mathcal{O}(\lambda^4). \tag{G12}$$

Since $\widehat{V}$ is diagonal, to cancel the block off-diagonal component $\widehat{T}$ at $\mathcal{O}(\lambda)$ in equation (G12) we require that S_0 satisfies

$$[S_0, \widehat{C}] = -\widehat{T}. \tag{G13}$$

Simplifying equation (G12) using equation (G13), we obtain

$$H_{\text{eff}} = \widehat{C} + \lambda\widehat{V} + \lambda^2\left\{\frac{1}{2}[S_0, \widehat{T}] + [S_0, \ \widehat{V}] + [S_1, \widehat{C}]\right\}$$

$$+ \lambda^3\left\{\frac{1}{3}[S_0, [S_0, \widehat{T}]] + \frac{1}{2}([S_1, \widehat{T}] + [S_0, [S_1, \widehat{C}]] + [S_0, [S_0, \widehat{V}]])\right.$$

$$\left. + [S_1, \widehat{V}] + [S_2, \widehat{C}]\right\} + \mathcal{O}(\lambda^4). \tag{G14}$$

To determine the block off-diagonal terms at $\mathcal{O}(\lambda^2)$ in equation (G14), we note that since $\widehat{C}$ and $\widehat{T}$ are block diagonal and block off-diagonal respectively, we can always choose S_0 in Eq. (G13) to be block off-diagonal. Thus, $[S_0, \widehat{T}]$ can have block diagonal terms, whereas $[S_0, \widehat{V}]$ is completely block off-diagonal. To cancel the block off-diagonal terms at $\mathcal{O}(\lambda^2)$ in Eq. (G14), we hence require that S_1 satisfies

$$[S_1, \widehat{C}] = -[S_0, \widehat{V}] - \frac{1}{2}([S_0, \widehat{T}] - \mathcal{P}[S_0, \widehat{T}]\mathcal{P}), \tag{G15}$$

where $\mathcal{P}$ is a projector that kills block off-diagonal components. That is, if a matrix X has both block diagonal and block off-diagonal components, $\mathcal{P}X\mathcal{P}$ (respectively, $(X - \mathcal{P}X\mathcal{P})$) is completely block diagonal (respectively, off-diagonal).

Simplifying the expression for H_{eff} in Eq. (G14) using Eq. (G15), we obtain

$$H_{\text{eff}} = \widehat{C} + \lambda \widehat{V} + \frac{\lambda^2}{2} \mathcal{P}[S_0, \widehat{T}]\mathcal{P}$$

$$+ \lambda^3 \left\{ \frac{1}{3}[S_0, [S_0, \widehat{T}]] + \frac{1}{2}[S_1, \widehat{T}] + [S_1, \widehat{V}] + [S_2, \widehat{C}] \right\} + \mathcal{O}(\lambda^4). \quad (G16)$$

In equation (G15), since the RHS is block off-diagonal and $\widehat{C}$ is block diagonal, S_1 can be chosen to be block off-diagonal.

Since S_0 and S_1 are block off-diagonal, $[S_1, \widehat{T}]$ and $[S_0, [S_0, \widehat{T}]]$ can have block diagonal components whereas $[S_1, \widehat{V}]$ is completely block off-diagonal. Moreover, in the Schrieffer-Wolff procedure, S_2 is chosen such that the term $[S_2, \widehat{C}]$ cancels block off-diagonal terms at $\mathcal{O}(\lambda^3)$. That is,

$$[S_2, \widehat{C}] = -[S_1, \widehat{V}] - \frac{1}{2}([S_1, \widehat{T}] - \mathcal{P}[S_1, \widehat{T}]\mathcal{P})$$

$$- \frac{1}{3}([S_0, [S_0, \widehat{T}]] - \mathcal{P}[S_0, [S_0, \widehat{T}]]\mathcal{P}). \quad (G17)$$

Thus, H_{eff} reads

$$H_{\text{eff}} = \widehat{C} + \lambda \widehat{V} + \frac{\lambda^2}{2} \mathcal{P}[S_0, \widehat{T}]\mathcal{P}$$

$$+ \lambda^3 \mathcal{P} \left(\frac{1}{2}[S_1, \widehat{T}] + \frac{1}{3}[S_0, [S_0, \widehat{T}]] \right) \mathcal{P} + \mathcal{O}(\lambda^4). \quad (G18)$$

Thus, we find that $H_{\text{eff}}^{(2)}$ and $H_{\text{eff}}^{(3)}$ are given by

$$H_{\text{eff}}^{(2)} = \frac{1}{2} \mathcal{P}[S_0, \widehat{T}]\mathcal{P}$$

$$H_{\text{eff}}^{(3)} = \mathcal{P} \left(\frac{1}{2}[S_1, \widehat{T}] + \frac{1}{3}[S_0, [S_0, \widehat{T}]] \right) \mathcal{P}. \quad (G19)$$

We now compute S_0 and S_1 in order to obtain the effective Hamiltonians $H_{\text{eff}}^{(2)}$ and $H_{\text{eff}}^{(3)}$. According to Eq. (G13), S_0 is determined by

$$[S_0, \widehat{C}] = -\widehat{T} = -(\widehat{T}_+ + \widehat{T}_-). \quad (G20)$$

We first compute some useful commutators:

$$[\widehat{T}_+, \widehat{T}_-] = 0, [\widehat{T}_+, \widehat{C}] = -\widehat{T}_+, [\widehat{T}_-, \widehat{C}] = \widehat{T}_-, \quad (G21)$$

Thus, equation (G13) is satisfied by choosing

$$S_0 = \widehat{T}_+ - \widehat{T}_-. \quad (G22)$$

Note that S_0 in equation (G22) is block off-diagonal and anti-Hermitian. Using Eqs. (G22) and (G21), we obtain

$$[S_0, \widehat{T}] = 0, H_{\text{eff}}^{(2)} = 0. \tag{G23}$$

S_1 is computed using Eq. (G15) and the relevant commutators read

$$[\widehat{T}_+, \widehat{V}_1] = \alpha_1 \sum_j (\hat{n}_{j-1} c_{j+1}^\dagger c_j - c_j^\dagger c_{j-1} \hat{n}_{j+1})$$

$$\equiv \alpha_1 (\widehat{O}_{+-} - \widehat{O}_{++})$$

$$[\widehat{T}_-, \widehat{V}_1] = \alpha_1 \sum_j (-\hat{n}_{j-1} c_j^\dagger c_{j+1} + c_{j-1}^\dagger c_j \hat{n}_{j+1})$$

$$\equiv \alpha_1 (\widehat{O}_{-+} - \widehat{O}_{--}),$$

$$[\widehat{T}_+, \widehat{V}_0] = \alpha_0 \sum_j (w_j - w_{j+1}) c_{j+1}^\dagger c_j$$

$$\equiv \alpha_0 (-\widehat{F}_+ + \widehat{B}_+)$$

$$[\widehat{T}_-, \widehat{V}_0] = \alpha_0 \sum_j (w_{j+1} - w_j) c_j^\dagger c_{j+1}$$

$$\equiv \alpha_0 (\widehat{F}_- - \widehat{B}_-)$$

$$\implies [S_0, \widehat{V}] = \alpha_1 (\widehat{O}_{+-} + \widehat{O}_{--} - \widehat{O}_{-+} - \widehat{O}_{++})$$

$$+ \alpha_0 (-\widehat{F}_+ - \widehat{F}_- + \widehat{B}_+ + \widehat{B}_-). \tag{G24}$$

where we have defined the operators

$$\widehat{O}_{++} = \sum_j c_j^\dagger c_{j-1} \hat{n}_{j+1}, \widehat{O}_{-+} = \widehat{O}_{++}^\dagger = \sum_j c_{j-1}^\dagger c_j \hat{n}_{j+1}$$

$$\widehat{O}_{+-} = \sum_j c_{j+1}^\dagger c_j \hat{n}_{j-1}, \widehat{O}_{--} = \widehat{O}_{+-}^\dagger = \sum_j c_j^\dagger c_{j+1} \hat{n}_{j-1},$$

$$\widehat{F}_+ = \sum_j w_{j+1} c_{j+1}^\dagger c_j, \widehat{F}_- = \widehat{F}_+^\dagger = \sum_j w_{j+1} c_j^\dagger c_{j+1},$$

$$\widehat{B}_+ = \sum_j w_j c_{j+1}^\dagger c_j, \widehat{B}_- = \widehat{B}_+^\dagger = \sum_j w_j c_j^\dagger c_{j+1}. \tag{G25}$$

Thus, according to equation (G15), S_1 should satisfy

$$[S_1, \widehat{C}] = -[S_0, \widehat{V}] = \alpha_1(-\widehat{O}_{+-} - \widehat{O}_{--} + \widehat{O}_{-+} + \widehat{O}_{++})$$
$$+ \alpha_0(\widehat{F}_+ + \widehat{F}_- - \widehat{B}_+ - \widehat{B}_-). \tag{G26}$$

We now show that S_1 can be chosen to be a linear superposition of $\widehat{O}_{\mu\nu}$, $\widehat{F}_\mu$ and $\widehat{B}_\mu$, where $\mu, \nu \in \{+, -\}$. The commutators $[\widehat{O}_{\mu\nu}, \widehat{C}]$, $[\widehat{F}_\mu, \widehat{C}]$ and $[\widehat{B}_\mu, \widehat{C}]$ read

$$[\widehat{O}_{++}, \widehat{C}] = -\widehat{O}_{++}, [\widehat{O}_{-+}, \widehat{C}] = \widehat{O}_{-+},$$
$$[\widehat{O}_{+-}, \widehat{C}] = -\widehat{O}_{+-}, [\widehat{O}_{--}, \widehat{C}] = \widehat{O}_{--},$$
$$[\widehat{F}_+, \widehat{C}] = -\widehat{F}_+, [\widehat{F}_-, \widehat{C}] = \widehat{F}_-,$$
$$[\widehat{B}_+, \widehat{C}] = -\widehat{B}_+, [\widehat{B}_-, \widehat{C}] = \widehat{B}_-. \tag{G27}$$

Thus, Eq. (G26) is satisfied by choosing

$$S_1 = \alpha_1(\widehat{O}_{+-} - \widehat{O}_{--} + \widehat{O}_{-+} - \widehat{O}_{++}) + \alpha_0(-\widehat{F}_+ + \widehat{F}_- + \widehat{B}_+ - \widehat{B}_-). \tag{G28}$$

Noting that $[S_0, \widehat{T}] = 0$, $H_{\text{eff}}^{(3)}$ in equation (G19) reads

$$H_{\text{eff}}^{(3)} = \frac{\lambda^3}{2} \mathcal{P}[S_1, \widehat{T}_+ + \widehat{T}_-]\mathcal{P}$$

$$= -\frac{\alpha_1 \lambda^3}{2} \mathcal{P}[\widehat{T}_+ + \widehat{T}_-, \widehat{O}_{+-} - \widehat{O}_{--} + \widehat{O}_{-+} - \widehat{O}_{++}]\mathcal{P}$$

$$- \frac{\alpha_0 \lambda^3}{2} \mathcal{P}[\widehat{T}_+ + \widehat{T}_-, -\widehat{F}_+ + \widehat{F}_- + \widehat{B}_+ - \widehat{B}_-]\mathcal{P}. \tag{G29}$$

We obtain the following commutators:

$$[\widehat{T}_+, \widehat{O}_{+-}] = \sum_j (-\hat{n}_j c_{j+1}^\dagger c_{j-1} + c_{j+2}^\dagger c_j \hat{n}_{j-1} - c_{j+1}^\dagger c_j c_{j-1}^\dagger c_{j-2}),$$

$$[\widehat{T}_+, \widehat{O}_{-+}] = \sum_j ((\hat{n}_j - \hat{n}_{j-1})\hat{n}_{j+1} + c_{j-1}^\dagger c_{j+2}^\dagger c_{j+1} c_j),$$

$$[\widehat{T}_+, \widehat{O}_{++}] = \sum_j (-c_j^\dagger \hat{n}_{j+1} c_{j-2} + c_{j+1}^\dagger c_{j-1} n_j + c_j^\dagger c_{j-1} c_{j+2}^\dagger c_{j+1}),$$

$$[\widehat{T}_-, \widehat{O}_{+-}] = [\widehat{T}_+^\dagger, \widehat{O}_{--}^\dagger] = -[\widehat{T}_+, \widehat{O}_{--}]^\dagger [\widehat{T}_-, \widehat{O}_{++}]$$
$$= [\widehat{T}_+^\dagger, \widehat{O}_{-+}^\dagger] = -[\widehat{T}_+, \widehat{O}_{-+}]^\dagger,$$

$$[\widehat{T}_-, \widehat{O}_{-+}] = [\widehat{T}_+^\dagger, \widehat{O}_{++}^\dagger] = -[\widehat{T}_+, \widehat{O}_{++}]^\dagger, [\widehat{T}_-, \widehat{O}_{--}]$$

$$= [\widehat{T}_+^\dagger, \widehat{O}_{+-}^\dagger] = -[\widehat{T}_+, \widehat{O}_{+-}]^\dagger,$$

$$[\widehat{T}_+, \widehat{F}_+] = \sum_j (w_j - w_{j+1}) c_{j+1}^\dagger c_{j-1}, [\widehat{T}_+, \widehat{F}_-] = \sum_j (w_j - w_{j+1}) \hat{n}_j,$$

$$[\widehat{T}_+, \widehat{B}_+] = \sum_j (w_{j-1} - w_j) c_{j+1}^\dagger c_{j-1}, [\widehat{T}_+, \widehat{B}_-] = \sum_j (w_{j-1} - w_j) \hat{n}_j,$$

$$[\widehat{T}_-, \widehat{F}_-] = [\widehat{T}_+^\dagger, \widehat{F}_+^\dagger] = -[\widehat{T}_+, \widehat{F}_+]^\dagger, [\widehat{T}_-, \widehat{F}_+] = [\widehat{T}_+^\dagger, \widehat{F}_-^\dagger] = -[\widehat{T}_+, \widehat{F}_-]^\dagger,$$

$$[\widehat{T}_-, \widehat{B}_-] = [\widehat{T}_+^\dagger, \widehat{B}_+^\dagger] = -[\widehat{T}_+, \widehat{B}_+]^\dagger, [\widehat{T}_-, \widehat{B}_+] = [\widehat{T}_+^\dagger, \widehat{B}_-^\dagger] = -[\widehat{T}_+, \widehat{B}_-]^\dagger.$$

$$\text{(G30)}$$

Note that

$$\mathcal{P}[\widehat{T}_+, \widehat{O}_{+-}]\mathcal{P} = \mathcal{P}[\widehat{T}_-, \widehat{O}_{--}]\mathcal{P} = 0, \mathcal{P}[\widehat{T}_+, \widehat{O}_{++}]\mathcal{P} = \mathcal{P}[\widehat{T}_-, \widehat{O}_{-+}]\mathcal{P} = 0,$$

$$\mathcal{P}[\widehat{T}_+, \widehat{F}_+]\mathcal{P} = \mathcal{P}[\widehat{T}_-, \widehat{F}_-]\mathcal{P} = 0, \mathcal{P}[\widehat{T}_+, \widehat{B}_+]\mathcal{P} = \mathcal{P}[\widehat{T}_-, \widehat{B}_-]\mathcal{P} = 0, \quad \text{(G31)}$$

since according to Eq. (G30), these terms change the energy of eigenstates of $\widehat{C}$ and are hence block off-diagonal. Simplifying Eq. (G29) using Eqs. (G30) and (G31), we obtain

$$H_{\text{eff}}^{(3)} = -\frac{\alpha_1 \lambda^3}{2} \mathcal{P}\{(-[\widehat{T}_+, \widehat{O}_{--}] + [\widehat{T}_+, \widehat{O}_{-+}]) + h.c.\}\mathcal{P}$$

$$-\frac{\alpha_0 \lambda^3}{2} \mathcal{P}\{([\widehat{T}_+, \widehat{F}_-] - [\widehat{T}_+, \widehat{B}_-]) + h.c.\}\mathcal{P}$$

$$= -\alpha_1 \lambda^3 \sum_j \{(c_j^\dagger c_{j+3}^\dagger c_{j+2} c_{j+1} + h.c.) + 2(\hat{n}_j \hat{n}_{j+1} - \hat{n}_j \hat{n}_{j+2})\}$$

$$-\alpha_0 \lambda^3 \sum_j (2w_j - w_{j-1} - w_{j+1}) \hat{n}_j \qquad \text{(G32)}$$

Finally, reintroducing the overall factor of E, the full effective Hamiltonian restricted to one center-of-mass sector is

$$H_{\text{eff}} = V_0 \sum_j \tilde{w}_j \hat{n}_j + \tilde{V}_1 \sum_j \hat{n}_j \hat{n}_{j+1} + \tilde{V}_2 \sum_j \hat{n}_j \hat{n}_{j+2}$$

$$-\frac{t^2 V_1}{E^2} \sum_j (c_j^\dagger c_{j+3}^\dagger c_{j+2} c_{j+1} + h.c.) + \mathcal{O}\left(\frac{t^3}{E^3}\right), \qquad \text{(G33)}$$

where we have omitted the term $E \sum_j j\hat{n}_j$ since it is a symmetry of the effective Hamiltonian and we have defined

$$\widetilde{w}_j \equiv \left(1 - \frac{2t^2}{E^2}\right) w_j + \frac{t^2}{E^2} \left(w_{j-1} + w_{j+1}\right), \widetilde{V}_1 \equiv V_1 \left(1 - \frac{2t^2}{E^2}\right), \widetilde{V}_2 \equiv \frac{2t^2 V_1}{E^2}.$$
$$\text{(G34)}$$

Acknowledgments

We thank Dan Arovas, Vedika Khemani, Alan Morningstar, Frank Pollmann, Gil Refael, Max Schultz, Shivaji Sondhi, Ruben Verresen and particularly David Huse for useful discussions. S.M. acknowledges the hospitality of the Laboratoire de Physique de l'Ecole Normale Supérieure, where parts of the manuscript were completed. A.P. acknowledges the hospitality of the Aspen Center of Physics, where part of this work was completed during a visit to the program "Realizations and Applications of Quantum Coherence in Non-Equlibrium Systems." The Aspen Center for Physics is supported by National Science Foundation grant PHY-1607611. A.P. is supported by a PCTS fellowship at Princeton University. This material is based in part (R.M.N.) upon work supported by Air Force Office of Sponsored Research under grant no. FA9550-17-1-0183. R.M.N. also acknowledges the hospitality of the KITP, where part of this work was done, during a visit to the program "Dynamics of Quantum Information." The KITP is supported in part by the National Science Foundation under grant PHY-1748958. B.A.B. and N.R. were supported by the Department of Energy Grant No. DE-SC0016239, the National Science Foundation EAGER Grant No. DMR 1643312, Simons Investigator Grant No. 404513, ONR Grant No. N00014-14-1-0330, the Packard Foundation, the Schmidt Fund for Innovative Research and a Guggenheim Fellowship from the John Simon Guggenheim Memorial Foundation.

References

[1] T. Kinoshita, T.Wenger and D. S.Weiss, *Nature (London)* **440**, 900 (2006).

[2] M. Gring, M. Kuhnert, T. Langen, T. Kitagawa, B. Rauer, M. Schreitl, I. Mazets, D. A. Smith, E. Demler and J. Schmiedmayer, *Science* **337**, 1318 (2012).

[3] M. Schreiber, S. S. Hodgman, P. Bordia, H. P. Luschen, M. H. Fischer, R. Vosk, E. Altman, U. Schneider and I. Bloch, *Science* **349**, 842 (2015).

[4] J. Smith, A. Lee, P. Richerme, B. Neyenhuis, P. W. Hess, P. Hauke, M. Heyl, D. A. Huse and C. Monroe, *Nature Phys.* **12**, 907 (2016).

[5] A. M. Kaufman, M. E. Tai, A. Lukin, M. Rispoli, R. Schittko, P. M. Preiss and M. Greiner, *Science* **353**, 794 (2016).

[6]   G. Kucsko, S. Choi, J. Choi, P. C. Maurer, H. Zhou, R. Landig, H. Sumiya, S. Onoda, J. Isoya, F. Jelezko, E. Demler, N. Y. Yao and M. D. Lukin, *Phys. Rev. Lett.* **121**, 023601 (2018).

[7]   J. M. Deutsch, *Phys. Rev. A* **43**, 2046 (1991).

[8]   M. Srednicki, *Phys. Rev. E* **50**, 888 (1994).

[9]   M. Rigol, V. Dunjko and M. Olshanii, *Nature (London)* **452**, 854 (2008).

[10]   A. Polkovnikov, K. Sengupta, A. Silva and M. Vengalattore, *Rev. Mod. Phys.* **83**, 863 (2011).

[11]   L. D'Alessio, Y. Kafri, A. Polkovnikov and M. Rigol, *Adv. Phys.* **65**, 239 (2016).

[12]   C. Gogolin and J. Eisert, *Rep. Prog. Phys.* **79**, 056001 (2016).

[13]   T. Mori, T. N. Ikeda, E. Kaminishi and M. Ueda, *J. Phys. B Atomic Mol. Phys.* **51**, 112001 (2018).

[14]   H. Kim, T. N. Ikeda and D. A. Huse, *Phys. Rev. E* **90**, 052105 (2014).

[15]   J. R. Garrison and T. Grover, *Phys. Rev. X* **8**, 021026 (2018).

[16]   P. W. Anderson, *Phys. Rev.* **109**, 1492 (1958).

[17]   I. V. Gornyi, A. D. Mirlin and D. G. Polyakov, *Phys. Rev. Lett.* **95**, 206603 (2005).

[18]   D. M. Basko, I. L. Aleiner and B. L. Altshuler, *Ann. Phys.* **321**, 1126 (2006).

[19]   M. Serbyn, Z. Papic and D. A. Abanin, *Phys. Rev. Lett.* **111**, 127201 (2013).

[20]   D. A. Huse, R. Nandkishore and V. Oganesyan, *Phys. Rev. B* **90**, 174202 (2014).

[21]   R. Nandkishore and D. A. Huse, *Ann. Rev. Conden. Matter Phys.* **6**, 15 (2015).

[22]   E. Altman and R. Vosk, *Ann. Revi. Conden. Matter Phys.* **6**, 383 (2015).

[23]   D. A. Abanin, E. Altman, I. Bloch and M. Serbyn, *Rev. Mod. Phys* **91**, 021001 (2019).

[24]   W. De Roeck and F. Huveneers, *Phys. Rev. B* **90**, 165137 (2014).

[25]   W. De Roeck and F. Huveneers, *Commun. Math. Phys.* **332**, 1017 (2014).

[26]   T. Grover and M. P. A. Fisher, *J. Stat. Mech. Theory Exp.* **2014**, P10010 (2014).

[27]   M. Schiulaz, A. Silva and M. Muller, *Phys. Rev. B* **91**, 184202 (2015).

[28]   Z. Papic, E. M. Stoudenmire and D. A. Abanin, *Ann. Phys.* **362**, 714 (2015).

[29]   N. Y. Yao, C. R. Laumann, J. I. Cirac, M. D. Lukin and J. E. Moore, *Phys. Rev. Lett.* **117**, 240601 (2016).

[30]   A. Smith, J. Knolle, D. L. Kovrizhin and R. Moessner, *Phys. Rev. Lett.* **118**, 266601 (2017).

[31]   A. Smith, J. Knolle, R. Moessner and D. L. Kovrizhin, *Phys. Rev. Lett* **119**, 176601 (2017).

[32]   A. A. Michailidis, M. Znidaric, M. Medvedyeva, D. A. Abanin, T. Prosen and Z. Papic, *Phys. Rev. B* **97**, 104307 (2018).

[33]   M. Brenes, M. Dalmonte, M. Heyl and A. Scardicchio, *Phys. Rev. Lett.* **120**, 030601 (2018).

[34]   G. Biroli, C. Kollath and A. M. Lauchli, *Phys. Rev. Lett.* **105**, 250401 (2010).

[35]   T. Mori, arXiv e-prints, arXiv:1609.09776 [condmat. stat-mech].

[36]   O. Vafek, N. Regnault and B. A. Bernevig, *Sci. Post Phys.* **3**, 043 (2017).

[37]   S. Moudgalya, N. Regnault and B. A. Bernevig, *Phys. Rev. B* **98**, 235156 (2018).

[38]   S. Moudgalya, S. Rachel, B. A. Bernevig and N. Regnault, *Phys. Rev. B* **98**, 235155 (2018).

[39]   I. Affleck, T. Kennedy, E. H. Lieb and H. Tasaki, *Commun. Math. Phys.* **115**, 477 (1988).

[40]   D. P. Arovas, *Phys. Lett. A* **137**, 431 (1989).

[41]   C.-J. Lin and O. I. Motrunich, *Phys. Rev. Lett.* **122**, 173401 (2019).

[42]   M. Schecter and T. Iadecola, *Phys. Rev. Lett.* **123**, 147201 (2019).

[43]   S. Chattopadhyay, H. Pichler, M. D. Lukin and W. W. Ho, arXiv:1910.08101 [quant-ph].

[44] T. Iadecola and M. Schecter, arXiv:1910.11350 [cond-mat.str-el].

[45] N. Shiraishi and T. Mori, *Phys. Rev. Lett* **119**, 030601 (2017).

[46] S. Ok, K. Choo, C. Mudry, C. Castelnovo, C. Chamon and T. Neupert, arXiv e-prints, arXiv:1901.01260 [cond-mat.other].

[47] H. Bernien, S. Schwartz, A. Keesling, H. Levine, A. Omran, H. Pichler, S. Choi, A. S. Zibrov, M. Endres, M. Greiner, V. Vuletic and M. D. Lukin, *Nature (London)* **551**, 579 (2017).

[48] C. Turner, A. Michailidis, D. Abanin, M. Serbyn and Z. Papic, *Nat. Phys.* **14**, 745 (2018).

[49] M. Schecter and T. Iadecola, *Phys. Rev. B* **98**, 035139 29 (2018).

[50] C. J. Turner, A. A. Michailidis, D. A. Abanin, M. Serbyn and Z. Papic, *Phys. Rev. B* **98**, 155134 (2018).

[51] S. Choi, C. J. Turner, H. Pichler, W. W. Ho, A. A. Michailidis, Z. Papic, M. Serbyn, M. D. Lukin and D. A. Abanin, *Phys. Rev. Lett.* **122**, 220603 (2019).

[52] W. W. Ho, S. Choi, H. Pichler and M. D. Lukin, *Phys. Rev. Lett* **122**, 040603 (2019).

[53] E. J. Heller, *Phys. Rev. Lett.* **53**, 1515 (1984).

[54] A. A. Michailidis, C. J. Turner, Z. Papic, D. A. Abanin and M. Serbyn, arXiv e-prints, arXiv:1905.08564 [quant-ph].

[55] V. Khemani, C. R. Laumann and A. Chandran, *Phys. Rev. B* **99**, 161101 (2019).

[56] F. M. Surace, P. P. Mazza, G. Giudici, A. Lerose, A. Gambassi and M. Dalmonte, arXiv e-prints, arXiv:1902.09551 [cond-mat.quant-gas].

[57] T. Iadecola, M. Schecter and S. Xu, arXiv e-prints, arXiv:1903.10517 [cond-mat.str-el].

[58] A. J. A. James, R. M. Konik and N. J. Robinson, *Phys. Rev. Lett* **122**, 130603 (2019).

[59] N. J. Robinson, A. J. A. James and R. M. Konik, *Phys. Rev. B* **99**, 195108 (2019).

[60] G. Magnico, M. Dalmonte, P. Facchi, S. Pascazio, F. V. Pepe and E. Ercolessi, arXiv:1909.04821 [quant-ph].

[61] K. Bull, I. Martin and Z. Papic, *Phys. Rev. Lett.* **123**, 030601 (2019).

[62] S. Moudgalya, B. A. Bernevig and N. Regnault, arXiv e-prints, arXiv:1906.05292 [cond-mat.str-el].

[63] A. Hudomal, I. Vasic, N. Regnault and Z. Papic, arXiv e-prints, arXiv:1910.09526 [quant-ph].

[64] C.-J. Lin, A. Chandran and O. I. Motrunich, arXiv eprints, arXiv:1910.07669 [cond-mat.quant-gas].

[65] S. Pai and M. Pretko, arXiv e-prints, arXiv:1903.06173 [cond-mat.stat-mech].

[66] B. Mukherjee, S. Nandy, A. Sen, D. Sen and K. Sengupta, arXiv:1907.08212 [quant-ph].

[67] A. Haldar, D. Sen, R. Moessner and A. Das, arXiv:1909.04064 [cond-mat.other].

[68] B. Olmos, M. Muller and I. Lesanovsky, *New J. Phys.* **12**, 013024 (2010).

[69] M. van Horssen, E. Levi and J. P. Garrahan, *Phys. Rev. B* **92**, 100305 (2015).

[70] Z. Lan, M. van Horssen, S. Powell and J. P. Garrahan, *Phys. Rev. Lett* **121**, 040603 (2018).

[71] B. Sun and F. Robicheaux, *New J. Phys.* **10**, 045032 (2008).

[72] B. Olmos, R. Gonzalez-Ferez and I. Lesanovsky, *Phys. Rev. A* **79**, 043419 (2009).

[73] B. Olmos, R. Gonzalez-Ferez, I. Lesanovsky and L. Velazquez, **45**, 325301 (2012).

[74] R. M. Nandkishore and M. Hermele, *Ann. Rev. Conden. Matt. Phys.* **10**, 295 (2019).

[75] C. Chamon, *Phys. Rev. Lett.* **94**, 040402 (2005).

[76] I. H. Kim and J. Haah, *Phys. Rev. Lett.* **116**, 027202 (2016).

[77] A. Prem, J. Haah and R. Nandkishore, *Phys. Rev. B* **95**, 155133 (2017).

[78] M. Pretko, *Phys. Rev. B* **95**, 115139 (2017).

[79] J. Sous and M. Pretko, arXiv e-prints, arXiv:1904.08424 [cond-mat.str-el].

[80] S. Pai and M. Pretko, arXiv e-prints, arXiv:1909.12306 [cond-mat.str-el].

[81] S. Pai, M. Pretko and R. M. Nandkishore, *Phys. Rev. X* **9**, 021003 (2019).

[82] O. Morsch and M. Oberthaler, *Rev. Mod. Phys.* **78**, 179 (2006).

[83] M. Znidaric, *Phys. Rev. Lett.* **110**, 070602 (2013).

[84] T. Iadecola and M. Znidaric, *Phys. Rev. Lett.* **123**, 036403 (2019).

[85] P. Sala, T. Rakovszky, R. Verresen, M. Knap and F. Pollmann, arXiv e-prints, arXiv:1904.04266 [cond-mat.str-el].

[86] V. Khemani and R. Nandkishore, arXiv e-prints, arXiv:1904.04815 [cond-mat.stat-mech].

[87] M. Schulz, C. A. Hooley, R. Moessner and F. Pollmann, *Phys. Rev. Lett.* **122**, 040606 (2019).

[88] E. van Nieuwenburg, Y. Baum and G. Refael, *Proc. Nat. Acad. Sci.* **116**, 9269 (2019).

[89] A. Seidel, H. Fu, D.-H. Lee, J. M. Leinaas and J. Moore, *Phys. Rev. Lett.* **95**, 266405 (2005).

[90] E. J. Bergholtz and A. Karlhede, *J. Stat. Mech. Theory Exp.* **2006**, L04001 (2006).

[91] E. J. Bergholtz and A. Karlhede, *Phys. Rev. B* **77**, 155308 (2008).

[92] C. H. Lee, Z. Papic and R. Thomale, *Phys. Rev. X* **5**, 041003 (2015).

[93] Z. Papic, *Phys. Rev. B* **90**, 075304 (2014).

[94] E. H. Rezayi and F. D. M. Haldane, *Phys. Rev. B* **50**, 17199 (1994).

[95] M. Nakamura, Z.-Y. Wang and E. J. Bergholtz, *Phys. Rev. Lett.* **109**, 016401 (2012).

[96] G. H. Wannier, *Rev. Mod. Phys* **34**, 645 (1962).

[97] D. Emin and C. F. Hart, *Phys. Rev. B* **36**, 7353 (1987).

[98] Z.-Y. Wang, S. Takayoshi and M. Nakamura, *Phys. Rev. B* **86**, 155104 (2012).

[99] M. Fremling, C. Repellin, J.-M. Stephan, N. Moran, J. K. Slingerland and M. Haque, *New J. Phys.* **20**, 103036 (2018).

[100] D. N. Page, *Phys. Rev. Lett* **71**, 1291 (1993).

[101] E. J. Bergholtz and A. Karlhede, *Phys. Rev. Lett.* **94**, 026802 (2005).

[102] E. H. Lieb and W. Liniger, *Phys. Rev.* **130**, 1605 (1963).

[103] D. Poilblanc, T. Ziman, J. Bellissard, F. Mila and G. Montambaux, *Europhys. Lett. (EPL)* **22**, 537 (1993).

[104] Y. Y. Atas, E. Bogomolny, O. Giraud and G. Roux, *Phys. Rev. Lett.* **110**, 084101 (2013).

[105] W. Beugeling, R. Moessner and M. Haque, *Phys. Rev. E* **89**, 042112 (2014).

[106] S. Bravyi, D. P. DiVincenzo and D. Loss, *Ann. Phys.* **326**, 2793 (2011).

[107] G. D. Tomasi, D. Hetterich, P. Sala and F. Pollmann, arXiv:1909.03073 [cond-mat.dis-nn].

[108] V. Khemani, M. Hermele and R. M. Nandkishore, arXiv:1910.01137 [cond-mat.stat-mech].

[109] S. R. Taylor, M. Schulz, F. Pollmann and R. Moessner, arXiv:1910.01154 [cond-mat.dis-nn].

[110] T. Rakovszky, P. Sala, R. Verresen, M. Knap and F. Pollmann, arXiv:1910.06341 [cond-mat.strel].

[111] A. de Pasquale, G. Costantini, P. Facchi, G. Florio, S. Pascazio and K. Yuasa, *Euro. Phys. J. Special Topics* **160**, 127 (2008).

[112] N. J. A. Sloane, in *Towards Mechanized Mathematical Assistants*, eds. M. Kauers, M. Kerber, R. Miner and W. Windsteiger (Springer, 2007), pp. 130–130.

Chapter 8

Classification of Strongly Disordered Topological Wires Using Machine Learning

Ye Zhuang*, Luiz H. Santos† and Taylor L. Hughes*

*Department of Physics and Institute of Condensed Matter Theory,
University of Illinois, Urbana, IL 61801, USA
†Department of Physics, Emory University,
Atlanta, GA 30322, USA

In this chapter, we apply the random forest machine-learning model to classify 1D topological phases when strong disorder is present. We show that using the entanglement spectrum as training features the model gives high classification accuracy. The trained model can be extended to other regions in phase space, and even to other symmetry classes on which it was not trained and still provides accurate results. After performing a detailed analysis of the trained model, we find that its dominant classification criteria capture degeneracy in the entanglement spectrum.

8.1 Introduction

Since the discovery of topological insulator phases, the problem of their classification has been an important subject. A classification table, predicated on the stability of a strong topological phase in the presence of disorder, was proposed for free fermion systems in the ten Altland–Zirnbauer symmetry classes [1, 2], though the table does not determine the explicit phases and phase diagrams of model systems. Many successful determinations of the phase diagrams of low-dimensional disordered topological phases have been made in symmetry classes A, AIII, and BDI using techniques based on, e.g., entanglement properties, level-spacing statistical analysis, and real-space topological indices [3–7]. In this chapter, we propose the use of a new technique based on the random forest (RF) machine-learning model to determine the phase diagrams of models of disordered topological phases.

Recent work has shown that machine learning can provide a new framework for solving problems in physics. In our context, promising developments have been achieved in applying machine-learning techniques to classifying phases of matter in condensed matter systems. Supervised learning has been used directly in characterizing phases in both classical spin systems [8–10] and quantum many-body systems [11–18]. Neural networks are the most widely used model to identify phases, especially topological phases of matter [19–22], and they are powerful models that have universal approximation capabilities [23, 24]. For example, Chern insulators and fractional Chern insulators can be classified by feeding quantum loop topography into a neural network [15]. Unfortunately, the black-box nature of neural networks makes it hard to interpret the trained models, and it is not easy to extract insightful physical intuition about the system under study. Additionally, the large number of hyper-parameters in a neural network can make it difficult to train.

In this chapter, we use the RF method as our machine-learning model to detect topological phases with strong disorder. RF is an ensemble method that is capable of representing complicated functions with much fewer parameters as compared with neural networks, and having more easily interpretable classification criteria once the model is trained [25, 26]. RF is a collection of decision trees, which can be understood as piecewise constant functions in feature space. An individual decision tree cannot make good predictions because, in general, predictions of decision trees have large variance. Averaging over decision trees reduces variance, making RF a popular method [27]. Some major advantages of RF are that it has few hyper-parameters, and is immune to problems such as over-fitting, collinearity, etc.

To train our RF model we will use the entanglement spectrum (ES) [28] of our physical system as our input data. We will focus on 1D systems where the ES has been widely used to characterize topological phases, and a robust degeneracy in the entanglement spectrum can serve as a general indicator for a topological phase [29]. To benchmark our machine-learning model, we will consider 1D free-fermion wires having chiral symmetry in the AIII class. We choose this system because the disordered phase diagram of this model has been carefully studied [6]. Here we find that the RF model, trained by the ES data generated from a small fraction of the phase diagram, can be generalized to the full phase space with high prediction accuracy. Furthermore, the RF model trained from the AIII class data shows high prediction ability for wires in symmetry class BDI as well. A detailed analysis reveals that the RF model is primarily capturing the degeneracy in the ES to make its classification, and may provide new routes to identify disordered topological phases from their entanglement properties.

8.2 Model

We start from the disordered chiral Hamiltonian in [6] defined on a one-dimensional chain with two sites A and B in one unit cell:

$$H = \sum_n \left[\frac{t_n}{2} c_n^\dagger \left(\sigma_x + i\sigma_y \right) c_{n+1} + h.c. \right], + \sum_n m_n c_n^\dagger \sigma_y c_n. \qquad (8.1)$$

where $c_n^\dagger = (c_{n,A}^\dagger, c_{n,B}^\dagger)$ are fermion creation operators in unit cell n. We have included disorder in both the hopping and mass terms, i.e., $t_n = 1 + W_1 \omega_1$, and $m_n = m + W_2 \omega_2$, where ω_1 and ω_2 are random variables generated from a uniform distribution on $[-0.5, 0.5]$, and W_1, W_2 represent the strengths of the disorder. The model preserves chiral symmetry $\mathcal{C} H \mathcal{C}^{-1} = -H$ with $\mathcal{C} = \sum_n c_n^\dagger \sigma_z c_n$.

In the clean limit, the system has translational symmetry and the Bloch Hamiltonian is

$$\mathcal{H}(k) = t \cos k \sigma_x + (t \sin k + m)\sigma_y. \qquad (8.2)$$

The chiral symmetry operator $\chi = \sigma_z$ anti-commutes with $\mathcal{H}(k)$, and one can use the topological winding number ν to identify the $\mathbb{Z}$ classification [30]. If we write the Bloch Hamiltonian in the form $\mathcal{H}(k) = d_x(k)\sigma_x + d_y(k)\sigma_y$, then ν is the number of times the unit vector $(\hat{d}_x, \hat{d}_y)$ travels around the origin as k traverses the whole Brillouin Zone. For example, when $|m| < |t|$, the system is in symmetry protected topological (SPT) phase with winding number $\nu = 1$, and when $|m| > |t|$ the system has $\nu = 0$.

When disorder is turned on, and in the limit that $W_2 \gg t$, the system is completely dimerized within individual unit cells, i.e., it is in the topologically trivial atomic limit. This result is independent of the value of m, and hence there must be a phase transition when W_2 is gradually increased from zero when $|m| < |t|$. One signature of the topological phase transition point is the divergence of the localization length of states at the Fermi level. Using this criterion, one can determine an analytic relation that is satisfied at a topological critical point [6].

$$\frac{|2 + W_1|^{1/W_1 + 1/2} |2m - W_1|^{m/W_1 - 1/2}}{|2 - W_2|^{1/W_2 - 1/2} |2m + W_2|^{m/W_2 + 1/2}} = 1. \qquad (8.3)$$

Indeed, using this relation one can determine the phase diagram of this model even in the presence of disorder.

8.3   Results

To generate training and testing data, we calculate the single-particle entanglement spectrum [31] using a central spatial cut of the lattice model. We use periodic boundary conditions on a chain of length $L = 400$ with $t = 1$. To be explicit, let us first focus on a line in the 3D phase space $\{m, W_1, W_2\}$ with $m = 0.5$ and $W_1 = 1$. For illustration, we plot the ES of single-disorder configuration for each value of W_2 in Fig. 8.1(a). The black vertical line indicates the theoretical transition point calculated from Eq. (8.3). We can clearly see double degeneracy at 0.5 in the region of weak disorder, which is a signature for SPT phases [29]. In the region of strong disorder, on the other hand, there are no such degeneracies in general, even though there may be accidental degeneracies induced by disorder.

In order to test the predictive power of the RF model, we first train the model in regions deep in the two phases using a set of test data based on the entanglement spectrum. We will further evaluate the RF model by testing if it can provide accurate predictions for other values of parameters different than those used to generate the training data. In particular, we test whether the model is capable of detecting the behavior of the system near the topological phase transition despite training it deep in the phases. We will indeed verify that the RF model can accurately detect disorder-induced phase transitions.

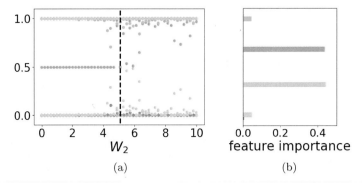

Fig. 8.1.   (a) Single-particle entanglement spectrum plotted vs. disorder strength W_2. The vertical black dashed line is the analytical transition point. Double degeneracy at 0.5 on the left-hand side indicates the nontrivial SPT phase. There may be accidental degeneracies in the trivial phase due to disorder. (b) Feature importance in our trained random forest model. The vertical axis is the average of the entanglement spectrum for each band in panel (a). We plot the feature importance of the bands in the horizontal direction. The high importance values for the middle two bands, i.e., the two that include the degenerate modes at 0.5, indicate their high influence on model predictions. The bands are colored-coded the same way in panels (a) and (b).

In order to implement the data training, 5000 *training* samples were generated with W_2 ranging from 0 to 4 and from 7 to 10, which correspond, respectively, to the topologically nontrivial and trivial phases. We intentionally skipped the region near the phase transition point $W_2 \approx 5$, in hopes that the RF model can locate it only with knowledge deep in the phases. *Test* data was generated separately over the whole range of W_2 from 0 to 10, and includes the transition region. In order to test the performance of the RF model with respect to other widely utilized algorithms, we trained three models using the same training and testing data: the linear model (LM), neural network (NN), and random forest, and we compare the predictions of the first two in relation to the latter. In our numerical analysis, the python package sklearn [32] was used for training and predicting.

We show the prediction results of the three models in Fig. 8.2. From left to right, the dots in the subfigures represent predicted probabilities of being in the topological phase from LM, NN, and RF models, respectively. We observe that, while most ground states are correctly classified for all the three models, two features stand out. First, the LM has a number of misclassified states in the region of strong disorder due to the simple linear assumption. The trained LM gives a linear relationship between the logarithm of the probability and the gap between the middle two entanglement bands. When the gap is small the model predicts the state is topologically nontrivial with high probability even when it is a trivial state. Second, while the NN and RF models give reliable predictions for strong and weak values of disorder, they classify with much higher variance near the phase boundary, which makes it hard to predict the resulting phase from a single-disorder configuration. This behavior may be expected for, near the transition, the (entanglement) spectral gap approaches zero, and the disorder causes strong fluctuations that make it difficult to correctly distinguish the intrinsic degeneracies of the ES on the topological side vs. the frequently encountered accidental degeneracies in the trivial phase near the phase boundary.

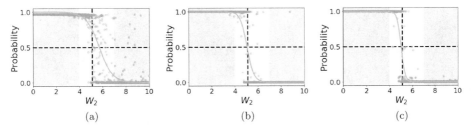

Fig. 8.2. (a) Predicted probability of being in the topological phase for the three models: (a) linear model, (b) neural network, and (c) random forest. The smooth curves are fits using Eq. (8.4).

The predicted critical point can be obtained for each model by fitting the predicted probability with the function

$$f(x) = \frac{1}{1 + e^{b+wx}}. \tag{8.4}$$

The fitted curves are shown in each of the subfigures in Fig. 8.2. To help identify the phase boundary we choose a cutoff value of 0.5, i.e., when the predicted probability is larger than 0.5, we say the state is in the topological phase; otherwise it is in the trivial phase. Therefore, the transition happens at the crossing point of the fitted curve and the horizontal dashed line at probability 0.5. For comparison, the black vertical dashed lines in each subfigure represent the true transition point. We see that both NN and RF models can predict the phase transition point with high accuracy, while the LM is not as accurate at predicting the critical point.

Quantitative assessments of the predictive properties of these three models can be obtained by evaluating accuracy and error. Accuracy is defined as the percentage of correctly predicted samples; higher accuracy means better prediction ability. We measure the error of the fitting by the cross entropy [33], which measures the closeness of two probability distributions p and q. The cross entropy is defined as (for discrete distributions)

$$H(p,q) = -\sum_x p(x) \log q(x), \tag{8.5}$$

which is the expectation value of $-\log q(x)$ for the random variable x following distribution p. Here p is the true probability distribution and q is the predicted probability distribution. For our problem, the distribution is discrete and has only two cases, topological and trivial, so the cross entropy reduces to just the log loss function for a binary classification problem

$$L(y, \hat{y}) = \frac{1}{n} \sum_{i=1}^{n} [-y_i \log \hat{y}_i - (1 - y_i) \log(1 - \hat{y}_i)], \tag{8.6}$$

where y_i is the true probability of being in the topological phase, and $\hat{y}_i$ is the predicted probability. Small errors indicate a better model, and indeed the LM and NN models are trained on training data to reduce the error. However, the RF model is trained based more on accuracy.

The accuracies of the LM, NN, and RF models on the test data are 0.923, 0.971, and 0.978, respectively. The corresponding errors are 0.202, 0.277, and 0.106. The LM has the lowest accuracy among the three, due to its simple linear assumption, while NN and RF models perform similarly. Nevertheless, we emphasize the use of the RF method has some advantages including the fact that it captures a high level of accuracy while requiring much fewer parameters than the NN model.

Another benefit of the RF model is the ability to interpret how it makes classification decisions. To illustrate this we can plot the feature importance of the model (Fig. 8.1(b)). Feature importance measures the number of splits in a tree that includes the feature [34], and higher feature importance means that the feature is more influential on prediction results. We use the ES as features in our model. For each band in the ES the feature importance is calculated (these bands are the ES states near 0, 1, and the two in the mid-gap region). To better understand the roles played by each band, we put Fig. 8.1(b) and Fig. 8.1(a) side-by-side. The vertical axis of Fig. 8.1(b) is the same as Fig. 8.1(a) with the bands represented by their averaged position from Fig. 8.1(a). The horizontal axis shows the importance of each feature. As shown in the figure, the middle two bands of the ES have the highest influence on predictions, which strongly indicates that the RF model focuses on the degeneracy of the ES to perform its classification decisions. Similar feature properties were also observed for our trained LM model, i.e., the coefficients of most features are nearly zero except for the features in the middle of the ES. We note that it is difficult to interpret coefficients of the NN model, so we have little that we can interpret about its behavior.

Let us move on to see how these methods can be extended. We trained our models with data deep in the topological and trivial phases, but a prior knowledge of the approximate phase transition point was needed to determine a reasonable parameter region to produce the training data. It would be more ideal if the RF model could also be used to find an unknown critical point as well. Indeed, this is possible if we use a scheme similar to the confusion scheme [16]. This method is a trial-and-test scheme that finds a point where two regions can be best distinguished by the model. This point is then the phase transition point. To use this scheme we choose a possible transition point at $W_2^{(c)}$ and calculate the prediction accuracy and error. If $W_2^{(c)}$ is the true transition point, we will produce high accuracy and low error. Otherwise, the accuracy will be low and error will be high. We plot the accuracy and error at different values of $W_2^{(c)}$ in Fig. 8.3 for the RF model. The $\wedge$ shape of the accuracy and the $\vee$ shape of the error are consistent with each other. These results suggest that the transition point is $W_2 \approx 5$, which is consistent with the analytic result indicated by the vertical dashed line. This shows that in principle we could have employed the confusion scheme to find the disorder-driven critical point in order to begin our model training, instead of knowing the exact critical point from an analytic calculation.

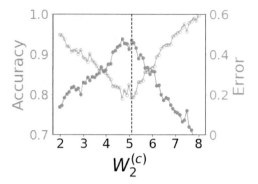

Fig. 8.3.   Accuracy and log loss error of predictions for the RF model with different test choices for transition points $W_2^{(c)}$. The point with the highest accuracy (lowest log loss) agrees well with the true transition point as anticipated from the confusion scheme.

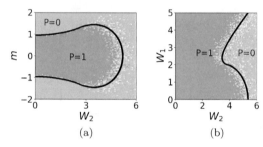

(a)   (b)

Fig. 8.4.   The predicted phase diagram of (a) $W_1 = 1.0$ and (b) $m = 0.5$. The black solid lines are theoretical phase boundaries. P is the predicted probability of being in the topological phase.

So far we have found that the trained model can locate the transition point with relatively high accuracy, even though the model is not given training data information near the phase boundary. Now we want to see if we can expand the region of applicability of our model to a wide range of phase space. For example, we can take two other cross-sections of the three-dimensional phase diagram: (i) fixed $W_1 = 1$ with varying (W_2, m) or (ii) fixed $m = 0.5$ with varying (W_2, W_1). The phase diagrams for these cross-sections are plotted in Figs. 8.4(a) and 8.4(b), respectively. The colormap indicates the predicted probability of being in the topological phase, and the exact phase boundaries are plotted as solid black lines. As can be seen, the RF model makes predictions with high confidence deep in the phases. When disorder is small, the predicted phase boundaries match very well with the exact ones, while there are some deviations near the phase boundaries at large disorder. This gives us confidence that our model generalizes to a broader range of parameters than those on which it has been trained.

Since the robust properties of the ES are characteristic features in topological phases, especially in 1D, we expect that the RF algorithm that we trained for the model in Eq. (8.1) can be applied to a much broader set of symmetry-protected topological phases. Furthermore, since the stable features of the ES — used to train our machine learning algorithm — rely on global symmetries of the system, we expect to be able to observe and quantify the breakdown of the method once symmetry-breaking effects are present.

As an example of the former, let us test the applicability of the RF model to another system. As an example, we apply our class AIII trained RF model to a disordered fermionic Kitaev chain in class BDI, whose Hamiltonian is given by [35]

$$H = \sum_n [t_n \, ib_n a_{n+1} + m_n \, ia_n b_n], \tag{8.7}$$

where a_n and b_n are Majorana fermions. We add disorder to the parameters $t_n = 1 + W_1\omega_1$ and $m_n = m + W_2\omega_2$ where t_n and m_n here can be interpreted as inter-cell and intra-cell Majorana coupling terms. In the clean limit corresponding to $W_1 = W_2 = 0$, this model is a one-dimensional topological superconductor with one (zero) isolated Majorana end state at each end for $|t| > |m|$ ($|t| < |m|$). For the test region, we chose the line $W = 2W_1 = W_2$ and $m = 0.5$. The prediction results for this model are shown in Fig. 8.5, where, similar to Fig. 8.2, the dots represent predicted probabilities of being in the topological phase. The orange curve is fitted using Eq. (8.4), and we find a predicted phase transition near $W = 4$. The black dots are winding numbers calculated in real space for the same system [6]. The transition point determined by the two methods are consistent with each other.

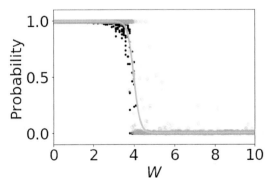

Fig. 8.5.   The predicted probability of being in the topological phase of Kitaev model in symmetry class BDI.

Now let us try to characterize some effects of symmetry breaking for the AIII model. To illustrate this we add a small σ_z term that breaks the chiral symmetry that protects the phase. We fix the other parameters as $m = 0.5$ and $W_1 = W_2 = 1$, so that the system is in topological phase when the symmetry is not broken, and add a term proportional to $\sum_n c_n^\dagger \sigma_z c_n$ to Eq. (8.1). Since the topological phase is protected by chiral symmetry, this term immediately breaks down the topological phase and the degeneracy in the entanglement spectrum is lifted. As the symmetry breaking term becomes stronger, the hitherto degenerate states in the middle of the entanglement spectrum are split away from each other, as indicated in Fig. 8.6(a). The resulting predictions made by the RF model are shown in Fig. 8.6(b), where the blue dots are the raw prediction probabilities. In the absence of symmetry breaking terms, the model confidently predicts a topological phase for these model parameters. On the other hand, as soon as the symmetry breaking strength is nonzero, the predicted probability immediately drops below 0.5, showing the sensitivity of the RF model to the removal of the degeneracy in the ES. Furthermore, this probability goes down gradually as the symmetry breaking strength increase.

Finally, we can compare the predictions of the trained RF model with the very simple classifier of just checking for the degeneracy of the mid-gap entanglement modes. To use the entanglement degeneracy to make predictions of the phase, we classify our data using a threshold and associate gaps in the ES smaller than 0.001 to the topological phase with probability 1, and larger gaps to be in the trivial phase with probability 1. We find that the predicted results using this simple method give accuracy 0.977, which is close to the accuracy of our random forest classifier. We plot the predictions from degeneracy in Fig. 8.7(a) with orange dots. Since we can only predict one or zero, we take the average of the predictions as the probability (green line).

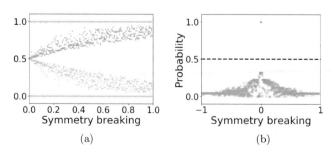

(a)   (b)

Fig. 8.6. (a) Entanglement spectrum with chiral symmetry breaking term. (b) The probability of being in the topological phase with symmetry breaking term added to the system. The model predicts that all configurations are in the trivial phase except when symmetry is preserved.

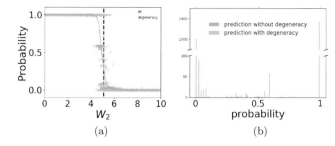

(a)   (b)

Fig. 8.7. (a) Predictions using the simple classifier that checks for degeneracy in the entanglement spectrum (orange dots) compared with random forest classifier (blue squares). The green line is average of predictions for the degeneracy method. The analytically calculated phase transition point is indicated by a vertical dashed line. (b) Distribution of predictions of the random forest model for two different testing datasets. When the test entanglement spectrum data has (no) double degeneracy, most predictions give probability one (zero) as being in the topological phase.

The predictions of the random forest model are shown with blue squares for comparison and the two predictions match extremely well.

We can dig a bit deeper into understanding how the RF model is classifying based on the ES degeneracy. We plot the distribution of random forest predictions in Fig. 8.7(b) for two sets of testing data: one has degeneracies in the ES while the other does not. Note that the y-axis is cut in the middle to reveal details for smaller y values. From the figure, we can see that if the ES has degeneracy the RF model almost always predicts the state as topological (probability > 0.5), while for nondegenerate ES, the model predicts the states as trivial (probability < 0.5) in most cases. Among the 3% incorrect predictions made by either the RF model or the simple degeneracy classifier, 84% are wrong by both models; 9% are wrong by the RF model but correct by degeneracy; and 7% are correct by the RF model but wrong by degeneracy. Therefore, the RF predictions are consistent with the predictions by degeneracies. So, after careful investigation we find that the RF model is making predictions based on the mid-gap degeneracy of the ES. We expect that the training process is essentially finding the best threshold for the gap size. The threshold ends up being about 0.0015, which is close to the value we set for our simple degeneracy classifier.

8.4 Summary

In summary, we applied the random forest model to classify disordered topological phases. Compared with the linear model, random forest gives better predictions. On the other hand, it preserves the easy interpretability of the linear model as compared to neural networks. Because of the generality of the entanglement spectrum, the model trained on a small training dataset can be generalized to test data in a larger phase space, and even to other models in different symmetry classes. A closer look at the RF model indicates that the model is capturing the degeneracy of the mid-gap entanglement spectrum modes and is very sensitive to any symmetry breaking which splits the modes.

Acknowledgments

This chapter is presented as a memorial tribute for Shoucheng Zhang and combines two of his recent interests of topological insulators and machine learning. TLH thanks him for the advice and support given during their long period of collaboration. YZ and TLH acknowledge support from the US National Science Foundation under grant DMR 1351895-CAR. LHS is supported by a faculty startup at Emory University.

References

[1] A. P. Schnyder, S. Ryu, A. Furusaki and A. W. W. Ludwig, Classification of topological insulators and superconductors in three spatial dimensions, *Phys. Rev. B* **78**, 195125 (2008).
[2] S. Ryu, A. P. Schnyder, A. Furusaki and A. W. W. Ludwig, Topological insulators and superconductors: tenfold way and dimensional hierarchy, *New J. Phys.* **12**(6), 065010 (2010).
[3] E. Prodan, T. L. Hughes and B. A. Bernevig, Entanglement spectrum of a disordered topological Chern insulator, *Phys. Rev. Lett.* **105**, 115501 (2010).
[4] J. Song, C. Fine and E. Prodan, Effect of strong disorder on three-dimensional Chiral topological insulators: Phase diagrams, maps of the bulk invariant, and existence of topological extended bulk states, *Phys. Rev. B* **90**, 184201 (2014).
[5] I. Mondragon-Shem and T. L. Hughes, Signatures of metal-insulator and topological phase transitions in the entanglement of one-dimensional disordered fermions, *Phys. Rev. B* **90**, 104204 (2014).
[6] I. Mondragon-Shem, T. L. Hughes, J. Song and E. Prodan, Topological criticality in the chiral-symmetric aiii class at strong disorder, *Phys. Rev. Lett.* **113**, 046802 (2014).
[7] J. Song and E. Prodan, AIII and BDI topological systems at strong disorder, *Phys. Rev. B*, **89**, 224203 (2014).
[8] G. Torlai and R. G. Melko, Learning thermodynamics with Boltzmann machines, *Phys. Rev. B* **94**, 165134 (2016).
[9] L. Wang, Discovering phase transitions with unsupervised learning, *Phys. Rev. B* **94**, 195105 (2016).
[10] J. Carrasquilla and R. G. Melko, Machine learning phases of matter, *Nat. Phys. Lett.*, **13**(5), 431 (2017).
[11] T. Ohtsuki and T. Ohtsuki, Deep learning the quantum phase transitions in random two-dimensional electron systems, *J. Phys. Soc. Jpn.* **85**, 123706 (2016).
[12] P. Broecker, J. Carrasquilla, R. G. Melko and S. Trebst, Machine learning quantum phases of mater beyond the fermion sign problem, *Sci. Rep.* **7**, 8823 (2017).

[13] K. Ch'ng, J. Carrasquilla, R. G. Melko and E. Khatami, Machine learning phases of strongly correlated fermions, *Phys. Rev. X* **7**, 031038 (2017).

[14] L. Huang and L. Wang, Accelerated Monte Carlo simulatons with restricted Boltzmann machines, *Phys. Rev. B*. **95**, 035105 (2017).

[15] Y. Zhang and E. A. Kim, Quantum loop topography for machine learning, *Phys. Rev. Lett.* **118**, 216401 (2017).

[16] E. P. L. van Nieuwenburg, Y.-H. Liu and S. D. Huber, Learning phase transitions by confusion, *Nat. Phys. Lett.* **13**, 435 (2017).

[17] T. Ohtsuki and T. Ohtsuki, Deep learning the quantum phase transitions in random electron systems: Applications to three dimensions, *J. Phys. Soc. Jpn.* **86**, 044708 (2017).

[18] T. Ohtsuki and T. Mano, Drawing phase diagrams of random quantum systems by deep learning the wave function, *J. Phys. Soc. Jpn.* **89**, 022001 (2020).

[19] P. Zhang, H. Shen and H. Zhai, Machine learning topological invariants with neural networks, *Phys. Rev. Lett.* **120**, 066401 (2018).

[20] D. Carvalho, N. A. García-Martínez, J. L. Lado and J. Fernández-Rossier, Real-space mapping of topological invariants using artifical neural networks, *Phys. Rev. B* **97**, 115453 (2018).

[21] N. Yoshioka, Y. Akagi and H. Katsura, Learning disordered topological phase by statistical recovery of symmetry, *Phys. Rev. B* **97**, 205110 (2018).

[22] N. Sun, J. Yi, P. Zhang, H. Shen and H. Zhai, Deep learning topological invariants of band insulators, *Phys. Rev. B* **98**, 085402 (2018).

[23] K. Hornik, M. Stinchcombe and H. White, Multilayer feedforward networks are universal approximators, *Neural Netw.* **2**, 359 (1989).

[24] K. Hornik, Approximation capabilities of multilayer feedforward network, *Neural Netw.* **4**, 251 (1991).

[25] L. Breiman, Random forests, *Machine Learn.* **45**, 5 (2001).

[26] J. Friedman, T. Hastie and R. Tibshirani, *The Elements of Statistical Learning*, Vol. 1 (Springer, New York, 2001).

[27] A. Liaw *et al.*, Cassification and regression by randomforest, *R News* **2**, 18 (2002).

[28] H. Liu and F. D. M. Haldane, Entanglement spectrum as a generalization of entanglement entropy: Identification of topological order in non-abelian fractional quantum hall effect states, *Phys. Rev. Lett.* **101** 010504 (2008).

[29] F. Pollmann, A. M. Turner, E. Berg and M. Oshikawa, Entanglement spectrum of a topological phase in one dimension, *Phys. Rev. B* **81**, 064439 (2010).

[30] A. P. Schnyder, S. Ryu and A. W. W. Ludwig, Lattice model of a three-dimensional topological singlet superconductor with time-reversal symmetry, *Phys. Rev. Lett.* **102**, 196804 (2009).

[31] I. Peschel, Calculation of reduced density matrices from correlation functions, *J. Phys. A: Math. Gen* **36**, L205 (2003).

[32] F. Pedregosa, G. Varoquauix, A. Gramfort, V. Michel, B. Thirion, O. Grisel, M. Blondel, P. Prettenhofer, R. Weiss, V. Dubourg, J. Vanderplas, A. Passos, D. Cournapeau, M. Brucher, M. Perrot and E. Duchesnay, Scikit-learn: Machine learning in Python, *J. Machine Learning Res.* **12**, 2825 (2011).

[33] I. Goodfellow, Y. Bengio and A. Courville, *Deep Learning* (MIT Press, 2016); http://www.deeplearningbook.org.

[34] L. Breiman, J. H. Friedman, R. A. Olshen and C. J. Stone, *Classification and Regression Trees*, The Wadsworth Statistics/Probability Series (Wadsworth and Brooks/Cole Advanced Books and Software, Monterey, CA, 1984).

[35] A. Yu. Kitaev, Unpaired Majorana fermions in quantum wires, *Phys. Uspekhi*, **44**, 131 (2001).

Chapter 9

Topological Physics with Mercury Telluride

Saquib Shamim[*,†,‡], Hartmut Buhmann[*,†,§] and Laurens W. Molenkamp[*,†,¶]

*Physikalisches Institut (EP3), Universität Würzburg,
Am Hubland, 97074 Würzburg, Germany
†Institute for Topological Insulators, Am Hubland,
97074 Würzburg, Germany
‡Saquib.Shamim@physik.uni-wuerzburg.de
§Hartmut.Buhmann@physik.uni-wuerzburg.de
¶Laurens.Molenkamp@physik.uni-wuerzburg.de

9.1 Introduction

Topology, traditionally associated with the study of geometric shapes of mathematical objects, is an active area of research in experimental and theoretical physics. Léon van Hove introduced topology in physics way back in 1953 to explain the frequency distribution of vibration in crystals [1]. The significance of topology in quantum condensed matter physics became more apparent when topological order was used to explain the precise quantization of transverse resistance in the quantum Hall effect [2,3]. The theoretical prediction and the subsequent experimental discovery of quantum spin Hall effect in 2007 led to a rapid growth of topological insulators as one of the most pursued research directions in physics [4–9]. Topological insulators have a bulk energy gap, which separates the lowest unoccupied band from the highest occupied band (conduction band and valence band in semiconductors), and conducting states at the surface. The gapless surface states are topologically protected by time-reversal symmetry. The topological classification of materials require the identification of a topological invariant. The topological invariant, e.g., Chern number, is characteristic to the topology of the system. In case of the integer quantum Hall effect, the Hall conductivity is quantized in integer multiples of e^2/h, $\sigma_{xy} = \nu e^2/h$, where the integer ν is the topological invariant of the system.

After the experimental realization of the quantized spin Hall conductance in HgTe quantum wells, concentrated efforts in experimental and theoretical condensed matter research have revealed two-dimensional and three-dimensional topological insulators hosted in various materials systems. Apart from bridging the gap between the mathematically abstract concept of topology, these systems have intrigued researchers due to many exciting properties like spin-momentum locking, protection against disorder and potential applications in topological quantum computation. Advances in material research have added more materials to the expanding topological family [10–13] and realized new topological states protected by additional symmetries such as topological crystalline insulators [14, 15] and topological superconductors.

This chapter is devoted to study of topological physics with HgTe. The advantage of HgTe is the high crystalline quality of the material. Molecular beam epitaxy grown HgTe-based two- and three-dimensional topological insulators have the highest mobility when compared to other topological materials. In addition, the ability to artificially engineer the band structure, by growing strained layers and alloying with other atoms, has allowed the investigation of a variety of novel topological phenomena in HgTe.

9.1.1 *Layout*

This chapter is organized as follows: In Sect. 9.2, we discuss the band structure of bulk HgTe, particularly the band inversion which becomes the defining property of topological insulators. Section 9.3 demonstrates HgTe quantum wells is the prototype material to realize the quantum spin Hall effect. This section also investigates the various properties of quantum spin Hall edge channels. Section 9.4 shows how the band structure of thick HgTe layers can be artificially engineered to realize exotic topological phases such as gapless Dirac-like surface states and bulk Dirac nodes. Finally, Sect. 9.5 concludes this chapter with a discussion on future directions and prospects of studying topology with HgTe.

9.2 Bandstructure of HgTe

The topological classification of materials is inherent to the band structure. The details of the band structure can substantially alter the electronic response of materials. An inverted band structure, due to relativistic corrections and spin–orbit interaction, is responsible for the topological

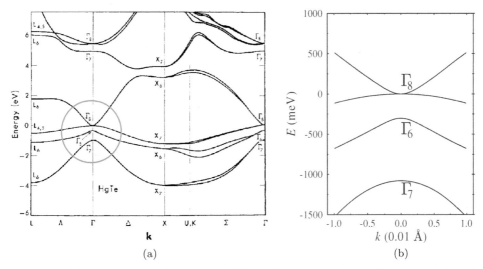

Fig. 9.1. (a) Electronic band structure of HgTe along the principal directions. Adapted from Ref. [16]. (b) Magnified view of the region enclosed by the circle in (a).

character of materials. In this section, we will focus on the band structure of HgTe with emphasis on how it can be tuned to realize a variety of topological systems such as 2D topological insulator, 3D topological insulator and Dirac/Weyl semimetal.

Bulk HgTe is a zinc blende semiconductor with a band structure similar to other zinc blende semiconductors (such as GaAs, Si or CdTe) with one important difference: The energetic ordering of the Γ_6- and Γ_8-bands is reversed in HgTe when compared to CdTe. For CdTe, the conduction and valence band are of Γ_6- and Γ_8-type, respectively. However, for HgTe, the relativistic correction to the Hamiltonian (due to heavy Hg atoms) leads to a rearrangement of Γ_6- and Γ_8-bands: The Γ_6-band now lies energetically below the doubly degenerate Γ_8-bands. The band structure of bulk HgTe along the symmetry directions is shown in Fig. 9.1(a). A magnified view of the band structure near the Γ-point is shown in Fig. 9.1(b). For an intrinsic HgTe crystal, the hole-like Γ_8-band is fully filled and forms the valence band while the electron-like Γ_8-band is empty and forms the conduction band. Hence, intrinsic HgTe is a zero-gap semiconductor and can be called a topological semimetal, which can be converted into a topological insulator by opening up a gap between the doubly degenerate Γ_8-bands. This is the starting point from which the band structure can be engineered using external parameters such as strain (compressive or tensile) or quantum confinement to realize a variety of topological systems.

9.3   HgTe as 2D topological insulator: Realization of quantum spin Hall effect

The theoretical prediction of the intrinsic spin Hall effect in 1971 and its experimental observation about three decades later generated significant interest among the condensed matter community [17–22]. The initial experimental observations of the spin Hall effect involved optical techniques [20,21] and the electrical detection of "extrinsic" spin Hall effect were reported for metallic nanostructures with ferromagnetic electrodes [23]. A clear demonstration of the ballistic "intrinsic" spin Hall effect was shown by our group in HgTe nanostructures [22]. An all-electrical manipulation of the spin in the absence of an external magnetic field has potential applications in spintronic devices. Additionally, the spin Hall effect laid the groundwork for the prediction of a new phenomena called the *quantum spin Hall* effect. The quantum spin Hall effect is a new state of matter that is topologically nontrivial and electrical conduction is solely due to helical edge channels in two-dimensional systems. Electrons with opposite spin propagate in opposite directions along the edges of the sample and lead to quantized conductance of $2e^2/h$ without any external magnetic field. In the absence of an external magnetic field, the time-reversal symmetry guarantees protection against backscattering, leading to dissipationless transport along the edges.

The quantum spin Hall effect was initially predicted for graphene [5] as well as spin–orbit coupled semiconductors [4,6]. The experimental realization of quantum spin Hall in graphene is deemed unrealistic because the spin–orbit interaction induced gap of $\sim$ few μeV is too small to be observable under common experimental conditions. However, after a collaborative exchange of Pfeuffer Jeschke's PhD thesis, Bernevig *et al.* show that it is possible to realize the quantum spin Hall effect in type III quantum wells of HgTe above a critical thickness [8, 24]. More specifically, at the critical thickness d_c, there is a topological phase transition between the insulating phase (for conventional quantum wells) and a topological phase hosting quantum spin Hall edge channels. This phase transition is inherently connected to the band structure since at precisely the same thickness d_c, the band structure of the HgTe quantum wells changes from normal to inverted type. The authors use an effective four-band model (called BHZ model) with E1($\pm$) and H1($\pm$) subbands, where the $\pm$ corresponds to spin-up and spin-down states. Figure 9.2(a) shows the energy of the E1- and H1-bands at $k = 0$ as a function of thickness of quantum well. The order of the bands reverses at the critical thickness d_c. The inset of Fig. 9.2(a) shows schematically the CdTe–HgTe–CgTe quantum well in the normal regime ($d < d_c$) and the inverted regime ($d > d_c$). In the inverted regime, the gap between the

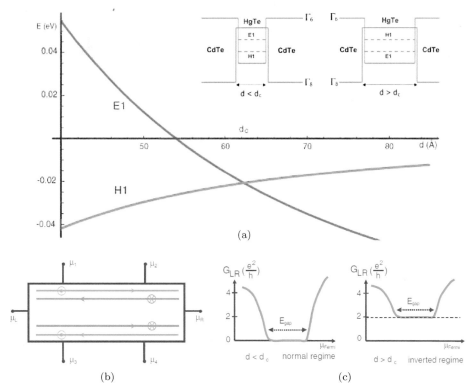

Fig. 9.2.   (a) Energy of the E1- and H1-bands at $k = 0$ as a function of the quantum well thickness d. The inset shows the schematic band structure of the CdTe–HgTe–CdTe quantum well for $d < d_c$ (left panel) and $d > d_c$ (right panel). (b) Experimental scheme of a six-terminal Hall bar to detect quantum spin Hall effect. The edge states of opposite spins are shown in green and purple. (c) Two-terminal conductance in the insulating regime (left panel) and the quantum spin Hall regime (right panel). Adapted from Ref. [8].

E1- and H1-bands reaches 15–20 meV (depending on the thickness of the quantum well) which is much greater than the $k_B T$ for $T < 4$ K and hence it is much more realistic to realize this experimentally compared to previous predictions of the quantum spin Hall effect. The prediction can be verified experimentally by making a six-terminal device in a Hall bar geometry (Fig. 9.2(c)) and measuring the longitudinal resistance for quantum wells in the normal and inverted regime. In the normal regime, the device should be in the conventional insulting regime characterized by zero conductance (left panel of Fig. 9.2(d)) while in the inverted regime, the device should support gapless helical edge channels characterized by quantum spin Hall conductance of $2e^2/h$ (right panel of Fig. 9.2(c)).

An experimental verification of the above prediction followed soon by our group, using HgTe quantum wells of various thickness sandwiched between (Hg,Cd)Te barriers [9]. All the layers are grown by molecular beam epitaxy

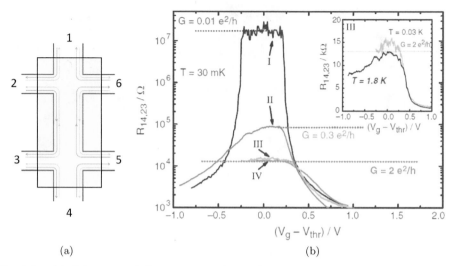

(a)   (b)

Fig. 9.3.   (a) Schematic of the six-terminal Hall bar used to detect quantum spin Hall effect in HgTe quantum wells. (b) The longitudinal resistance $R_{14,23}$ of quantum wells of various thickness and device dimensions. Adapted from Ref. [9].

and the quantum wells show mobility $\sim 10^5$ cm^{2}V^{-1}s^{-1}. The wafers are patterned into six-terminal Hall bar devices by electron beam lithography and dry etching with Ar$^+$ ions. A schematic of the device is shown in Fig. 9.3(a). The four terminal resistance is measured at zero magnetic field and at low temperatures (30 mK) using ac lock-in technique. Figure 9.3(b) shows the resistance $R_{14,23} = V_{23}/I_{14}$ for four devices: I (20 × 13.3 μm^2 device from a 5.5 nm quantum well), II (20 × 13.3 μm^2 device from a 7.3 nm quantum well), III (1 × 1 μm^2 device from a 7.3 nm quantum well) and IV (1×0.5 μm^2 device from a 7.3 nm quantum well). Device I has a normal band structure and exhibits an insulating state characterized by zero conductance when the chemical potential is tuned to the bulk gap. Devices II, III and IV have an inverted band structure and show a finite conductance even when the chemical potential is in the bulk gap. For micron-sized devices (III and IV), the conductance is independent of the width of the Hall bar and is quantized to $2e^2/h$. The inset of Fig. 9.3(b) shows the conductance quantization for two Hall bars at 30 mK and 1.8 K. This was the first experimental evidence of the quantum spin Hall effect.

9.3.1   *Nonlocality and spin polarization of quantum spin Hall edge channels*

Soon after the discovery of quantum spin Hall effect, Roth *et al.* [25] showed further evidence of helical edge states in HgTe quantum wells by measuring

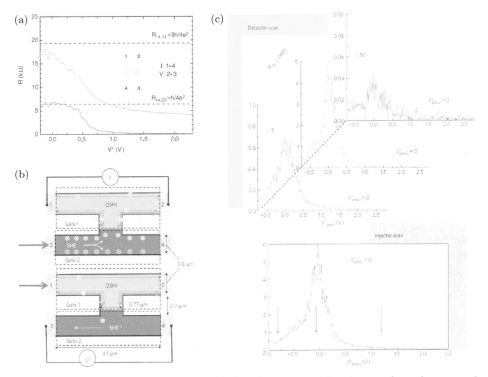

Fig. 9.4. (a) The nonlocal resistance (red) and two-terminal resistance (green) measured in an H-bar device (schematic in the inset) at 1.8 K. The dashed lines indicate the value expected from Landau–Buttiker calculation. Adapted from Ref. [25]. (b) Schematic of the H-bar device used to detect the spin polarization of the quantum spin Hall edge states. The yellow colored region indicates quantum spin Hall regime while the green colored region indicates metallic regime. (c) Nonlocal resistance data at 1.8 K, corresponding to the measurement scheme shown in (b). Adapted from Ref. [26].

nonlocal transport in the quantum spin Hall regime. The authors fabricated multi-terminal devices from inverted HgTe quantum wells and measured the nonlocal voltage across contacts which are spatially separated from the current path. The inset of Fig. 9.4(a) shows the schematic of an H-bar device used for nonlocal measurements. The current is applied along the contacts 1–4 while the voltage is measured across 2–3. In the trivial case, the voltage across 2–3 would be zero since the current flowing along the contacts is zero. However, in case of helical edge channels, the expected nonlocal resistance can be calculated using the Landau–Buttiker formalism. Figure 9.4(a) shows the measured nonlocal resistance and comparison with the value expected from Landau–Buttiker formalism (dashed lines). The agreement between the experiment and theory unambiguously confirmed the quantum spin Hall effect in HgTe quantum wells.

Apart from the conductance quantization and nonlocal transport, the quantum spin Hall edge channels are spin polarized. Brüne *et al.* [26] demonstrated an all-electrical detection of the spin polarization of the quantum spin Hall edge channels. Figure 9.4(b) shows the device schematic and the measurement principle. The authors patterned the mesa into an H-bar whose two legs could be independently tuned into the metallic (denoted by green color) or the quantum spin Hall regime (denoted by the yellow color) by means of split gate. A current injected into the metallic leg causes opposite spins to move towards the opposite edges of the leg due to intrinsic spin Hall effect (already shown to exist in inverted HgTe quantum wells). A nonlocal voltage then develops across the leg induced in the quantum spin Hall regime. This happens due to coupling of the spin-polarized quantum spin Hall states and the spin in the metallic legs with same polarization. In the absence of spin polarization of the quantum spin Hall edges states, the nonlocal signal would be zero. Similarly, current could be injected along the quantum spin Hall leg and a nonlocal voltage would be detected in the metallic due to the inverse spin Hall effect. Figure 9.4(c) shows the injector and detector scan and the presence of finite nonlocal signal establishes the spin-polarized nature of the quantum spin Hall edge channels.

9.3.2 *Refined fabrication process: Chemical wet etching*

The conductance quantization due to quantum spin Hall effect was observed only for microscopic devices. All these devices were fabricated by dry-etching using high-energy Ar^+ ions which is detrimental to the sample quality. A clear indication of this is seen when the mobility of devices of different dimensions are compared. Figure 9.5(a) shows that the mobility of the devices fabricated by dry etching decreases from 5×10^5 $cm^2V^{-1}s^{-1}$ for macroscopic devices to 4×10^4 $cm^2V^{-1}s^{-1}$ for microscopic devices. This drastic reduction in mobility is associated with the dry-etching-induced damage to the mesa. Since the microstructures are most important for quantum spin Hall studies as well as further experiments on these devices by integrating with superconductors, it became essential to develop an alternate fabrication procedure, which is not detrimental to the sample quality. A chemical wet-etching process, developed by Bendias *et al.* [27], preserves the inherent crystal quality of the pristine material by using an aqueous solution of $KI:I_2:HBr$ to etch the mesa. The resulting devices are of superior quality as evidenced by higher mobility even in wet-etched microstructures. Figure 9.5(b) shows the mobility of devices with dimensions ranging from few hundreds μm to a few μm. A high mobility 4×10^5 $cm^2V^{-1}s^{-1}$

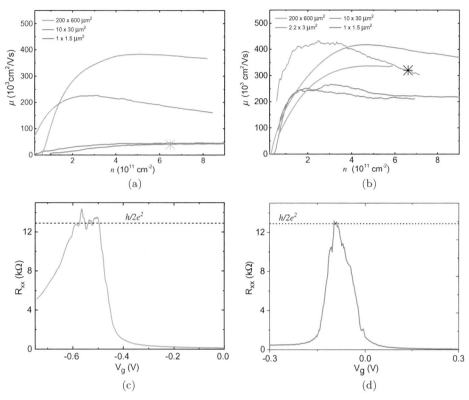

Fig. 9.5. The mobility μ as a function of density at 4.2 K for Hall bars of various dimensions for (a) dry-etched device and (b) wet-etched device. (c) and (d) The longitudinal resistance as a function of gate voltage for two micron-sized Hall bars from two different wafers. The black dashed lines indicate quantized value expected in the quantum spin Hall regime. The devices in (c) and (d) were measured at 4.2 K and 1.4 K, respectively. Adapted from Ref. [27].

even for microstructures establishes wet-etching as a superior alternative to dry-etching. The authors also confirmed that the devices fabricated from the new process show the quantized conductance due to quantum spin Hall effect. The quantized spin Hall conductance is shown for two devices fabricated from different wafers in Figs. 9.5(c) and 9.5(d). The reproducible fluctuations of the resistance in the quantum spin Hall regime are observed for multiple devices but the exact origin of these fluctuations is not yet known. Theoretical investigations have suggested the influence of inelastic backscattering due to charge puddles (common in narrow-gap semiconductors) [28–30]. A report by Lunczer *et al.* [31] has established that charge puddles-induced potential fluctuations lead to deviation of the conductance from the quantized value. Recently, Strunz *et al.* [32] have used the wet-etch technique to realize quantum point contacts in a topological material. The cleaner mesa edges enabled the authors to study the interaction between the pristine helical edge channels when they are brought in close proximity to each other. The authors show the appearance of an interaction-induced 0.5-anomaly in the conductance plateaus of a quantum point contact in a topological material.

9.4   HgTe as 3D topological insulator

In this section, we review bulk HgTe as a three-dimensional (3D) topological insulator. For thicker HgTe layers sandwiched between (Hg,Cd)Te barriers, when the confinement energies are small such that the layers can be considered three-dimensional, one expects that band inversion leads to Dirac-like dispersion emerging from the 2D surface states. ARPES measurements on a fully relaxed 1 μm thick HgTe film confirm the presence of massless surface states (Fig. 9.6(a)). The Dirac-like dispersion stems from the surface state formed due to inversion of Γ_6- and Γ_8-bands. Though the surface states exist, in the absence of a bulk band gap, thicker HgTe films are more accurately described as a zero-gap semiconductor or a semimetal. However, a tensile strain can open up a bulk band gap, turning HgTe into a 3D topological insulator and making the surface states accessible in transport experiments.

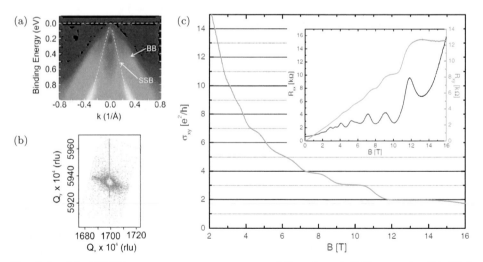

Fig. 9.6.   (a) ARPES measurements on a 1 μm fully relaxed HgTe sample. (b) High resolution X-ray diffraction measurements showing the reciprocal space map of a 70 nm HgTe film under tensile strain. (c) The Hall conductivity σ_{xy} of a 70 nm HgTe film at 50 mK. The inset shows the longitudinal and transverse resistance, R_{xx} and R_{xy} respectively as a function of perpendicular magnetic field. Adapted from Ref. [34].

(a)

(b)

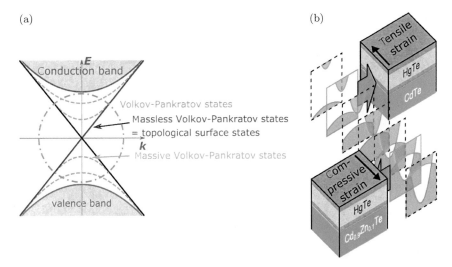

Fig. 9.7. (a) A schematic representation of the band structure at the interface of semiconductors with mutually inverted bands. (b) Evolution of band structure of HgTe films as a function of tensile strain and compressive strain. Adapted from Ref. [33].

9.4.1 *Bandstructure of HgTe: Effect of strain*

Figure 9.7(b) shows the evolution of band structure of HgTe as a function of strain [33]. Application of a strain lifts the degeneracy of the Γ_8-bands. The strain can either be tensile or compressive, the two types leading to the formation of two different classes of systems. A tensile strain opens a topological band gap and the conduction happens due to massless surface states. A compressive strain also breaks the degeneracy of Γ_8-bands, which now move in opposite directions (as compared to tensile strain) leading to linear crossing points (see the far right-band structure is Fig. 9.7(b)). The former case is the well-known example of a 3D topological insulator while the latter case is the manifestation of bulk Dirac crossings.

9.4.2 *Surface states of HgTe 3D TI*

Brüne *et al.* [34] have used the above concept to realize a 3D TI in HgTe films under tensile strain. The authors used CdTe substrates to grow 70 nm thick HgTe films by molecular beam epitaxy. The 0.3% lattice constant mismatch between HgTe and CdTe results in strained HgTe films when the thickness is less than the critical thickness ($\sim$200 nm) at which the layer are fully relaxed. Hence the authors used a 70 nm film for their investigations of HgTe films as a 3D topological insulator where a bulk gap $\sim$20 meV ensures that surface states can be accessed in transport experiments at low temperatures. The high resolution X-ray diffraction map shown in Fig. 9.6(b) confirms that

70 nm films are fully strained. Figure 9.6(c) shows the magnetotransport measurements of a 600×200 μm^2 Hall bar at 50 mK. The occurrence of odd filling factor ($\nu = 9, 7, 5$) Landau level at low magnetic fields indicates the presence of zero mode Landau level confirming that the transport stems from two independent Dirac-like surface states.

9.4.3   *Volkov–Pankratov states*

Band inversion is the key to the topological character of materials. However, one must always be careful to consider the formation of additional conducting states which form at the interface of materials with inverted bands with respect to each other. The interface of two materials with mutually inverted bands was studied by Volkov and Pankratov [35] almost four decades ago and their observations are summarized in Fig. 9.7(a). In addition to the bulk valence and conduction band, there are two distinct types of surface states: Massless surface states with Dirac-like dispersion (solid black lines) and massive surface states, also called massive Volkov–Pankratov states (dashed magenta lines). The massless surface states are topologically nontrivial while the massive Volkov–Pankratov states are trivial. Recently, Inhofer *et al.* [36] demonstrated the existence of massive Volkov–Pankratov states in HgTe 3D topological insulators using high-frequency compressibility measurements. The presence of massive surface states acquires additional significance for the recently discovered class of Weyl/Dirac semimetals where the coexistence of nontrivial massless surface states, trivial massive surface states and bulk conduction complicates the scenario and could lead to inaccurate conclusions.

9.4.4   *Coexistence of Dirac nodes and Volkov–Pankratov states in bulk HgTe*

In this section, we focus on the realization of Dirac/Weyl nodes in compressively strained HgTe. As discussed previously, the three kinds of states, the topological surface states, massive Volkov–Pankratov states and Weyl/Dirac crossing, can coexist when an interface is made from material with mutually inverted bands. Hence, any claims of demonstration of bulk Dirac nodes must also account for the conduction from the topological surface as well as the massive Volkov–Pankratov states. Recently, our group has shown a clear demonstration of the interplay between the Dirac nodes, topological surface states and the massive Volkov–Pankratov states in compressively strained HgTe layers [33]. The high crystalline quality of the molecular beam epitaxy grown layers guarantees extremely low intrinsic carrier density which

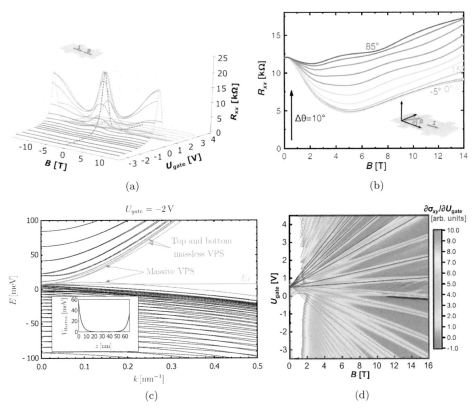

Fig. 9.8. (a) The longitudinal resistance R_{xx} as a function of magnetic field applied parallel to the direction of current for different gate voltages V_g at 2 K. (b) $R_x x$ as a function of magnetic field applied at various angles θ with respect to the current (see schematic in inset) at 0.3 K. (c) $k \cdot p$-band structure of a 66 nm compressively strained HgTe film including the bulk inversion asymmetry terms and Hartree terms. (d) Landau-level fan diagram showing $\delta\sigma_{xy}/\delta U_{gate}$ as a function of gate voltage U_{gate} and perpendicular magnetic field B at 20 mK. The black and magenta lines indicate the contribution from massless topological surface states and the massive Volkov–Pankratov states. Adapted from Ref. [33].

allowed the authors to tune the chemical potential exactly to the Dirac nodes and study the resulting bulk conduction. Figure 9.8(a) shows the longitudinal resistance as a function magnetic field parallel to the direction of current flowing through the sample. For the chemical potential tuned to the Dirac node, a negative magentoresistance of $\sim$60% is observed. The negative magnetoresistance decreases as the chemical potential is moved away from the Dirac node in either direction (by means of a gate voltage). A negative magnetoresistance is a hallmark of the chiral anomaly associated with Weyl nodes. Further, as shown in Fig. 9.8(b), the negative magnetoresistance decreases as the angle between the applied magnetic field and the current increases in confirmation with the expected behavior from the chiral anomaly.

When the chemical potential is moved away from the Dirac nodes, magnetotransport measurements in perpendicular magnetic fields demonstrate clear evidence of the coexistence of topological surface states and massive Volkov–Pankratov states. The transverse resistance shows accurate quantization to the Klitzing constant in perpendicular magnetic fields implying conduction from 2D surface states. The Landau fan chart in Fig. 9.8(d) shows the contribution from topological surface states (black lines) and massive Volkov–Pankratov states (magenta lines). A 6×6 $\boldsymbol{k} \cdot \boldsymbol{p}$-band structure calculation taking into account the compressive strain as well as Hartree potential (inset of Fig. 9.8(c)) confirms the presence of massless surface states and massive Volkov–Pankratov states as shown in Fig. 9.8(c).

9.5   Conclusion and outlook

In this chapter, we have shown that HgTe provides an excellent platform to emulate a variety of topological phenomena. The high crystalline quality of the materials grown by molecular beam epitaxy guarantees extremely high mobility as compared to other topological materials. This has enabled the realization of various topological states by engineering the band structure using various parameters such as thickness, strain and doping. There are various promising directions of research to be explored in HgTe, some of which are enumerated below:

(1) Combining quantum point contacts in high quality QSH material (like HgTe) with s-wave superconductors provides an opportunity to test the recent prediction of realization of parafermions [37].

(2) The effect of strong Coulomb interactions between the helical edge states is relatively unexplored experimentally. Strunz *et al.* [32] have reported a novel 0.5-anomaly in quantum point contacts in HgTe as a consequence of interaction among the helical edge channels. By bringing two helical edges in close proximity to each other either laterally (by nanofabrication) or vertically (using molecular beam epitaxy to grow a double quantum well), one should be able to study the Coulomb drag among the helical edge channels [38, 39].

(3) Though the extraordinary electrical conductivity of helical edge channels has been unambiguously demonstrated by various experiments, measurements of thermal conductivity of the these states remain elusive. By using high-frequency Johnson noise thermometry, one can determine the limits of thermal conductivity of helical edges channels.

References

[1] L. Van Hove, The occurrence of singularities in the elastic frequency distribution of a crystal, *Phys. Rev.* **89**, 1189 (1953); doi: 10.1103/PhysRev.89.1189; URL https:// link.aps.org/doi/10.1103/PhysRev.89.1189.

[2] D. J. Thouless, M. Kohmoto, M. P. Nightingale and M. den Nijs, Quantized Hall conductance in a two-dimensional periodic potential, *Phys. Rev. Lett.* **49**(6), 405 (1982); doi: 10.1103/PhysRevLett.49.405; URL https://link.aps.org/doi/10.1103/ PhysRevLett.49.405.

[3] X.-G. Wen, Topological orders and edge excitations in fractional quantum Hall states, *Adv. Phys.* **44**(5), 405 (1995); doi: 10.1080/00018739500101566; URL https://doi.org/ 10.1080/00018739500101566.

[4] B. A. Bernevig and S.-C. Zhang, Quantum spin Hall effect, *Phys. Rev. Lett.* **96**, 106802 (2006); doi: 10.1103/PhysRevLett.96.106802; URL https://link.aps.org/doi/ 10.1103/PhysRevLett.96.106802.

[5] C. L. Kane and E. J. Mele, Quantum spin Hall effect in graphene, *Phys. Rev. Lett.* **95**(22), 226801 (2005); doi: 10.1103/PhysRevLett.95.226801.

[6] X.-L. Qi, Y.-S. Wu and S.-C. Zhang, Topological quantization of the spin hall effect in two-dimensional paramagnetic semiconductors, *Phys. Rev. B* **74**, 085308 (2006); doi: 10.1103/PhysRevB.74.085308; URL http://link.aps.org/doi/10.1103/PhysRevB. 74.085308.

[7] D. N. Sheng, Z. Y. Weng, L. Sheng and F. D. M. Haldane, Quantum spin-Hall effect and topologically invariant chern numbers, *Phys. Rev. Lett.* **97**, 036808 (2006); doi: 10.1103/PhysRevLett.97.036808; URL https://link.aps.org/doi/10.1103/ PhysRevLett.97.036808.

[8] B. A. Bernevig, T. L. Hughes and S.-C. Zhang, Quantum spin Hall effect and topological phase transition in HgTe quantum wells, *Science* **314**(5806), 1757 (2006); doi: 10.1126/science.1133734; URL http://science.sciencemag.org/content/ 314/5806/1757.

[9] M. König, S. Wiedmann, C. Brüne, A. Roth, H. Buhmann, L. W. Molenkamp, X.-L. Qi and S.-C. Zhang, Quantum spin Hall insulator state in HgTe quantum wells, *Science* **318**(5851), 766 (2007); doi: 10.1126/science.1148047; URL http:// www.sciencemag.org/content/318/5851/766.abstract.

[10] I. Knez, R.-R. Du and G. Sullivan, Evidence for helical edge modes in inverted InAs/GaSb quantum wells, *Phys. Rev. Lett.* **107**, 136603 (2011); doi: 10.1103/ PhysRevLett.107.136603; URL https://link.aps.org/doi/10.1103/PhysRevLett.107. 136603.

[11] Y. Xu, I. Miotkowski, C. Liu, J. Tian, H. Nam, N. Alidoust, J. Hu, C.-K. Shih, M. Z. Hasan and Y. P. Chen, Observation of topological surface state quantum Hall effect in an intrinsic three-dimensional topological insulator, *Nat. Phys.* **10**(12), 956 (2014); URL https://www.nature.com/articles/nphys3140.

[12] F. Reis, G. Li, L. Dudy, M. Bauernfeind, S. Glass, W. Hanke, R. Thomale, J. Schäfer and R. Claessen, Bismuthene on a SiC substrate: A candidate for a high-temperature quantum spin Hall material, *Science* **357**(6348), 287 (2017); doi: 10.1126/science. aai8142; URL http://science.sciencemag.org/content/357/6348/287.

[13] S. Wu, V. Fatemi, Q. D. Gibson, K. Watanabe, T. Taniguchi, R. J. Cava and P. Jarillo-Herrero, Observation of the quantum spin Hall effect up to 100 Kelvin in a monolayer crystal, *Science* **359**(6371), 76 (2018); doi: 10.1126/science.aan6003; URL http:// science.sciencemag.org/content/359/6371/76.

[14] L. Fu, Topological crystalline insulators, *Phys. Rev. Lett.* **106**, 106802 (2011); doi: 10. 1103/PhysRevLett.106.106802; URL https://link.aps.org/doi/10.1103/PhysRevLett. 106.106802.

[15] Y. Tanaka, Z. Ren, T. Sato, K. Nakayama, S. Souma, T. Takahashi, K. Segawa and Y. Ando, Experimental realization of a topological crystalline insulator in SnTe, *Nat. Phys.* **8**(11), 800 (2012); URL https://www.nature.com/articles/nphys2442.

[16] D. J. Chadi, J. P. Walter, M. L. Cohen, Y. Petroff and M. Balkanski, Reflectivities and electronic band structures of CdTe and HgTe, *Phys. Rev. B* **5**, 3058 (1972); doi: 10.1103/PhysRevB.5.3058; URL https://link.aps.org/doi/10.1103/PhysRevB.5.3058.

[17] M. Dyakonov and V. Perel, Current-induced spin orientation of electrons in semiconductors, *Phys. Lett. A* **35**(6) 459 (1971); doi: https://doi.org/10.1016/0375-9601(71)90196-4; URL http://www.sciencedirect.com/science/article/pii/0375960171901964.

[18] M. Dyakonov and V. Perel, Possibility of orienting electron spins with current, *Soviet J. Exp. Theoret. Phys. Lett.* **13**, 467 (1971).

[19] J. E. Hirsch, Spin Hall effect, *Phys. Rev. Lett.* **83**, 1834 (1999); doi: 10.1103/PhysRevLett.83.1834; URL https://link.aps.org/doi/10.1103/PhysRevLett.83.1834.

[20] Y. K. Kato, R. C. Myers, A. C. Gossard and D. D. Awschalom, Observation of the spin Hall effect in semiconductors, *Science* **306**(5703), 1910 (2004); doi: 10.1126/science.1105514; URL https://science.sciencemag.org/content/306/5703/1910.

[21] J. Wunderlich, B. Kaestner, J. Sinova and T. Jungwirth, Experimental observation of the spin-Hall effect in a two-dimensional spin–orbit coupled semiconductor system, *Phys. Rev. Lett.* **94**, 047204 (2005); doi: 10.1103/PhysRevLett.94.047204; URL https://link.aps.org/doi/10.1103/PhysRevLett.94.047204.

[22] C. Brüne, A. Roth, E. G. Novik, M. König, H. Buhmann, E. M. Hankiewicz, W. Hanke, J. Sinova and L. W. Molenkamp, Evidence for the ballistic intrinsic spin Hall effect in HgTe nanostructures, *Nature Phys.* **6**(6), 448 (2010); doi: 10.1038/nphys1655; URL http://dx.doi.org/10.1038/nphys1655.

[23] S. O. Valenzuela and M. Tinkham, Direct electronic measurement of the spin Hall effect, *Nature* **442**(7099), 176 (2006).

[24] A. Pfeuffer-Jeschke. PhD thesis, Würzburg University, Würzburg, Germany (2000).

[25] A. Roth, C. Brüne, H. Buhmann, L. W. Molenkamp, J. Maciejko, X.-L. Qi and S.-C. Zhang, Nonlocal transport in the quantum spin Hall state, *Science* **325**(5938), 294 (2009); doi: 10.1126/science.1174736; URL http://www.sciencemag.org/content/325/5938/294.abstract.

[26] C. Brune, A. Roth, H. Buhmann, E. M. Hankiewicz, L. W. Molenkamp, J. Maciejko, X.-L. Qi and S.-C. Zhang, Spin polarization of the quantum spin Hall edge states, *Nature Phys.* **8**(6), 486 (2012); doi: 10.1038/nphys2322; URL http://dx.doi.org/10.1038/nphys2322.

[27] K. Bendias, S. Shamim, O. Herrmann, A. Budewitz, P. Shekhar, P. Leubner, J. Kleinlein, E. Bocquillon, H. Buhmann and L. W. Molenkamp, High mobility HgTe microstructures for quantum spin Hall studies, *Nano Lett.* **18**(8), 4831 (2018); doi: 10.1021/acs.nanolett.8b01405; URL https://doi.org/10.1021/acs.nanolett.8b01405.

[28] T. L. Schmidt, S. Rachel, F. von Oppen and L. I. Glazman, Inelastic electron backscattering in a generic helical edge channel, *Phys. Rev. Lett.* **108**, 156402 (2012); doi: 10.1103/PhysRevLett.108.156402; URL https://link.aps.org/doi/10.1103/PhysRevLett.108.156402.

[29] J. I. Väyrynen, M. Goldstein and L. I. Glazman, Helical edge resistance introduced by charge puddles, *Phys. Rev. Lett.* **110**, 216402 (2013); doi: 10.1103/PhysRevLett.110.216402; URL https://link.aps.org/doi/10.1103/PhysRevLett.110.216402.

[30] J. I. Väyrynen, M. Goldstein, Y. Gefen and L. I. Glazman, Resistance of helical edges formed in a semiconductor heterostructure, *Phys. Rev. B* **90**, 115309 (2014); doi: 10.1103/PhysRevB.90.115309; URL https://link.aps.org/doi/10.1103/PhysRevB.90.115309.

[31] L. Lunczer, P. Leubner, M. Endres, V. L. Müller, C. Brüne, H. Buhmann and L. W. Molenkamp, Approaching quantization in macroscopic quantum spin Hall devices through gate training, *Phys. Rev. Lett.* **123**, 047701 (2019); doi: 10.1103/PhysRevLett.123.047701; URL https://link.aps.org/doi/10.1103/PhysRevLett.123.047701.

[32] J. Strunz, J. Wiedenmann, C. Fleckenstein, L. Lunczer, W. Beugeling, V. L. Müller, P. Shekhar, N. T. Ziani, S. Shamim, J. Kleinlein, H. Buhmann, B. Trauzettel and L. W. Molenkamp, Interacting topological edge channels, preprint (2019); arXiv:1905.08175; URL https://arxiv.org/abs/1905.08175.

[33] D. M. Mahler, J.-B. Mayer, P. Leubner, L. Lunczer, D. Di Sante, G. Sangiovanni, R. Thomale, E. M. Hankiewicz, H. Buhmann, C. Gould and L. W. Molenkamp, Interplay of Dirac nodes and Volkov–Pankratov surface states in compressively strained HgTe, *Phys. Rev. X* **9**, 031034 (2019); doi: 10.1103/PhysRevX.9.031034; URL https://link.aps.org/doi/10.1103/PhysRevX.9.031034.

[34] C. Brüne, C. X. Liu, E. G. Novik, E. M. Hankiewicz, H. Buhmann, Y. L. Chen, X. L. Qi, Z. X. Shen, S. C. Zhang and L. W. Molenkamp, Quantum Hall effect from the topological surface states of strained bulk HgTe, *Phys. Rev. Lett.* **106**(12), 126803 (2011); doi: 10.1103/PhysRevLett.106.126803.

[35] B. A. Volkov and O. A. Pankratov, Two-dimensional massless electrons in an inverted contact, *JETP Lett.* **42**(4), 178 (1985); *Pis'ma Zh. Eksp. Teor. Fiz.* **42**, 145 (1985).

[36] A. Inhofer, S. Tchoumakov, B. A. Assaf, G. Fève, J. M. Berroir, V. Jouffrey, D. Carpentier, M. O. Goerbig, B. Plaçais, K. Bendias, D. M. Mahler, E. Bocquillon, R. Schlereth, C. Brüne, H. Buhmann and L. W. Molenkamp, Observation of Volkov–Pankratov states in topological HgTe heterojunctions using high-frequency compressibility, *Phys. Rev. B* **96**, 195104 (2017); doi: 10.1103/PhysRevB.96.195104; URL https://link.aps.org/doi/10.1103/PhysRevB.96.195104.

[37] C. Fleckenstein, N. T. Ziani and B. Trauzettel, F_4 parafermions in weakly interacting superconducting constrictions at the helical edge of quantum spin Hall insulators, *Phys. Rev. Lett.* **122**, 066801 (2019); doi: 10.1103/PhysRevLett.122.066801; URL https://link.aps.org/doi/10.1103/PhysRevLett.122.066801.

[38] P. Sternativo and F. Dolcini, Tunnel junction of helical edge states: Determining and controlling spin-preserving and spin-flipping processes through transconductance, *Phys. Rev. B* **89**, 035415 (2014); doi: 10.1103/PhysRevB.89.035415; URL https://link.aps.org/doi/10.1103/PhysRevB.89.035415.

[39] Y. Tanaka and N. Nagaosa, Two interacting helical edge modes in quantum spin hall systems, *Phys. Rev. Lett.* **103**, 166403 (2009); doi: 10.1103/PhysRevLett.103.166403; URL https://link.aps.org/doi/10.1103/PhysRevLett.103.166403.

Chapter 10

Topology and Interactions in InAs/GaSb Quantum Wells

Rui-Rui Du

School of Physics, Peking University, Beijing 100871, China

rrd@pku.edu.cn

Shou-Cheng invited me to Fudan in the summer of 2008 as he told me that he has some very interesting idea to discuss. I had been interested in looking for excitonic condensation in InAs/GaSb quantum wells and was working on high quality materials for experiments. I mentioned this to Shou-Cheng during previous meetings. In Fundan, Shou-Cheng and Xiao-Liang spent three days explaining to me how this material can become a topological insulator. I understood that Chao-Xing Liu in Shou-Cheng's group had worked out with his collaborators the theory of quantum spin Hall effect in inverted type II semiconductor quantum wells.

Shou-Cheng was already known to be a bright young theoretician since early years. When I joined the group of DanTsui and Stormer to do experiments on the fractional quantum Hall effect (FQHE) in 1991, the field was experiencing a sea changing in theoretical understanding of the quantum fluid and its excitations. The composite fermion model had just emerged, which explains the transport in the vicinity of the half filling at the lowest Landau level and the FQHE states around $\nu = 1/2$. On the experimental side, Hong-Wen Jiang before me in the group had just discovered a puzzling metallic state at exactly 1/2 [1] around the same time of Bob Willet's surface acoustic wave experiments on the $\nu = 1/2$ state [2].

Shou-Cheng and his collaborator published a short paper explaining "the $v = 1/2$ anomaly" by invoking the Chem–Simons fermions [3], which I felt very illuminating. This and other work on the quantum Hall effect by Shou-Cheng and collaborators, the most famous ones being "global phase diagram" [4] and "four-dimensional quantum Hall effect" [5], have certainly influenced my own experimental research. Our interests intersect again on the topological insulators. In fact it was Shou-Cheng who introduced me into this field, and his theoretical thoughts and many discussions have guided our experiments. Looking back, this has been a long and exciting journey. This brief summary is a tribute to Shou-Cheng's science and his friendship.

10.1 Introduction

The rise of topological insulator and topological superconductor [6, 7] is a most important development in condensed matter physics in the last 15 years or so. The quantum spin Hall insulator (QSHI) in InAs/GaSb quantum wells (QWs) was proposed in 2008 by Liu *et al.* [8]. In this semiconductor, heterostructure electron–hole bilayer naturally occurs due to the unique broken-gap band alignment of InAs and GaSb [9]. In particular, the conduction band of InAs is lower than the valence band of GaSb, which results in charge transfer between two layers, and the emergence of coexisting 2D sheets of electrons and holes, as shown in Fig. 10.1(a) [10]. The Fermi energies of the electron and hole subbands can be adjusted by changing the thickness of InAs and GaSb layers, or by gate voltages [11], resulting in topologically trivial and nontrivial energy spectra for narrower wells and wider wells, respectively.

Topology in the InAs/GaSb was revealed when considering the inverted band structures [8, 12]. In the topologically nontrivial regime, electron–hole subbands cross in the momentum space and due to the quantum tunneling between the wells, electron and hole states hybridize, opening an inverted energy gap, Δ, termed minigap. It has been proposed [8] that at the edge of the system, this inverted gap must close before the normal gap in nearby native oxides or vacuum can open, creating a linearly dispersing helical edge spectrum, as depicted in Fig. 10.1(b). The term "helical" refers to the property that the edge states appear as a pair, with one branch moving to the left with spin up, and the other moving to the right with spin down. This spin-momentum locking property provides a topological protection such that the motion of electrons in the edge states would be free of back scatterings, a cause for energy dissipation. Further research

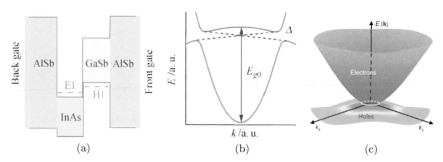

Fig. 10.1. Band structure and band dispersion of inverted InAs/ GaSb QWs. The degree of band inversion is given by $E_{g0} = H1 - E1$, the energy difference between the first subband of holes and the first subband of electrons. The bands cross in momentum space at a carrier density n_{cross}, opening a minigap Δ due to band hybridization. Adapted from [10]

on this material will help to find way to transmit quantum information by spins, with little dissipation of energy as compared to the today's computer technology.

10.1.1 *Experiments revealing bulk gap prior to topological insulators*

Transport measurements of the mini gap were first reported by US NRL group (1997) [13] and Cambridge group (1998) [14], respectively. Those earlier transport studies of inverted InAs/GaSb QWs, which have been only conducted on larger Hall bar samples, have confirmed a hybridization gap in the bulk, although all data show a residual conductivity down to liquid helium temperature.

10.1.2 *Theoretical model*

The quantum spin Hall physics in inverted InAs/GaSb QWs is based on the now famous Bernevig–Hughes–Zhang (BHZ) model [8, 12]: the Hamiltonian

$$H = H_0 + H_{\text{BIA}} + H_{\text{SIA}}, \tag{10.1}$$

$$H_0 = \varepsilon(k)I_{4\times4} + \begin{pmatrix} M(k) & Ak_+ & 0 & 0 \\ Ak_- & -M(k) & 0 & 0 \\ 0 & 0 & M(k) & -Ak_- \\ 0 & 0 & -Ak_+ & -M(k) \end{pmatrix}, \tag{10.2}$$

where $I_{4\times4}$ is the 4×4 identity matrix, $M(k) = M_0 + M_2k^2$ and $\varepsilon(k) = C_0 + C_2k^2$. This is simply the Hamiltonian used by BHZ. Two different atoms in each unit cell break bulk inversion symmetry and leads to additional terms. When projected onto the lowest subbands, the bulk inversion asymmetric (BIA) term is

$$H_{\text{BIA}} = \begin{pmatrix} 0 & 0 & \Delta_e k_+ & -\Delta_0 \\ 0 & 0 & \Delta_0 & \Delta_h k_- \\ \Delta_e k_- & \Delta_0 & 0 & 0 \\ -\Delta_0 & \Delta_h k_+ & 0 & 0 \end{pmatrix}. \tag{10.3}$$

and the structure inversion asymmetric (SIA) term reads

$$H_{\text{SIA}} = \begin{pmatrix} 0 & 0 & i\xi_e k_- & 0 \\ 0 & 0 & 0 & 0 \\ -i\xi_e^* k_+ & 0 & 0 & 0 \\ 0 & 0 & 0 & 0 \end{pmatrix}. \tag{10.4}$$

Here the SIA term is the electron k-linear Rashba term; the heavy-hole k-cubic Rashba term is neglected. The parameters $\Delta_h, \Delta_e, \Delta_0, \xi_e$ depend on the quantum well geometry.

10.2   Quantum spin hall effect

10.2.1   *Quantum spin hall effect in dilute InAs/GaSb QWs*

Experimentally, it is essential to understand the following two aspects specifically for observing the quantum spin Hall effect (QSHE) in InAs/GaSb. Firstly, the degree of band inversion, E_{g0}, can be tuned by the widths of the QWs and gate voltages, and it has dramatic influences on the bulk transport properties. In the deeply inverted regime where typically n_{cross} above 2×10^{11} cm^{-2}, there always exist considerable residual states in the hybridization gap; thus, the bulk of InAs/GaSb QWs is not truly insulating, which limits the studies and applications of the QSHE. It is understood theoretically [15] that the residual conductivity in the hybridization gap goes as $\sigma_0 \approx \frac{e^2}{h} \frac{E_{g0}}{\Delta}$, hence QWs with shallower inversion should result in a more insulating bulk state. In the shallowly inverted regime ($n_{\text{cross}} \lesssim 1 \times 10^{11}$ cm^{-2}), the bulk is insulating to a high degree [16]. Secondly, in order to observe the quantized edge conductance, the Hall bar length L should satisfy $L \ll \lambda_\varphi$, where λ_φ is the edge coherent length and its value is on the order of 4 microns in typical InAs/GaSb QWs shown in Fig. 10.3. In a large sample $L \gg \lambda_\varphi$, edge mode resistance can be estimated as $(L/\lambda_\varphi)(h/e^2)$ [16].

We have first observed the signatures of the QSHE, namely, the edge modes emerging in the inverted QWs, in 2011 [17]. By 2015, we reported the work on quantized helical edge conductivity in high quality InAs/GaSb

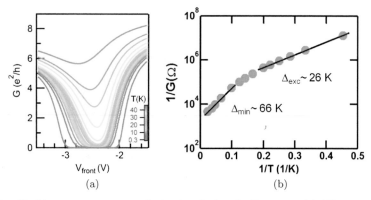

Fig. 10.2.   Corbino measurement of the insulating bulk state. (a) The temperature-dependent conductance traces measured in a Corbino disk are displayed. (b) The Arrhenius plot shows a thermally activated transport in the bulk state. Adapted from [16].

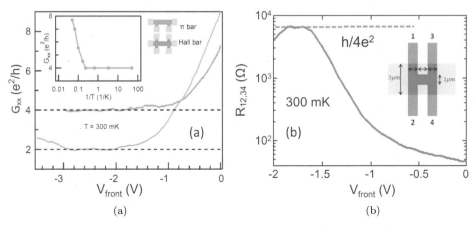

Fig. 10.3. (a) Wide conductance plateaus quantized to $2e^2/h$ and $4e^2/h$, respectively, for two device configurations shown in inset, both having length 2 μm and width 1 μm. (b) Quantized edge conductivity in an H-bar sample. Adapted from [16].

QWs with shallowly inverted bands [16], as shown in Figs. 10.2 and 10.3. For the bulk state, the gap energy Δ is quantitatively determined from conductivity measurements in a Corbino disk, shown in Figs. 10.2(a) and 10.2(b), as a function of temperature. The energy gap is deduced by fitting the conductance: $G_{xx} \propto \exp(-\Delta/2k_B T)$, where Δ is the energy required to create a pair of electron–hole over the gap and k_B is the Boltzmann constant. At higher temperatures, a gap value $\Delta \approx 66$ K is deduced, consistent with a hybridization-induced minigap. As the temperature is further reduced below ~ 10 K, the conductance continues to vanish exponentially with a different slope, indicating the opening of a gap $\Delta_{exc} \approx 26$ K in the energy spectrum; a wide plateau of near-zero conductivity emerges in this regime. In hindsight, we now understand that the Δ_{exc} is an excitonic gap opening in a dilute electron–hole double layer [18–20]. Robust helical edge states were observed as wide conductance plateaus precisely quantized to $2e^2/h$ in mesoscopic Hall samples, and to $4e^2/h$ in mesoscopic H-bar samples, as shown in Fig. 10.3. The edge current in the QSHE was imaged by SQUID measurements by Stanford group [21].

We would like to note the following characteristics of the QSHE observed in InAs/GaSb QWs. First, the Fermi velocity of the edge mode $\nu_F \sim 1.5 \times 10^4$ m/s here is at least 1 order of magnitude smaller than that of GaAs 2D electron gas (2DEG) or HgTe/CdTe QW ($\nu_F \sim 5.5 \times 10^5$ m/s) due to the fact that the gap opens at a finite wave vector k_{cross} instead of the zone center. Remarkably, the edge scattering time, i.e., $\tau = \lambda_\varphi/\nu_F = 2\lambda_\varphi \cdot k_{\mathrm{cross}}/\Delta \approx$ 200 ps (approaching that of the ultrahigh-mobility 2DEG in GaAs), appears to be extremely long regardless of the disordered bulk. Such a small ν_F should indicate that the edge transport is in a strong electron–electron interaction regime where the one-dimensional (1-D) Luttinger liquid may become the ground state, as we shall discuss. Second, it is found that the quantized conductance plateaus persist under strong external magnetic fields, in contrast with the theoretical expectations for time-reversal-symmetry (TRS) protected helical edge states [8]. This experimental finding has not been fully explained by existing theoretical models, including the proposals based on the InAs/GaSb band structure calculations which show that the Dirac point of the edge states is hidden below the valance band [22, 23]. Recent theories have also proposed that coulomb interactions of electron–hole pairs dominate over hybridization effects in such a dilute limit, leading to the possibility of a two-dimensional excitonic ground state [19, 20] accompanied by exotic edge states. The experimental results for topological insulator in InAs/GaSb QWs have been subsequently reported by a number of groups [24–29].

10.2.2 *Quantum spin Hall effect in large gap InAs/GaInSb QWs*

While the quantized edge transport has been observed in micrometer-size samples of shallowly inverted InAs/GaSb, our results have shown that electron interaction effect in the edge states may be relevant. From an experimental perspective, it is desirable to develop a large gap, plain vanilla QSHI with properties dominated by single-particle physics. Ideally, to some degree the interaction effects may be set in by tuning experimental parameters such as ν_F.

The physical idea guiding the construction of a large-gap QSHI in InAs/GaSb QWs is strain-engineering [30–32]. By alloying GaSb (the lattice constant is about 6.1 Å) with InSb (6.4 Å), because of the strain in the growth plane, the energy of the conduction band (CB) in InAs shifts downward while the energy level of the valence band (VB) in $Ga_{1-x}In_xSb$ splits into the heavy hole (HH) level and light hole (LL) level, respectively, where the energy of the HH level is higher than the original top VB in GaSb. Based on the calculated band shifts, one can design inverted band structure with narrower QWs in strained-layer $InAs/Ga_{1-x}In_xSb$ (Fig. 10.4), comparing to unstrained InAs/GaSb. The hybridization-induced gap should increase in such narrower QWs primarily due to the enhanced overlap of electron and hole wavefunctions.

We have realized a large gap QSHI in strained-layer $InAs/Ga_{1-x}In_xSb$ QWs. Experimentally measured gap values for several strained-layer QWs are listed in Table 10.1. We reported in 2017 a gap value $\Delta \sim 250$ K

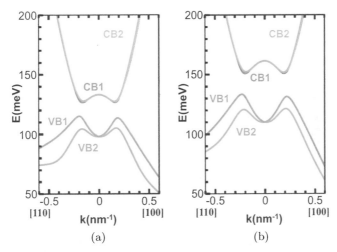

Fig. 10.4. Calculated band dispersions of the strained-layer InAs/Ga$_{1-x}$In$_x$Sb QWs, respectively for (a) the InAs/Ga$_{0.80}$In$_{0.20}$Sb (8.7 nm/4 nm) QWs, and (b) InAsGa$_{0.68}$In$_{0.32}$Sb (8 nm/4 nm) QWs; CB1, VB1 and CB2, VB2 are bands of different spin component. Adapted from [32].

Table 10.1. List of strained-layer InAs/Ga$_{1-x}$In$_x$Sb QWs and experimentally measured respective gap values.

Composition	$x = 0$	$x = 0.20$	$x = 0.32$	$x = 0.40$
Gap value	66 K [16]	120K [32]	250K [32]	400K [33]

for the InAs/Ga$_{0.68}$In$_{0.32}$Sb QWs [32]. Recently, NTT group has reported [33] a gap value $\Delta \sim 400$ K in InAs/Ga$_{0.6}$In$_{0.4}$Sb QWs, This is one order of magnitude enhancement comparing to unstrained InAs/GaSb QWs. Strained-layer and narrower QWs results in more insulating hybridization gaps at low temperatures even when the n_{cross} is larger than 3×10^{11} cm^{-2}. Remarkably, we found that the edge coherent length λ_φ is correlated with ν_F, which could be well controlled by lattice strain and the gate voltages, see Fig. 10.5.

Perhaps more importantly, edge modes in large-gap QSHI clearly manifest TRS protected properties correlating to their increasing ν_F [32] As shown in Fig. 10.6, contrary to the weak response of slow-moving edge modes in unstrained InAs/GaSb QWs, the helical edge conductance of the edge modes in larger-gap InAs/Ga$_{0.68}$In$_{0.32}$Sb QWs decreases under either perpendicular or in-plane magnetic fields, indicating the opening of mass gaps in the edge states. We have systematically studied [34] the low-temperature transport properties of strained InAs/Ga$_{0.68}$In$_{0.32}$Sb QWs, where the helical edge state is weakly interacting (Luttinger parameter $K \sim 0.43$). Our results

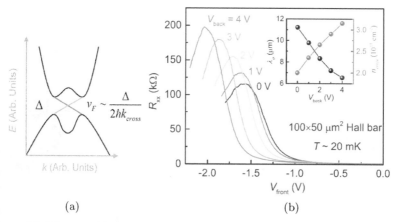

(a)   (b)

Fig. 10.5.   (a) Schematic drawing of band dispersions (both bulk states and edge states) is shown for the InAs/Ga$_{1-x}$In$_x$Sb QWs. The group velocity of the edge modes is determined by the ratio $\Delta/2hk_{cross}$. (b) Helical edge transport in strained-layer InAs/Ga$_{0.75}$In$_{0.25}$Sb QWs as a function of the gate voltage. The edge coherence length λ_φ increases with decreasing n_{cross} values, or with increasing ν_F. The inset shows the λ_φ and the n_{cross} at different back-gate bias V$_{back}$; the λ_φ reaches 11 μm in this sample. Adapted from [32].

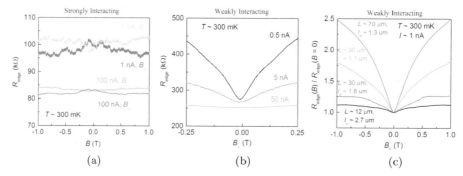

(a)   (b)   (c)

Fig. 10.6.   Magnetoresistance of helical edge states. (a) R_{edge} of a 30 μm $\times$ 10 μm Hall bar made from shallowly inverted InAs/GaSb QWs with strongly interacting edge states, as a function of magnetic fields. (b) R_{edge} of a 30 μm $\times$ 10 μm Hall bar made from a strained-layer InAs/Ga$_{0.68}$In$_{0.32}$Sb QWs with weakly interacting edge states, as a function of $B_\perp$. (c) Normalized edge magnetoresistance as a function of $B_\parallel$, for several samples made from InAs/Ga$_{0.68}$In$_{0.32}$Sb QWs with different L and l_ϕ. The aspect ratio (length/width) of all four samples is 3. Adapted from [34].

indicate that although the electron–electron interaction still exists here in helical edge states, it becomes less relevant. Thus the system follows the behavior of a TRS-protected QSHI based on the single-particle picture. Present experiment provides a clear comparison of the physical properties between the strongly interacting and the weakly interacting helical edge states. Recently, we have also observed anomalous conductance oscillations in the hybridization gap [35] of InAs/GaSb OWs as well as strained layer InAs/Ga$_{1-x}$In$_x$Sb QWs, and the QSH to QH topological phase transition under a strong perpendicular field.

10.3 Topological excitonic insulator

We have also explored a new experimental direction of exciton formation and condensation in InAs/GaSb QWs, the same material hosting the quantum spin Hall effect [18–20]. The ideal of excitonic insulator can be traced back to the 1960s, when it was predicted [36–43] that Coulomb interactions in an electron–hole (e–h) coexisting system can make the normal semimetallic state unstable against the spontaneous appearance of excitons, or bound e–h pairs, inducing a phase transition into an insulator, called the excitonic insulator (EI) or Bardeen–Cooper–Schrieffer (BCS)-like excitonic condensation.

10.3.1 *Excitonic insulator in electron–hole bilayers*

InAs/GaSb QWs exhibit unique inverted band structure with finite overlap of the conduction and valance bands, allowing the coexistence of spatially separated electrons and holes without photoexcitation, which offers a natural setting for equilibrium excitons and consequent formation of condensates including the EI phase. Electrons are located in the InAs QW and holes are located in the GaSb QW, and thus, they are spatially separated in real space. The average separation can be defined as one half of the thickness of the double-QW structure. In this system, signatures of magneto-excitons were reported in 1997 [44].

Our devices were made from inverted InAs/GaSb QWs. By dual gates, we could reach the charge neutrality point (CNP), i.e., $n_0 = p_0$. The lowest n_0 in our devices was $\sim 5.5 \times 10^{10}$ cm^{-2}. The average inter-exciton in-plane distance, $2r_{\mathrm{avg}}$, corresponding to $n_0 \sim 5.5 \times 10^{10}$ cm^{-2}, defined through $1 = n_0 \cdot \pi r_{\mathrm{avg}}^2$, is ~ 48 nm. This value should be compared with the effective Bohr radius, a_B, which is estimated to be ~ 30 nm within a simple effective-mass approximation using the following parameters: electron effective mass $m_e^* \sim 0.032m_0$ ($m_0 = 9.11 \times 10^{-31}$ kg), hole effective mass $m_h^* \sim 0.136m_0$, dielectric constant $\epsilon \sim 15$, and interlayer distance between the centers of the electron and hole wells $d \sim 10$ nm. Therefore, we have $r_{\mathrm{avg}}/a_B \sim 0.8$, indicating strong wavefunction overlap between excitons, a situation reminiscent of Cooper pairs.

10.3.2 *Optical spectroscopy and transport results for topological excitonic insulator*

We performed low-temperature THz transmission spectroscopy experiments [18] on a device covered by a 5 mm × 5 mm semi-transparent gate, in a frequency (energy) range of 0.25–2.4 THz ($\sim$ 1–10 meV). A transmittance spectrum when the system is at the CNP at 1.4 K is shown in Fig. 10.7c. Here, two absorption lines (or transmission dips) are present in the spectrum, including Line A at $\sim$ 2 meV, and Line B at 7.3 meV, consistent with what we expect from Figs. 10.2(a) and 10.2(b). We interpret Line A as coming from pair-breaking excitation near the Fermi level and Line B from excitation near $k = 0$. What we are measuring is the joint density of states of the pair-breaking excitation process, which is characterized by the spectrum $E(k)$, not a direct extraction of the gap. It should be noted that these features are observed only when n_0 is low ($\sim 5.5 \times 10^{10}\,\text{cm}^{-2}$); they disappear in an electron- or hole-dominating regime, or in a high $n_0 (\sim > 10^{11}\,\text{cm}^{-2})$ case. In addition to the fine structure in Joint density of states (Fig. 10.7(b)), inhomogeneous broadening due to random potential fluctuations would contribute to the Line B linewidth. Moreover, disorder has stronger effect to low k states that have lower energy, contributing to a broader line. Also, random potential fluctuations are expected to enhance the exciton density at low k.

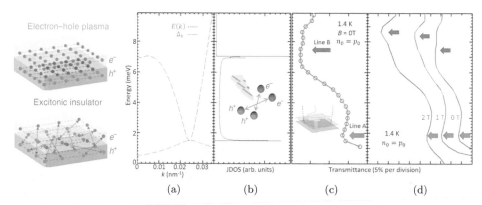

Fig. 10.7. Excitonic insulator in dilute inverted InAs/GaSb QWs. Left: Schematic drawings for e–h plasma (or a semimetal) at high temperatures, and an excitonic insulator ground state formed below critical temperature, where electrons and holes are weakly bound, like Cooper pairs. Right: Pair-breaking excitation spectra of excitonic insulator. (a) Gap function $\Delta(k)$ (red dashed line) and the pair-breaking energy $E(k)$ (blue dashed line) of the exciton as a function of k. (b) Joint density of states as a function of energy. (c) Transmission spectrum at the charge neutral point at temperature 1.4 K and zero magnetic field. (d) Transmittance spectra at various perpendicular magnetic fields. Adapted from [18].

The interpretation of the lines as from the pair-breaking excitations across the BCS-like gap of an EI state is also supported by their temperature dependence [18]. When a magnetic field perpendicular to the QWs, $B_\perp$, is applied at 1.4 K, the amplitudes of the absorption lines increase and Line B slightly blue-shifts with the magnetic field, as shown in Fig. 10.7(d). The strengthened absorption under $B_\perp$ is likely due to the fact that a $B_\perp$ makes e–h pairs more tightly bound, hence more stable against dissociation. Overall, the data not only confirms that the lines have the same origin but also provides optical spectroscopic evidence for the spontaneous formation of an EI gap through the Coulomb attraction between spatially separated electrons and holes.

In the low n_0 regime comparable to the case here, the quantization plateau of helical edge states has been previously observed [16] and taken as the evidence of the QSH effect. Moreover, the plateaus were found to be quite robust under the variance of external parameters such as in-plane magnetic field. Such a robustness of helical edge states cannot be explained by existing single-particle theory concerning two-dimensional (2D) topological insulators. Pikulin and Hyart subsequently proposed [19] the emergence of an unconventional EI ground state in the inverted InAs/GaSb QWs, providing a plausible explanation for these experimental observations. In their model, the interplay between an excitonic ground state and interlayer tunneling, which is naturally existing in the present InAs/GaSb structure (i.e., without a tunneling barrier between the two layers) can lead to a p-wave EI with topologically protected edge states. To explore the nontrivial topological properties of the condensate, we performed nonlocal transport measurements in a mesoscopic H-bar device at low n_0, and confirm the existence of quantized edge states [18]. Altogether, the optical and transport results suggest that an EI spontaneously emerges in the bulk of dilute InAs/GaSb QWs with helical edge modes propagating along the perimeters, consistent with the observed unusual QSH properties in InAs/ GaSb within the picture of the theoretically proposed topological EI.

10.4   1D helical Luttinger liquid

To explore the electron–electron interaction physics in helical edge states in shallowly inverted QWs, we performed the ultralow temperature (down to 7 mK) transport measurements, and observed a helical Luttinger-liquid in the edge of InAs/GaSb quantum spin Hall insulator [45]. This is the first experimental observations of a Luttinger-liquid state in topological materials. The Luttinger liquid describes an 1D interacting electron system.

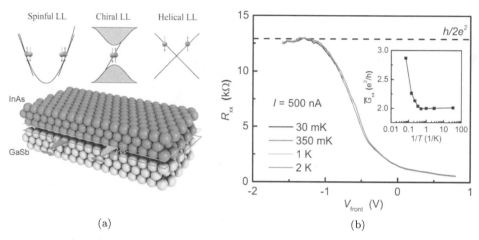

(a)   (b)

Fig. 10.8.   (a) Family of Luttinger liquids: schematic drawing of energy dispersions for spinful LL, chiral LL, and helical LL, respectively, and schematic drawing of electronic transport in a helical LL. (b) Resistance for a mesoscopic device (edge length 1.2 μm) taken at various temperatures with a high-bias 500 nA excitation current. Quantized resistivity plateau of $h/2e^2$ persists from 30 mK to 2 K. Adapted from [45].

During the past 20 years, theoretical and experimental studies [46–54] of Luttinger liquid have become a frontier of condensed matter physics. Carbon nanotubes, as well as fractional quantum Hall edge states, have been experimentally studied as spinful Luttinger liquid and chiral Luttinger liquid, respectively (Fig. 10.8).

Transport of electrons in InAs/GaSb edge states has been shown to exhibit Luttinger-liquid behavior, where the resistance of the edge states increases following a power-law with the decreasing of temperature and bias voltage (Fig. 10.9). The underlying physics, often referred as "helical Luttinger liquid", is quite distinct from the known types of Luttinger liquids due to the helical nature (i.e. spin-momentum locking) of the QSHI edge states [55–66]. The results indicate that strong electron–electron interaction effects could play important roles in symmetry-protected topological matters, influencing their electronic properties in a fundamental way.

In summary, we observe a strong suppression of the helical edge conductance in InAs/GaSb QWs at low temperature and low bias voltage, which suggests that strong electron–electron interactions in the helical edges should lead to a correlated electronic insulator phase in the limit of $T \sim 0$ and vanishing bias voltage. Because of the fact that the bulk gaps (and hence the v_F of edge states) in InAs/GaSb materials can be engineered

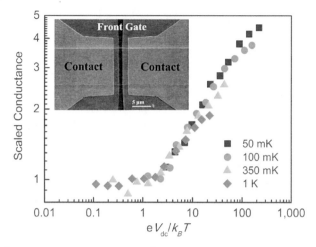

Fig. 10.9. Scaling properties of differential conductance measured in helical edge of the device shown in Fig. 10.8(b). The inset shows the SEM image of the device. The main plot illustrates all the measured dI/dV data points except the saturation region collapse onto a single curve by scaling the energy eV_{dc}/K_BT. The result is consistent with a Luttinger parameter $K \sim 0.21$. Adapted from [45].

by molecular beam epitaxy growth and gating architectures, the electron–electron interactions can be fine-tuned, leading to a well-controlled model system for studies of 1D electronic and spin correlation physics.

10.5 Future perspectives

We have briefly reviewed the progresses made in the understanding of topology and interactions in a clean semiconductor materials system made of InAs/GaAs. On the one hand, the inverted bands made it possible to study the symmetry protected topological state, namely the quantum spin Hall phase consisting of the gaped bulk and the gapless helical edge modes. On the other hand, the electron–electron interactions inherent in two-dimensional electron–hole heterostructures must be considered in order to understand the emergent phases, for example, the spontaneous formation of excitonic ground states in the 2D bulk, and helical Luttinger liquid in the edge modes. The interplay between topology and interaction should give rise to a rich phase diagram in the InAs/GaSb QWs as proposed by Xue and MacDonald [20].

By strain engineering, we have demonstrated a QSHI in InAs/GaInSb QWs clearly manifesting TRS protection, which shows a larger hybridization gap and a longer edge coherent length compared to that in InAs/GaSb

QWs. Moreover, the bulk of the InAs/GaInSb QWs is highly insulating at low temperatures and the edge coherent length may be well controlled by the gates; data show that the edge states can be gapped out by applying magnetic fields.

The realization of an excitonic insulator state in the InAs/GaSb QWs paves the way for further studying excitonic condensations in great detail as well as depth (e.g., BCS–BEC crossover, and excitonic BEC) in an equilibrium electron–hole system without optical pumping. With an additional AlGaSb barrier inserted between the electron and hole layers [67, 68], counter-flow studies would be enabled to explore excitonic BEC and superfluidity [69, 70]. Moreover, the inverted band structure of the current system brings in topological nature to the EI state, which will allow one to study 2D interacting topological insulators in a highly controllable manner. For example, by adjusting a thin AlGaSb tunneling barrier or utilizing strained-layer InAs/GaInSb QWs, the symmetry of the order parameter can be controlled by tuning the interplay of intra- and inter-layer interactions. The spin states of the condensations can be explored as well.

One prevailing feature of the InAs/GaSb system is that the helical edge modes are in a strongly interacting regime, making it an ideal model system for studies of correlation effect and many-particle quantum phases in a controlled manner. Luttinger liquid in a small K parameter range around 0.25 should be carefully studied with improved sample quality and innovative measurement techniques. Future experiments such as quantum point contact [57] and noise correlation [71, 72] measurements could, in principle, reveal the microscopic physical processes inside such strongly interacting helical edge states. It is well known that [73–75] the QSHI helical edge states coupled with superconductors can support Majorana zero modes. More interestingly, the presence of interactions promotes these Majorana modes splitting into Z_4 parafermionic modes [76–78]. The Josephson junction mediated by interacting QSHI edge states creates a pair of parafermions, yielding a novel 8π-Josephson effect reflecting the tunneling processes of $e/2$ charge quasiparticles between superconductors. These are the bases for platforms which are promising for universal, topologically-protected quantum computation [79]. Further studies of interaction effects on the helical edge states in the InAs/GaSb system would be necessary to advance in this direction. With semiconductor technology, it can be expected that the materials will be further refined to reveal intrinsic electron–electron interaction physics in the simplest 1D electronic system.

The International Workshop on Majorana Fermions in Solid States, June 3–7, 2013, Peking University

Acknowledgments

I would like to acknowledge my former and current graduate students Ivan Knez, Lingjie Du, Tingxin Li, Xiaoyang Mu, Jie Zhang, Xingjun Wu, Bingbing Tong, Xiaoxue Liu, Zongdong Han, Shiqi Yao, Chi Zhang for their indispensable contributions to this endeavour, Kai Chang and Wenkai Lou for many contributions in theoretical modeling and numerical calculations, Gerard Sullivan and Amal Ikhlassi for expert sample preparation by molecular beam epitaxy, Eric Spanton, Kathryn Moler, Xinwei Li, Junichiro Kono, Loah Stevens, Douglas Natelson, Pengjie Wang, Hailong Fu, Kate Schreiber, Gábor Csáthy, Xi Lin for experimental collaborations.

References

[1] H. W. Jiang, H. L. Stormer, D. C. Tsui, L. N. Pfeiffer and K. W. West, Transport anomalies in the lowest Landau level of two-dimensional electrons at half-filling, *Phys. Rev. B* **40**, 12013(R) (1989).

[2] R. L. Willett, M. A. Paalanen, R. R. Ruel, K. W. West, L. N. Pfeiffer and D. J. Bishop, Anomalous sound propagation at $\nu = 1/2$ in a 2D electron gas: Observation of a spontaneously broken translational symmetry? *Phys. Rev. Lett.* **65**, 112 (1990).

[3] V. Kalmeyer and S.-C. Zhang, Metallic phase of the quantum Hall system at even-denominator filling fractions, *Phys. Rev. B* **46**, 9889(R) (1992).

[4] S. Kivelson, D. H. Lee and S. C. Zhang, Global phase diagram in the quantum Hall effect, *Phys. Rev. B* **46**, 2223 (1992).

[5] S.-C. Zhang and J. Hu, A four-dimensional generalization of the quantum hall effect, *Science* **294**, 823 (2001).

[6] M. Z. Hasan and C. L. Kane, Colloquium: Topological insulators, *Rev. Mod. Phys.* **82**, 3045 (2010).

[7] X. L. Qi and S. C. Zhang, Topological insulator and superconductors, *Rev. Mod. Phys.* **83**, 1057 (2011).

[8] C. X. Liu, T. L. Hughes, X. L. Qi, K. Wang and S. C. Zhang, Quantum spin hall effect in inverted type-II semiconductors, *Phys. Rev. Lett.* **100**, 236601 (2008).

[9] H. Kroemer, The 6.1 Å family (InAs, GaSb, AlSb) and its heterostructures: A selective review *Physica E* **20**, 196 (2004).

[10] I Knez and R. R. Du, Quantum spin Hall effect in inverted InAs/GaSb quantum wells, *Frontiers Phys.* **7**, 200 (2012).

[11] Y. Naveh, B. Laikhtman, Band-structure tailoring by electric field in a weakly coupled electron–hole system, *Appl. Phys. Lett.* **66**, 1980(1995).

[12] B. A. Bernevig, T. L. Hughes and S. C. Zhang, Quantum spin Hall effect and topological phase transition in HgTe quantum wells, *Science* **314**, 1757 (2006).

[13] M. J. Yang, C. H. Yang, B. R. Bennett and B. V. Shanabrook, Evidence of a hybridization gap in "semimetallic" InAs/GaSb systems, *Phys. Rev. Lett.* **78**, 4613 (1997).

[14] L. J. Cooper, N. K. Patel, V. Drouot, E. H. Linfield, D. A. Ritchie and M. Pepper, Resistance resonance induced by electron–hole hybridization in a strongly coupled InAs/GaSb/AlSb heterostructure, *Phys. Rev. B* **57**, 11915 (1998).

[15] Y. Naveh and B. Laikhtman, Magnetotransport of coupled electron–holes, *Europhys. Lett.* **55**, 545 (2001).

[16] L. Du, I. Knez, G. Sullivan and R. R. Du, Robust helical edge transport in gated InAs/GaSb bilayers, *Phys. Rev. Lett.* **114**, 096802 (2015).

[17] I. Knez, R. R. Du and G. Sullivan, Evidence for helical edge modes in inverted InAs/GaSb quantum wells, *Phys. Rev. Lett.* **107**, 136603 (2011).

[18] L. Du, X. Li, W. Lou, G. Sullivan, K. Chang, J. Kono and R.-R. Du, Evidence for a topological excitonic insulator in InAs/GaSb bilayers, *Nature Commun.* **8**, 1971 (2017).

[19] D. I. Pikulin and T. Hyart, Interplay of exciton condensation and the quantum spin Hall effect in InAs/GaSb bilayers, *Phys. Rev. Lett.* **112**, 176403 (2014).

[20] F. Xue and A. H. MacDonald, Time-reversal symmetry-breaking nematic insulators near quantum spin hall phase transitions, *Phys. Rev. Lett.* **120**, 186802 (2018).

[21] E. M. Spanton, K. C. Nowack, L. J. Du, G. Sullivan, R. R. Du and K. A. Moler, Images of edge current in InAs/GaSb quantum wells, *Phys. Rev. Lett.* **113**, 026804 (2014).

[22] C.-A. Li, S.-B. Zhang and S.-Q. Shen, Hidden edge Dirac point and robust quantum edge transport in InAs/GaSb quantum wells, *Phys. Rev. B* **97**, 045420 (2018).

[23] R. Skolasinski, D. I. Pikulin, J. Alicea and M. Wimmer, Robust helical edge transport in quantum spin Hall quantum wells, *Phys. Rev. B* **98**, 201404(R) (2018).

[24] W. Pan, J. F. Klem, J. K. Kim, M. Thalakulam, M. J. Cich and S. K. Lyo, Chaotic quantum transport near the charge neutrality point in inverted type-II InAs/GaSb field-effect transistors, *Appl. Phys. Lett.* **102**, 033504 (2013).

[25] K. Suzuki, Y. Harada, K. Onomitsu and K. Muraki, Gatecontrolled semimetal-topological insulator transition in an InAs/GaSb heterostructure, *Phys. Rev. B* **91**, 245309 (2015).

[26] S. Mueller, A. N. Pal, M. Karalic, T. Tschirky, C. Charpentier, W. Wegscheider, K. Ensslin and T. Ihn, Nonlocal transport via edge states in InAs/GaSb coupled quantum wells, *Phys. Rev. B* **92**, 081303(R) (2015).

[27] F. Qu, A. J. A. Beukman, S. Nadj-Perge, M.Wimmer, B.M. Nguyen, W. Yi, J. Thorp, M. Sokolich, A. A. Kiselev, M. J. Manfra, C. M. Marcus and L. P. Kouwenhoven, Electric and magnetic tuning between the trivial and topological phases in InAs/GaSb double quantum wells. *Phys. Rev. Lett.* **115**, 036803 (2015).

[28] F. Couëdo, H. Irie, K. Suzuki, K. Onomitsu and K. Muraki, Single-edge transport in an InAs/GaSb quantum spin Hall insulator, *Phys. Rev. B* **94**, 035301 (2016).

[29] F. Nichele, H. J. Suominen, M. Kjaergaard, C. M. Marcus, E. Sajadi, J. A. Folk, F. Qu, A. J. A. Beukman and F. K. de Vries, Edge transport in the trivial phase of InAs/GaSb, *New J. Phys.* **18**, 083005 (2016).

[30] D. L. Smith and C. Mailhiot, Proposal for strained type II superlattice infrared detectors, *J. Appl. Phys.* **62**, 2545 (1987).

[31] T. Akiho, F. Couëdo, H. Irie, K. Suzuki, K. Onomitsu and K. Muraki, Engineering quantum spin Hall insulator by strainedlayer hereterostructures, *Appl. Phys. Lett.* **109**, 192105 (2016).

[32] L. Du, T. Li, W. Lou, X. Wu, X. Liu, Z. Han, C. Zhang, G. Sullivan, A. Ikhlassi, K. Chang and R. R. Du, Tuning edge states in strained-layer InAs/GaInSb quantum spin hall insulators, *Phys. Rev. Lett.* **119**, 056803 (2017).

[33] H. Irie, T. Akiho, F. Couëdo, K. Suzuki, K. Onomitsu and K. Muraki, Energy gap tuning and gate-controlled topological phase transition in InAs/In$_x$Ga$_{1-x}$Sb composite quantum wells, Preprint (2008); arXiv:2008.00664 (cond-mat).

[34] T. Li, P. Wang, G. Sullivan, X. Lin and R.-R. Du, Low-temperature conductivity of weakly interacting quantum spin Hall edges in strained-layer InAs/GaInSb, *Phy. Rev. B* **96**, 241406(R) (2017).

[35] Z. Han, T. Li, L. Zhang, G. Sullivan and R.-R. Du, Anomalous conductance oscillations in the hybridization gap of InAs/GaSb quantum wells, *Phys. Rev. Lett.* **123**, 126803 (2019).

[36] N. F. Mott, The transition to the metallic state, *Phil. Mag.* **6**, 287 (1961).

[37] R. S. Knox, Theory of excitons, *Solid State Phys. Suppl.* **5**, 100 (1963).

[38] L. V. Keldysh and Y. V. Kopaev, Possible instability of the semimetallic state toward coulomb interaction, *Fiz. Tverd. Tela* **6**, 2791 (1964).

[39] D. Jérome, T. M. Rice and W. Kohn, Excitonic insulator, *Phys. Rev.* **158**, 462 (1967).

[40] B. I. Halperin and T. M. Rice, Possible anomalies at a semimetal-semiconductor transition, *Rev. Mod. Phys.* **40**, 755 (1968).

[41] Yu. E. Lozovik and Y. I. Yudson, Feasibility of superfluidity of paired spatially separated electrons and holes; a new superconductivity mechanism, *JETP Lett.* **22**, 274 (1975).

[42] P. B. Littlewood and X. Zhu, Possibilities for exciton condensation in semiconductor quantum-well structures, *Phys. Scripta T* **68**, 56 (1996).

[43] Y. Naveh and B. Laikhtman, Excitonic instability and electric-field-induced phase transition towards a two-dimensional exciton condensate, *Phys. Rev. Lett.* **77**, 900 (1996).

[44] J. Kono, *et al.*, Far-infrared magneto-optical study of two-dimensional electrons and holes in InAs/AlGaSb quantum wells, *Phys. Rev. B* **55**, 1617 (1997).

[45] T. X. Li, P. J. Wang, H. L. Fu, L. J. Du, K. A. Schreiber, X. Y. Mu, X. X. Liu, G. Sullivan, G. A. Csáthy, X. Lin and R. R. Du, Observation of a helical luttinger liquid in quantum spin hall edges, *Phys. Rev. Lett.* **115**, 136804 (2015).

[46] F. D. M. Haldane, 'Luttinger liquid theory' of onedimensional quantum fluids: I. Properties of the Luttinger model and their extension to the general 1D interacting spinless Fermi gas, *J. Phys. C* **14**, 2585 (1981).

[47] C. L. Kane and M. P. A. Fisher, Transport in a one-channel luttinger liquid, *Phys. Rev. Lett.* **68**, 1220 (1992).

[48] A. Furusaki and N. Nagaosa, Kondo effect in a Tomonaga–Luttinger liquid, *Phys. Rev. Lett.* **72**, 892 (1994).

[49] M. Bockrath, D. H. Cobden, J. Lu, A. G. Rinzler, R. E. Smalley, L. Balents and P. L. McEuen, Luttinger-liquid behaviour in carbon nanotubes, *Nature (London)* **397**, 598 (1999).

[50] Z. Yao, H.W. Ch. Postma, L. Balents and C. Dekker, Carbon nanotube intramolecular junctions, *Nature (London)* **402**, 273 (1999).

[51] B. Gao, A. Komnik, R. Egger, D. C. Glattli and A. Bachtold, Evidence for Luttinger-liquid behavior in crossed metallic single-wall nanotubes, *Phys. Rev. Lett.* **92**, 216804 (2004).

[52] E. Levy, A. Tsukernik, M. Karpovski, A. Palevski, B. Dwir, E. Pelucchi, A. Rudra, E. Kapon and Y. Oreg, Luttinger-liquid behavior in weakly disordered quantum wires, *Phys. Rev. Lett.* **97**, 196802 (2006).

[53] A. Yacoby, H. L. Stormer, N. S. Wingreen, L. N. Pfeiffer, K. W. Baldwin and K. W. West, Nonuniversal conductance quantization in quantum wires, *Phys. Rev. Lett.* **77**, 4612 (1996).

[54] A. M. Chang, L. N. Pfeiffer and K. W. West, Observation of chiral luttinger behavior in electron tunneling into fractional quantum hall edges, *Phys. Rev. Lett.* **77**, 2538 (1996).

[55] C. J. Wu, B. A. Bernevig and S. C. Zhang, Helical liquid and the edge of quantum spin hall systems, *Phys. Rev. Lett.* **96**, 106401 (2006).

[56] C. K. Xu and J. E. Moore, Stability of the quantum spin Hall effect: Effects of interactions, disorder, and $\mathbb{Z}_2$ topology, *Phys. Rev. B* **73**, 045322 (2006).

[57] J. C. Y. Teo and C. L. Kane, Critical behavior of a point contact in a quantum spin Hall insulator, *Phys. Rev. B* **79**, 235321 (2009).

[58] J. Maciejko, C. X. Liu, Y. Oreg, X. L. Qi, C. J. Wu and S. C. Zhang, Kondo effect in the helical edge liquid of the quantum spin hall state, *Phys. Rev. Lett.* **102**, 256803 (2009).

[59] T. L. Schmidt, S. Rachel, F. von Oppen and L. I. Glazman, Inelastic electron backscattering in a generic helical edge channel, *Phys. Rev. Lett.* **108**, 156402 (2012).

[60] A. Rod, T. L. Schmidt and S. Rachel, Spin texture of generic helical edge states, *Phys. Rev. B* **91**, 245112 (2015).

[61] N. Lezmy, Y. Oreg and M. Berkooz, Single and multiparticle scattering in helical liquid with an impurity, *Phys. Rev. B* **85**, 235304 (2012).

[62] F. Crépin, J. C. Budich, F. Dolcini, P. Recher and B. Trauzettel, Renormalization group approach for the scattering off a single Rashba impurity in a helical liquid, *Phys. Rev. B* **86**, 121106(R) (2012).

[63] T. L. Schmidt, S. Rachel, F. von Oppen and L. I. Glazman, Inelastic electron backscattering in a generic helical edge channel, *Phys. Rev. Lett.* **108**, 156402 (2012).

[64] N. Kainaris, I. V. Gornyi, S. T. Carr and A. D. Mirlin, Conductivity of a generic helical liquid, *Phys. Rev. B* **90**, 075118 (2014).

[65] J. I. Väyrynen, M. Goldstein, Y. Gefen and L. I. Glazman, Resistance of helical edges formed in a semiconductor heterostructure, *Phys. Rev. B* **90**, 115309 (2014).

[66] Y.-Z. Chou, A. Levchenko and M. S. Foster, Helical quantum edge gears in 2D topological insulators, *Phys. Rev. Lett.* **115**, 186404 (2015).

[67] X. J. Wu, W. K. Lou, K. Chang, G. Sullivan and R. R. Du, Resistive signature of excitonic coupling in an electron-hole double layer with a middle barrier, *Phys. Rev. B* **99**, 085307 (2019).

[68] X. J. Wu, W. K. Lou, K. Chang, G. Sullivan, A. Ikhlassi and R. R. Du, Electrically tuning many-body states in a Coulomb-coupled InAs/InGaSb double layer, *Phys. Rev. B* **100**, 165309 (2019).

[69] J. P. Eisenstein and A. H. MacDonald, Bose–Einstein condensation of excitons in bilayer electron systems, *Nature (London)* **432**, 691 (2004).

[70] J.-J. Su and A.H. MacDonald, How to make a bilayer exciton condensate flow, *Nature Phys.* **4**, 799 (2008).

[71] X.-L. Qi, T. L. Hughes and S.-C. Zhang, Fractional charge and quantized current in the quantum spin Hall state, *Nature Phys.* **4**, 273 (2008).

[72] L. A. Stevens, T. Li, R.-R. Du and D. Natelson, Noise processes in InAs/Ga(In)Sb Corbino structures, *Appl. Phys. Lett.* **115**, 052107 (2019).

[73] I. Knez, R. R. Du and G. Sullivan, Andreev reflection of helical edge modes in InAs/GaSb quantum spin Hall insulator, *Phys. Rev. Lett.* **109**, 186603 (2015).

[74] S. Hart, H. C. Ren, T. Wagner, P. Leubner, M. Mühlbauer, C. Brüne, H. Buhmann, L. W. Molenkamp and A. Yacoby, Induced superconductivity in the quantum spin Hall edge, *Nature Phys.* **10**, 638 (2014).

[75] V. S. Pribiag, A. Beukman, F. M. Qu, M. C. Cassidy, C. Charpentier, W. Wegscheider and L. P. Kouwenhoven, *Nature Nanotechnol.* **10**, 593 (2015).

[76] F. Zhang and C. L. Kane, Time-reversal-invariant $\mathbb{Z}_4$ fractional josephson effect, *Phys. Rev. Lett.* **113**, 036401 (2014).

[77] C. P. Orth, R. P. Tiwari, T. Meng and T. L. Schmidt, Non-Abelian parafermions in time-reversal-invariant interacting helical systems, *Phys. Rev. B* **91**, 081406(R) (2015).

[78] Y. Peng, Y. V. Aviv, P. W. Brouwer, L. I. Glazman and F. von Oppen, Parity anomaly and spin transmutation in quantum spin Hall Josephson junctions, *Phys. Rev. Lett.* **117**, 267001 (2016).

[79] J. Alicea, New directions in the pursuit of Majorana fermions in solid state systems, *Rep. Prog. Phys.* (2012).

Chapter 11

First Principle Calculation of the Effective Zeeman's Couplings in Topological Materials

Zhida Song[*,†], Song Sun[*,†], Yuanfeng Xu[*,†], Simin Nie[‡],
Hongming Weng[*,§], Zhong Fang[*,§] and Xi Dai[*,¶,‖]

[*]*Beijing National Laboratory for Condensed Matter Physics
and Institute of Physics, Chinese Academy of Sciences,
Beijing 100190, China*
[†]*University of Chinese Academy of Sciences,
Beijing 100049, China*
[‡]*Department of Materials Science and Engineering,
Stanford University, Stanford, CA 94305, USA*
[§]*Collaborative Innovation Center of Quantum Matter,
Beijing, 100084, China*
[¶]*Department of Physics, Hong Kong University of Science
and Technology, Clear Water Bay, Kowloon 999077, Hong Kong*
[‖]*dai@ust.hk*

In this chapter, we propose a first principle calculation method for the effective Zeeman's coupling based on the second order perturbation theory and apply it to a number of topological materials. For Bi metal and Bi_2Se_3, our numerical results are in good agreement with the experimental data; for Na_3Bi, TaN, and $ZrTe_5$, the structure of the multi-band Zeeman's couplings are discussed; especially, we discuss the impact of Zeeman's coupling on the Fermi surface's topology in Na_3Bi in details.

11.1 Introduction

The Landé g-factor, or the form of effective Zeeman effect in particular materials, has been an old topic in the condensed matter physics, which determines how a pair of otherwise degenerate states split under the external magnetic field and can lead to a series of consequences in magnetic responses such as the Van Vleck paramagnetism in insulators [1], the Pauli paramagnetism in metals [2], as well as the frequency splitting of quantum oscillations in metals or doped semiconductors [3]. The widely used theory about the origin of the effective Landé g-factor in materials was developed in 1950s by Luttinger [4] and Cohen [5], according to which, the Zeeman's coupling is not only contributed by the spin angular momentums but also the orbital angular momentums of the Bloch states. As clarified by Chang and Niu in a semiclassical picture, such an orbital contribution can be interpreted as the self-rotation effect of the quasiparticle wave packet [6, 7]. However, although the theory and physical picture of the effective Zeeman's coupling have already been established for a long time, there are still no reported first principle calculations of g-factor based on the density functional theory (DFT), which motivates us to make a first try. In this work, we present a successful first principle calculation method of the effective Zeeman's coupling parameters within the projector augmented-wave (PAW) formulation of the DFT.

In the traditional semiconductors described by the Fermi liquid theory, the responses of the system to the external magnetic field can be well described by two sets of parameters, the effective mass and Landé g-factor tensors. In traditional metals and semi-conductors, the Zeeman effect only splits the two degenerate states and modifies the Landau-level behavior of the free electrons under an external magnetic field. While, for topological materials with strong spin–orbit coupling (SOC), the Zeeman effect will also generate Berry's curvature near the Fermi energy, which will affect various transport properties of the materials. The consequences of the Berry curvature have been studied thoroughly in various theoretical and experimental works, including the weak localization [8], the negative magnetoresistance [9–13], the nonlocal transport [14], the anomalous quantum oscillation [15], and the anomalous coupling with pseudo-scalar phonon [16], etc. However, the role of the Zeeman's coupling in these transport phenomena has not attracted much attentions until recently [17], which makes it timely and important to develop a first principle method to calculate the g-factor tensor for realistic material systems.

The paper is organized as the following. In Sect. 11.2, the theory of effective Zeeman's coupling will be introduced in the framework of the second-order quasi-degenerate perturbation theory. In Sect. 11.3, the first principle calculation method is introduced and applied to a few topological materials. In Sect. 11.4, a most direct application of our data, i.e., the Fermi surface topology of Dirac semimetal under Zeeman's coupling, is discussed. In the end, we give a brief summary in Sect. 11.5.

11.2 Theory

In this section, we will give a short review about the quasi-degenerate perturbation theroy [18–20] and show how the effective Zeeman's coupling emerges as a gauge invariant second-order perturbation term.

In semiconductors and semimetals, the low-energy physics usually involves only a small part of the Brillouin zone, i.e., the neighborhoods around a few special wave vectors, and so can be well described by the $k \cdot p$ Hamiltonians around these points. For each of the wave vectors $\mathbf{K}$, we can write the $k \cdot p$ Hamiltonian $e^{-i(\mathbf{K}+\mathbf{k}) \cdot \mathbf{r}} \hat{H} e^{i(\mathbf{K}+\mathbf{k}) \cdot \mathbf{r}}$ on the periodic parts of Bloch wavefunctions as [20]

$$H_{nn'}(\mathbf{k}) = \delta_{nn'} \left[\epsilon_n + \frac{\hbar^2 \mathbf{k}^2}{2m_e} \right] + \frac{\hbar}{m_e} \boldsymbol{\pi}_{nn'} \cdot \mathbf{k},$$

where ϵ_n is the nth band energy at $\mathbf{K}$, and $\boldsymbol{\pi}_{nn'} = \langle \psi_{n\mathbf{K}} | \hat{\mathbf{p}} + \frac{1}{2m_e c^2} (\hat{\mathbf{s}} \times \nabla V) | \psi_{n'\mathbf{K}} \rangle$ is the momentum element with spin–orbit coupling (SOC) correction, with $\hat{\mathbf{p}}$ being the canonical momentum operator, $\hat{\mathbf{s}}$ being the spin operator, V being the scalar potential in crystal, m_e being the electron mass, c being the light speed, and $|\psi_{n'\mathbf{K}}\rangle$ being the n'th Bloch wavefunction. Since the excitations near the Fermi-level dominate the low-energy physics, we only include a small number of bands near the Fermi level. However, a direct cutoff in the Hilbert space is a too rough approximation because the high-energy subspace couples to the low-energy subspace through the off-diagonal terms in the Hamiltonian. To solve this problem, in the quasi-degenerate perturbation theory the coupling Hamiltonian is treated as a small quantity and an unitary transformation is constructed to decouple the two subspaces. With the second-order approximation, the transformed Hamiltonian in the low-energy subspace can be derived as

$$H_{mm'} = \delta_{mm'} \left[\epsilon_m + \frac{\hbar^2 \mathbf{k}^2}{2m_e} \right] + \frac{\hbar}{m_e} \boldsymbol{\pi}_{mm'} \cdot \mathbf{k}$$

$$+ \frac{\hbar^2}{2m_e^2} \sum_l' \sum_{ij} \left[\frac{1}{\epsilon_m - \epsilon_l} + \frac{1}{\epsilon_{m'} - \epsilon_l} \right] \pi_{ml}^i \pi_{lm'}^j k^i k^j, \qquad (11.1)$$

where m, m' are the band indexes in the low-energy subspace, and the summation over l is limited within the high-energy subspace. In the presence of magnetic field, according to the Peierls substitution the momentums $\hbar k^i$ should be replaced by the kinetic momentum operators $-i\hbar \partial^i + eA^i$, which are not commutative with each other. Thus, the product $(-i\hbar \partial^i + eA^i)(-i\hbar \partial^j + eA^j)$ substituting $\hbar^2 k^i k^j$ can be decomposed into a gauge-dependent symmetric component $\frac{1}{2}\{-i\hbar \partial^i + eA^i, -i\hbar \partial^j + eA^j\}$ and a gauge invariant anti-symmetric component $\frac{1}{2}[-i\hbar \partial^i + eA^i, -i\hbar \partial^j + eA^j]$ which equals to $-\frac{i\hbar e}{2} \sum_k \epsilon^{ijk} B^k$ [4, 5, 21]. Here the charge of electron is $-e$, A^i is the vector potential, ϵ^{ijk} is the Levi–Civita tensor and the curly and square brackets represent the anti-commutator and the commutator respectively. Finally, the total Hamiltonian under magnetic field can be summarized as a gauge dependent part

$$\hat{H}^{kp}_{mm'} = \delta_{mm'}\epsilon_m + \frac{\hbar}{m_e}\boldsymbol{\pi}_{mm'} \cdot \left(-i\nabla + \frac{e}{\hbar}\mathbf{A}\right)$$
$$+ \sum_{ij} M^{ij}_{mm'}\left(-i\partial^i + \frac{e}{\hbar}A^i\right)\left(-i\partial^j + \frac{e}{\hbar}A^j\right), \qquad (11.2)$$

and a gauge invariant part, i.e., the effective Zeeman's coupling

$$\hat{H}^Z_{mm'} = \mu_B \frac{1}{\hbar}(\mathbf{L}_{mm'} + 2\mathbf{s}_{mm'}) \cdot \mathbf{B}. \qquad (11.3)$$

Here $M^{ij}_{mm'}$ is the symmetrized rank-2 tensor describing the inverse effective mass

$$M^{ij}_{mm'} = \delta_{mm'}\delta_{ij}\frac{\hbar^2}{2m_e} + \frac{\hbar^2}{2m_e^2}\sum_l{}' \left[\frac{1}{\epsilon_m - \epsilon_l} + \frac{1}{\epsilon_{m'} - \epsilon_l}\right] \times \frac{\pi^i_{ml}\pi^j_{lm'} + \pi^j_{ml}\pi^i_{lm'}}{2},$$
$$(11.4)$$

$\mathbf{L}_{mm'}$ is the effective orbital momentum contributed by the anti-symmetrized rank-2 tensor as

$$L^k_{mm'} = -\frac{i\hbar}{2m_e}\sum_l{}'\sum_{ij} \left[\frac{1}{\epsilon_m - \epsilon_l} + \frac{1}{\epsilon_{m'} - \epsilon_l}\right]\epsilon_{ijk}\pi^i_{ml}\pi^j_{lm'}, \qquad (11.5)$$

$\mu_B = \frac{e\hbar}{2m_e}$ is the Bohr magneton, and $\frac{2}{\hbar}\mu_B\mathbf{s}\cdot\mathbf{B}$ is the bare Zeeman's coupling from Schrödinger's equation.

In the above derivation, we would like to emphasize two points. The first point is that both the form and value of the g-factor tensor themselves are not the observable quantities, because it depends on the low-energy subspace chosen by us. Therefore, in principle only the resulting final Landau-level spectrum or other transport properties are measurable quantities. The second point is that such a quasi-degenerate perturbation theory is well defined only if there is a relatively large gap, compared to other energy scales such as Fermi energy, between the low- and high-energy subspaces.

11.3   First principle calculations

In this section, we calculate the second-order $k \cdot p$ models and the effective Zeeman's couplings for a few typical topological materials by Eqs. (11.4) and (11.5), in which the momentum and spin matrix elements are computed with the Vienna *ab-initio* simulation package (VASP). Technical details about the computation method within the projector augmented-wave (PAW) formulation are given in Appendix A. Several commonly used types of exchange and correlation potentials are implemented in the method, i.e., the local density approximation (LDA) [22, 23], the generalized gradient approximation (GGA) [24], and the LDA correlation plus modified Becke and Johnson (mBJ) exchange potential [25], and at least 300 bands is calculated for the summation over l in Eqs. (11.4) and (11.5).

11.3.1 *Two bands models*

The g-factor of Schrödinger electron is a dimensionless scalar that characterizes the ratio between electron's magnetic moment and spin moment, which equals to 2 in vacuum. In semiconductors where both inversion and time-reversal (TR) symmetry are present, the conduction or valance bands are doubly dengenerate and are well described by a quasi-Schrödinger's equation with renormalized mass and g-factor. Generally speaking, according to the anisotropy that may occur in real materials, both the effective mass and effective g-factor should be 3×3 tensor now. Here we define the effective g-factor tensor as the expansion coefficients of the Zeeman's coupling Hamiltonian on Pauli's matrices

$$H^Z_{mm'} = \mu_B \frac{1}{2} \sum_i g_{ij} \sigma^j_{mm'} B^i, \tag{11.6}$$

which is reduced to $g_{ij} = 2\delta_{ij}$ in vacuum.

In the following, we present two examples, Bi metal and Bi_2Se_3 whose g-factor has been measured in experiments [26, 27] or studied in previous theoretical work [21], to verify the validity of our method.

11.3.1.1 *Hole pocket in Bi metal*

Bismuth has played an important role in the topological materials because of its large SOC, which, in Ref. [28], is also believed to be responsible for the large anisotropic Zeeman's coupling of the valence bands in its elemental crystal. The elemental bismuth has the space group $R\bar{3}m$ and is a typical semimetal with a hole pocket at T point and three equivalent electron pockets at L point [29]. Here we only focus on the hole-pocket bands, which form a two-dimensional Irreducible representation (IR) $E_{\frac{3}{2}u}$ of the little group D_{3d} at T point due to the time reversal symmetry. Choosing the two bases as $|P\frac{3}{2}\rangle$, $|P\bar{\frac{3}{2}}\rangle$, we find that all the mass and g-factor elements, except $m_{xx} = m_{yy} = m_\perp$, $m_{zz} = m_\parallel$, $g_{zz} = g_\parallel$, are zero. In fact, the absence of the correction to g_{xx} and g_{yy}, or, the absence of off-diagonal elements in H^Z, is guaranteed by the angular momentum conservation condition $\langle P\frac{3}{2}|L_x \pm$

Table 11.1.   Effective g-factor and mass tensors for the two bands models of conduction bands at Γ of Bi_2Se_3 and valance bands at T of Bi. Both g and m^* are dimensionless numbers. The energy gaps shown here are the direct gaps at these high symmetry points.

		LDA	GGA	mBJ	Exp
Bi hole	Gap(eV)	0.272	0.298	0.453	0.18 to 0.41 [29]
	$g_\parallel$	79.40	73.83	55.31	63.2 [26]
	$m^*_\parallel$	−0.592	−0.604	−0.626	−0.69 to −0.702 [29]
	$m^*_\perp$	−0.0389	−0.0447	−0.0895	−0.064 [29]
Bi_2Se_3	Gap(eV)	0.521	0.435	0.236	
	$g_\parallel$	18.4	21.76	41.80	32 [27]
	$g_\perp$	16.37	17.86	26.18	23 [27]
	$m^*_\parallel$	0.851	0.488	0.238	
	$m^*_\perp$	2.61	0.420	0.077	0.124 [27]

$iL_y|P\frac{\bar{3}}{2}\rangle = 0$, which indicates that the Zeeman effect is nonlinear for in-plane field direction. Numerical results are summarized in Table 11.1. It shows that the experimental g-factor and mass lie between the LDA/GGA and mBJ values, which is very reasonable because LDA and GGA usually underestimate while mBJ may overestimate the band gap.

11.3.1.2 Bi_2Se_3

The second system we study is Bi_2Se_3 with the space group $R\bar{3}m$, which has been a famous material in the past few years as the first large gap three-dimensional $\mathbb{Z}_2$ topological insulator [30, 31]. Besides its nontrivial topology, Bi_2Se_3 also has nontrivial large g-factors in its conduction bands, which are doubly degenerate and form the IR $E_{\frac{1}{2}u}$ of the little group D_{3d} at Γ point [27]. Taking the gauge in which the bases transform as $|P\frac{1}{2}\rangle$ and $|P\frac{\bar{1}}{2}\rangle$ (see [32]), we find that due to the symmetry constraints the effective mass and g-factor tensor have very concise forms: $m_{xx} = m_{yy} = m_\perp$, $m_{zz} = m_\parallel$, $g_{xx} = g_{yy} = g_\perp$, $g_{zz} = g_\parallel$, and the rest of components are zero. Results comparable with experimental data are obtained, as shown in Table 11.1, which again shows that the experimental values lie between the LDA/GGA and mBJ values. An interesting observation is that the mBJ gap here is smaller than the LDA/GGA gap, which is different with the tendency in bismuth, because the gap in Bi_2Se_3 is the typical SOC type rather than the charge transfer type.

11.3.2 *Four bands models*

In topological semimetals with gapless nodes or topological insulators near critical point, due to the zero or extremely small band gap, there is no longer a well-defined second-order quasi-degenerate perturbation theory for a two-band subspace. Therefore, the effective model should include both the conduction and valence bands, and consequently becomes four-band. We will show that, such a multi-band Zeeman's coupling reveals much more fruitful physical consequences.

11.3.2.1 Na_3Bi

A typical instance of Dirac semimetal is Na_3Bi with space group $P6_3/mmc$, where the two Dirac nodes are generated by the crossings of two doubly degenerate bands, i.e., the $\pm\frac{3}{2}$ and $\pm\frac{1}{2}$ states forming the $E_{\frac{3}{2}}$ and $E_{\frac{1}{2}}$ IRs of the little group C_{6v} along z-axis [33,34]. Here the little group C_{6v} is crucial for the existence of Dirac nodes, as the crossing between different IRs is protected by the rotational symmetry and the double degeneracy of each IR is guaranteed by the vertical mirrors. Expanding the $k \cdot p$ Hamiltonian around one of the Dirac node, say, $(00k_c)$ for example, we get the effective $k \cdot p$ model as

$$H^{kp}(\mathbf{k}) = C(\mathbf{k}) + \begin{pmatrix} -M(\mathbf{k}) & -v_\perp k_- - \gamma_1 k_z k_- & \gamma_2 k_-^2 & 0 \\ * & M(\mathbf{k}) & 0 & \gamma_2 k_-^2 \\ * & * & M(\mathbf{k}) & v_\perp k_- + \gamma_1 k_z k_- \\ * & * & * & -M(\mathbf{k}) \end{pmatrix},$$

(11.7)

and the effective Zeeman's coupling as

$$H^Z = \mu_B \begin{pmatrix} g_\parallel^{\frac{3}{2}} B_z & g_\perp' B_- & 0 & 0 \\ * & g_\parallel^{\frac{1}{2}} B_z & g_\perp^{\frac{1}{2}} B_- & 0 \\ * & * & -g_\parallel^{\frac{1}{2}} B_z & g_\perp' B_- \\ * & * & * & -g_\parallel^{\frac{3}{2}} B_z \end{pmatrix},$$

(11.8)

where the basis set is chosen as $|\frac{3}{2}\rangle$, $|\frac{1}{2}\rangle$, $|\bar{\frac{1}{2}}\rangle$, $|\bar{\frac{3}{2}}\rangle$, and the quantities in the above equations are defined as $k_\pm = k_x \pm i k_y$, $B_\pm = B_x \pm i B_y$, $M(\mathbf{k}) = v_z k_z + M_\parallel k_z^2 + M_\perp(k_x^2 + k_y^2)$, and $C(\mathbf{k}) = v_0 k_z + C_\parallel k_z^2 + C_\perp(k_x^2 + k_y^2)$, respectively. As shown in Eq. (11.8), such a matrix form Zeeman's coupling not only splits the bands within each Weyl subblock but also couples the two Weyl subblocks together, leading to some exotic Fermi surface structure, as will be discussed in the next section. The calculated model parameters are summarized in Table 11.2.

Table 11.2. Parameters in the Zeeman's couplings and effective $k \cdot p$ Hamiltonians for the four bands models. The parameters g defining the Zeeman's couplings are dimensionless numbers, while the parameters Δ, v, C (M, γ), i.e., the coefficients of constants, $\mathcal{O}(k)$ terms, and $\mathcal{O}(k^2)$ terms, are in units of eV, eV·Å, and eV·Å^2, respectively. Since the self-consistent LDA + mBJ potential heavily overestimates the corrections on band inversions in Na$_3$Bi and bismuth, here we fix the MBJ parameter as $c_{\mathrm{MBJ}} = 0.93$ and $c_{\mathrm{MBJ}} = 1.14$ for Na$_3$Bi and bismuth respectively to recover the HSE band inversion [33] and experimental band gap [29].

Na$_3$Bi				TaN			ZrTe$_5$				Bi electron			Unit
LDA	GGA	mBJ		LDA	GGA	mBJ	LDA	GGA	mBJ		LDA	GGA	mBJ	
5.78	5.84	6.36	$g^{3/2}_{\parallel}$	−0.25	−0.28	−0.36	−0.12	−0.04	0.08	g^p_x	−2.73	−2.70	−2.56	1
2.90	2.96	3.41	$g^{1/2}_{\parallel}$	1.84	1.96	2.45	12.24	11.63	9.66	g^p_y	5.90	5.94	5.11	
4.40	4.45	4.50	$g^{1/2}_{\perp}$	0.54	0.57	0.66	−5.19	−4.56	−2.22	g^p_z	1.08+0.29i	1.05+0.25i	0.55+0.19i	
3.10	3.12	3.06	$g'_{\perp}$	0.25	0.23	0.16	0.64	0.67	0.77	g^s_x	0.025	0.054	0.033	
							−2.42	−2.89	−6.45	g^s_y	5.98	5.99	5.98	
							−0.55	−0.61	−0.93	g^s_z	2.64−1.98i	2.60−1.98i	1.83−2.47i	
1.41	1.47	1.70	Δ	2.74	2.68	2.48	0.0113	0.0118	0.0323	Δ	0.113	0.102	0.013	eV
1.63	1.68	1.87	v_0	−0.72	−0.71	−0.66	−1.88	−2.14	−3.91	v_x	1.24−5.51i	1.22−5.52i	0.80−5.89i	eVÅ
1.99	1.99	2.12	$v_{\parallel}$	3.08	3.14	3.34	0.43	0.38	0.13	v_y	−0.43−0.71i	−0.42−0.68i	−0.29−0.41i	
			$v_{\perp}$				1.55	1.66	2.07	v_z	−1.81+4.48i	−1.75+4.52i	−1.17+4.66i	
1.90	2.13	2.24	$C_{\parallel}$	3.20	3.28	3.66	34.44	31.08	5.91	C^p_x	24.54	25.40	23.74	eVÅ^2
5.97	6.26	8.70	$C_{\perp}$	−2.39	−2.36	−1.97	−9.64	−8.90	−5.19	C^p_y	−5.95	−5.87	−6.50	
4.21	4.45	4.42	$M_{\parallel}$	−10.37	−10.44	−10.15	−9.52	−8.84	−0.12	C^p_z	10.50	10.55	9.86	
0.99	0.97	0.60	$M_{\perp}$	6.28	6.53	7.87	−28.54	−35.42	−49.83	C^s_x	−13.76	−14.40	−14.39	
14.22	14.36	14.34	γ_1	0.68	0.60	0.40	4.06	4.08	3.27	C^s_y	4.10	4.11	2.99	
7.64	7.89	9.85	γ_2	−1.01	−1.03	−0.92	8.82	7.26	−6.94	C^s_z	−8.76	−8.61	−11.14	
										C^p_{yz}	3.02	3.05	3.62	
										C^s_{yz}	−4.60	−4.63	−4.98	

11.3.2.2 *Tantalum nitride*

In the above model, the presence of quartic degenerate Dirac nodes is protected by the little group C_{6v}. Then a natural question to ask is that what if the system has a lower symmetry, where, for example, the C_6 symmetry is broken? An example is the θ-phase TaN with space group $P\bar{6}m2$, where the little group along the z-axis is reduced to C_{3v} and the Dirac nodes split into two triply degenerate nodes [35]. Specifically, the $\pm\frac{1}{2}$ states having the C_3 eigenvalues $e^{\mp i\frac{\pi}{3}}$ are still degenerate along the z-axis due to the presence of vertical mirrors; while the degeneracy of $\pm\frac{3}{2}$ states is no longer guaranteed since their C_3 eigenvalues are the same (-1). For convenience, here we choose the basis set $|\frac{1}{2}\rangle$, $|\bar{\frac{1}{2}}\rangle$, $|\frac{3}{2}\rangle$, $|\bar{\frac{3}{2}}\rangle$ at the high symmetry point A $= (00\pi)$, and get the effective $k \cdot p$ model and Zeeman's coupling as

$$
H^{kp}(\mathbf{k}) = \begin{pmatrix}
\frac{\Delta}{2} + C^{\frac{1}{2}}(\mathbf{k}) & 0 & \gamma_1 k_z k_+ & v_\perp k_+ + \gamma_2 k_-^2 \\
* & \frac{\Delta}{2} + C^{\frac{1}{2}}(\mathbf{k}) & v_\perp k_- - \gamma_2 k_+^2 & \gamma_1 k_z k_- \\
* & * & -\frac{\Delta}{2} + C^{\frac{3}{2}}(\mathbf{k}) & v_\| k_z \\
* & * & * & -\frac{\Delta}{2} + C^{\frac{3}{2}}(\mathbf{k})
\end{pmatrix},
$$
$$(11.9)$$

$$
H^Z = \mu_B \begin{pmatrix}
g_\|^{\frac{1}{2}} B_z & g_\perp^{\frac{1}{2}} B_- & g'_\perp B_+ & 0 \\
* & -g_\|^{\frac{1}{2}} B_z & 0 & -g'_\perp B_- \\
* & * & g_\|^{\frac{3}{2}} B_z & 0 \\
* & * & * & -g_\|^{\frac{3}{2}} B_z
\end{pmatrix},
\qquad (11.10)
$$

where the quadratic terms are defined as $C^{\frac{1}{2}/\frac{3}{2}}(\mathbf{k}) = C_\|^{\frac{1}{2}/\frac{3}{2}} k_z^2 + C_\perp^{\frac{1}{2}/\frac{3}{2}}(k_x^2 + k_y^2)$ and all the model parameters are summarized in Table 11.2. Although the symmetry here is different with Na$_3$Bi, we find that to linear order of magnetic field, the Zeeman's coupling shares the same form.

11.3.2.3 *ZrTe$_5$*

The transition-metal pentatelluride ZrTe$_5$ with space group *Cmcm* can be thought as the stacking of quantum spin Hall layers with medium strength van der Waals interlayer bonding. As a consequence, this material is very close to the critical point between weak and strong topological insulator phases [36,37]. Since the band gap is very sensitive to the external pressure and lattice constants, whether the experimental phase is in strong or weak topological phase is still to be determined experimentally [38–44]. Probably because of this, ZrTe$_5$ has some very nontrivial transport behaviors, such as "anomalous" Hall effect [45] and negative magnetoresistance [40]. As discussed in [17], a vital issue relevant to these transport properties is the interplay between Zeeman's coupling and Berry curvature. To study the impact of Zeeman's coupling, we pick up the conductance and valance bands at Γ, which form the IRs $E_{\frac{1}{2}u}$ and $E_{\frac{1}{2}g}$ of the little group D_{2h}, as the basis to build the effective model. Taking the bases order as $|P\frac{1}{2}\rangle$, $|P\bar{\frac{1}{2}}\rangle$, $|S\frac{1}{2}\rangle$, $|S\bar{\frac{1}{2}}\rangle$, we get

$$
H^{kp}(\mathbf{k}) =
\begin{pmatrix}
\frac{\Delta}{2} + C^p(\mathbf{k}) & 0 & v_z k_z & v_x k_x - i v_y k_y \\
* & \frac{\Delta}{2} + C^p(\mathbf{k}) & v_x k_x + i v_y k_y & -v_z k_z \\
* & * & -\frac{\Delta}{2} + C^s(\mathbf{k}) & 0 \\
* & * & * & -\frac{\Delta}{2} + C^s(\mathbf{k})
\end{pmatrix},
$$
$$(11.11)$$

$$
H^Z = \mu_B
\begin{pmatrix}
g_z^p B_z & g_x^p B_x - i g_y^p B_y & 0 & 0 \\
* & -g_z^p B_z & 0 & 0 \\
* & * & g_z^s B_z & g_x^s B_x - i g_y^s B_y \\
* & * & * & -g_z^s B_z
\end{pmatrix},
\qquad (11.12)
$$

where the quadratic terms are defined as $C^{p/s}(\mathbf{k}) = C_x^{p/s} k_x^2 + C_y^{p/s} k_y^2 + C_z^{p/s} k_z^2$. Different from the two models above, the presence of inversion symmetry guarantees the absence of off-block diagonal Zeeman's couplings. The first principle parameters summarized in Table 11.2 are calculated with the optimized crystal structure described in [37].

11.3.2.4 Electron pocket in bismuth

In the previous section, we have calculated the effective mass and the g-factor of the hole pocket at the T point in bismuth. To complete the discussion, now we consider the electron pocket at L point, where the conduction and valance bands form two-dimensional IRs $E_{\frac{1}{2}g}$ and $E_{\frac{1}{2}u}$ of the little group C_{2h}, respectively, due to the TR symmetry. Although the Fermi level cuts only the conduction bands, as the energy gap between conduction and valance bands is very small and comparable with the Fermi energy [29], the effective model should include both of them. For convenience, among the three equivalent L points, we pick the L point locating at the $k_x = 0$ plane, where the C_2 axis is along the x-axis. Choose the basis set as $|S\frac{1}{2}\rangle$, $|S\bar{\frac{1}{2}}\rangle$, $|P\frac{1}{2}\rangle$, $|P\bar{\frac{1}{2}}\rangle$, we get

$$
H^{kp}(\mathbf{k}) =
\begin{pmatrix}
\frac{\Delta}{2} + C^s(\mathbf{k}) & 0 & v_x k_x & v_y k_y - i v_z k_z \\
* & \frac{\Delta}{2} + C^s(\mathbf{k}) & v_y^* k_y + i v_z^* k_z & -v_x^* k_x \\
* & * & -\frac{\Delta}{2} + C^p(\mathbf{k}) & 0 \\
* & * & * & -\frac{\Delta}{2} + C^p(\mathbf{k})
\end{pmatrix}
\tag{11.13}
$$

and

$$
H^Z = \mu_B
\begin{pmatrix}
g_x^s B_x & g_y^s B_y - i g_z^s B_z & 0 & 0 \\
* & -g_x^s B_x & 0 & 0 \\
* & * & g_x^p B_x & g_y^p B_y - i g_z^p B_z \\
* & * & * & -g_x^p B_x
\end{pmatrix},
\tag{11.14}
$$

where the quadratic terms are defined as $C^{p/s}(\mathbf{k}) = C_x^{p/s} k_x^2 + C_y^{p/s} k_y^2 + C_z^{p/s} k_z^2 + C_{yz}^{p/s} k_y k_z$. Since the C_{2h} group can be obtained by just removing the two C_2-axes in the horizontal plane in the D_{2h} group, this model shares similar form with the ZrTe$_5$ model, except that the lower symmetry here makes all the off-diagonal parameters, i.e., $v_{x/y/z}$ and $g_{y/z}^{s/p}$, no longer real but complex. The calculated parameters are summarized in Table 11.2, where, in order to fix the gauge freedom, we have absorbed phases of g_y^s and g_y^p into basis and keep them to be real numbers.

11.4   Discussions

One of the greatest interesting points of the topological semimetals with gapless nodes or the topological insulators near critical points may be their novel responses under magnetic field due to the Berry curvature correction on the quasiparticle dynamics. Such responses include weak localization [8], negative magnetoresistance [10–13], anomalous quantum oscillation [15], and anomalous coupling with pseudo-scalar phonon [16], etc. However, a related important issue, i.e., the Zeeman's coupling, which is recently revealed to play an important role in the magnetoresistance in topological insulators [17], is usually neglected in previous theoretical works. Now, with the calculated g-factor tensors obtained above, the magnetic responses can be studied with the correction of Zeeman's coupling, where the interplay between Berry curvature and Zeeman's coupling would certainly lead to interesting new physics.

Here we only show a most direct application of our results: the Fermi surface topology of Dirac semimetal under Zeeman's coupling. The intuitive picture that the Zeeman's coupling would split the double degenerate Fermi surface in Dirac semimetal into two Fermi surfaces with nonzero Chern numbers was firstly proposed to understand some magneto-transport experiments [45]. However, no serious calculation about such splitting has been done because of the lacking of reliable g-factors. Now we present a full evolution of the Fermi surface around a single Dirac point in Na$_3$Bi with respect to the strength and direction of the magnetic field. Since there are only two relevant energy scales, i.e., the Fermi energy ϵ_F and the Zeeman's splitting $\mu_B|\mathbf{B}|$, if we presume the clean limit and keep only k linear terms in Eq. (11.7), the model can be rescaled such that the Fermi surface topology will only depend on the dimensionless parameter $\mu_B\mathbf{B}/\epsilon_F$. Our results are summarized as a "phase diagram" shown in Fig. (11.1), where different "phases" are distinguished by the carrier types (electron or hole) and Chern numbers of the Fermi surfaces, and the magnetic field is applied only within the xz-plane since all the vertical planes are equivalent with each other. With different magnetic field, various of phases, including the phase with one trivial electron pocket, the phase with two separate nontrivial electron pockets, the phase with two nested nontrivial electron pockets, and the phase with a nontrivial electron pocket plus a nontrivial hole pocket, can be achieved, leading to a fruitful phase diagram as shown in Fig. 11.1.

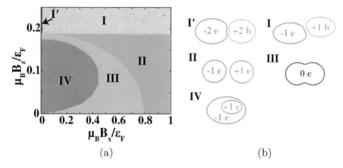

(a)   (b)

Fig. 11.1.   "Phase diagram" of the Fermi surface topology around a single Dirac point in Na$_3$Bi under the Zeeman's coupling. To get this phase diagram, the GGA parameters in Table 11.2 are adopted. In (a), a phase diagram consist of five phases are plotted with respect to the dimensionless parameter $\mu_B\mathbf{B}/\epsilon_F$, and in (b) diagrammatic sketches of the Fermi surface topologies for the five phases are plotted. Here the blue, gray, and red circles represent the right-hand, trivial, and left-hand Fermi surfaces, respectively, where the Chern number of each Fermi surface is defined by the wavefunctions on the occupied side. The solid and dashed circles represent the electron and hole pockets, respectively. As mentioned in the text, we have presumed the clean limit and neglect k^2 terms, without which the two splitted Fermi surfaces have a nodal line crossing at the $k_z = 0$ plane if the magnetic field is applied along z. However, the presence of infinite small k^2 terms will open such crossings and leave the two separated Fermi surfaces with ± 2 Chern numbers, leading to the I$'$ phase.

11.5 Summary

In summary, a first principle method to calculate the effective g-factor tensor in solid has been proposed, and the coupling parameters in a few topological materials have been calculated and discussed. In the framework of quasi-degenerate perturbation theory, we define the effective Zeeman's coupling in magnetic fields as the gauge invariant part of the second-order perturbation Hamiltonian and express all the parameters in it by the momentum and spin matrix elements among the Bloch states, which can be computed within the PAW formulation of the DFT in this work. To verify the validity of this method, the Landé g-factor of the two-band systems, i.e., the hole pocket in bismuth and the conduction bands in Bi_2Se_3, whose Landé g-factor have been measured in experiments, are calculated and the results fit quite well with the experiments. Furthermore, we also derive, discuss, and calculate the effective Zeeman's couplings for a few multi-band models in topological materials such as Na_3Bi, TaN, and $ZrTe_5$. As a simple application of our method, a full evolution of the Fermi surface topology in Dirac semimetals under Zeeman's coupling with respect to the strength and direction of magnetic fields is obtained and discussed based on the calculations of Na_3Bi. Further theoretical investigations for other applications, such as the interplay of the Zeeman's coupling and the Berry phase in magnetotransport experiments, and the role of multi-band Zeeman's coupling in quantum oscillations will be discussed in the future works.

Appendix A. PAW formulation of the matrix elements

In the PAW formulation adopted in VASP, the matrix element of operator $\hat{O}$ can be calculated as [46, 47]

$$
\begin{aligned}
O_{nn'} &= \langle \psi_{n\mathbf{K}} | \hat{O} | \psi_{n'\mathbf{K}} \rangle \\
&= \langle \widetilde{\psi}_{n\mathbf{K}} | \hat{O} | \widetilde{\psi}_{n'\mathbf{K}} \rangle \\
&\quad + \sum_{a\mu\nu} \sum_{\zeta\zeta'} \langle \widetilde{\psi}_{n\mathbf{K}} | \widetilde{p}_{a\mu}\zeta \rangle O^a_{\mu\zeta,\nu\zeta'} \langle \widetilde{p}_{a\nu}\zeta' | \widetilde{\psi}_{n'\mathbf{K}} \rangle,
\end{aligned}
\tag{A1}
$$

where $|\widetilde{\psi}_{n\mathbf{K}}\rangle$ is the pseudo-Bloch wavefunction, $|\widetilde{p}_{a\mu}\zeta\rangle$ is a projector wavefunction at the ath atom consisting of a real space projector wavefunction $\widetilde{p}_{a\mu}(\mathbf{r})$ and a spinor wavefunction $|\zeta\rangle$ ($\zeta = \uparrow\downarrow$), and $O^a_{\mu\nu}$ is the PAW matrix of $\hat{O}$ in the ath atom's augmentation sphere. Here the pseudo-Bloch wavefunction is spanned by the plane waves

$$
\psi_{n\mathbf{K}}(\mathbf{r}\zeta) = \sum_{\mathbf{G}} c^{n\mathbf{K}}_{\zeta,\mathbf{G}} e^{i(\mathbf{K}+\mathbf{G})\cdot\mathbf{r}},
\tag{A2}
$$

and the PAW matrix of $\hat{O}$ is defined as

$$O^a_{\mu\zeta,\nu\zeta'} = \langle\phi_{a\mu}\zeta|\hat{O}|\phi_{a\nu}\zeta'\rangle - \langle\widetilde{\phi}_{a\mu}\zeta|\hat{O}|\widetilde{\phi}_{a\nu}\zeta'\rangle, \tag{A3}$$

with $\phi_{a\mu}(\mathbf{r})$ being the all electron partial wavefunction, and $\widetilde{\phi}_{a\mu}(\mathbf{r})$ the pseudo-partial wavefunction, both of which are stored as an angular part and a radial part

$$\phi_{a\mu}(\mathbf{r}) = Y^{m_\mu}_{l_\mu}(\widehat{\mathbf{r} - \mathbf{R}_a})R_{a\mu}(|\mathbf{r} - \mathbf{R}_a|), \tag{A4}$$

$$\widetilde{\phi}_{a\mu}(\mathbf{r}) = Y^{m_\mu}_{l_\mu}(\widehat{\mathbf{r} - \mathbf{R}_a})\widetilde{R}_{a\mu}(|\mathbf{r} - \mathbf{R}_a|), \tag{A5}$$

where $\mathbf{R}_a$ is the location of the ath atom. Both the plane wave coefficients $c^{n\mathbf{K}}_{\zeta,\mathbf{G}}$ and the projection coefficients $\langle\widetilde{p}_{a\mu}\zeta|\widetilde{\psi}_{n\mathbf{K}}\rangle$ are calculated in the VASP code. To calculate the matrix element of $\hat{O}$ we only need to calculate the pseudo-wavefunction contributions, i.e., the first term in Eq. (A1), and the PAW matrix elements separately and then substitute them back to Eq. (A1).

The spin elements are very direct to calculate. The pseudo-wavefunction contribution is

$$\langle\widetilde{\psi}_{n\mathbf{K}}|\hat{\mathbf{s}}|\widetilde{\psi}_{n'\mathbf{K}}\rangle = \frac{\hbar}{2}\sum_{\mathbf{G}\zeta\zeta'}\boldsymbol{\sigma}_{\zeta,\zeta'}c^{n\mathbf{K}*}_{\zeta,\mathbf{G}}c^{n'\mathbf{K}}_{\zeta',\mathbf{G}}, \tag{A6}$$

and the PAW matrix is

$$\mathbf{s}^a_{\mu\zeta,\nu\zeta'} = \delta_{l_\mu l_\nu}\delta_{m_\mu m_\nu}\frac{\hbar}{2}\sum_{\zeta\zeta'}\boldsymbol{\sigma}_{\zeta,\zeta'}\int dr \cdot r^2 R^*_{a\mu}R_{a\nu}$$

$$- \text{ the integral with } \widetilde{R}, \tag{A7}$$

where the integral over r is evaluated on a logarithmic radial grid. Nevertheless, the calculation of the momentum element is a bit more complicated. According to Eq. (A1), the momentum element can be directly written as

$$\boldsymbol{\pi}_{nn'} = \langle\widetilde{\psi}_{n\mathbf{K}}|\hat{\mathbf{p}}|\widetilde{\psi}_{n'\mathbf{K}}\rangle + \frac{1}{2m_ee^2}\langle\widetilde{\psi}_{n\mathbf{K}}|\hat{\mathbf{s}}\times\nabla V|\widetilde{\psi}_{n'\mathbf{K}}\rangle$$

$$+ \sum_{a\mu\nu}\sum_{\zeta\zeta'}\langle\widetilde{\psi}_{n\mathbf{K}}|\widetilde{p}_{a\mu}\zeta\rangle[\langle\phi_{a\mu}|\hat{\mathbf{p}}|\phi_{a\nu}\rangle - \langle\widetilde{\phi}_{a\mu}|\hat{\mathbf{p}}|\widetilde{\phi}_{a\nu}\rangle]\langle\widetilde{p}_{a\nu}\zeta'|\widetilde{\psi}_{n'\mathbf{K}}\rangle$$

$$+ \frac{1}{2m_ee^2}\sum_{a\mu\nu}\sum_{\zeta\zeta'}\langle\widetilde{\psi}_{n\mathbf{K}}|\widetilde{p}_{a\mu}\zeta\rangle[\langle\phi_{a\mu}\zeta|\hat{\mathbf{s}}\times\nabla V|\zeta'\phi_{a\nu}\rangle$$

$$- \langle\widetilde{\phi}_{a\mu}\zeta|\hat{\mathbf{s}}\times\nabla V|\widetilde{\phi}_{a\nu}\zeta'\rangle]\langle\widetilde{p}_{a\nu}\zeta'|\widetilde{\psi}_{n'\mathbf{K}}\rangle. \tag{A8}$$

Please note that the second term in the first line and the second term in the third line will cancel each other because in VASP the SOC effect is considered only within the augmentation spheres, in which the pseudo-Bloch wavefunctions equal to their projections on pseudo-partial wavefunctions, or equivalently $|\widetilde{\psi}_{n\mathbf{K}}\rangle = \sum_{a\mu\zeta} |\widetilde{\phi}_{a\mu}\zeta\rangle\langle\widetilde{p}_{a\nu}\zeta|\widetilde{\psi}_{n\mathbf{K}}\rangle$. Therefore, the momentum element matrix elements can be simplified to

$$
\begin{aligned}
\boldsymbol{\pi}_{nn'} &= \langle\widetilde{\psi}_{n\mathbf{K}}|\hat{\mathbf{p}}|\widetilde{\psi}_{n'\mathbf{K}}\rangle \\
&+ \sum_{a\mu\nu}\sum_{\zeta\zeta'}\langle\widetilde{\psi}_{n\mathbf{K}}|\widetilde{p}_{a\mu}\zeta\rangle\boldsymbol{\pi}^{a'}_{\mu\zeta,\nu\zeta'}\langle\widetilde{p}_{a\nu}\zeta'|\widetilde{\psi}_{n'\mathbf{K}}\rangle,
\end{aligned}
\tag{A9}
$$

where the psuedo-wavefunction contribution is

$$
\langle\widetilde{\psi}_{n\mathbf{K}}|\hat{\mathbf{p}}|\widetilde{\psi}_{n'\mathbf{K}}\rangle = \sum_{\mathbf{G}\zeta}\hbar\,(\mathbf{K}+\mathbf{G})\,c^{n\mathbf{K}*}_{\zeta,\mathbf{G}}c^{n'\mathbf{K}}_{\zeta,\mathbf{G}},
\tag{A10}
$$

and $\boldsymbol{\pi}^{a'}_{\mu\zeta,\nu\zeta'}$ is defined as

$$
\begin{aligned}
\boldsymbol{\pi}^{a'}_{\mu\zeta,\nu\zeta'} &= \delta_{\zeta\zeta'}\langle\phi_{a\mu}|\hat{\mathbf{p}}|\phi_{a\mu}\rangle - \delta_{\zeta\zeta'}\langle\widetilde{\phi}_{a\mu}|\hat{p}^i|\widetilde{\phi}_{a\mu}\rangle \\
&+ \frac{\hbar}{2m_e e^2}\cdot\frac{\hbar}{2}\boldsymbol{\sigma}_{\zeta,\zeta'}\times\langle\phi_{a\mu}|\nabla V|\phi_{a\mu}\rangle.
\end{aligned}
\tag{A11}
$$

All the integrals between partial wavefunctions can be decomposed as an angular part and a radial part, such as

$$
\begin{aligned}
\langle\phi_{a\mu}|\hat{\mathbf{p}}|\phi_{a\mu}\rangle &= -i\hbar\int d\Omega\, Y^{m_\mu *}_{l_\mu}\nabla Y^{m_\nu}_{l_\nu}\int dr\cdot r^2 R^*_{a\mu}R_{a\nu} \\
&- i\hbar\int d\Omega\, Y^{m_\mu *}_{l_\mu}\frac{\mathbf{r}}{r}Y^{m_\nu}_{l_\nu}\int dr\cdot r^2 R^*_{a\mu}\partial_r R_{a\nu}
\end{aligned}
\tag{A12}
$$

and

$$
\langle\phi_{a\mu}|\nabla V|\phi_{a\mu}\rangle \approx \int d\Omega\, Y^{m_\mu *}_{l_\mu}\frac{\mathbf{r}}{r}Y^{m_\nu}_{l_\nu}\int dr\cdot r^2 R^*_{a\mu}\partial_r V R_{a\nu}.
\tag{A13}
$$

In practice, the angular integrals concerning with spherical harmonics are evaluated with the help of the code in asa.F of VASP.

Acknowledgments

This work is supported by the National Natural Science Foundation of China, the National 973 program of China (Grant No. 2013CB921700) and the "Strategic Priority Research Program (B)" of the Chinese Academy of Sciences (Grant No. XDB07020100).

References

[1] J. H. Van Vleck, *The theory of electric and magnetic susceptibilities* (Clarendon Press, 1932).

[2] N. W. Ashcroft and N. D. Mermin, *Solid State Physics* (Holt, Rinehart and Winston, 1976).

[3] D. Shoenberg, *Magnetic Oscillations in Metals* (Cambridge University Press, 2009).

[4] J. M. Luttinger, Quantum theory of cyclotron resonance in semiconductors: General theory, *Phys. Rev.* **102**(4), 1030 (1956); doi: 10.1103/PhysRev.102.1030; URL http://link.aps.org/doi/10.1103/PhysRev.102.1030.

[5] M. H. Cohen and E. I. Blount, The g-factor and de haas-van alphen effect of electrons in bismuth, *Philoso. Mag.* **5**(50), 115 (1960); doi: 10.1080/14786436008243294; URL http://dx.doi.org/10.1080/14786436008243294.

[6] M.-C. Chang and Q. Niu, Berry phase, hyperorbits, and the Hofstadter spectrum: Semiclassical dynamics in magnetic Bloch bands, *Phys. Rev. B* **53**(11), 7010 (1996); doi: 10.1103/PhysRevB.53.7010; URL http://link.aps.org/doi/10.1103/PhysRevB.53.7010.

[7] M.-C. Chang and Q. Niu, Berry curvature, orbital moment, and effective quantum theory of electrons in electromagnetic fields, *J. Phy.: Conden. Matter* **20**(19), 193202 (2008); doi: 10.1088/0953-8984/20/19/193202; URL http://iopscience.iop.org/0953-8984/20/19/193202/cites.

[8] H.-Z. Lu and S.-Q. Shen, Weak antilocalization and localization in disordered and interacting Weyl semimetals, *Phys. Rev. B* **92**(3) (2015); doi: 10.1103/PhysRevB.92.035203; URL http://arxiv.org/abs/1411.2686.

[9] H. B. Nielsen and M. Ninomiya, The Adler–Bell–Jackiw anomaly and Weyl fermions in a crystal, *Phys. Lett. B* **130**(6), 389 (1983); doi: 10.1016/0370-2693(83)91529-0; URL http://www.sciencedirect.com/science/article/pii/0370269383915290.

[10] A. Burkov, Chiral anomaly and diffusive magnetotransport in Weyl metals, *Phys. Rev. Lett.* **113**(24), 247203 (2014); doi: 10.1103/PhysRevLett.113.247203; URL http://link.aps.org/doi/10.1103/PhysRevLett.113.247203.

[11] A. A. Burkov, Negative longitudinal magnetoresistance in Dirac and Weyl metals, *Phys. Rev. B* **91**(24), 245157 (2015); doi: 10.1103/PhysRevB.91.245157; URL http://link.aps.org/doi/10.1103/PhysRevB.91.245157.

[12] D. T. Son and B. Z. Spivak, Chiral anomaly and classical negative magnetoresistance of Weyl metals, *Phys. Rev. B* **88**(10), 104412 (2013); doi: 10.1103/PhysRevB.88.104412; URL http://link.aps.org/doi/10.1103/PhysRevB.88.104412.

[13] J. Xiong, S. K. Kushwaha, T. Liang, J. W. Krizan, M. Hirschberger, W. Wang, R. J. Cava and N. P. Ong, Evidence for the chiral anomaly in the Dirac semimetal Na_3Bi, *Science* **350**(6259), 413 (2015); doi: 10.1126/science.aac6089; URL http://science.sciencemag.org/content/350/6259/413.

[14] S. Parameswaran, T. Grover, D. Abanin, D. Pesin and A. Vishwanath, Probing the chiral anomaly with nonlocal transport in three-dimensional topological semimetals, *Phys. Rev. X* **4**(3), 031035 (2014); doi: 10.1103/PhysRevX.4.031035; URL http://link.aps.org/doi/10.1103/PhysRevX.4.031035.

[15] C. Wang, H.-Z. Lu and S.-Q. Shen, Anomalous phase shift of quantum oscillations in 3d topological semimetals, *Phys. Rev. Lett.* **117**(7), 077201 (2016); doi: 10.1103/PhysRevLett.117.077201; URL https://link.aps.org/doi/10.1103/PhysRevLett.117.077201.

[16] Z. Song, J. Zhao, Z. Fang and X. Dai, Detecting the Chiral magnetic effect by lattice dynamics in Weyl semimetals, *Phys. Rev. B* **94**(21), 214306 (2016);

doi: 10.1103/PhysRevB.94.214306; URL http://link.aps.org/doi/10.1103/PhysRevB. 94.214306.

[17] X. Dai, Z. Z. Du and H.-Z. Lu, Negative magnetoresistance without Chiral anomaly in topological insulators, Preprint (2017); arXiv:1705.02724 [cond-mat]; URL http:// arxiv.org/abs/1705.02724; arXiv: 1705.02724.

[18] P.-O. Löwdin, A note on the quantum-mechanical perturbation theory, *J. Chem. Phys.* **19**(11), 1396 (1951); URL http://dx.doi.org/10.1063/1.1748067.

[19] J. M. Luttinger and W. Kohn, Motion of electrons and holes in perturbed periodic fields, *Phys. Rev.* **97**(4), 869, (1955); doi: 10.1103/PhysRev.97.869; URL http://link. aps.org/doi/10.1103/PhysRev.97.869.

[20] R. Winkler, *Spin–Orbit Coupling Effects in Two-Dimensional Electron and Hole Systems* (Springer Science & Business Media, 2003).

[21] Y. Fuseya, Z. Zhu, B. Fauqué, W. Kang, B. Lenoir and K. Behnia, Origin of the large anisotropic *g* factor of holes in bismuth, *Phys. Rev. Lett.* **115**(21), 216401 (2015); doi: 10.1103/PhysRevLett.115.216401; URL https://link.aps.org/doi/ 10.1103/PhysRevLett.115.216401.

[22] W. Kohn and L. J. Sham, Self-consistent equations including exchange and correlation effects, *Phys. Rev.* **140**, A1133 (1965); doi: 10.1103/PhysRev.140.A1133; URL https://link.aps.org/doi/10.1103/PhysRev.140.A1133.

[23] D. M. Ceperley and B. J. Alder, Ground state of the electron gas by a stochastic method, *Phys. Rev. Lett.* **45**, 566 (1980); doi: 10.1103/PhysRevLett.45.566; URL https://link.aps.org/doi/10.1103/PhysRevLett.45.566.

[24] J. P. Perdew, K. Burke and M. Ernzerhof, Generalized gradient approximation made simple, *Phys. Rev. Lett.* **77**(18), 3865 (1996); doi: 10.1103/PhysRevLett.77.3865; URL http://link.aps.org/doi/10.1103/PhysRevLett.77.3865.

[25] F. Tran and P. Blaha, Accurate band gaps of semiconductors and insulators with a semilocal exchange-correlation potential, *Phys. Rev. Lett.* **102**(22), 226401 (2009); doi: 10.1103/PhysRevLett.102.226401; URL http://link.aps.org/doi/10.1103/ PhysRevLett.102.226401.

[26] H. R. Verdún and H. D. Drew, Far-infrared magnetospectroscopy of the hole pocket in bismuth. I. Band-structure effects, *Phys. Rev. B* **14**(4), 1370 (1976); doi: 10.1103/PhysRevB.14.1370; URL http://link.aps.org/doi/10.1103/PhysRevB. 14.1370.

[27] H. Köhler and E. Wöchner, The g-factor of the conduction electrons in Bi_2Se_3, *Phys. Stat. Sol.* (*b*) **67**(2), 665 (1975); doi: 10.1002/pssb.2220670229; URL http:// onlinelibrary.wiley.com/doi/10.1002/pssb.2220670229/abstract.

[28] J. Alicea and L. Balents, Bismuth in strong magnetic fields: Unconventional Zeeman coupling and correlation effects, *Phys. Rev. B* **79**(24), 241101 (2009); doi: 10.1103/PhysRevB.79.241101; URL http://link.aps.org/doi/10.1103/PhysRevB.79. 241101.

[29] Y. Liu and R. E. Allen, Electronic structure of the semimetals Bi and Sb, *Phys. Rev. B* **52**(3), 1566–1577 (1995); doi: 10.1103/PhysRevB.52.1566; URL http://link.aps. org/doi/10.1103/PhysRevB.52.1566.

[30] H. Zhang, C.-X. Liu, X.-L. Qi, X. Dai, Z. Fang and S.-C. Zhang, Topological insulators in Bi_2Se_3, Bi_2Te_3 and Sb_2Te_3 with a single Dirac cone on the surface, *Nature Phys.* **5**(6), 438 (2009); doi: 10.1038/nphys1270; URL http://www.nature. com/nphys/journal/v5/n6/abs/nphys1270.html.

[31] Y. Zhang, K. He, C.-Z. Chang, C.-L. Song, L.-L. Wang, X. Chen, J.-F. Jia, Z. Fang, X. Dai, W.-Y. Shan, S.-Q. Shen, Q. Niu, X.-L. Qi, S.-C. Zhang, X.-C. Ma and Q.-K. Xue, Crossover of the three-dimensional topological insulator Bi_2Se_3 to the

two-dimensional limit, *Nature Phys.* **6**(8), 584 (2010); doi: 10.1038/nphys1689; URL http://www.nature.com/nphys/journal/v6/n8/full/nphys1689.html.

[32] S. L. Altmann and P. Herzig, *Point-Group Theory Tables* (Clarendon Press, 1994).

[33] Z. Wang, Y. Sun, X.-Q. Chen, C. Franchini, G. Xu, H. Weng, X. Dai and Z. Fang, *Phys. Rev. B* **85**(19), 195320 (2012); doi: 10.1103/PhysRevB.85.195320; URL http://link.aps.org/doi/10.1103/PhysRevB.85.195320.

[34] Z. K. Liu, B. Zhou, Y. Zhang, Z. J. Wang, H. M. Weng, D. Prabhakaran, S.-K. Mo, Z. X. Shen, Z. Fang, X. Dai, Z. Hussain and Y. L. Chen, Discovery of a three-dimensional topological Dirac semimetal, Na$_3$Bi, *Science* **343**(6173), 864 (2014); doi: 10.1126/science.1245085; URL http://science.sciencemag.org/content/343/6173/864.

[35] H. Weng, C. Fang, Z. Fang and X. Dai, *Phys. Rev. B* **93**(24), 241202 (2016); doi: 10.1103/PhysRevB.93.241202; URL https://link.aps.org/doi/10.1103/PhysRevB.93.241202.

[36] H. Weng, X. Dai and Z. Fang, *Phys. Rev. X* **4**(1), 011002 (2014); doi: 10.1103/PhysRevX.4.011002; URL http://link.aps.org/doi/10.1103/PhysRevX.4.011002.

[37] Z. Fan, Q.-F. Liang, Y. B. Chen, S.-H. Yao and J. Zhou, *Sci. Rep.* **7** (2017); doi: 10.1038/srep45667; URL http://www.ncbi.nlm.nih.gov/pmc/articles/PMC5379478/.

[38] R. Y. Chen, Z. G. Chen, X.-Y. Song, J. A. Schneeloch, G. D. Gu, F. Wang and N. L. Wang, Magnetoinfrared spectroscopy of Landau levels and Zeeman splitting of three-dimensional massless dirac fermions in ZrTe$_5$, *Phys. Rev. Lett.* **115**, 176404 (2015); doi: 10.1103/PhysRevLett.115.176404; URL https://link.aps.org/doi/10.1103/PhysRevLett.115.176404.

[39] X. Yuan *et al.*, Observation of quasi-two-dimensional dirac fermions in ZrTe$_5$, *NPG Asia Mater.* **8**(11), e325 (2016).

[40] Q. Li, D. E. Kharzeev, C. Zhang, Y. Huang, I. Pletikosić, A. V. Fedorov, R. D. Zhong, J. A. Schneeloch, G. D. Gu and T. Valla, Chiral magnetic effect in ZrTe$_5$, *Nature Phys.* **advance online publication**, (2016); doi: 10.1038/nphys3648; URL http://www.nature.com/nphys/journal/vaop/ncurrent/full/nphys3648.html.

[41] Y. Zhang, C. Wang, L. Yu, G. Liu, A. Liang, J. Huang, S. Nie, Y. Zhang, B. Shen, J. Liu, H. Weng, L. Zhao, G. Chen, X. Jia, C. Hu, Y. Ding, S. He, L. Zhao, F. Zhang, S. Zhang, F. Yang, Z. Wang, Q. Peng, X. Dai, Z. Fan, Z. Xu, C. Chen and X. J. Zhou, Electronic evidence of temperature-induced Lifshitz transition and topological nature in ZrTe$_5$, Preprint (2016).

[42] G. Manzoni, L. Gragnaniello, G. Autès, T. Kuhn, A. Sterzi, F. Cilento, M. Zacchigna, V. Enenkel, I. Vobornik, L. Barba, F. Bisti, P. Bugnon, A. Magrez, V. N. Strocov, H. Berger, O. V. Yazyev, M. Fonin, F. Parmigiani and A. Crepaldi, Evidence for a strong topological insulator phase in ZrTe$_5$, *Phys. Rev. Lett.* **117**, 237601 (2016); doi: 10.1103/PhysRevLett.117.237601; URL https://link.aps.org/doi/10.1103/PhysRevLett.117.237601.

[43] X.-B. Li, W.-K. Huang, Y.-Y. Lv, K.-W. Zhang, C.-L. Yang, B.-B. Zhang, Y. B. Chen, S.-H. Yao, J. Zhou, M.-H. Lu, L. Sheng, S.-C. Li, J.-F. Jia, Q.-K. Xue, Y.-F. Chen and D.-Y. Xing, Experimental observation of topological edge states at the surface step edge of the topological insulator ZrTe$_5$, *Phys. Rev. Lett.* **116**, 176803 (2016); doi: 10.1103/PhysRevLett.116.176803; URL https://link.aps.org/doi/10.1103/PhysRevLett.116.176803.

[44] R. Wu, J.-Z. Ma, S.-M. Nie, L.-X. Zhao, X. Huang, J.-X. Yin, B.-B. Fu, P. Richard, G.-F. Chen, Z. Fang, X. Dai, H.-M. Weng, T. Qian, H. Ding and S. H. Pan, Evidence for topological edge states in a large energy gap near the step edges on the surface of ZrTe$_5$, *Phys. Rev. X* **6**, 021017 (2016); doi: 10.1103/PhysRevX.6.021017; URL https://link.aps.org/doi/10.1103/PhysRevX.6.021017.

[45] T. Liang, Q. Gibson, M. Liu, W. Wang, R. J. Cava and N. P. Ong, Anomalous Hall effect in ZrTe$_5$, *Nature Phys.* **14**(5), 451 (2018); arXiv:1612.06972 [cond-mat]; URL http://arxiv.org/abs/1612.06972. arXiv: 1612.06972.

[46] P. E. Blöchl, Projector augmented-wave method, *Phys. Rev. B* **50**(24), 17953 (1994); doi: 10.1103/PhysRevB.50.17953; URL http://link.aps.org/doi/10.1103/PhysRevB.50.17953.

[47] G. Kresse and D. Joubert, From ultrasoft pseudopotentials to the projector augmented-wave method, *Phys. Rev. B* **59**(3), 1758 (1999); doi: 10.1103/PhysRevB.59.1758; URL http://link.aps.org/doi/10.1103/PhysRevB.59.1758.

Chapter 12

Anomaly Inflow and the η-Invariant

Edward Witten* and Kazuya Yonekura[†]

*School of Natural Sciences, Institute for Advanced Study,
Einstein Drive, Princeton, NJ 08540, USA
[†]Department of Physics, Tohoku University,
Sendai 980-8578, Japan

Perturbative fermion anomalies in spacetime dimension d have a well-known relation to Chern–Simons functions in dimension $D = d + 1$. This relationship is manifested in a beautiful way in "anomaly inflow" from the bulk of a system to its boundary. Along with perturbative anomalies, fermions also have global or nonperturbative anomalies, which can be incorporated by using the η-invariant of Atiyah, Patodi, and Singer instead of the Chern–Simons function. Here we give a nonperturbative description of anomaly inflow, involving the η-invariant. This formula has been expected in the past based on the Dai–Freed theorem, but has not been fully justified. It leads to a general description of perturbative and nonperturbative fermion anomalies in d dimensions in terms of an η-invariant in D dimensions. This η-invariant is a cobordism invariant whenever perturbative anomalies cancel.

12.1   Introduction

Shoucheng Zhang made many outstanding contributions in condensed matter physics. He started his career in particle physics and was always very interested in the interplay between condensed matter physics and relativistic physics. He made numerous lasting contributions in areas that combine ideas of the two fields.

Fermion anomalies — originally the one-loop triangle anomaly of Adler, Bell, and Jackiw [1,2] — have been important in particle physics since their discovery, and have proved to be important in condensed matter physics. In particular, they played a role in some of Shoucheng Zhang's important contributions. The lecture of one of us at the Shoucheng Zhang Memorial Workshop was devoted to a particular question about fermion anomalies [3]. (The question is briefly described at the end of Sec. 12.4.1.) The conference organizers suggested that written contributions to the proceedings could be on broader themes related to the topics of the lectures. Here, we will describe a nonperturbative approach to fermion anomaly inflow.

A rather well-known fact about perturbative fermion anomalies is that the perturbative anomaly in spacetime dimension d is related to a Chern–Simons function in dimension $D = d + 1$ [4–6]. This idea has a beautiful manifestation in the idea of anomaly inflow [7], which is important in both particle physics and condensed matter physics. The classic example in condensed matter physics is the integer quantum Hall effect. The worldvolume of a quantum Hall sample has spacetime dimension $D = 3$ and its boundary has dimension $d = 2$. The "edge mode" of a quantum Hall system is — in relativistic language — a massless electrically charged chiral fermion. The coupling of this mode to electromagnetism has an anomaly — analogous to the original triangle anomaly. In the bulk of a quantum Hall system, the effective action for the gauge field A of electromagnetism includes a multiple of the Chern–Simons coupling $\int A \wedge \mathrm{d}A$. This coupling is not gauge-invariant on a manifold with boundary, and that failure of gauge invariance cancels the quantum anomaly of the edge mode.

Along with perturbative anomalies, fermions can also have global or nonperturbative anomalies that are not visible in perturbation theory [8]. The relation between perturbative anomalies in d dimensions and a Chern–Simons function in $d + 1$ dimensions can be generalized to include nonperturbative anomalies [9]. One just has to replace the Chern–Simons function by the η-invariant of Atiyah, Patodi, and Singer (APS) [10]. The η-invariant is equivalent in perturbation theory to a Chern–Simons function, but it contains additional nonperturbative information. The relation between anomalies and the η-invariant motivated a mathematical result that we will call the Dai–Freed theorem [11]. See [12] for an explanation of the Dai–Freed theorem from a physical point of view. The Dai–Freed theorem makes it possible to get more precise results about fermion anomalies than were available otherwise; see, for example, [13, Sec. 2.2].

Since anomaly inflow is an important aspect of understanding perturbative fermion anomalies, and since fermions do have nonperturbative anomalies, one would like to formulate anomaly inflow in a way that incorporates nonperturbative anomalies as well as perturbative ones. The Dai–Freed theorem strongly suggests how to do this. The basic idea is to replace the Chern–Simons function with an η-invariant defined with APS boundary conditions. The resulting nonperturbative formula for anomaly inflow (Eq. (12.52) below) was described in [14]. However, this formula has not been fully justified. The main goal of the present chapter is to do this.

We present our nonperturbative approach to fermion anomaly inflow in Sec. 12.2. We take a viewpoint that is natural in modern condensed matter physics. An arbitrary, possibly anomalous, fermion system in dimension d can be regarded as the boundary state of an anomaly-free gapped fermion system in dimension $D = d + 1$. Then, the anomaly inflow formula is simply obtained, in principle, by integrating out the gapped fermions in the bulk. Since it is already known that integrating out a massive fermion gives a Chern–Simons term perturbatively [15] and an η-invariant nonperturbatively [16], it is fairly clear that this process will give an answer of a familiar kind but with Chern–Simons replaced with η. Less obvious is why the η-invariant should be computed with APS boundary conditions, as suggested by the Dai–Freed theorem. Explaining this is one of the main tasks of Sec. 12.2. To avoid extraneous details, we begin Sec. 12.2 assuming that the bulk fermion is a Dirac fermion, but then we extend the discussion to apply to the anomalies of a completely general relativistic fermion system. (In a somewhat similar way, though presented more abstractly, the Dai–Freed theorem was generalized recently to an arbitrary relativistic fermion system in the appendix of [17].)

In Sec. 12.3, we explain that the nonperturbative formula for anomaly inflow implies a general description of perturbative and nonperturbative fermion anomalies in dimension d in terms of an exponentiated η-invariant in dimension $d + 1$. This generalizes the global anomaly formula obtained in [9] (where d was assumed to be even and only orientable manifolds were considered), and also gives more precise information about the fermion path integral in anomaly-free examples. In Sec. 12.4, we use the description via the η-invariant to analyze potential global anomalies in various examples in dimensions $d = 1, 2, 3, 4$. One of the examples we consider involves a celebrated contribution by Shoucheng Zhang [18], involving the electromagnetic θ angle in the bulk of a $3 + 1$-dimensional topological insulator. Another example we consider is the Standard Model of particle physics.

Though in the present chapter we consider only fermion anomalies, we should note that in modern condensed matter and particle physics, anomalies in bosonic systems play a role as well. In particular, in condensed matter physics, a symmetry protected topological (SPT) phase of matter is a gapped phase with some global symmetry group G, such that the phase is in some sense topologically nontrivial but would become trivial if the symmetry is explicitly broken. A version of anomaly inflow is important in understanding

SPT phases [19–21], with group cohomology playing the role that for fermion anomalies is played by the η-invariant. In all cases, the anomaly in dimension d is related to an "invertible topological field theory" in dimension $d+1$. For a general description of this point of view about anomalies, see [22, 23].

12.2   A precise formula for anomaly inflow

In this section, we study massive fermions on a manifold Y with boundary W. W will have dimension d, while Y has dimension $D = d + 1$. The goal is to get a precise formula for anomaly inflow from Y to W.

12.2.1   *Massive Dirac fermion with local boundary condition*

We begin with a massive Dirac fermion Ψ on a D-dimensional manifold Y. A Dirac fermion is just a fermion field Ψ whose components are all charged under a global or gauge U(1) symmetry. The conjugate field $\overline{\Psi}$ carries opposite U(1) charges. The assumption of a U(1) symmetry is certainly nongeneric, but we begin with this case in order to explain the main idea of the derivation without potentially distracting details.

The discussion will be quite general; D may be even or odd, Y may be orientable or unorientable, and Ψ may be coupled to gauge fields as well as to the geometry of Y. We assume only that Y is endowed with the appropriate structure (such as a spin or pin structure, a spin$_c$ structure, etc.) so that a suitable action for Ψ exists, of the usual form for a Dirac fermion:

$$I = -\int_Y \mathrm{d}^D x \sqrt{g}\, \overline{\Psi}(\slashed{D}_Y + m)\Psi. \tag{12.1}$$

Ψ is coupled to the Riemannian metric g of Y and possibly to a background gauge field A. The Dirac operator is defined in the usual way, $\slashed{D}_Y = \sum_{\mu=1}^{D} \gamma^\mu D_\mu$, with $\{\gamma_\nu, \gamma_\nu\} = 2g_{\mu\nu}$. On a manifold without boundary, the operator $\mathcal{D}_Y = i\slashed{D}_Y$ is self-adjoint.

Now suppose that the manifold Y has boundary $W = \partial Y$, with a metric that looks like a product near the boundary. Thus, near the boundary the metric of Y is $\mathrm{d}s_Y^2 = \mathrm{d}\tau^2 + \mathrm{d}s_W^2$, where $\mathrm{d}s_W^2$ is a metric on W, and τ parameterizes the normal direction. We can normalize τ to vanish along W and to be negative away from W. We assume this product description is valid at least in a range $(-\epsilon, 0]$.

We impose on Ψ the following local boundary condition:

$$\mathsf{L} : (1 - \gamma^\tau)\Psi|_{\tau=0} = 0, \tag{12.2}$$

where γ^τ is the gamma matrix in the direction τ. Before proceeding with the analysis, we will say a word about this boundary condition.

The boundary condition (12.2) is elliptic but not self-adjoint.[a] Ellipticity of the boundary condition is necessary and sufficient to ensure that the Dirac operator with this boundary condition has the properties needed to make Euclidean field theory well-defined. For example, on a compact manifold, the Dirac operator with an elliptic boundary condition has at most a finite-dimensional space of zero-modes, and, after removing those zero-modes, it has a Green's function with the usual properties, such as short distance singularities of a standard form. Technically, the condition for ellipticity is as follows. Suppose that $W = \mathbb{R}^d$ and $Y = \mathbb{R}^d \times \mathbb{R}_- = \mathbb{R}^{d+1}_-$, where $\mathbb{R}_-$ is a half-line parameterized by $\tau \leq 0$. A boundary condition on the Dirac equation at $\tau = 0$ is elliptic if, after dropping from the equation lower order terms (notably mass terms and couplings to background fields), or equivalently after taking the momentum along $\mathbb{R}^d$ to be extremely large, the equation and its adjoint have no solutions that satisfy the boundary condition and vanish for $\tau \to -\infty$. The local boundary condition (12.2) has this property, because any solution of the free Dirac equation on $\mathbb{R}^{d+1}_-$ that has nonzero momentum along the boundary and vanishes for $\tau \to -\infty$ has components with $\gamma^\tau = -1$ as well as components with $\gamma^\tau = +1$.

A boundary condition on the Euclidean Dirac operator is self-adjoint if, with this boundary condition, $i\slashed{D}_Y$ is self-adjoint on a manifold with boundary. As usual, a self-adjoint operator has a well-defined eigenvalue problem with real eigenvalues. The boundary condition (12.2) is not self-adjoint; an attempt to prove self-adjointness will fail because of a surface term at $\tau = 0$. A self-adjoint boundary condition could not lead to the chiral, potentially anomalous physics that we are interested in. In Euclidean signature, self-adjointness of a local boundary condition is not a physical requirement. The actual condition that a local boundary condition should satisfy in order to be physically sensible is the following. Suppose that one analytically continues W to Lorentz signature, with time coordinate x^0. One wants the boundary condition to be such that the Hamiltonian that propagates a state in the x^0 direction is self-adjoint. One can verify that this is true for the boundary condition L, using the fact that $\{\gamma^0, \gamma^\tau\} = 0$.

---

[a]Self-adjoint local boundary conditions are also possible, of course, at least in some cases. For recent analysis of a self-adjoint local boundary condition that leads to no localized boundary mode (but may explicitly violate symmetries such as time-reversal or reflection symmetry), see, e.g., [24–26].

The action (12.1) can be written near the boundary as

$$I = - \int_Y d^D x \sqrt{g} \, \overline{\Psi} \gamma^\tau \left(\frac{\partial}{\partial \tau} + \mathcal{D}_W + \gamma^\tau m \right) \Psi. \tag{12.3}$$

where[b]

$$\mathcal{D}_W = \sum_{\mu \neq \tau} \gamma^\tau \gamma^\mu D_\mu \tag{12.4}$$

is a self-adjoint Dirac operator on the boundary W. Notice that γ^τ and $\mathcal{D}_W$ anti-commute with each other. If d is even, the operator γ^τ measures what is usually called the "chirality" of a fermion that propagates along W. We will call γ^τ a chirality operator in any case, though for odd d this is not standard terminology.

Our discussion so far makes sense for either sign of m, but in what follows, given our choice of boundary condition, the interesting case is $m < 0$. For $m < 0$, the Dirac equation $(\slashed{D} + m)\Psi = 0$ has a mode localized near the boundary and given by

$$\Psi = \chi \exp(|m|\tau), \quad (1 - \gamma^\tau)\chi = 0, \quad \mathcal{D}_W \chi = 0, \tag{12.5}$$

where χ is a fermion field on W. Since it vanishes exponentially for $\tau \ll 0$, this mode is localized along W. Since it satisfies $\mathcal{D}_W \chi = 0$, it propagates along W as a massless fermion. Finally, as it obeys $\gamma^\tau \chi = \chi$, it is a chiral fermion along W. For $m > 0$, there is no such boundary-localized mode.

To quantize the fermion field Ψ, some sort of regulator is needed. A useful choice for our purposes is a simple Pauli–Villars regulator, defined by adding a very massive field of opposite statistics, satisfying the same Dirac equation (possibly with a different mass parameter) and the same boundary condition. We take the Pauli–Villars regulator field to have a positive mass parameter, since we do not want the regulator field to have a low-energy mode propagating along the boundary, which would be quite unphysical.

Now we want to compute the partition function of the above massive fermion Ψ on the manifold Y with the boundary condition L. For this purpose, it is useful to think of τ as a Euclidean time coordinate, and the boundary $W = \partial Y$ as a time-slice. In this point of view, the path integral over Y gives a physical state vector which we denote as $|Y\rangle$. This is a state vector in the Hilbert space $\mathcal{H}_W$ of W. The boundary condition

---

[b] $\mathcal{D}_W$ differs from the usual Dirac operator $i \sum_{\mu=1}^d \gamma^\mu D_\mu$ of W only by the use of a different set of gamma matrices $(-i\gamma^\tau\gamma^\mu$ instead of $\gamma^\mu)$ that obey the same algebra and lead to the same rotation generators $\frac{1}{4}[\gamma_\mu, \gamma_\nu]$. So the two operators are equivalent.

L also corresponds to a state vector $|L\rangle \in \mathcal{H}_W$, which we will later discuss explicitly. Then, the partition function on Y with the boundary condition L is just given by

$$Z(Y,\mathsf{L}) = \langle \mathsf{L}|Y\rangle. \tag{12.6}$$

The important point is now as follows. We take the mass $|m|$ to be very large so that its Compton wavelength $1/|m|$ is much shorter than the typical scale of the manifolds Y and W; equivalently, we take the length scale of Y and W to be much larger than $1/|m|$. In this limit, there is a large mass gap in the Hilbert space $\mathcal{H}_W$. Moreover, the path integral on a cylindrical region $(-\epsilon, 0] \times W$ gives the Euclidean time evolution $e^{-\epsilon H}$ where H is the Hamiltonian. If the mass gap is large so that $\epsilon|m| \gg 1$, the factor $e^{-\epsilon H}$ plays the role of the projection operator to the ground state $|\Omega\rangle$:

$$e^{-\epsilon H} \simeq |\Omega\rangle\langle\Omega| \quad (\epsilon|m| \gg 1). \tag{12.7}$$

This is valid up to errors which are exponentially suppressed by $e^{-\epsilon|m|}$, which we neglect. In this limit, $|Y\rangle$ is proportional to the ground state,

$$|Y\rangle \propto |\Omega\rangle, \tag{12.8}$$

where we may assume $\langle\Omega|\Omega\rangle = 1$. So we can rewrite the partition function as

$$Z(Y,\mathsf{L}) = \langle \mathsf{L}|\Omega\rangle\langle\Omega|Y\rangle. \tag{12.9}$$

In this way, we can split the bulk contribution $\langle\Omega|Y\rangle$ and the boundary contribution $\langle \mathsf{L}|\Omega\rangle$.

The ground state $|\Omega\rangle$ has a phase ambiguity, and hence the splitting into $\langle\Omega|Y\rangle$ and $\langle \mathsf{L}|\Omega\rangle$ also has this ambiguity. In the context of the Dai–Freed theorem [11], a physical interpretation of this ambiguity is as follows [12]. The space of ground states is a one-dimensional subspace of the Hilbert space $\mathcal{H}_W$. We denote this one-dimensional subspace as $\mathcal{L}_W \subset \mathcal{H}_W$. Now suppose that we make an adiabatic change of background fields such as, e.g., the metric g of W or an external gauge field A. The ground state changes adiabatically, and evolves with a Berry phase. Let $\mathcal{W}$ be the parameter space of the background fields. The existence of such a Berry phase means that we may get a nontrivial holonomy when going around a loop in $\mathcal{W}$. The one-dimensional spaces $\mathcal{L}_W$ at each point of $\mathcal{W}$ combine into a rank one complex vector bundle (a complex line bundle) over $\mathcal{W}$ with nontrivial holonomies determined by the Berry phases. Then instead of $\langle\Omega|Y\rangle$ and $\langle \mathsf{L}|\Omega\rangle$, which do not have well-defined phases, we can define the following:

$$|\Omega\rangle\langle\Omega|Y\rangle \in \mathcal{L}_W, \quad \langle L|\Omega\rangle\langle\Omega| \in \mathcal{L}_W^{-1}. \tag{12.10}$$

These formulas are well-defined, since they do not depend on the phase of the state $|\Omega\rangle$. But instead of a number $\langle\Omega|Y\rangle$ we get a vector $|\Omega\rangle\langle\Omega|Y\rangle \in \mathcal{L}_W$, and likewise instead of a number $\langle L|\Omega\rangle$ we get a vector $\langle L|\Omega\rangle\langle\Omega| \in \mathcal{L}_W^{-1}$. All this is related to the following fact [27]: the partition function of the chiral fermion χ on the boundary W is naturally understood as a section of a line bundle $\mathcal{L}_W^{-1}$ (called the determinant line bundle) rather than a complex number. Also, the factor $\langle\Omega|Y\rangle$ is an exponentiated η-invariant that can be regarded as an element of $\mathcal{L}_W$ [11]. See [12] for more details on these points, from a physical perspective.

Somewhat surprisingly, however, as long as we are in a situation in which the Dirac operator on W generically has no zero-mode,[c] it is possible to avoid introducing the line bundle $\mathcal{L}_W$. One can avoid the line bundle $\mathcal{L}_W$ by using instead of $|\Omega\rangle$ a state which has no phase ambiguity. As long as the boundary Dirac operator $\mathcal{D}_W$ has no zero modes, we can consider the global Atiyah–Patodi–Singer (APS) boundary condition [10], which we denote as APS. This boundary condition will be described explicitly in Sec. 12.2.2. The path integral on Y with boundary condition APS corresponds to a state vector $|\mathsf{APS}\rangle \in \mathcal{H}_W$, which we will also describe explicitly in Sec. 12.2.2. Because $|Y\rangle$ is a multiple of $|\Omega\rangle$, we can rewrite (12.9) in the following longer but useful form:

$$Z(Y, L) = \frac{\langle L|\Omega\rangle\langle\Omega|\mathsf{APS}\rangle}{|\langle\mathsf{APS}|\Omega\rangle|^2} \cdot \langle\mathsf{APS}|Y\rangle. \tag{12.11}$$

In this formula, the inner product $\langle\mathsf{APS}|Y\rangle$ is the partition function of the massive fermion Ψ on Y with the APS boundary condition. In Sec. 12.2.2, we show that at long distances (that is, if Y and W are large compared to $1/|m|$), we have, modulo nonuniversal factors that do not affect the analysis of anomalies,[d]

$$\langle\mathsf{APS}|Y\rangle = \exp(-i\pi\eta_D), \tag{12.12}$$

---

[c]The case that $\mathcal{D}_W$ does generically have a zero-mode will be analyzed in Sec. 12.2.5. As explained there, this case is somewhat more exotic since, for example, it does not arise if W is connected.

[d]These are factors that can be removed by adding to the action the integral of a gauge invariant function of the background fields g, A. For example, the ground-state energy per unit volume of the Ψ field is a nonuniversal effect that can be removed by adding to the action a multiple of the volume of Y.

where η_D is the APS η-invariant (defined with APS boundary conditions along $W = \partial Y$) for the Dirac fermion Ψ; η_D will be introduced in Sec. 12.2.2. The remaining factor $\langle \mathsf{L}|\Omega\rangle\langle\Omega|\mathsf{APS}\rangle/|\langle\mathsf{APS}|\Omega\rangle|^2$ in $Z(Y, \mathsf{L})$ will be analyzed in Sec. 12.2.3, and we will see that (modulo nonuniversal factors) it equals $|\mathrm{Det}\,\mathcal{D}_W^+|$, the absolute value of the path integral on W for the massless chiral fermion χ. (The superscript $+$ in $\mathcal{D}_W^+$ means that here $\mathcal{D}_W$ is taken to act only on the field χ of positive chirality.) It is essential that an absolute value comes in here. In general, the path integral of the boundary mode χ might be anomalous, so the corresponding determinant, without taking an absolute value, may not be well-defined. However, an anomaly always affects only the phase of the path integral, so the absolute value $|\mathrm{Det}\,\mathcal{D}_W^+|$ is always anomaly-free.

Combining these claims about the various factors in $Z(Y, \mathsf{L})$, we get the following formula for the partition function on Y with boundary condition L:

$$Z(Y, \mathsf{L}) = |\mathrm{Det}\,\mathcal{D}_W^+|\exp(-i\pi\eta_D). \tag{12.13}$$

We claim that (for fermions charged under a $\mathrm{U}(1)$ symmetry) this is the general statement of anomaly inflow, including all perturbative and nonperturbative anomalies.

If one is only interested in perturbative anomalies, then Eq. (12.13) can be replaced by a slightly simpler and probably more familiar version:

$$Z(Y, \mathsf{L}) \stackrel{?}{=} \mathrm{Det}\,\mathcal{D}_W^+ \, \exp(-i\,\mathrm{CS}(g, A)). \tag{12.14}$$

Here, $\mathrm{CS}(g, A)$ is a Chern–Simons interaction that depends on the background fields g, A on Y. In perturbation theory, this formula is a satisfactory expression of the idea of anomaly inflow [7]. The idea behind this formula is that the path integral of the boundary fermion is the determinant $\mathrm{Det}\,\mathcal{D}_W^+$, and integrating out the massive fermion Ψ on Y induces the Chern–Simons coupling $\mathrm{CS}(g, A)$. When one proves the gauge-invariance of $\mathrm{CS}(g, A)$, one encounters a surface term, as a result of which $\mathrm{CS}(g, A)$ is not gauge-invariant on a manifold with boundary. This failure of gauge invariance of $\mathrm{CS}(g, A)$ on a manifold with boundary precisely matches the perturbative anomaly of a chiral fermion on the boundary [4–6], as a result of which the formula (12.14) is satisfactory as far as perturbative anomalies are concerned.

In general, however, in a topologically nontrivial situation, the result of integrating out a massive Dirac fermion, even on a manifold without boundary, is not $\exp(-i\,\mathrm{CS}(g, A))$ but $\exp(-i\pi\eta_D)$ [14,16]. This fact will be reviewed in Sec. 12.2.2. The two expressions are equivalent in perturbation theory around Euclidean space, that is in perturbation theory around $Y = \mathbb{R}^D$, $g_{\mu\nu} = \delta_{\mu\nu}$, and $A = 0$. But in general, the description by $\exp(-i\,\mathrm{CS}(g, A))$ misses many topological subtleties that are important in an understanding of global or nonperturbative anomalies and anomaly inflow and that are properly described by $\exp(-i\pi\eta_D)$. This is true for all D, but perhaps it is most obvious if D is even. If D is even, there are no conventional Chern–Simons couplings, and in perturbation theory around Euclidean space, one sees nothing. But η_D can still be nontrivial and describes anomaly inflow.

Bearing in mind that $\exp(-i\,\mathrm{CS})$ should be replaced by $\exp(-i\pi\eta_D)$ if we want a nonperturbative description, let us ask what might be the correct general formula for anomaly inflow, replacing (12.14). Part of the Dai–Freed theorem [11], which has been elucidated from a physical point of view in [12], is an abstract definition of $\exp(-i\pi\eta_D)$ (with APS boundary conditions along $W = \partial Y$) as a section of the line bundle $\mathcal{L}_W$. In that framework, the obvious candidate for improving on (12.14) is $Z(Y, \mathsf{L}) = \mathrm{Det}\,\mathcal{D}_W^+ \cdot \exp(-i\pi\eta_D)$. Neither factor is gauge-invariant separately, but the Dai–Freed theorem shows that the product is well-defined and completely free of any perturbative or nonperturbative anomaly.

This formulation is correct but unnecessarily abstract. As long as the operator $\mathcal{D}_W$ generically has no zero-mode, it is possible to simply define η_D as a number; this is the route that we will follow.[e] Then the formula takes the form of Eq. (12.13). In this formulation, the two factors are separately gauge-invariant, but they are not separately physically sensible. Indeed, because of the absolute value, $|\mathrm{Det}\,\mathcal{D}_W^+|$ does not vary smoothly near a value of background fields g, A at which $\mathcal{D}_W$ has a zero-mode (and therefore $\mathrm{Det}\,\mathcal{D}_W^+ = 0$), while $\exp(-i\pi\eta_D)$ does not vary smoothly near such a point because the definition of η_D breaks down. The Dai–Freed theorem says that the product in (12.13) is always smoothly varying.

Thus Eq. (12.13) is the natural generalization of the more familiar but imprecise formula (12.14). But to our knowledge a full justification of Eq. (12.13) has not previously been given. We will address this issue in Secs. 12.2.3 and 12.2.4, after first explaining the basic facts about the η-invariant.

---

[e]Fundamentally, it is not necessary to introduce the line bundle $\mathcal{L}_W$ because as long as $\mathcal{D}_W$ has no zero-mode, $\mathrm{Det}\,\mathcal{D}_W^+$ is nonzero and trivializes $\mathcal{L}_W$. Relative to this trivialization, $\mathrm{Det}\,\mathcal{D}_W^+$ is positive (so it is replaced by its absolute value $|\mathrm{Det}\,\mathcal{D}_W^+|$) and $\exp(-i\pi\eta_D)$ becomes a number. To understand what follows, it is not necessary to think in such abstract terms.

12.2.2 *Integrating out a fermion*

Though our main interest in the present chapter is in integrating out a massive fermion, it is also of interest to know what happens if one integrates out a massless fermion. It turns out to be useful to consider this case first.

We consider a Dirac fermion Ψ on a D-manifold Y, initially without boundary. Ψ has a self-adjoint Dirac operator $\mathcal{D}_Y = i\slashed{D}_Y$. It has real eigenvalues that we denote λ_k.

Formally, the path integral of Ψ is $\text{Det}\,\mathcal{D}_Y = \prod_k \lambda_k$. This of course needs to be regulated. We do so with the aid of a Pauli–Villars regulator of mass M. A regulated version of the determinant is[f]

$$\text{Det}\,\mathcal{D}_Y = \prod_k \frac{\lambda_k}{\lambda_k + iM}. \tag{12.15}$$

Now let us determine the phase or argument of $\text{Det}\,\mathcal{D}_Y$. This is ill-defined if any of the λ_k vanish, so we assume they are nonvanishing. Assuming $M > 0$ and $M \gg |\lambda_k|$, we see that the argument of $\lambda_k/(\lambda_k + iM)$ is approximately $-\frac{i\pi}{2}\text{sign}(\lambda_k)$. So the argument of $\text{Det}\,\mathcal{D}_Y$ is a regularized version of $-\frac{i\pi}{2}\sum_k \text{sign}(\lambda_k)$.

The Atiyah–Patodi–Singer η-invariant is a regularized version of $\sum_k \text{sign}(\lambda_k)$. The precise regularization does not matter. We can take, for example,[g]

$$\eta_D = \lim_{\epsilon \to 0^+} \sum_k \exp(-\epsilon|\lambda_k|)\text{sign}(\lambda_k). \tag{12.16}$$

The Pauli–Villars regularization in (12.15) gives a different regularization of $\sum_k \text{sign}(\lambda_k)$ (it gives a regularization since the argument of $\lambda_k/(\lambda_k + iM)$ vanishes for $|\lambda_k| \to \infty$). The two regularizations are equivalent in the limit $M \to \infty$ or $\epsilon \to 0^+$.

The phase of the path integral for a massless fermion (with a positive regulator mass) is therefore $\exp(-i\pi\eta_D/2)$ [16]. This is a refinement of a well-known perturbative calculation, expanding around Euclidean space with a flat metric and $A = 0$, in which one finds a Chern–Simons coupling [15]. The nonperturbative statement with η_D can be related to the perturbative

---

[f] Actually, to completely regularize the determinant, one needs several Pauli–Villars fields with different masses and statistics. The details do not really modify the following derivation. The same remark applies whenever we consider expressions that require regularization.

[g] The subscript D in η_D is for Dirac; for a Dirac fermion Ψ, we define η_D by summing over modes of Ψ, ignoring the modes of the conjugate field $\overline{\Psi}$. We will shortly introduce a related invariant η that is defined by summing over all fermion modes regardless of charge.

statement based on a Chern–Simons coupling by using the APS index theorem [10] for a manifold with boundary. Recall first that if X is a $D + 1$-manifold without boundary, then the original Atiyah–Singer index theorem [28] gives a formula for the index $\mathcal{I}$ of a Dirac operator on X as the integral over X of a certain differential form[h] Φ_{d+2} on X: $\mathcal{I} = \int_X \Phi_{d+2}$. If instead X has a nonempty boundary Y, the APS index theorem gives a more general formula

$$\mathcal{I} = \int_X \Phi_{d+2} - \frac{\eta_D}{2}, \qquad (12.17)$$

where η_D is the η-invariant of a suitable fermion field on Y. (Roughly, a fermion field Ψ on X reduces along $Y = \partial X$ to two copies of a fermion field on Y, and η_D is here the η-invariant of one such copy. For more detail, see Sec. 12.3.2.) In perturbation theory around Euclidean space, one can assume that a suitable X exists with $\partial X = Y$ (if $Y = \mathbb{R}^D$, X can be a half-space in $\mathbb{R}^{D+1}$). Moreover in perturbation theory, one does not see the integer $\mathcal{I}$. Under these conditions, Eq. (12.17) reduces to $\eta_D/2 = \int_X \Phi_{d+2}$, so one can think of $\eta_D/2$ as a boundary term related to the characteristic class Φ_{d+2}, or in other words as a Chern–Simons interaction in a generalized sense. In a topologically nontrivial situation, X may not exist and if it does the integer $\mathcal{I}$ cannot necessarily be neglected. So one has to describe the phase of the path integral for a massless fermion as $\exp(-i\pi\eta_D/2)$, not as the exponential of a Chern–Simons coupling.

Now let us consider integrating out a massive fermion, still on a D-manifold Y without boundary. In this case, the path integral is $\mathrm{Det}(\mathcal{D}_Y + im)$. This formally equals $\prod_k (\lambda_k + im)$. A regularized version is $\prod_k (\lambda_k + im)/(\lambda_k + iM)$. This can be factored as a ratio of two factors that we have already studied:

$$\prod_k \frac{\lambda_k}{\lambda_k + iM} \prod_{k'} \frac{\lambda_{k'} + im}{\lambda_{k'}}. \qquad (12.18)$$

The resulting phase is trivial if m and M have the same sign. Indeed, in the previous derivation, only the sign of M was relevant, so if m and M have the same sign, we can set $m = M$. But then the product in Eq. (12.18) is simply 1. The interesting case is therefore the "topologically nontrivial" case in which the physical fermion and the regulator have mass parameters with opposite signs. For example (as in Sec. 12.2.1), we may take $m < 0$ and

---

[h]Concretely, for the case that Ψ transforms in a representation V of some gauge group, $\Phi_{d+2} = \widehat{A}(R) \, \mathrm{Tr}_V \exp(F/2\pi)$, where R is the Riemann tensor, $\widehat{A}(R)$ is a certain polynomial in R, and F is the gauge field strength.

$M > 0$. Then Eq. (12.18) is the product of a factor we have already studied divided by the complex conjugate of the same factor. The argument of the path integral is hence $\exp(-i\pi\eta_D)$.

This derivation has been so general that it also applies if Y has a boundary with a self-adjoint boundary condition, the role of self-adjointness being to ensure that the Dirac eigenvalues are real. In particular, an example of a self-adjoint boundary condition is the global boundary condition of Atiyah, Patodi, and Singer [10]. This accounts for the claim that $\langle \text{APS}|Y \rangle = \exp(-i\pi\eta_D)$ (Eq. (12.12)).

But what is the APS boundary condition? It is not a local boundary condition but a rather subtle global one. The original APS definition (in the notation of the present paper) was that Ψ, restricted to $W = \partial Y$, is required to be a linear combination of eigenfunctions of $\mathcal{D}_W$ with positive eigenvalue.[i] For our purposes, however, the most useful way to describe the APS boundary condition is to say that $|\text{APS}\rangle$ is the state $|\Omega_{m=0}\rangle$ that would be the fermion ground state if one sets $m = 0$. This can be regarded as a state in the $m \neq 0$ Hilbert space (though it is certainly not the ground state of the $m \neq 0$ Hamiltonian); this is made explicit in Sec. 12.2.3. Because of our assumption that the operator $\mathcal{D}_W$ has no zero-mode, the $m = 0$ ground state $|\Omega_{m=0}\rangle$ is uniquely determined up to phase.[j] After Pauli–Villars regularization, there is no phase ambiguity in the state $|\text{APS}\rangle$, because in defining $|\text{APS}\rangle$, we use the same massless ground state $|\Omega_{m=0}\rangle$, with the same phase, for the physical fermion and for the regulator. The choice of phase cancels between the physical fermion field and the regulator.

We conclude with three remarks that are important in generalizations.

First, what happens if we replace the local boundary condition L of Eq. (12.2) with the opposite boundary condition L' defined by $(1 + \gamma^\tau)\Psi|_W = 0$? In this case, we get no boundary-localized mode if we set $m < 0$, but for $m > 0$, we get a boundary-localized mode with the opposite

---

[i]This boundary condition does not have a sensible continuation to the case that W has Lorentz signature. The properties of the APS boundary condition such as nonlocality and the absence of a continuation to Lorentz signature motivated some study of the APS theorem in a physical context [29]. See also [30] for a new treatment of the APS index theorem. For an early use of the APS boundary condition in the physics literature in a different context, see [31]. Notice that the local boundary condition L is completely physically sensible, and we just introduce the APS boundary condition as a way to compute the partition function (12.6).

[j]If $\mathcal{D}_W$ has zero-modes, then quantizing these modes gives a space of degenerate ground states of the $m = 0$ theory and the simple definition of the APS state and boundary condition breaks down. As a result, a more elaborate definition of APS boundary conditions is needed in that situation. This case is treated in Sec. 12.2.5.

chirality to that in Eq. (12.5). In quantizing this theory, we have to take $M < 0$, since we do not want the regulator field to have a massless mode propagating on the boundary. Reversing the sign of both m and M relative to the above derivation has the effect of complex-conjugating the product in Eq. (12.18), so now the phase is $\exp(i\pi\eta_D)$. In other words, anomaly inflow for a boundary mode of one chirality requires a phase $\exp(-i\pi\eta_D)$, and anomaly inflow for the opposite chirality requires a complex conjugate phase $\exp(i\pi\eta_D)$. The fact that the sign of the inflow depends on the chirality of the boundary mode is a standard result, usually deduced perturbatively in terms of Chern–Simons couplings rather than an η-invariant.

Second, in this derivation, we have assumed that Ψ carries a U(1) gauge or global symmetry. A "Dirac fermion" is usually understood as a fermion field such that Ψ has positive U(1) charge and its conjugate $\overline{\Psi}$ (which appeared in the original action (12.1) though we have not had to discuss it subsequently) has negative U(1) charge. The Dirac operator $\mathcal{D}_Y$ could be taken to act on fermions of either positive or negative charge, that is, on either Ψ or $\overline{\Psi}$. We defined η_D by summing over eigenvalues of $\mathcal{D}_Y$ acting on Ψ only. Obviously, we could have defined a similar invariant η by summing over all eigenvalues of $\mathcal{D}_Y$, acting on either Ψ or $\overline{\Psi}$. The relation between η and η_D is simply $\eta = 2\eta_D$, since complex conjugation can be used to show that the eigenvalues of $\mathcal{D}_Y$ acting on Ψ or on $\overline{\Psi}$ are the same.[k] In the absence of a U(1) symmetry, there is no distinction between Ψ and $\overline{\Psi}$, so the only natural definition involves a sum over all eigenvalues of $\mathcal{D}_Y$. In other words, to express our results in a way that generalizes naturally to an arbitrary fermion system, we should use η rather than η_D. In terms of η, the effect of integrating out a massive fermion (with $m < 0$, $M > 0$) is a factor $\exp(-i\pi\eta/2)$. This is the formula that we will use when we drop the assumption of U(1) symmetry, starting in Sec. 12.2.4.

Finally, in the above derivation, we assumed that $\mathcal{D}_Y$ has no zero-modes, so that all λ_k are nonzero. We did this because we started with a massless fermion; the path integral of a massless fermion vanishes when there is a zero-mode, and does not have a well-defined phase. However, a zero-mode of $\mathcal{D}_Y$ causes no problem in defining the phase that is induced by integrating out a *massive* fermion. Taking $m = -M$, each mode contributes to the regularized fermion path integral a factor $(\lambda - iM)/(\lambda + iM)$, which is -1 if $\lambda = 0$. So in including zero-modes, we want each such mode to give a factor

---

[k]We could have made the whole derivation thinking of $\mathcal{D}_Y$ as an operator acting on $\overline{\Psi}$ rather than Ψ, leading to an η-invariant defined in terms of the eigenvalues of $\mathcal{D}_Y$ acting on $\overline{\Psi}$. So η-invariants defined using eigenvalues of $\mathcal{D}_Y$ acting on Ψ or on $\overline{\Psi}$ must be equal.

of -1. For this, we define

$$\text{sign}(\lambda) = \begin{cases} 1 & \text{if } \lambda \geq 0, \\ -1 & \text{if } \lambda < 0, \end{cases} \tag{12.19}$$

and then we define η_D (and $\eta = 2\eta_D$) precisely as in (12.16). Then the phase induced by integrating out a fermion with $m < 0$, $M > 0$ is $\exp(-i\pi\eta_D)$ or $\exp(-i\pi\eta/2)$, whether zero-modes are present or not. We should warn the reader, however, that this definition of η or η_D including the zero-modes is not completely standard. In the original APS paper [10] (and many subsequent mathematical references), such a sum with the zero-modes included is called 2ξ rather than η; η is defined precisely as in Eq. (12.16), but summing over nonzero modes only.

12.2.3   *The chiral fermion partition function as a state overlap*

To complete the proof of the formula (12.13) for the partition function, we must calculate the state overlaps $\langle L|\Omega\rangle$ and $\langle APS|\Omega\rangle$ that appear in Eq. (12.11). We will see that $\langle L|\Omega\rangle$ is essentially the partition function of the boundary chiral fermion χ.[1]

They are computed by a straightforward quantization of the fermion Ψ on the space W. We make a Wick rotation $\tau \to it$, and the Hamiltonian is easily read off from the action (12.1) as

$$H = \int_W d^d x \sqrt{g}\, \Psi^\dagger (\mathcal{D}_W + \gamma^\tau m)\Psi, \tag{12.20}$$

where we have used the usual relation $\overline{\Psi} = \Psi^\dagger \gamma^\tau = \Psi^\dagger(i\gamma^t)$ (for Dirac fermions).

The operators γ^τ and $\mathcal{D}_W$ anti-commute, so we can make a mode expansion of the following kind. All modes of the Ψ field on W can be expressed as linear combinations of pairs of modes $(\psi_{+,a}, \psi_{-,a})$ which satisfy

$$\gamma^\tau \psi_{\pm,a} = \pm\psi_{\pm,a}, \quad \mathcal{D}_W \psi_{+,a} = \lambda_a \psi_{-,a}, \quad \mathcal{D}_W \psi_{-,a} = \lambda_a \psi_{+,a}. \tag{12.21}$$

---

[1]A similar phenomenon that a state vector in $d+1$-dimensions is related to a partition function in d-dimensions was also observed in a different system [32]. There, the d-dimensional theory is a free U(1) Maxwell theory with $d = 4$ and the $D = d + 1$-dimensional theory is a gauge theory which flows to a topological field theory in the low-energy limit. Their result might be interpreted as anomaly inflow of the global anomaly of the Maxwell theory under S-duality [33–35].

Here λ_a are the positive square roots of the eigenvalues of $(\mathcal{D}_W)^2$. Since we assume that $\mathcal{D}_W$ has no zero-mode, all modes have $\lambda_a > 0$ and are paired in the above fashion. We expand

$$\Psi = \sum_a (A_{+,a}\psi_{+,a} + A_{-,a}\psi_{-,a}). \tag{12.22}$$

The Hamiltonian is given by

$$H = \sum_a (A_{+,a}^\dagger, A_{-,a}^\dagger) \begin{pmatrix} m & \lambda_a \\ \lambda_a & -m \end{pmatrix} \begin{pmatrix} A_{+,a} \\ A_{-,a} \end{pmatrix}. \tag{12.23}$$

The operators $A_{\pm,a}$ are annihilation operators, and $A_{\pm,a}^\dagger$ are the corresponding creation operators.

We can consider each value of a separately, so we omit $\sum_a$ (or $\prod_a$, depending on the context) until the end of this section. After defining

$$(\cos(2\theta_a), \sin(2\theta_a)) = \frac{(m, \lambda_a)}{\sqrt{m^2 + \lambda_a^2}}, \tag{12.24}$$

the Hamiltonian is easily diagonalized as

$$H = \sqrt{\lambda_a^2 + m^2}(B_{1,a}^\dagger B_{1,a} - B_{2,a}^\dagger B_{2,a}), \tag{12.25}$$

where

$$B_{1,a} = \cos\theta_a A_{+,a} + \sin\theta_a A_{-,a}, \quad B_{2,a} = -\sin\theta_a A_{+,a} + \cos\theta_a A_{-,a}. \tag{12.26}$$

The ground state $|\Omega\rangle$ is specified by the conditions $B_{1,a}|\Omega\rangle = 0$ and $B_{2,a}^\dagger|\Omega\rangle = 0$, or $0 = \langle\Omega|B_{1,a}^\dagger = \langle\Omega|B_{2,a}$.

The local boundary condition L given in (12.2) requires the condition that $A_{-,a} = 0$ at the boundary, while $A_{+,a}$ is free. Then the corresponding state $|\mathsf{L}\rangle$ is specified by the condition that $\langle\mathsf{L}|A_{-,a} = 0$ and $\langle\mathsf{L}|A_{+,a}^\dagger = 0$. (See [12] for more details about the relation between boundary conditions and state vectors.) The APS boundary state $|\mathsf{APS}\rangle$ is the ground state with m set to 0, which corresponds to $\theta_a = \pi/4$ for all a and hence $\cos\theta_a = \sin\theta_a$. So $|\mathsf{APS}\rangle$ is characterized by $\langle\mathsf{APS}|(A_{-,a} - A_{+,a}) = 0$ and $\langle\mathsf{APS}|(A_{-,a} + A_{+,a})^\dagger = 0$.

These conditions leave us free to multiply the states $\langle\mathsf{L}|$ and $\langle\mathsf{APS}|$ by arbitrary nonzero complex numbers. We could constrain them to have unit norm, but this leaves an undetermined phase. However, the ambiguities do not matter. This is because we will eventually take the ratio between the negative mass theory $m < 0$ (for the physical fermion) and the positive mass theory $m > 0$ (for the Pauli–Villars regulator). The ambiguities cancel out in the ratio, because the boundary conditions $\langle L|$ and $\langle APS|$ do not depend on the mass parameter. The ground state $|\Omega\rangle$ has a phase ambiguity, which is related to the appearance of the line bundle $\mathcal{L}_W$, as discussed in Sec. 12.2.1. This ambiguity is physically unavoidable because of Berry phases when background fields are varied. But this ambiguity cancels out in the product $\langle L|\Omega\rangle\langle\Omega|APS\rangle$. Thus overall normalization factors will always cancel out and we can use any explicit state vectors satisfying the above conditions.

Let $|E\rangle$ be a state vector with

$$A_{+,a}|E\rangle = A_{-,a}|E\rangle = 0. \tag{12.27}$$

Then we can realize the relevant state vectors as

$$|\Omega\rangle = B_{2,a}^\dagger|E\rangle, \quad \langle L| = \langle E|A_{-,a}, \quad \langle APS| = \langle E|(A_{-,a} - A_{+,a}). \tag{12.28}$$

From these expressions, we easily get

$$\langle APS|\Omega\rangle = \cos\theta_a + \sin\theta_a \to 1 \quad (\lambda_a \ll |m|), \tag{12.29}$$

and

$$\langle L|\Omega\rangle = \cos\theta_a \to \begin{cases} 1 & (m > 0, \quad \lambda_a \ll |m|), \\ \lambda_a/(2|m|) & (m < 0, \quad \lambda_a \ll |m|). \end{cases} \tag{12.30}$$

Essentially, $\cos\theta_a$ for $m < 0$ is the eigenvalue λ_a normalized by $2|m|$ as long as $\lambda_a \ll |m|$. But $|m|$ plays the role of a regulator. In the limit $\lambda_a/|m| \to \infty$, we have $\cos\theta_a \to 1/\sqrt{2}$, independent of a or m. Upon taking the ratio between the theories with $m < 0$ and $m > 0$, the factors of $\cos\theta_a$ associated to eigenvalues with $|\lambda_a| \gg m$ cancel out, and hence the ultraviolet is regularized. Therefore, after taking the ratio, we finally get

$$\frac{\langle L|\Omega\rangle\langle\Omega|APS\rangle}{|\langle APS|\Omega\rangle|^2} = \prod_a \left(\frac{\lambda_a}{2|m|}\right)_{\text{reg}} = |\text{Det}(\mathcal{D}_W^+)| \tag{12.31}$$

where the subscript "reg" means that the product is regularized by $\cos\theta_a$ in the way just mentioned. Note that the determinant of the nonchiral Dirac operator $\mathcal{D}_W$ would have a factor of λ_a^2 for each pair of modes, while in Eq. (12.31), there is just one factor of λ_a for each pair. That is why the right-hand side of Eq. (12.31) is $|\text{Det}\,\mathcal{D}_W^+|$ (where as before $\mathcal{D}_W^+$ is the chiral Dirac operator on W), not $\text{Det}\,\mathcal{D}_W$. Note that the λ_a are all positive, and that $|\text{Det}\,\mathcal{D}_W^+|$ is the same as $|\text{Det}\,\mathcal{D}_W|^{1/2}$.

Combining this result with Eq. (12.11) and with what we learned in Sec. 12.2.2, it follows that the total partition function of the bulk massive fermion

Ψ with the boundary condition L is given by

$$Z(Y, \mathsf{L}) = |\mathrm{Det}(\mathcal{D}_W^+)| \exp(-\pi \mathrm{i} \eta_D). \tag{12.32}$$

as we claimed in Eq. (12.13), and as suggested by the Dai–Freed theorem [14, 36].

12.2.4 *The general case*

We have so far described anomaly inflow for a fermion all of whose components are charged under a U(1) symmetry. Here we want to generalize the construction to describe anomaly inflow for an arbitrary set of relativistic boundary fermions. For this, we will start with an arbitrary relativistic fermion field χ on the d-manifold W, and then find a massive fermion system on Y that is related to χ by anomaly inflow. This problem has also been studied recently in the appendix of [17], in the context of generalizing the Dai–Freed theorem to an arbitrary system of relativistic fermions.

We will ultimately work in Euclidean signature, but the necessary conditions to define a physically sensible theory are most naturally stated in Lorentz signature. The field χ at each point in W will take values in a vector space S that will be a representation of a group that includes Lorentz symmetries possibly together with some gauge or global symmetries. In the Lorentz group, we possibly include disconnected components, related to time-reversal and reflection symmetry; also the symmetry group that acts on the fermions is really a $\mathbb{Z}_2$ central extension of a product of the Lorentz group with a group of gauge or global symmetries. The central $\mathbb{Z}_2$ subgroup is generated by the operator $(-1)^{\mathsf{F}}$ that distinguishes bosons and fermions. This $\mathbb{Z}_2$ central extension may be a spin group, a pin group, or a refinement such as spin$_c$. We treat all cases uniformly.[m]

We will denote the components of χ generically as χ^a, where a runs over a basis of S. The only assumption that we make about χ is that it has a first order action of the general form

$$\frac{\mathrm{i}}{2} \int \mathrm{d}^d x \sqrt{g} \sum_{a,b} \chi^a \sigma_{ab}^\mu D_\mu \chi^b, \tag{12.33}$$

with some matrices σ_{ab}^μ, which obey constraints that will be described in a moment. We do not make any assumption about S beyond whatever is

---

[m]The symmetry group of a relativistic fermion system in any dimension $d \geq 2$ is as just described. For $d = 1$, relativity loses its force and there are more general possibilities. We will run into this in Sec. 12.4.1.

needed to ensure the existence of a physically sensible action of this form. We have not written a mass term in Eq. (12.33) and the more interesting case (since anomalies are possible) is that the symmetries do not allow a mass term. Possible mass terms and possible additional couplings, such as couplings to scalar fields, do not modify the following analysis in any essential way.

Because χ satisfies fermi statistics, and in view of the possibility of integration by parts, only the symmetric part of σ^μ really contributes to the action, so we can assume that $\sigma^\mu_{ab} = \sigma^\mu_{ba}$. In Lorentz signature, the χ^a are real (after quantization, they become hermitian operators) and the matrices σ^μ are real, so as to make the action and the Hamiltonian real.[n] Also, in Lorentz signature, one wants σ^0 to be a positive-definite matrix for unitarity (positivity of σ^0 ensures that if $\beta = \sum_a r_a \chi^a$ with real coefficients r_a, then after quantization, $\{\beta, \beta\} > 0$, consistent with positivity of the Hilbert space inner product). Here σ^0 is the time component of σ^μ in any local Lorentz frame. (In some particular local Lorentz frame, one can make a linear transformation of the χ^a to set $\sigma^0 = 1$, but this statement would then only hold in that particular frame.)

The Dirac equation is

$$\sigma^\mu_{ab} D_\mu \chi^b = 0. \tag{12.34}$$

The left-hand side of this equation is not valued in the same vector space S in which χ takes values, but rather in the dual space $\widetilde{S}$. The duality is clear from the fact that the action (which is a pairing of χ^a with the left-hand side of Eq. (12.34)) exists. The σ^μ are not gamma matrices, since they map the vector space S not to itself, but rather to a dual space $\widetilde{S}$. Fields valued in S and $\widetilde{S}$ will be represented by spinor fields χ^a or $\widetilde{\chi}_a$ with "index up" or "index down."

To get a dispersion relation of the appropriate form for relativistic fermions, the matrices σ^μ should satisfy the following. Let $g_{\mu\nu}$ be the spacetime metric (in a local Lorentz frame), and define $\overline{\sigma}_\mu = g_{\mu\mu} \sigma_\mu^{-1}$, with no summation over μ. (Thus in Euclidean signature with $g_{\mu\nu} = \delta_{\mu\nu}$, $\overline{\sigma}_\mu$ is

---

[n]The real and imaginary parts of a complex field are real, so there is no loss of generality in assuming the χ^a to be real. If there are gauge and global symmetries, then their generators acting on χ are real matrices. Although σ^μ and the gauge and global symmetry generators are all real, in general one cannot write $S = S' \otimes S''$ with S' and S'' being real vector spaces such that the σ^μ are bilinear forms on S' and the gauge and global symmetry generators act only on S''. The existence or not of such a decomposition will play no role in the present paper.

just the inverse of σ_μ.) Since σ_μ maps S to $\widetilde{S}$, its inverse matrix σ_μ^{-1} maps $\widetilde{S}$ to S, and therefore $\overline{\sigma}_\mu$ does the same. Then we want

$$\overline{\sigma}_\mu \sigma_\nu + \overline{\sigma}_\nu \sigma_\mu = 2g_{\mu\nu}. \tag{12.35}$$

This equation makes sense: the σ_μ map S to $\widetilde{S}$ and the $\overline{\sigma}_\mu$ map in the opposite direction, so the product maps S to itself and can be a c-number. From Eq. (12.35) we can deduce that χ obeys the expected dispersion relation, since Eq. (12.34) implies (in flat space) that

$$0 = (\overline{\sigma}^\nu \partial_\nu)(\sigma^\mu \partial_\mu)\chi = g^{\mu\nu}\partial_\mu \partial_\nu \chi. \tag{12.36}$$

Therefore we require (12.35) as a condition on σ^μ. Note that the Clifford algebra of Eq. (12.35) implies Lorentz invariance with generators $\sigma_{\mu\nu} = \frac{1}{4}(\overline{\sigma}_\mu \sigma_\nu - \overline{\sigma}_\nu \sigma_\mu)$. These generators map χ to χ, so the original theory of χ only was in fact Lorentz-invariant.

As an example of this, in two dimensions with Euclidean signature, σ_1 and σ_2 can be the 1×1 matrices 1 and i. Then $\overline{\sigma}_1$ and $\overline{\sigma}_2$ are 1 and $-$i. The fermion χ has a single component, with definite chirality. In Lorentz signature, χ is called a Majorana–Weyl fermion. In this example, the conjugate fermion $\widetilde{\chi}$ that is introduced in a moment has opposite chirality.

We now introduce a second spinor field $\widetilde{\chi}_a$ that transforms in the dual representation $\widetilde{S}$. (In a particular example, the original representation might be self-dual so it may happen that χ and $\widetilde{\chi}$ actually transform the same way. We proceed the same way whether this is so or not. Note that even if χ and $\widetilde{\chi}$ transform the same way, a mass term for χ might be forbidden by fermi statistics, so anomalies may be possible.) For $\widetilde{\chi}$ we can always write an action

$$-\frac{i}{2}\int d^d x \sqrt{g} \sum_{a,b} \widetilde{\chi}_a \overline{\sigma}^{\mu\,ab} D_\mu \widetilde{\chi}_b, \tag{12.37}$$

since the matrices $\overline{\sigma}^\mu$ map $\widetilde{S}$ back to S. Here we have put a minus sign in the action relative to Eq. (12.33), because $\overline{\sigma}^0 = -(\sigma^0)^{-1}$ is negative-definite in Lorentz signature. We write $\mathcal{D}_W^+ = -\sigma^\mu D_\mu$ and $\mathcal{D}_W^- = \overline{\sigma}^\mu D_\mu$.

Now we can combine χ and $\widetilde{\chi}$ to a fermi field $\Psi = \begin{pmatrix} \chi \\ \widetilde{\chi} \end{pmatrix}$. Acting on Ψ, we define the matrices

$$\gamma_\mu = \begin{pmatrix} 0 & \overline{\sigma}_\mu \\ \sigma_\mu & 0 \end{pmatrix}. \tag{12.38}$$

These are gamma matrices $\{\gamma_\mu, \gamma_\nu\} = 2g_{\mu\nu}$ by virtue of Eq. (12.35).

The action for Ψ is the sum of the actions for χ and for $\widetilde{\chi}$ plus a possible mass term:

$$- \mathrm{i}m \int \mathrm{d}^d x \sqrt{g}\, (\widetilde{\chi}, \chi) \tag{12.39}$$

where $(\widetilde{\chi}, \chi) = \widetilde{\chi}_a \chi^a$. We can introduce an antisymmetric bilinear form $\langle,\rangle$ on the space $S \oplus \widetilde{S}$, invariant under all symmetries, by

$$\langle \Psi_1, \Psi_2 \rangle = (\widetilde{\chi}_1)_a (\chi_2)^a - (\chi_1)^a (\widetilde{\chi}_2)_a. \tag{12.40}$$

One can verify that

$$\langle \gamma^\mu \Psi_1, \Psi_2 \rangle = -\langle \Psi_1, \gamma^\mu \Psi_2 \rangle. \tag{12.41}$$

The action can be written in the manifestly Lorentz-invariant form:

$$-\frac{\mathrm{i}}{2} \int \mathrm{d}^d x \sqrt{g}\, \langle \Psi, (\gamma^\mu D_\mu + m)\Psi \rangle. \tag{12.42}$$

The identify (12.41), together with integration by parts and the anti-symmetry of $\langle,\rangle$, can be used to verify that the equation of motion derived from this action is the expected $(\slashed{D} + m)\Psi = 0$.

Crucially, it is also possible to add another dimension. One adds a new coordinate τ and a new gamma matrix

$$\gamma^\tau = \begin{pmatrix} 1 & 0 \\ 0 & -1 \end{pmatrix} \tag{12.43}$$

that obeys $(\gamma^\tau)^2 = 1$ and anti-commutes with the previous gamma matrices. One can verify the analog of (12.41):

$$\langle \gamma^\tau \Psi_1, \Psi_2 \rangle = -\langle \Psi_1, \gamma^\tau \Psi_2 \rangle. \tag{12.44}$$

With a new gamma matrix, one can extend d-dimensional Lorentz symmetry to $D = (d+1)$-dimensional Lorentz symmetry, the new generators being $\frac{1}{4}[\gamma^\tau, \gamma^\mu]$. The form $\langle,\rangle$ possesses D-dimensional Lorentz symmetry.[o]

---

[o]To check invariance under the additional generators, the identity we need is $\langle \gamma^\tau \gamma^\mu \Psi_1, \Psi_2 \rangle + \langle \Psi_1, \gamma^\tau \gamma^\mu \Psi_2 \rangle = 0$, which follows immediately from (12.41) and (12.44). Note also that if the original d-dimensional theory has a time-reversal or reflection symmetry, the D-dimensional theory has the same symmetry. This follows from Lorentz invariance in D dimensions together with the discrete symmetry in d dimensions.

Then there can be a new term in the action

$$-i \int d^d x d\tau \sqrt{g}(\widetilde{\chi}, D_\tau \chi) = -\frac{i}{2} \int d^d x d\tau \sqrt{g}\langle \Psi, \gamma^\tau D_\tau \Psi\rangle. \qquad (12.45)$$

Combining (12.42) (or rather its integral over τ) and (12.45), we construct the action of the massive fermion Ψ in $D = d + 1$ dimensions

$$-\frac{i}{2} \int d^{d+1} x \sqrt{g}\,\langle \Psi, (\slashed{D} + m)\Psi\rangle. \qquad (12.46)$$

From this action, one can derive the expected D-dimensional Dirac equation $(\slashed{D} + m)\Psi = 0$ (all facts used to prove this statement in dimension d also hold in dimension $D = d + 1$). The action and the Dirac equation have manifest D-dimensional Lorentz invariance. Therefore, this action can be defined on any D-manifold Y that is endowed with all the structures[P] (such as orientations, spin structures, gauge bundles) that were needed to define the original fermion field χ on W.

With the d-manifold W still understood to have Lorentz signature, the original d gamma matrices (12.38) are all real, and of course γ^τ as defined in Eq. (12.43) is also real. So all gamma matrices are real, and therefore the Lorentz generators are also real. χ was real to begin with, and $\widetilde{\chi}$, transforming as the dual to χ, is also real. So there is a physically sensible theory of a real Ψ field in D dimensions with Lorentz signature. This is important in order to make the formula we will get for anomaly inflow physically meaningful.

Thus, on a completely general D-manifold with Lorentz signature, Ψ can be considered real. What happens when we go to Euclidean signature? Then Ψ is generally no longer real, but in fact it becomes pseudoreal.[q] We rotate to Euclidean signature by a Wick rotation of W, say $x^0 \to -ix_E^0$, where x^0 parameterizes a time direction in W. The corresponding transformation of gamma matrices is $\gamma^0 = -i\gamma_E^0$. In Lorentz signature, all gamma matrices were real, so in Euclidean signature, they are all real except for one, namely

---

[P]By a spin structure or gauge bundle on Y, we always mean a spin structure or gauge bundle on Y that extend the spin structure and gauge bundle of W. Along ∂Y, a specific isomorphism is chosen between the spin structure and gauge bundle of Y and those of W.
[q]In some cases, Ψ can also be given a real structure in Euclidean signature. This happens if Ψ has $2k$ components, and the full symmetry group including rotations and gauge and global symmetries is a subgroup of $U(k)$. This group is a subgroup of $Sp(2k)$ (corresponding to the pseudoreal structure of Ψ described in the text) and of $O(2k)$ (corresponding to an additional real structure). When Ψ does have a real structure, this can be used to simplify the analysis of anomalies. However, we will focus on the pseudoreal structure that Ψ carries universally.

γ_E^0. Let $*$ be complex conjugation. Evidently, $*$ anti-commutes with γ_E^0 and commutes with other gamma matrices. Accordingly,

$$\mathsf{C} = *(-\mathrm{i}\gamma_E^0) = *\gamma^0 \tag{12.47}$$

anti-commutes with all gamma matrices. C therefore commutes with the D-dimensional rotation generators $\frac{1}{4}[\gamma_\mu, \gamma_\nu]$, and of course it commutes with all gauge and global symmetry generators (which are real and commute with the gamma matrices). So C commutes with all symmetries. C is antilinear and satisfies

$$\mathsf{C}^2 = -1. \tag{12.48}$$

The existence of an antilinear operator C that commutes with all symmetries and obeys $\mathsf{C}^2 = -1$ means that Ψ is in a pseudoreal representation of the symmetry group. (This statement follows more abstractly from the existence of the antisymmetric bilinear form $\langle, \rangle$ that preserves all symmetries. But that reasoning would not give an explicit construction of C.) Since C is antilinear and anti-commutes with gamma matrices, it commutes with the self-adjoint Dirac operator $\mathcal{D}_Y = \mathrm{i}\sum_{\mu=1}^D \gamma^\mu D_\mu$. Moreover, the fact that C is invariant under all symmetries means that the construction can be made on an arbitrary D-manifold Y that admits all the relevant structures, though we started by singling out a particular "time" direction.[r]

Using the antilinear operator C that was just introduced, we can define a hermitian form on the space of Ψ fields,

$$(\Psi_1, \Psi_2) = \langle \mathsf{C}\Psi_1, \Psi_2 \rangle, \quad (\Psi_1, \Psi_2)_Y = \int_Y \mathrm{d}^D x \sqrt{g}\, (\Psi_1, \Psi_2). \tag{12.49}$$

Calling $(\ ,\ )$ a hermitian form means that it is linear in the second variable and antilinear in the first. This hermitian form is clearly invariant under all symmetries, since it was constructed from the invariant ingredients $\langle, \rangle$ and C. Explicitly if $\Psi = \begin{pmatrix} \chi \\ \tilde{\chi} \end{pmatrix}$, then $\mathsf{C}\Psi = \begin{pmatrix} \overline{\sigma}^0 \tilde{\chi}^* \\ \sigma^0 \chi^* \end{pmatrix}$, where $*$ is complex conjugation. So

$$(\Psi, \Psi) = \sigma_{ab}^0 \chi^{a*} \chi^b - \overline{\sigma}^{0ab} \tilde{\chi}_a^* \tilde{\chi}_b. \tag{12.50}$$

Since σ^0 is positive-definite and $\overline{\sigma}^0$ is negative-definite, we see that (Ψ, Ψ) is positive-definite. Moreover, the identities that have been described imply

---

[r]Concretely, to make this construction in curved spacetime, one uses the vierbein formalism that is anyway necessary to define fermions in curved spacetime, and one uses the above formulas in a locally Euclidean frame. Because C commutes with the rotation generators in any locally Euclidean frame, the definitions of C in different locally Euclidean frames are compatible.

that, assuming Y has no boundary, $\mathcal{D}_Y = i\gamma^\mu D_\mu$ is self-adjoint with respect to the hermitian form $(\ ,\ )_Y$:

$$(\mathcal{D}_Y \Psi_1, \Psi_2)_Y = (\Psi_1, \mathcal{D}_Y \Psi_2)_Y. \tag{12.51}$$

Though Ψ is not real on a general D-manifold with Euclidean signature, it can be considered real on a D-manifold that is presented with a given factorization as $W \times \mathbb{R}$, where W has Euclidean signature (and we only care about the behavior of Ψ under symmetries that preserve this factorization). One way to explain this statement is to observe that whether the $\mathbb{R}$ direction is Lorentzian or Euclidean does not affect the behavior under symmetries of W, and if the $\mathbb{R}$ direction is Lorentzian, we are on a Lorentz signature manifold, so Ψ can be considered real. Since the point is important, we will give another explanation. The vector spaces S and $\widetilde{S}$ in which χ and $\widetilde{\chi}$ take values were dual to each other when W has Lorentz signature, and this duality persists after continuation to Euclidean signature. But in Euclidean signature, the symmetry group — a double cover of the group of rotations and possible gauge or global symmetries — is compact. The dual of a representation of a compact Lie group is isomorphic to the complex conjugate representation. This means that when W has Euclidean signature, as far as symmetries of W are concerned, S and $\widetilde{S}$ can be regarded as complex conjugate vector spaces, while in Lorentz signature they were each real and were not related to each other by complex conjugation. Hence χ and $\widetilde{\chi}$ transform as complex conjugates and Ψ can be considered real and takes values in a real vector space S. Informally $\mathsf{S} = S \oplus \widetilde{S}$ (the precise statement is that S is a real vector space whose complexification has such a decomposition).

We stress that there are two senses in which Ψ might be considered real. In the starting point, in Lorentz signature, Ψ is naturally real. This is important for showing that our study of anomaly inflow applies to a physically sensible quantum field theory. In Lorentz signature, for Ψ to be real just means that χ and $\widetilde{\chi}$ are both real. But in our calculations, we will be in Euclidean signature, and then it is useful that a different real structure can be defined in the case of a D-manifold that locally is a product $\mathbb{R} \times W$. In this second real structure, $\widetilde{\chi}$ is proportional to the complex conjugate of χ. The relationship between them is described more precisely presently.

The Dirac operator $\mathcal{D}_W = \sum_{\mu=1}^{d} \gamma^\tau \gamma^\mu D_\mu$ (see footnote b for its relation to the usual Dirac operator $i\sum_{\mu=1}^{d} \gamma^\mu D_\mu$) can be written $\mathcal{D}_W = \begin{pmatrix} 0 & \mathcal{D}_W^- \\ \mathcal{D}_W^+ & 0 \end{pmatrix}$, where $\mathcal{D}_W^+ = -\sigma^\mu D_\mu$, $\mathcal{D}_W^- = \overline{\sigma}^\mu D_\mu$ are the Dirac operators for $\chi, \widetilde{\chi}$ that we had in the beginning before going to D dimensions.

Now consider the theory of the Ψ field on a D-manifold Y with boundary W, with the local boundary condition L defined by $\widetilde{\chi}|_W = 0$. For $m < 0$, and treating Y as a product near the boundary, Ψ has a boundary-localized mode given by the ansatz of Eq. (12.5). In particular, $\widetilde{\chi}$ vanishes identically (in the approximation that $Y = W \times \mathbb{R}_-$), and χ vanishes exponentially fast away from W. The effective theory for this boundary-localized mode is the original purely d-dimensional theory (12.33) for χ only. This means that anomaly inflow from Y, generated by integrating out the massive field Ψ, will cancel any anomaly of the original d-dimensional theory of χ.

To describe this anomaly inflow precisely, we would like to generalize to this situation the formula (12.13) for $Z(Y, \mathsf{L})$. It is not difficult to guess the generalization. First of all, integrating out the massive field Ψ will generate a factor of $\exp(-i\pi\eta/2)$. (When there is no U(1) symmetry, we have to use η, defined by summing over all eigenvalues of $\mathcal{D}_Y$, rather than $\eta_D = \eta/2$, defined by summing only over eigenvalues of positive charge. This was explained at the end of Sec. 12.2.2.) Also, by fermi statistics, the kinetic operator $\mathcal{D}_W^+ = -\sigma^\mu D_\mu$ of χ can be viewed as an antisymmetric matrix. The path integral of χ is its Pfaffian, $\mathrm{Pf}(\mathcal{D}_W^+)$. When there is a U(1) symmetry carried by all the fermions, this Pfaffian can be viewed as the determinant of a kinetic operator that acts on a smaller set of fields (those of positive U(1) charge), but in the absence of a U(1) symmetry, the path integral is best understood as a Pfaffian. So the natural analog of Eq. (12.13) is

$$Z(Y, \mathsf{L}) = |\mathrm{Pf}(\mathcal{D}_W^+)| \exp(-i\pi\eta/2), \qquad (12.52)$$

and we will aim to justify this formula by adapting the derivation of Sec. 12.2.3.

The first step is straightforward. $Z(Y, \mathsf{L})$ can be expressed in terms of inner products by the same logic as before, leading to Eq. (12.11). Here $\langle \mathsf{APS}|Y \rangle$ is known from the arguments of Sec. 12.2.2, and we will slightly modify the arguments of Sec. 12.2.3 to compute the inner products $\langle \mathsf{L}|\Omega \rangle$ and $\langle \mathsf{APS}|\Omega \rangle$.

In general, when one quantizes fermions, to get a Hilbert space with a positive-definite inner product, the fermions that are being quantized carry a real structure. (If complex fermion fields are present, one can take their real and imaginary parts.) Real fermion fields become hermitian operators after quantization. And on the space of real fermion fields, there is a positive-definite inner product that appears in the canonical anti-commutation relations. So in order to quantize the Ψ field on W, we want to describe explicitly the real structure that is appropriate if $Y = \mathbb{R} \times W$, and the natural positive-definite inner product in this real structure.

From our definitions, it follows immediately that if W has Lorentz signature with time coordinate x^0, and γ^0 is the corresponding gamma matrix (defined in Eq. (12.38)), then $-\langle \Psi_1, \gamma^0 \Psi_2 \rangle = \sigma^0_{ab} \chi_1^a \chi_2^b - \overline{\sigma}^{0\,ab} (\widetilde{\chi}_1)_a (\widetilde{\chi}_2)_b$. Since σ^0 is positive-definite and $\overline{\sigma}^0$ is negative-definite, it follows that the inner product $(\Psi_1, \Psi_2)_0 = -\langle \Psi_1, \gamma^0 \Psi_2 \rangle$ is positive-definite, as long as χ and $\widetilde{\chi}$ are real. We actually want to take W to be Euclidean and make a Wick rotation $\tau = it$ of the coordinate orthogonal to W. Lorentz invariance of the pairing $\langle \Psi_1, \Psi_2 \rangle$ means, of course, that any statement with x^0 viewed as a time coordinate and τ as a space coordinate has an analog if the roles are reversed. Thus, setting $\gamma^t = -i\gamma^\tau$, and now assuming W to be Euclidean, we define the inner product

$$(\Psi_1, \Psi_2)_t = -\langle \Psi_1, \gamma^t \Psi_2 \rangle = \langle \Psi_1, i\gamma^\tau \Psi_2 \rangle, \qquad (12.53)$$

where γ^τ was defined in Eq. (12.43). This will be positive-definite if we place on Ψ the appropriate Wick-rotated reality condition. What reality condition will do the job? The positivity of $(\,,\,)_t$ is ensured if we impose on Ψ a reality condition such that

$$(\Psi_1, \Psi_2)_t = (\Psi_1, \Psi_2), \qquad (12.54)$$

where the right-hand side is the positive definite hermitian inner product introduced in Eq. (12.49). Notice that the left-hand side is linear in Ψ_1, while the right-hand side is antilinear in Ψ_1, so this equation imposes a certain reality condition on Ψ_1. Explicitly, since $(\Psi_1, \Psi_2)_t = \langle -i\gamma^\tau \Psi_1, \Psi_2 \rangle$ and $(\Psi_1, \Psi_2) = \langle \mathsf{C} \Psi_1, \Psi_2 \rangle$, the reality condition is given by

$$i\gamma^\tau \mathsf{C} \Psi = \Psi. \qquad (12.55)$$

This is a consistent reality condition, since C and γ^τ anti-commute and hence $(i\gamma^\tau \mathsf{C})^2 = -\mathsf{C}^2 = 1$. More explicitly, in terms of χ and $\widetilde{\chi}$, Eq. (12.55) becomes $\widetilde{\chi}_a = -i\sigma^0_{ab}(\chi^b)^*$ and $\chi^a = i\overline{\sigma}^{0\,ab}(\widetilde{\chi}_b)^*$, where $*$ is complex conjugation.[s] This is consistent with the claim that when W has Euclidean signature, $\widetilde{\chi}$ and χ transform as complex conjugates.

---

[s]This way of writing the reality condition looks noncovariant, but it actually is covariant, since Eq. (12.55) is a manifestly covariant version. As always, formulas such as $\widetilde{\chi}_a = -i\sigma^0_{ab}(\chi^b)^*$ are written in a locally Euclidean frame. With the representation of the gamma matrices that led to the explicit definition (12.47) of C, some rotation generators of the tangent space of W are real and some are imaginary; when this is taken into account, the formula $\widetilde{\chi}_a = -i\sigma^0_{ab}(\chi^b)^*$ is covariant under a change of the locally Euclidean frame, as is the definition of C. A different representation of the gamma matrices (in which γ_τ is imaginary and the gamma matrices of W are real) would make the relation between $\widetilde{\chi}$ and χ^* look more natural while obscuring the relationship to the starting point, which was a physically sensible theory of the χ field in Lorentz signature.

We define $\widetilde{\chi}$ as

$$\widetilde{\chi}_a = \sigma^0_{ab}(\chi^b)^*. \tag{12.56}$$

Then the reality condition is

$$\widetilde{\chi} = -i\overline{\chi}. \tag{12.57}$$

We impose this condition in canonical quantization of the system when W has Euclidean signature and $t = -i\tau$ is the time coordinate. In particular, the product $\overline{\chi}\chi = \overline{\chi}_a\chi^a = (\chi^b)^*\sigma^0_{ab}\chi^b$ is positive-definite and invariant under all symmetries on W. The matrix σ^0 plays the role of a hermitian metric on S, invariant under all symmetries.

Using $\widetilde{\chi} = -i\overline{\chi}$, the action after the Wick rotation $\tau \to it$ is

$$I = \int d^d x dt \sqrt{g} \left[i\overline{\chi}\partial_t \chi - m\overline{\chi}\chi + \frac{i}{2}\chi\sigma^\mu D_\mu\chi + \frac{i}{2}\overline{\chi}\,\overline{\sigma}^\mu D_\mu\overline{\chi} \right], \tag{12.58}$$

where $\overline{\chi}\partial_t\chi = \overline{\chi}_a\partial_t\chi^a$, $\chi\sigma^\mu D_\mu\chi = \chi^a\sigma^\mu_{ab}D_\mu\chi^b$ and so on. The Hamiltonian is

$$H = \int d^d x \sqrt{g} \left[m\overline{\chi}\chi + \frac{i}{2}\chi D^+_W\chi + (\text{h.c.}) \right], \tag{12.59}$$

where $D^+_W = -\sigma^\mu D_\mu$ and (h.c.) is the hermitian conjugate of $\frac{i}{2}\chi D^+_W\chi$.

We work in the subspace in which $(D^+_W)^\dagger(D^+_W)\chi = \lambda_a^2\chi$. We can assume that $\lambda_a > 0$, since until Sec. 12.2.5, we assume there are no zero-modes. The existence of the action $\int \chi D^+_W\chi$, consistent with fermi statistics, means that the differential operator D^+_W can be regarded as an antisymmetric matrix. Thus we can put it in a block diagonal form with 2×2 blocks of the form $\lambda_a\epsilon_{ij}$, where ϵ_{ij} is the totally antisymmetric 2×2 matrix with $\epsilon_{12} = 1$. We can normalize the corresponding orthonormal eigenmodes ψ_1 and ψ_2 such that

$$\int d^d x \sqrt{g}\,\overline{\psi}^i\psi_j = \delta^i_j, \quad \int d^d x \sqrt{g}\,\psi_i D^+_W\psi_j = i\lambda_a\epsilon_{ij}, \tag{12.60}$$

where $\overline{\psi}^i_a = \sigma^0_{ab}(\psi^b_i)^*$. The phase of $\int d^d x \sqrt{g}\,\psi_1 D_W\psi_2$ can be freely chosen by rotating the phases of ψ_1 and ψ_2, and we have chosen it to be i to slightly simplify the later equations.

Using these modes, we expand χ and $\overline{\chi}$ as

$$\chi = A_+\psi_1 + A^\dagger_-\psi_2, \quad \overline{\chi} = A^\dagger_+\overline{\psi}^1 + A_-\overline{\psi}^2. \tag{12.61}$$

The reason for the notation A_+ and $A^\dagger_-$ for the coefficients of the expansion of χ is to make the Hamiltonian below to be of the same form as in the case of a Dirac fermion studied in Sec. 12.2.3. By canonical quantization, the above action gives the anti-commutation relations $\{A_+, A_+^\dagger\} = 1$, $\{A_-, A_-^\dagger\} = 1$, with others zero. The Hamiltonian is

$$H = m(A_+^\dagger A_+ - A_-^\dagger A_-) + \lambda_a(A_-^\dagger A_+ + A_+^\dagger A_-)$$

$$= (A_+^\dagger, A_-^\dagger) \begin{pmatrix} m & \lambda_a \\ \lambda_a & -m \end{pmatrix} \begin{pmatrix} A_+ \\ A_- \end{pmatrix}. \tag{12.62}$$

This is the same as we had for a Dirac fermion.

The boundary conditions have basis-independent characterizations as follows. We denote the Hamiltonian with the parameters m, λ_a as $H(m, \lambda_a)$. The ground state $|\Omega\rangle$ is the lowest energy state of the Hamiltonian $H(m, \lambda_a)$. The APS boundary condition $|\text{APS}\rangle$ is the lowest energy state of the Hamiltonian $H(0, \lambda_a)$ (i.e., the Hamiltonian of the massless theory $m = 0$). The state $|\text{L}\rangle$ defined by the local boundary condition L is the lowest energy state of the Hamiltonian $H(|m|, 0)$ (i.e., the Hamiltonian of the positive mass theory in the limit $\lambda_a/|m| \to 0$).

Now the problem reduces to the case of the Dirac fermion studied in the previous section. The Hamiltonian as well as the boundary conditions are completely the same. The only change is that the infinite product over a now gives a Pfaffian:

$$\frac{\langle \text{L}|\Omega\rangle\langle\Omega|\text{APS}\rangle}{|\langle\text{APS}|\Omega\rangle|^2} = \prod_a \left(\frac{\lambda_a}{2|m|}\right)_{\text{reg}} = |\text{Pf}(\mathcal{D}_W^+)|. \tag{12.63}$$

Indeed, Eq. (12.60) shows that $|\text{Pf}(\mathcal{D}_W^+)|$ is a regularized version of $\prod_a \lambda_a$.

Together with the fact that $\langle \text{APS}|Y\rangle$ is the path integral of the massive fermion field Ψ on Y with APS boundary conditions, and so is equal to $\exp(-i\pi\eta_Y/2)$, Eq. (12.63) is what we need to justify Eq. (12.52).

12.2.5   *Treatment of zero-modes*

So far we have assumed that the boundary Dirac operator $\mathcal{D}_W$ has no zero-modes. We have justified the anomaly inflow formula (12.52) under this assumption. As long as there are no zero-modes for generic background fields g, A, this formula gives a good characterization of anomaly inflow. (The formula remains valid if the background fields are varied so that zero-modes appear, since the left- and right-hand sides of Eq. (12.52) both vanish in that case.)

We need a new derivation for the case that for arbitrary background fields, $\mathcal{D}_W$ has zero-modes. This might happen because of a nonzero index or mod 2 index that implies the existence of zero-modes.[t] Actually, in the anomaly inflow problem, we consider a d-manifold W that by definition is the boundary of some Y. The index and the mod 2 index of the Dirac operator on W are cobordism invariants, so they vanish.[u] But it may happen that W is the union of disconnected components W_i. In that case, although the overall Dirac index and mod 2 index on W will vanish, on individual components there may be a nonvanishing index or mod 2 index. That is the situation in which generically (and in fact always) $\mathcal{D}_W$ has zero-modes. Though this situation may at first sight seem rather esoteric, it actually plays a role in understanding relatively simple examples, as we will see in Sec. 12.4. Therefore, it is important to generalize the anomaly inflow formula (12.52) to cover this situation.

In doing so, we will use the fact that there is always a generic number of zero-modes, the minimum number allowed by any index or mod 2 index theorem. Let b_i be the generic number of zero-modes of the chiral Dirac operator $\mathcal{D}_W^+$ on W_i. Then summing over all components, the generic number of zero-modes of the chiral Dirac operator $\mathcal{D}_W^+$ on W is $b = \sum_i b_i$. Since (when W has Euclidean signature), complex conjugation exchanges χ and $\widetilde{\chi}$, the number of zero-modes of the opposite chirality Dirac operator $\mathcal{D}_W^-$ is also b. In the anomaly inflow problem, the number b is always even, because the

---

[t]The mod 2 index ζ_W of the Dirac operator $\mathcal{D}_W^+$ is the number of zero-modes of χ mod 2. As we will explain in Sec. 12.4.1, fermi statistics imply that it is a topological invariant. When $\zeta_W \neq 0$, $\mathcal{D}_W^+$ must have a zero-mode. (By complex conjugation, $\mathcal{D}_W^-$ has the same number of zero-modes as $\mathcal{D}_W^+$.) An ordinary Dirac index can likewise imply the existence of zero-modes. For example, in $d = 4$, suppose that χ is a Majorana fermion coupled to gravity only. Then the number of positive chirality zero-modes of χ minus the number of negative-chirality zero-modes of χ is the Dirac index $\mathcal{I}$; when it is nonzero, again $\mathcal{D}_W^+$ must have zero-modes.

[u]Cobordism invariance of the mod 2 index ζ_W of $\mathcal{D}_W^+$ is an easy consequence of the anomaly inflow construction. Because the equation $(\slashed{D}_Y + m)\Psi = 0$ on Y with the local boundary condition L that leads to anomaly inflow can be derived from an action consistent with fermi statistics, the number ζ_Y of zero-modes of this equation mod 2 is a deformation invariant and in particular independent of m (see Sec. 12.4.1 for this argument). Taking $m \gg 0$, the equation $(\slashed{D}_Y + m)\Psi = 0$ has no approximate solutions on Y that satisfy the boundary condition, so $\zeta_Y = 0$. Taking $m \ll 0$, the approximate solutions of the equation are the same as the zero-modes of χ on W, so $\zeta_W = \zeta_Y$ and hence $\zeta_W = 0$. Note that zero-modes of $\mathcal{D}_W^+$ may not correspond to *exact* zero-modes of $\slashed{D}_Y + m$, but small corrections to the spectrum do not affect the number of zero-modes of an antisymmetric matrix mod 2, and so do not affect the statement that $\zeta_W = \zeta_Y$. For the ordinary index $\mathcal{I}$, the Atiyah–Singer theorem gives a formula $\mathcal{I} = \int_W \Phi$ for some characteristic class Φ. If $W = \partial Y$ and the structures needed to define Φ extend over Y then $\mathcal{I} = \int_W \Phi = \int_Y d\Phi = 0$, so $\mathcal{I}$ is likewise invariant under cobordism. This statement can also be proved without knowing the index formula by adapting the proof we explained for the mod 2 index.

mod 2 index is a cobordism invariant and W is assumed to be the boundary of some Y. So we set $b = 2\nu$.

Generalizing Eq. (12.52), we will describe a formula that characterizes anomaly inflow as long as $\mathcal{D}_W^+$ has the generic number 2ν of zero-modes. When the background fields are varied so that additional zero-modes appear, the formula will remain valid but will become a trivial identity $0 = 0$.

We denote orthonormal bases of the spaces of zero-modes of $\mathcal{D}_W^+$ as ψ_i, where $i = 1, \ldots, 2\nu$. By definition, the ψ_i are modes of the positive chirality field χ. The zero-modes for the negative chirality field $\tilde{\chi}$ are $\overline{\psi}^i = \sigma^0(\psi_i)^*$. We impose the orthonormality condition

$$\int \mathrm{d}^d x \sqrt{g}\, \overline{\psi}^i \psi_j = \delta^i_j. \tag{12.64}$$

We neglect nonzero modes because their treatment is completely the same as before. We expand $\Psi = \begin{pmatrix} \chi \\ \tilde{\chi} \end{pmatrix}$ as

$$\chi = \sum_{i=1}^{2\nu} A^i \psi_i, \quad \tilde{\chi} = -\mathrm{i} \sum_{i=1}^{2\nu} \overline{A}_i \overline{\psi}^i. \tag{12.65}$$

When the theory is formulated on the Euclidean signature manifold W or on $W \times \mathbb{R}$ (where $\mathbb{R}$ parameterizes the "time" in the Hamiltonian framework), the reality condition on Ψ is $\tilde{\chi} = -\mathrm{i}\overline{\chi}$ and hence $\overline{A}_i$ is the complex conjugate of A^i. Upon quantization, they become hermitian conjugate operators, the nonzero canonical anti-commutators being

$$\{A^i, \overline{A}_j\} = \delta^i_j. \tag{12.66}$$

The Hamiltonian is

$$H = \int \mathrm{d}^d x \sqrt{g}\, m\overline{\chi}\chi = m \sum_{i=1}^{2\nu} \overline{A}_i A^i. \tag{12.67}$$

If the mass parameter is positive $m > 0$, the ground state is specified by $A^i|\Omega\rangle = 0$, while if it is negative $m < 0$, the condition is $\overline{A}_i|\Omega\rangle = 0$. The local boundary condition L is $\langle \mathsf{L}|\overline{A}_i = 0$, which is the same as the one for the ground state of the positive mass theory.

When we formulate the theory on a general Euclidean signature manifold Y, Ψ is no longer real[v] and we consider the Dirac operator $\mathcal{D}_Y$ acting on a complex-valued field Ψ. If Y has boundary W, it would be meaningful

---

[v]In fact, the reality condition (12.55) cannot be extended into the bulk of Euclidean Y, because this condition involves $(\Psi_1, \Psi_2)_t = \langle \gamma^t \Psi_1, \Psi_2 \rangle$ which is not invariant under D-dimensional Lorentz transformations in Y. It is invariant only under d-dimensional Lorentz transformations in W.

to say that Ψ is real along the boundary, but neither the local boundary condition L nor the APS boundary condition, which we describe shortly, imposes such a constraint. We have already noted that L constrains $\overline{A}_i$ but not A^i along the boundary, and similarly the APS boundary condition will put a constraint on $\overline{A}_i$ and A^i that is not consistent with Ψ being real along the boundary. Quantum mechanically, the constraint will mean that certain nonhermitian linear combinations of A^i and $\overline{A}_i$ annihilate the state $|\text{APS}\rangle$ (similarly to the fact that the nonhermitian operators $\overline{A}_i$ annihilate the state $\langle \text{L}|$). Note that because of the canonical anti-commutation relations (12.66), there is no state that is annihilated by a hermitian linear combination of the A^i and $\overline{A}_j$, since the square of such an operator is strictly positive. So no reasonable boundary condition can make Ψ real along the boundary at the classical level.

Let us proceed to the discussion of the APS boundary condition. When zero-modes are present, the APS boundary condition is subtle. Remember that in the absence of zero-modes, the state $|\text{APS}\rangle$ associated with the APS boundary condition is the ground state of the Ψ field for $m = 0$. When the operator $\mathcal{D}_W$ has zero-modes, the $m = 0$ theory does not have a unique ground state (even up to an overall scalar multiple) but a nontrivial space of ground states, obtained by quantizing the zero-modes. Hence, to define an APS boundary condition, it is necessary to pick a particular state (or more precisely, a particular one-dimensional subspace) in the space of ground states. A completely arbitrary choice will not do, as we want the APS boundary condition to be such that the path integral of the Ψ field on Y with APS boundary conditions will generate the usual phase $\exp(-i\pi\eta_Y/2)$. In order for this to happen, the equation $(\mathcal{D}_Y + im)\Psi = 0$ should be the equation of motion derived from the fermion action, and in addition $\mathcal{D}_Y$ must be self-adjoint. These were inputs to the derivation in Sec. 12.2.2 (the first condition is needed so that the fermion path integral is the Pfaffian of $\mathcal{D}_Y + im$, and the second makes the eigenvalues of $\mathcal{D}_Y$ real, as assumed in the derivation). Each of these requirements is nontrivial.

First, we ask whether the equation of motion derived from the usual Dirac action $\int_Y d^D x \sqrt{g} \langle \Psi, (\slashed{D} + m)\Psi \rangle$ is the Dirac equation $(\mathcal{D}_Y + im)\Psi = 0$, or whether the equation of motion contains additional delta function terms supported on the boundary. To avoid such terms, we need

$$\langle \Psi_1, \mathcal{D}_Y \Psi_2 \rangle_Y = \langle \mathcal{D}_Y \Psi_1, \Psi_2 \rangle_Y \tag{12.68}$$

for c-number (commuting) fermion fields Ψ_1, Ψ_2, where

$$\langle \Psi_1, \Psi_2 \rangle_Y = \int d^D x \sqrt{g} \, \langle \Psi_1, \Psi_2 \rangle. \tag{12.69}$$

To prove Eq. (12.68), we have to integrate by parts, and we encounter a surface term proportional to

$$- \int d^d x \sqrt{g} \, \langle \Psi_1, \gamma^t \Psi_2 \rangle = \int d^d x \sqrt{g} \, (\Psi_1, \Psi_2)_t, \qquad (12.70)$$

where $(\, , \,)_t$ was introduced in Eq. (12.53). We want this expression to vanish whenever Ψ_1 and Ψ_2 satisfy the APS boundary condition. Expanding

$$\Psi_1 = \begin{pmatrix} \chi_1 \\ \tilde{\chi}_1 \end{pmatrix} = \sum_{i=1}^{2\nu} \begin{pmatrix} A^i \psi_i \\ -i \overline{A}_i \overline{\psi}^{\, i} \end{pmatrix}, \quad \Psi_2 = \begin{pmatrix} \chi_2 \\ \tilde{\chi}_2 \end{pmatrix} = \sum_{i=1}^{2\nu} \begin{pmatrix} B^i \psi_i \\ -i \overline{B}_i \overline{\psi}^{\, i} \end{pmatrix}, \qquad (12.71)$$

we get

$$\int d^d x \sqrt{g} \, (\Psi_1, \Psi_2)_t = \sum_{i=1}^{2\nu} (A^i \overline{B}_i + \overline{A}_i B^i), \qquad (12.72)$$

where we have used the orthonormality condition (12.64). As noted earlier, along W, Ψ is not constrained to be real, so in this formula, A^i and $\overline{A}_j$ (and similarly B^i and $\overline{B}_j$) should be understood as independent complex variables.

Now let us look at the condition for self-adjointness of $\mathcal{D}_Y$. When Y has no boundary, $\mathcal{D}_Y$ is hermitian with respect to the hermitian form $(\, , \,)_Y$ that was introduced in Eq. (12.49): $(\Psi_1, \mathcal{D}_Y \Psi_2)_Y = (\mathcal{D}_Y \Psi_1, \Psi_2)_Y$. When Y has a boundary, integration by parts generates a surface term $\int d^d x \sqrt{g} \, \langle \mathsf{C} \Psi_1, \gamma^\tau \Psi_2 \rangle$. From the definition of C given in Eq. (12.47) along with $\sigma^0 (\psi_i)^* = \overline{\psi}^{\, i}$, $\overline{\sigma}^0 (\overline{\psi}^{\, i})^* = -\psi_i$, we get

$$\Psi_1 = \sum_{i=1}^{2\nu} \begin{pmatrix} A^i \psi_i \\ -i \overline{A}_i \overline{\psi}^{\, i} \end{pmatrix} \implies \mathsf{C} \Psi_1 = \sum_{i=1}^{2\nu} \begin{pmatrix} -i (\overline{A}_i)^* \psi_i \\ (A^i)^* \overline{\psi}^{\, i} \end{pmatrix}. \qquad (12.73)$$

Therefore, the surface term is given by

$$\int d^d x \sqrt{g} \, \langle \mathsf{C} \Psi_1, \gamma^\tau \Psi_2 \rangle = \sum_{i=1}^{2\nu} \left((A^i)^* B^i - (\overline{A}_i)^* \overline{B}_i \right). \qquad (12.74)$$

For the surface terms (12.72) and (12.74) to vanish, we impose a boundary condition of the following sort. Let $J_{ij} = -J_{ji}$ be a matrix that is antisymmetric and also unitary. Then we impose

$$\overline{A}_i = \sum_j J_{ij} A^j. \qquad (12.75)$$

(with the same condition on B^i, $\overline{B}_i$, since Ψ_1 and Ψ_2 take values in the same space). The antisymmetry of the matrix J guarantees that the term

(12.72) vanishes and hence the Pfaffian of $\mathcal{D}_Y$ is well-defined. The unitarity of J guarantees that the term (12.74) vanishes and hence the eigenvalues are real. For the existence of an antisymmetric unitary J, it is crucial that the total mod 2 index is zero and hence the index i takes the values $i = 1, \ldots, 2\nu$. If the number of zero modes were odd, an antisymmetric matrix necessarily would have a zero eigenvalue and could not be unitary.

There is no unique way to choose the matrix J_{ij}. That leads to the nonuniqueness of APS boundary conditions in the presence of zero modes. We allow any choice of J. But we note that it is always possible to choose a basis of zero modes ψ_i to put J in the standard form

$$J_{2i-1,2i} = -J_{2i,2i-1} = -1, \quad \text{other } J_{ij} = 0. \tag{12.76}$$

Quantum mechanically, A^i and $\overline{A}_j$ become operators that satisfy the canonical anti-commutation relations (12.66). The condition (12.75) means that the state $|\text{APS}\rangle$ satisfies

$$\langle \text{APS}| \left(\overline{A}_i - \sum_j J_{ij} A^j \right) = 0. \tag{12.77}$$

The operators $\overline{A}_i - \sum_j J_{ij} A^j$ are a maximal set of anti-commuting operators constructed from the zero-modes, so these constraints are consistent and uniquely determine a linear combination of the ground states (up to a scalar multiple). On the other hand, not every linear combination of the ground states satisfies a condition of this form. The above detailed analysis has singled out a preferred class of ground states. There is, however, no way to avoid making a choice of the matrix J. For nonzero modes, there was no need to make such a choice; one simply says that the state $|\text{APS}\rangle$ is annihilated by the negative energy modes of Ψ.

Let us go back to the Hamiltonian framework on W. Let $|E\rangle$ be a state vector with $A_i|E\rangle = 0$, $i = 1, \ldots, 2\nu$. For notational simplicity we focus on a block of four operators $(A^{2k-1}, A^{2k}, \overline{A}_{2k-1}, \overline{A}_{2k})$ (with some fixed value of k) and omit $\prod_k$ in the following equations. Then we have

$$\langle \text{L}| = \langle E|, \quad \langle \text{APS}| = \langle E|(-\overline{A}_{2k} + A^{2k-1})(\overline{A}_{2k-1} + A^{2k}), \tag{12.78}$$

and

$$|\Omega\rangle = \begin{cases} |E\rangle, & m > 0, \\ \overline{A}_{2k}\overline{A}_{2k-1}|E\rangle, & m < 0. \end{cases} \tag{12.79}$$

From this, we get

$$\langle \text{APS}|\Omega\rangle = 1 \tag{12.80}$$

and

$$\langle L|\Omega\rangle = \begin{cases} 1, & m > 0, \\ 0, & m < 0. \end{cases} \tag{12.81}$$

This result could have been anticipated. For $m > 0$, there are no localized chiral fermions on the boundary and hence nothing happens. On the other hand, when $m < 0$, we have a localized chiral fermion whose partition function $\mathrm{Pf}(\mathcal{D}_W^+)$ vanishes because of the zero-modes.

To get something nonzero for $m < 0$, we have to insert an operator. We choose some operator O that is a function of the field χ on $W = \partial Y$. Then instead of a vacuum path integral, we compute a path integral with an insertion of O. We write $\langle O\rangle_{Y,L}$ for an unnormalized path integral on Y with this insertion, and with boundary condition L. Before going on, let us note that $\langle O\rangle_{Y,L}$ is completely well-defined and anomaly-free. It is an observable in the D-dimensional theory of the massive field Ψ; because this field is massive, it admits Pauli–Villars regularization and is free of any anomaly.

We can get an illuminating formula for $\langle O\rangle_{Y,L}$ as follows. Reasoning as in the derivation of Eq. (12.9), in the long distance limit we have $\langle O\rangle_{Y,L} = \langle L|O|\Omega\rangle\langle\Omega|Y\rangle$. Choosing an APS boundary condition and repeating the derivation that led to Eq. (12.11) gives

$$\langle O\rangle_{Y,L} = \frac{\langle L|O|\Omega\rangle\langle\Omega|\mathrm{APS}\rangle}{|\langle \mathrm{APS}|\Omega\rangle|^2} \cdot \langle \mathrm{APS}|Y\rangle. \tag{12.82}$$

In other words, the only change from the previous derivation is that $\langle L|\Omega\rangle$ is replaced by $\langle L|O|\Omega\rangle$.

It does not matter very much which O we pick, as long as $\langle L|O|\Omega\rangle$ is generically nonzero. A minimal choice is to take $O = \chi(x_1)\chi(x_2)\dots\chi(x_{2\nu})$ for points $x_1, x_2, \dots, x_{2\nu} \in W$. The zero-mode part of χ is $\chi(x) = \sum_i A^i\psi_i(x)$. The matrix element $\langle L|\chi(x_1)\dots\chi(x_{2\nu})|\Omega\rangle$ is a product of a zero-mode factor and a factor coming from nonzero modes. The factor that comes from nonzero modes can be analyzed precisely as in Secs. 12.2.3 and 12.2.4 and is $|\mathrm{Pf}'(\mathcal{D}_W^+)|$, where Pf' is the Pfaffian in the space orthogonal to the zero-modes. The zero-mode factor is

$$\langle L|\chi(x_1)\dots\chi(x_{2\nu})|\Omega\rangle_0 = \sum_\sigma \mathrm{sign}(\sigma)\prod_{i=1}^{2\nu}\psi_i(x_{\sigma(i)}), \tag{12.83}$$

where the sum is over all permutations σ of 2ν numbers, and we have restored $\prod_i$ in the notation. This follows from the formulas given earlier for the states

$|\mathsf{L}\rangle$ and $|\Omega\rangle$. Combining the two factors,

$$\langle\mathsf{L}|\chi(x_1)\dots\chi(x_{2\nu})|\Omega\rangle = |\mathrm{Pf}'(\mathcal{D}_W^+)|\sum_\sigma \mathrm{sign}(\sigma)\prod_{i=1}^{2\nu}\psi_i(x_{\sigma(i)})\,. \tag{12.84}$$

The bulk contribution is exactly the same as before, namely

$$\langle\mathsf{APS}|Y\rangle = \exp\left(-\mathrm{i}\pi\eta_Y/2\right), \tag{12.85}$$

where η_Y is computed using the chosen APS boundary conditions. In fact, we have defined this boundary condition so that $\mathcal{D}_Y$ has all the properties that were needed in Sec. 12.2.2 for the derivation of Eq. (12.85).

Combining these results, we get

$$\langle\chi(x_1)\dots\chi(x_{2\nu})\rangle_{Y,\mathsf{L}} = \frac{\langle\mathsf{L}|\chi(x_1)\dots\chi(x_{2\nu})|\Omega\rangle\langle\Omega|\mathsf{APS}\rangle}{|\langle\mathsf{APS}|\Omega\rangle|^2}\langle\mathsf{APS}|Y\rangle$$

$$= |\mathrm{Pf}'(\mathcal{D}_W^+)|\exp\left(-\mathrm{i}\pi\eta_Y/2\right)\sum_\sigma\mathrm{sign}(\sigma)\prod_{i=1}^{2\nu}\psi_i(x_{\sigma(i)})\,. \tag{12.86}$$

This is the result with a minimal operator insertion that gives a nonzero result. We could also consider some other choice of O. The only consequence is to modify the factor $\langle\mathsf{L}|O|\Omega\rangle$. The anomaly inflow factor $\exp(-\mathrm{i}\pi\eta_Y/2)$ is unchanged. This is important because — in the context of the derivation explained in Sec. 12.3 — it will imply that the anomaly of the path integral of the original χ field on W does not depend on what operator insertion is made. This is equally true whether $\nu = 0$ — the case assumed in Secs. 12.2.3 and 12.2.4 — or $\nu > 0$, as analyzed here. The only difference is that for $\nu = 0$, the minimal choice of O is $O = 1$, so in the previous analysis, we did not introduce O explicitly.

We conclude with one last remark. One might worry that the above result depends on the choice of a basis of zero modes ψ_i, because if we transform the basis as $\psi_i \to \sum_j \psi_j U_i^j$ with a unitary matrix U_j^i, the factor $\sum_\sigma\mathrm{sign}(\sigma)\prod_{i=1}^{2\nu}\psi_i(x_{\sigma(i)})$ is multiplied by $\det U$. However, if we change the basis in this way, we are also changing the APS boundary condition, since we have fixed J to the standard form (12.76). The Dai–Freed theorem [11] states that if we change the APS boundary condition in this way, the exponentiated η-invariant $\exp(-\mathrm{i}\pi\eta_Y/2)$ changes by $\det U^{-1}$ so that the above correlation function is independent of U. In this way the final result is independent of any choice of basis vectors or APS boundary condition. Alternatively, since in this derivation we knew at the beginning that $\langle O\rangle_{Y,\mathsf{L}}$ is well-defined, and the APS boundary condition was only introduced as a way to calculate it, one may see the above result as a physical proof of this part of the Dai–Freed theorem [12].

Of course, the result for $\langle O \rangle_{Y,\mathsf{L}}$ will in general depend on the choice of Y and of its spin structure. Here is a special case — an important one, though unfortunately somewhat technical to describe. In general, if W has s connected components $W_1, W_2, \ldots, W_s$, and we are given a spin structure on Y whose restriction to the boundary is isomorphic to some given spin structures on $W_1, W_2, \ldots, W_s$, then there are 2^s such isomorphisms,[w] as a given isomorphism on any given component can be multiplied by $(-1)^{\mathsf{F}}$. According to the definition of footnote p in Sec. 12.2.4, the choice of such an isomorphism is part of the definition of a spin structure on Y. However, an overall $(-1)^{\mathsf{F}}$ gauge transformation on Y, changing the sign of all fermions, would "flip" the isomorphism on each of the W_i. So spin structures on Y come in groups of 2^{s-1} that differ only by the chosen isomorphisms with the given spin structures on $W_1, W_2, \ldots, W_s$. Suppose that we "flip" the isomorphism on one of the boundary components, say W_1. This is gauge-equivalent to changing the APS boundary condition on Y without changing the bases of zero-modes on W, so it multiplies $\exp(-i\pi\eta_Y/2)$ by $\det U^{-1}$, with no compensating factor of $\det U$. Here U is the matrix that acts as -1 on zero-modes supported on W_1, and $+1$ on zero-modes supported on other components. As a result, $\langle O \rangle_{Y,\mathsf{L}}$ is multiplied by $(-1)^{\zeta_1}$, where ζ_1 is mod 2 index on W_1 (the number of χ zero-modes on W_1, mod 2). If $(-1)^{\zeta_1}$ can be nontrivial, then the product of an odd number of χ fields on W_1 can have an expectation value in the original d-dimensional theory on W_1. Since χ is odd under $(-1)^{\mathsf{F}}$, this represents an anomaly in $(-1)^{\mathsf{F}}$ symmetry. This particular anomaly is reproduced in the D-dimensional formalism by the spin structure dependence that was just described.

12.3   The anomaly

In Sec. 12.2, we have obtained a general formula describing anomaly inflow for an arbitrary fermion field χ on a manifold W. Implicit in this formula, as we will now explain, is a description of the anomaly of χ. The anomaly inflow involved an η-invariant on a manifold Y with boundary W, and this will enable us to express the anomaly in d dimensions in terms of an η-invariant in dimension $D = d + 1$.

---

[w]We assume that a specific isomorphism between the gauge bundle of ∂Y and that of W has been fixed. In general, if the gauge group has a nontrivial center, then $\langle O \rangle_{Y,\mathsf{L}}$ may depend on the choice of this isomorphism.

The fact that an anomaly in d dimensions is naturally related to some quantity in $d + 1$ dimensions is familiar for perturbative anomalies. The most familiar version of the statement is that the perturbative anomaly in d dimensions is related to a Chern–Simons function in $d + 1$ dimensions [4–6]. Our point here is to explain that nonperturbative or global fermion anomalies can be incorporated in this statement by just replacing the Chern–Simons function with η.

The relation between the global fermion anomaly in d dimensions and an η-invariant in $d + 1$ dimensions was originally found in [9]. The original derivation involved computing what one might call the holonomy of a Berry connection on the determinant (or Pfaffian) line bundle and expressing this in terms of η. Our derivation here, inspired by the Dai–Freed theorem [11,12] (which itself was partly inspired by the computation in [9]), is more general as it applies in any dimension, even or odd, and on any manifold, orientable or not, on which a fermion system can be defined. (The Dai–Freed theorem has been extended to this more general context in the appendix to [17].) Also, our derivation gives a more precise answer than was sought in the original work. When there is no anomaly, we determine the actual phase of the fermion path integral on W. In early work on anomalies, one aimed to show that the overall phase of the path integral was not affected by any inconsistency or anomaly, but one did not aim to get a formula for that phase. To determine the phase is much more precise than just showing that the phase can be defined. For the use of the Dai–Freed theorem and related ideas to define the absolute phase of a path integral in particular cases, see [13, Sec. 2.2]; [37, Sec. 2]; [38].

12.3.1 *Defining the phase of the path integral*

Let us start with a fermion field χ on a d-manifold W, and try to define the corresponding partition function. If χ can have a bare mass, the partition function can always be defined using Pauli–Villars regularization. Even if a bare mass is not possible, Pauli–Villars regularization can always be used to define the absolute value of the fermion partition function. But in general, if a bare mass is not possible, there is no simple direct way to define the phase of the fermion partition function, and there may be an anomaly that makes it impossible to get a satisfactory definition of this phase.

However, from Sec. 12.2, we know how to define the partition function of a modified system if W is the boundary of some manifold Y over which all the structures (such as a spin or pin structure and possibly a gauge bundle) needed to define the original fermion field χ on W have been extended. We regard χ as a boundary mode of a massive fermion field Ψ on Y, whose partition function is

$$Z(Y, \mathsf{L}) = |\mathrm{Pf}(\mathcal{D}_W^+)| \exp\left(-\mathrm{i}\pi\eta_Y/2\right), \tag{12.87}$$

where we put the subscript Y on η to indicate that it is computed on the manifold Y. This is the partition function of a combined system consisting of a massless fermion on W with anomaly inflow from a massive fermion in bulk.

If the expression (12.87) actually does not depend on the choice of Y, we can regard it as a definition of the path integral for the χ field on W. So let us investigate the dependence on Y. (If a suitable Y does not exist at all, then it is necessary to generalize the procedure. We postpone this issue for the moment.) To investigate the dependence on Y, let Y' be another manifold with the same boundary $\partial Y' = W$. The ratio between the partition functions $Z(Y, \mathsf{L})$ and $Z(Y', \mathsf{L})$ is given by $\exp(-\mathrm{i}\pi(\eta_Y - \eta_{Y'})/2)$. Let $\overline{Y}$ be the closed manifold[x] which is constructed by gluing Y and the orientation reversal of Y' along their common boundary W. The gluing theorem for η [11] says that we have

$$\exp\left(-\mathrm{i}\pi(\eta_Y - \eta_{Y'})/2\right) = \exp\left(-\mathrm{i}\pi\eta_{\overline{Y}}/2\right). \tag{12.88}$$

The physical interpretation of this gluing theorem is as follows. The partition function of the massive fermion on the closed manifold $\overline{Y}$ is represented in the path integral formulation as

$$Z(\overline{Y}) = \langle Y'|Y\rangle, \tag{12.89}$$

where $|Y\rangle$, $|Y'\rangle$ are physical states on the Hilbert space $\mathcal{H}_W$ as introduced in Sec. 12.2.1. As explained in that section, these states are proportional to the ground state $|Y\rangle \propto |\Omega\rangle$, $|Y'\rangle \propto |\Omega\rangle$, and hence we can write

$$\langle Y'|Y\rangle = \frac{\langle Y'|\mathsf{APS}\rangle\langle \mathsf{APS}|Y\rangle}{|\langle \mathsf{APS}|\Omega\rangle|^2}. \tag{12.90}$$

By computing as in Sec. 12.2.2, we get the gluing formula (12.88). The universal phase in the numerator on the right-hand side is $\exp(-\mathrm{i}\pi(\eta_Y - \eta_{Y'})/2)$ (note that reversing the orientation of Y' reverses the sign of $\eta_{Y'}$) and the denominator is positive. The universal phase on the left-hand side is $\exp(-\mathrm{i}\pi\eta_{\overline{Y}}/2)$.

---

[x]A closed manifold is a compact manifold without boundary. In what follows, $\overline{Y}$ is always a closed D-manifold, and Y is a D-manifold that might have a boundary.

Therefore, the dependence of the partition function $Z(Y, \mathsf{L})$ on Y is characterized by

$$\frac{Z(Y, \mathsf{L})}{Z(Y', \mathsf{L})} = \exp(-i\pi\eta_{\overline{Y}}/2). \qquad (12.91)$$

If $\Upsilon_{\overline{Y}} = \exp(-i\pi\eta_{\overline{Y}}/2)$ is always equal to 1 for any closed manifold $\overline{Y}$, $Z(Y, \mathsf{L})$ does not depend on the choice of Y. In that case, $Z(Y, \mathsf{L})$ can serve as a definition of the path integral $Z(W)$ for the original fermion system on W, which therefore is completely anomaly-free. If instead $\Upsilon_{\overline{Y}}$ is nontrivial, then $Z(Y, \mathsf{L})$ does depend on Y and cannot serve as a satisfactory definition of $Z(W)$. However, we should ask if there is some other definition that we should use instead of $Z(Y, \mathsf{L})$. This will be discussed in Sec. 12.3.3. The conclusion will be to show that the anomaly can be eliminated by using a better definition if and only if $\eta_{\overline{Y}}$ can be written as a local integral over $\overline{Y}$.

In the above derivation, we have implicitly assumed that the operator $\mathcal{D}_W$ has no zero-modes, so that the anomaly can be probed by studying the partition function $Z(Y, \mathsf{L})$. When that is not the case, to get something nonzero, one has to consider a path integral $\langle O \rangle_{Y, \mathsf{L}}$ with some operator insertion, as analyzed in Sec. 12.2.5, and the APS boundary condition is no longer unique. However, the above derivation is still valid in this more general situation. The only properties of the APS boundary condition that are needed in the derivation are that the universal phase of $\langle \mathrm{APS}|Y \rangle$ is $\exp(-i\pi\eta_Y/2)$, and that the matrix element $\langle \mathrm{APS}|\Omega \rangle$ is nonzero. The APS boundary condition was chosen in Sec. 12.2.5 to ensure the first property, and the second property also holds in general (Eq. (12.80)).

When $\exp(-i\pi\eta_{\overline{Y}}/2) = 1$ for any closed manifold $\overline{Y}$ that carries the appropriate structures, this tells us in a very strong sense that the original fermion system on W was anomaly-free. Not only is there no anomaly that would prevent us from determining the phase of the path integral of the original fermion system, but we have actually determined this phase in Eq. (12.87). In Secs. 12.3.2 and 12.3.3, we will explore this result more fully. But first we fill a gap in the explanation that we have given so far.

In this derivation, we assumed that W is the boundary of some manifold Y over which all relevant structures are extended. What happens if a suitable Y does not exist? This actually does not mean that the original theory on W cannot be defined. It means that if it is possible to define this theory, then the definition is not unique. The condition under which it is possible to define the theory is the same as before: $\exp(-i\pi\eta_{\overline{Y}}/2)$ should equal 1 for any closed manifold $\overline{Y}$ carrying the appropriate structures (or a slight generalization of this described in Sec. 12.3.3).

The nonuniqueness arises for the following reason. The very fact that W is not the boundary of any Y over which the appropriate structures extend means that there is a nontrivial cobordism group Γ whose elements are d-manifolds carrying such structures modulo those that are boundaries. The group operation in Γ is disjoint union of manifolds. Any homomorphism φ from Γ to $U(1)$ gives the partition function of an "invertible" topological field theory [38–41]. This is a purely d-dimensional theory whose partition function on a manifold W is $\varphi(W)$. When Γ is nontrivial, we cannot expect to uniquely determine the partition function of the original fermion theory on W by any general arguments, because any definition that is consistent with all general principles of quantum field theory could always be modified by multiplying it by $\varphi(W)$. In general, different regularizations of the same theory will give results that differ by such a factor.

We can proceed as follows, as in [36, Sec. 2.6]. Rather than being abstract, we will assume that $\Gamma = \mathbb{Z}_k$ for some integer k. This means that there is some manifold W_0 such that W_0 is not the boundary of any suitable Y, but the disjoint union of k copies of W_0 is such a boundary. Writing W' for this disjoint union, we determine the partition function $Z(W')$ from the formula (12.87). Since W' is the disjoint union of k copies of W_0, we interpret $Z(W')$ as $Z(W_0)^k$. Now we define $Z(W_0)$ as $(Z(W'))^{1/k}$, with some choice of the kth root. It is in this choice of a kth root that the nonuniqueness enters. Now given any W, since W_0 generates the cobordism group, there is some integer r such that W'', defined as the disjoint union of W with r copies of W_0, is the boundary of some Y. Thus we can use Eq. (12.87) to define $Z(W'')$. We interpret this as $Z(W)Z(W_0)^r$, since W'' is a disjoint union of W with r copies of W_0, and we define $Z(W) = Z(W'')/Z(W_0)^r$.

In this construction, changing the kth root in the definition of $Z(W_0)$ will multiply $Z(W)$ for any d-manifold W by $\varphi(W)$, for some $\varphi : \Gamma \to U(1)$. Since W_0 was assumed to generate Γ, φ is completely determined by $\varphi(W_0)$.

12.3.2   Topological field theory and cobordism

To understand the global anomaly more deeply, we have to return to the APS index theorem [10], which was briefly introduced in Sec. 12.2.2.

Suppose that the $d + 1$-manifold $\overline{Y}$ is the boundary of a $d + 2$-manifold X. Suppose also that we are given a fermion field Ψ on $\overline{Y}$ with a self-adjoint Dirac operator $\mathcal{D}_{\overline{Y}} = i \sum_{\mu=1}^{d+1} \gamma^\mu D_\mu$, and that all structures (such as spin structures and gauge bundles) needed to define $\mathcal{D}_{\overline{Y}}$ have been extended over X. The APS index theorem relates $\eta_{\overline{Y}}$, the η-invariant of $\mathcal{D}_{\overline{Y}}$, to the index $\mathcal{I}$ of a certain Dirac operator $\mathcal{D}_X$ on X. $\mathcal{D}_X$ is defined using the same doubling procedure that we used in Sec. 12.2.4 to go from W to $\overline{Y}$. In brief, we introduce a second copy $\widetilde{\Psi}$ of Ψ and a combined field $\widehat{\Psi} = \begin{pmatrix} \Psi \\ \widetilde{\Psi} \end{pmatrix}$. Acting on $\widehat{\Psi}$, we define $d+2$ gamma matrices rather as in Eqs. (12.38) and (12.43):

$$\Gamma^\mu = \begin{pmatrix} 0 & \gamma^\mu \\ \gamma^\mu & 0 \end{pmatrix}, \quad \Gamma^\tau = \begin{pmatrix} 0 & -i \\ i & 0 \end{pmatrix}. \tag{12.92}$$

This enables us to define a Dirac operator $\mathcal{D}_X = i(\Gamma^\mu D_\mu + \Gamma^\tau D_\tau)$ on $\overline{Y} \times \mathbb{R}_-$. By Lorentz invariance of the construction, this operator can be defined on any X with boundary $\overline{Y}$ such that the relevant structures on $\overline{Y}$ have been extended over X.

There is one very important difference from the previous case. In Sec. 12.2.4, we started with a d-dimensional fermion field χ that did not necessarily have a self-adjoint Dirac operator. Doubling involved introducing a dual field $\widetilde{\chi}$ that did not necessarily transform the same way as χ under the symmetry group. As a result, in general the nonzero blocks in Eq. (12.38) are not equal; they consist of distinct matrices σ_μ and $\overline{\sigma}_\mu$. But in the present discussion, we started with a $d+1$-dimensional fermion field that by hypothesis does have a self-adjoint Dirac operator, and the additional field $\widetilde{\Psi}$ that we introduced in the doubling was just a second copy of Ψ. Accordingly the nonzero blocks in the definition of Γ^μ are equal. This gave more freedom in the definition of Γ^τ than we had in the previous case, and we have taken advantage of that freedom in making a convenient choice of Γ^τ. The extra freedom also means that in addition to Γ^τ, we can define a "chirality" operator:

$$\overline{\Gamma} = \begin{pmatrix} 1 & 0 \\ 0 & -1 \end{pmatrix}. \tag{12.93}$$

Because $\overline{\Gamma}$ anti-commutes with all gamma matrices Γ^μ and Γ^τ, it anti-commutes with the Dirac operator $\mathcal{D}_X = i(\Gamma^\mu D_\mu + \Gamma^\tau D_\tau)$. Hence we can define the index $\mathcal{I}$ of $\mathcal{D}_X$: the number of zero-modes of $\mathcal{D}_X$ with $\overline{\Gamma} = 1$ minus the number with $\overline{\Gamma} = -1$. This did not have an analog when we went from d dimensions to $d+1$ dimensions.

We stress that this index can in general be nonzero in any dimension, even or odd. The Atiyah–Singer index theorem implies that the index of an elliptic operator on an odd-dimensional manifold without boundary vanishes, but this is not so on an odd-dimensional manifold X with boundary $\overline{Y}$. In general, the index of $\mathcal{D}_X$ can be nonzero in any dimension.

So far, everything makes sense for any $d+1$-dimensional fermion field Ψ with a self-adjoint Dirac operator $\mathcal{D}_{\overline{Y}}$. But we are actually interested in the case that Ψ was itself defined by the doubling procedure starting from some fermion field χ in d dimensions. In that case, there is an antilinear operator C, defined in Eq. (12.47), that anti-commutes with all gamma matrices while commuting with all symmetries and with the Dirac operator $\mathcal{D}_{\overline{Y}}$. After another round of doubling, the Dirac operator $\mathcal{D}_X$ on X has an antilinear symmetry $\overline{\mathsf{C}}$ that also obeys $\overline{\mathsf{C}}^2 = -1$. The definition of Γ^τ was chosen to make the definition simple:

$$\overline{\mathsf{C}} = \begin{pmatrix} \mathsf{C} & 0 \\ 0 & \mathsf{C} \end{pmatrix}. \tag{12.94}$$

Given that C anti-commutes with the γ^μ, $\overline{\mathsf{C}}$ anti-commutes with Γ^μ and Γ^τ, and hence commutes with $\mathcal{D}_X = i(\Gamma^\mu D_\mu + \Gamma^\tau D_\tau)$.

The existence of an antilinear transformation $\overline{\mathsf{C}}$ that commutes with $\mathcal{D}_X$ and squares to -1 implies that all eigenvalues of $\mathcal{D}_X$ have even multiplicity, by a kind of Kramers doubling of eigenvalues. In particular, the index $\mathcal{I}$ of $\mathcal{D}_X$ is even. We will see shortly why this is important.

Now let us look at the APS index formula for $\mathcal{I}$:

$$\mathcal{I} = \int_X \Phi_{d+2} - \frac{\eta_{\overline{Y}}}{2}. \tag{12.95}$$

If $\overline{Y}$ is empty, this reduces to the usual Atiyah–Singer index formula $\mathcal{I} = \int_X \Phi_{d+2}$. Φ_{d+2} can be expressed in terms of the gauge field strength and the Riemann tensor, as already remarked in footnote h of Sec. 12.2.2. However, for us, the most useful characterization of Φ_{d+2} is that it is the anomaly $d + 2$-form of the d-dimensional fermion χ. This follows directly from the APS index formula. The anomaly $d + 2$-form Φ_{d+2} is by definition the polynomial in the Riemann tensor and the gauge field strength whose associated Chern–Simons $d + 1$-form is related by anomaly inflow to the perturbative anomaly of χ. But the formula (12.95) shows that modulo the integer $\mathcal{I}$ (which plays no role in perturbation theory) and provided X exists (which can be assumed in perturbation theory) $\eta_{\overline{Y}}/2$ is the Chern–Simons form associated to Φ_{d+2}. Moreover, we have learned in Sec. 12.2 that $\eta_{\overline{Y}}/2$ is related by anomaly inflow to the perturbative (and nonperturbative) anomaly of χ. So Φ_{d+2} is the anomaly $d + 2$-form.

Therefore, Φ_{d+2} vanishes if and only if the original theory of the χ field in d dimensions is free of perturbative anomalies. For example, this is always the case for odd d; in odd dimensions, there are no perturbative anomalies, and the Atiyah–Singer index theorem shows that $\Phi_{d+2} = 0$ for odd d. For even d, perturbative anomalies are possible, of course, but many interesting theories — such as the Standard Model of particle physics — are free of them. These are again theories with $\Phi_{d+2} = 0$.

When there is no perturbative anomaly, there may still be a global anomaly. As we discussed in Sec. 12.3.1, the anomaly is governed by $\exp(-i\pi\eta_{\overline{Y}}/2)$ for closed $d+1$-manifolds $\overline{Y}$. This can definitely be nontrivial even when there is no perturbative anomaly. However, from the APS index formula, we can reach an important conclusion about the function $\exp(-i\pi\eta_{\overline{Y}}/2)$ when $\Phi_{d+2} = 0$. Suppose that $\overline{Y}$ is the boundary of a $d+2$-manifold X over which the relevant structures extend. The index formula then reduces to $\eta_{\overline{Y}} = -2\mathcal{I}$. Since $\mathcal{I}$ is an even integer, $\eta_{\overline{Y}}$ is an integer multiple of 4, and therefore in this situation $\exp(-i\pi\eta_{\overline{Y}}/2) = 1$. In other words, we have learned that when there is no perturbative anomaly, the function $\exp(-i\pi\eta_{\overline{Y}}/2)$ that governs the global anomaly is a cobordism invariant: it is trivial on any $\overline{Y}$ that is the boundary of some X. It may be a nontrivial cobordism invariant; if $\overline{Y}$ is not the boundary of any X, it may happen that $\exp(-i\pi\eta_{\overline{Y}}/2) \neq 1$. This is precisely the case that the original theory of the χ field in dimension d has a nontrivial global anomaly.

As a special case of what we have said, $\exp(-i\pi\eta_{\overline{Y}}/2)$ is a topological invariant, unchanged in any continuous deformations of metrics and gauge fields. Cobordism invariance is much stronger than topological invariance.

The APS index formula and the associated cobordism invariance of the global anomaly when perturbative anomalies vanish was used in [43] (without using the term "cobordism") for certain applications to string theory. The framework was more restrictive than that of the present paper: in relating anomalies to η-invariants, d was assumed to be even and all manifolds were assumed orientable. Nowadays, there is a more general framework for understanding the role of cobordism invariance. Any gapped theory with no topological order reduces at long distances (modulo nonuniversal terms that can be removed by local counterterms) to an "invertible" topological quantum field theory whose partition function is a cobordism invariant. This was conjectured in [39], and proved under some axioms of locality and unitarity in [40,41]. A system of free massive fermions coupled to background fields is an example of a gapped system with no topological order, and our discussion establishes directly the claim about cobordism invariance at long distances for such a system.

12.3.3   *Deformation classes and anomalies*

In Sec. 12.3.1, assuming that $\Upsilon_{\overline{Y}} = \exp(-i\pi\eta_{\overline{Y}}/2)$ is trivial for every closed D-manifold $\overline{Y}$, we have found a satisfactory partition function for the purely

d-dimensional theory of the χ field,[y] namely $Z_W^0 = |\mathrm{Pf}\,\mathcal{D}_W^+|\exp(-i\pi\eta_Y/2)$, where Y is any appropriate manifold with boundary W. We call this Z_W^0 (and not just Z_W) because we will consider some more general possibilities in a moment.

What happens if $\Upsilon_{\overline{Y}}$ is not always trivial? Then the formula Z_W^0 is not a satisfactory definition for the partition function of the χ field as a purely d-dimensional theory, but we should ask if this formula can be improved. Let us try some other definition $Z_W = |\mathrm{Pf}\,\mathcal{D}_W^+|\exp(-i\pi\eta_Y/2)\exp(iQ_Y)$, where to begin with Q_Y is some unknown function of the background fields g, A on Y (we can assume that Q is real as the anomaly only affects the phase of the partition function).

In order for Z_W to be independent of the background fields g, A on Y once Y has been chosen, Q_Y must be unaffected by any variation of the background fields on Y that leaves them fixed along W. (Therefore, when presently we discuss the variation of Q_Y under a change of g, A along W, it will not matter how g, A are being changed away from W.) For Z_W to be independent of the choice of Y, we need $\exp(iQ_Y)$ to satisfy conditions similar to those that were used in the previous analysis of Z_W^0: $\exp(iQ_Y)$ must satisfy the same gluing law (12.88) as $\exp(-i\pi\eta_Y/2)$, and for a closed D-manifold $\overline{Y}$, we need

$$\exp(-i\pi\eta_{\overline{Y}}/2)\exp(iQ_{\overline{Y}}) = 1. \qquad (12.96)$$

These conditions could be trivially satisfied with $\exp(iQ_Y) = \exp(i\pi\eta_Y/2)$. However, we need more. Since Z_W^0 is the partition function of a physically sensible system consisting of the χ field plus massive (D-dimensional) regulator degrees of freedom, it is manifestly physically sensible (at least in a D-dimensional sense) and we did not have to discuss what properties make it physically sensible. An important aspect is the following. Consider making a small variation of the background fields g, A along W. The variation of the logarithm of the partition function of χ should be given by the one-point function of the stress tensor T or the current operator J of χ. In any theory free of perturbative anomaly, these one-point functions $\langle T \rangle$ and $\langle J \rangle$ can be regularized in a way consistent with all physical principles, including conservation of T and J. (If there is a global anomaly, it may appear when one tries to integrate $\delta \log Z/\delta g$ and $\delta \log Z/\delta A$ to determine Z.) The regularization is unique up to the possibility of adding to T and J

---

[y] Refinements discussed in Sec. 12.3.1, where $\mathcal{D}_W$ has zero-modes (so that the path integral measure cannot be characterized by a partition function) or W is not a boundary (leading to a slightly more elaborate discussion), can be straightforwardly included in the following. We omit them for brevity.

a function of the background fields that is local, gauge-invariant, and conserved. The derivation that led to the formula Z_W^0 for the partition of the χ field plus massive degrees of freedom makes it manifest that $\delta \log Z_W^0/\delta g$ and $\delta \log Z_W^0/\delta A$ are related in the expected way to $\langle T \rangle$ and $\langle J \rangle$.

What happens if we replace Z_W^0 with $Z_W = Z_W^0 \exp(iQ_Y)$? Since $\log Z_W = \log Z_W^0 + iQ_Y$, this shifts $\delta \log Z/\delta g$ and $\delta \log Z/\delta A$ by $i\delta Q_Y/\delta g$ and $i\delta Q_Y/\delta A$. For Z_W to be a physically sensible candidate formula for a partition function of the χ field, it must be possible to interpret $i\delta Q_Y/\delta g$ and $i\delta Q_Y/\delta A$ as contributions to T and J. So $\delta Q_Y/\delta g$ and $\delta Q_Y/\delta A$ must be gauge-invariant, local (and conserved) functions of the background fields along W.

For this to be the case, Q_Y must be the integral over Y of some local operator Φ: $Q_Y = \int_Y \Phi$. Φ must be such that Q_Y respects all symmetries of the theory, including possible time-reversal or reflection symmetries. Equation (12.96) tells us that in the case of a closed manifold $\overline{Y}$,

$$\exp(-i\pi\eta_{\overline{Y}}/2) = \exp\left(-i\int_{\overline{Y}} \Phi\right), \tag{12.97}$$

so $\exp\left(-i\int_{\overline{Y}} \Phi\right)$ is a topological invariant and in fact a cobordism invariant. For a local operator Φ to have that property, Φ must be a D-form constructed as a gauge-invariant polynomial in the Riemann tensor R and the gauge field strength F, plus a possible exact form $d\Lambda$ (here Λ is a gauge-invariant $(D-1)$-form, locally constructed from g, A). Adding $d\Lambda$ to Φ will modify Z_W by $Z_W \to Z_W \exp(i\int_Y d\Lambda) = Z_W \exp(i\int_W \Lambda)$. This is equivalent to adding to the action of the original theory on W a c-number term $-i\int_W \Lambda$ constructed from the background fields. That does not affect the consistency of the theory, so an exact term in Φ is not important. So in short, the case that Φ can help in eliminating an anomaly is that Φ is a polynomial in R and F. Φ is supposed to be a D-form, so this is only possible if D is even. For such a Φ, $\int_{\overline{Y}} \Phi$ is a characteristic class.

Whenever $\Upsilon_{\overline{Y}} = \exp(-i\pi\eta_{\overline{Y}}/2)$ can be expressed as in Eq. (12.97) in terms of a characteristic class $\int_{\overline{Y}} \Phi$, we can define a purely d-dimensional partition function for the field χ. Under these assumptions,

$$Z_W = |\text{Pf } \mathcal{D}_W^+| \exp(-i\pi\eta_Y/2) \exp\left(i\int_Y \Phi\right) \tag{12.98}$$

depends only on W and not on Y.[z]

---

[z] Actually, there is one last requirement to make the definition (12.98) physically sensible: Φ should be odd under reflections, to ensure reflection positivity of the theory. Accordingly, the case that Φ is the polynomial in R related to the Euler characteristic is not satisfactory.

When Y has boundary W, $CS(g, A) = \int_Y \Phi$ is a generalized Chern–Simons function of the fields g, A on W, and is independent of how those fields are extended over Y (and of the choice of Y) modulo the values of $\int_{\overline{Y}} \Phi$ for closed manifolds $\overline{Y}$. (This is proved by using the analog of Eq. (12.88) with $\pi \eta_Y / 2$ replaced by $\int_Y \Phi$.) If $\int_{\overline{Y}} \Phi$ takes values in $2\pi \mathbb{Z}$, then $\exp(i\, CS(g, A))$ depends only on the background fields on W, and not on anything about Y. In that case, we say that $CS(g, A)$ is a properly normalized Chern–Simons action; it makes sense as a purely d-dimensional coupling. However, Eq. (12.97) tells us that whenever $\Upsilon_{\overline{Y}}$ is nontrivial to begin with, $\int_{\overline{Y}} \Phi$ is not valued in $2\pi \mathbb{Z}$ and $CS(g, A)$ is not a properly normalized Chern–Simons action. It does not make sense by itself as a d-dimensional action; rather, it is being used to cancel an anomaly. The classic example of this situation is the "parity anomaly" for odd d [15], in which $CS(g, A)$ is $\frac{1}{2}$ of a properly normalized Chern–Simons coupling.[aa] We will discuss that case from the present point of view in Sec. 12.4.2.

Whenever there is a Φ that satisfies (12.97), it is not unique, because we can always shift $\Phi \to \Phi + \Phi'$, where $\int_{\overline{Y}} \Phi'$ is a characteristic class normalized to take values in $2\pi \mathbb{Z}$. This will have the effect of shifting the effective action on W by a properly quantized, physically sensible Chern–Simons function of the background fields. This shift can be interpreted as the result of using a different regularization of the underlying theory.

Now let us discuss the same subject from the point of view of the massive theory in the D-dimensional bulk. The massive Ψ field has a single ground state when quantized on any $(D-1)$-manifold, so at long distances the theory of the Ψ field becomes an "invertible topological field theory." The partition function of this invertible topological field theory on a closed manifold $\overline{Y}$ is the cobordism invariant $\exp(-i\pi \eta_{\overline{Y}}/2)$. Let us ask what are the moduli of this theory as an invertible topological field theory. In general, a first-order deformation with small parameter ε of any quantum field theory multiplies the partition function by $\exp\left(i\varepsilon \int_{\overline{Y}} \langle \mathcal{O} \rangle\right)$ for some local operator $\mathcal{O}$. In an invertible topological field theory, $\mathcal{O}$ is just a function of background fields g, A (since there are no other local operators), and so $\langle \mathcal{O} \rangle$ reduces to the classical value of $\mathcal{O}$ in the given background fields. For $\exp\left(i\varepsilon \int_{\overline{Y}} \mathcal{O}\right)$ to be a topological invariant (and in fact a cobordism invariant), the exponent must be a characteristic class. In other words, the moduli of an invertible

---

Reflection positivity implies that the Euler characteristic should appear in the Euclidean signature effective action with a real coefficient, not an imaginary one.
[aa]Somewhat similar are fractional Chern–Simons counterterms that arise in certain situations [42].

topological field theory precisely correspond to the possibility of multiplying the partition function by $\exp\left(i\int_Y \Phi\right)$, where as before Φ is a gauge-invariant polynomial in the curvature and field strength.

We are free here to take $\Phi = \sum_i t_i \Phi_i$, where Φ_i are a basis of possible gauge-invariant polynomials and t_i are arbitrary real numbers. In particular, we can interpolate continuously between any given Φ and $\Phi = 0$. Therefore, Eq. (12.97) is precisely the condition under which the invertible topological field theory with partition function $\exp(-i\pi\eta_{\overline{Y}}/2)$ can be deformed, through a family of invertible topological field theories, to a trivial theory. In other words, the anomaly of the original theory in d dimensions is really associated not with the invertible topological field theory associated to $\exp(-i\pi\eta_{\overline{Y}}/2)$ but with the deformation class of this theory. There is no anomaly if and only if this invertible topological field theory is deformable to a trivial one.

This result is natural from the point of view of condensed matter physics. In the discussion of symmetry protected topological (SPT) phases [19–21], we consider two phases to be equivalent if they can be continuously deformed to each other without any phase transition while preserving the relevant symmetries. In particular, since the coefficients with which the polynomials Φ_i appear in the effective action can be varied continuously, these coefficients are not universal and are not part of the classification of SPT phases. The SPT theory of the massive field Ψ is considered nontrivial if the invertible topological field theory associated to it, with partition function $\exp(-i\pi\eta_{\overline{Y}}/2)$, is not deformable to a trivial theory. As we have seen, this is the case in which the theory of the χ field cannot be defined as a purely d-dimensional theory.

12.4   Examples in dimensions $d = 1, 2, 3, 4$

In this section, we consider some examples of the use of the η-invariant to analyze global anomalies in dimensions $d = 1, 2, 3, 4$.

The cases of odd d have a different flavor, because for odd d there are no perturbative anomalies, and all anomalies are global from the beginning. We will consider the odd d cases first.

To keep our examples simple, we will primarily work on orientable manifolds only; in other words, we will generally not incorporate time-reversal or reflection symmetry. When d is odd, the global anomaly is related to an η-invariant in an even dimension $D = d + 1$. On an even-dimensional orientable manifold, the η-invariant reduces to a more familiar invariant — an index or a mod 2 index. Therefore, our odd d examples could be described more directly in terms of the index or the mod 2 index rather than the

η-invariant. (See [14, 43] for such a treatment of the $d = 1$ and $d = 3$ examples, respectively.) However, it seems useful to explain that all examples can be deduced from the η-invariant.

In even d, we consider examples in which perturbative anomalies cancel, and analyze the global anomalies using the η-invariant.

12.4.1 $d = 1$

In $d = 1$, we consider a system of n real fermions with a classical $O(n)$ symmetry and a simple action

$$I = \int dt \frac{i}{2} \sum_{j=1}^{n} \chi_j \frac{d}{dt} \chi_j. \tag{12.99}$$

For simplicity, to begin with we take n even, do not incorporate time-reversal symmetry, and consider background $SO(n)$ gauge fields only. We briefly explain at the end what happens if one relaxes one of those conditions.

A compact 1-manifold W will have to be a circle. The circle has two possible spin structures, as χ could be either periodic or antiperiodic in going around the circle. We will refer to these spin structures as Ramond (R) or Neveu–Schwarz (NS).

The Hamiltonian derived from the action I simply vanishes. Though we could add additional terms to the action to get a nonzero Hamiltonian, instead we will turn on a background $SO(n)$ gauge field on the circle. We write U for the holonomy of this gauge field. In its canonical form, U is block diagonal with 2×2 blocks of the form

$$\begin{pmatrix} \cos\theta_k & \sin\theta_k \\ -\sin\theta_k & \cos\theta_k \end{pmatrix}, \quad k = 1, \ldots, n/2. \tag{12.100}$$

As this formula makes clear, shifting any of the θ_k by an integer multiple of 2π does not change U and so is a gauge transformation of the background gauge field.

Upon quantization, the χ_k satisfy $\{\chi_k, \chi_{k'}\} = \delta_{kk'}$ so (up to a factor of $\sqrt{2}$) they are gamma matrices. Hence the group that acts on the quantum Hilbert space $\mathcal{H}$ is not the classical symmetry group $SO(n)$ but its double cover $\mathrm{Spin}(n)$. The fact that the group that acts quantum mechanically is a double cover of the classical symmetry group is an anomaly of sorts. To see that this can be understood as an anomaly in the conventional sense — an ill-definedness of the path integral — we consider the path integral on a circle in the presence of a background gauge field with holonomy U.

The path integral with antiperiodic boundary conditions for the fermions (NS spin structure) computes $\text{Tr}_{\mathcal{H}} U$. Using the explicit description of $\mathcal{H}$ as a spinor representation of $SO(n)$, we can write a formula for this trace:

$$\text{Tr}_{\mathcal{H}} U = \prod_{k=1}^{n/2} 2\cos(\theta_k/2). \tag{12.101}$$

Here we see the anomaly: shifting any one of the θ_k by 2π is a gauge transformation of the background gauge field, but it changes the sign of the path integral.

The path integral with periodic boundary conditions for the fermions (R spin structure) computes $\text{Tr}_{\mathcal{H}} (-1)^{\text{F}} U$. Again we can write an explicit formula:

$$\text{Tr}_{\mathcal{H}} (-1)^{\text{F}} U = \pm \prod_{k=1}^{n/2} 2i \sin(\theta_k/2). \tag{12.102}$$

Again we see the anomaly, but now there is a new ingredient: the overall sign of the path integral depends on an arbitrary choice. One way to explain this fact is the following. The operator $(-1)^{\text{F}}$ is characterized by the fact that it anti-commutes with the elementary fermions, and its square is 1. But these conditions do not determine the overall sign of the operator $(-1)^{\text{F}}$, and without more input there is no natural way to fix this sign. We can define $(-1)^{\text{F}}$ up to sign by the product $\chi_1 \chi_2 \cdots \chi_n$, but as the χ_k anti-commute with each other, there is no way to determine the overall sign of this expression without knowing something about how the χ_k should be ordered. We run into the same issue from the point of view of path integrals. Consider the system of n real fermions χ_k with periodic spin structure in the special case $U = 1$. Each of the χ_k has a zero-mode; let us call these modes χ_k^0. The measure for the fermion zero-modes is, up to sign, $d\chi_1^0 d\chi_2^0 \cdots d\chi_n^0$, But again, the sign of this measure depends on how we order the $d\chi_k^0$.

Now let us look at this anomaly from the perspective of the η-invariant. Applied to a 1-component real fermion in dimension $d = 1$, the doubling procedure of Sec. 12.2.4 produces a 2-component real fermion field Ψ in dimension $D = d + 1 = 2$. So starting with the n fields $\chi_1, \ldots, \chi_n$ in dimension 1 transforming in the fundamental representation of $O(n)$, we get in two dimensions an n-component Majorana fermion field Ψ in the fundamental representation of $O(n)$.

To study the anomaly by our general procedure, we regard the circle W (at least in the NS case; see below for the R case) as the boundary of a two-manifold Y, over which the spin structure of W is extended. Since we do not incorporate time-reversal symmetry as part of the discussion, we can consider the original circle W to be oriented, and then in the procedure of Sec. 12.2.4, we consider only oriented Y. The anomaly is given by $\exp(-i\pi\eta_{\overline{Y}}/2)$, where $\eta_{\overline{Y}}$ is the η-invariant of the self-adjoint Dirac operator $\mathcal{D}_{\overline{Y}}$ of the Ψ-field on a closed manifold $\overline{Y}$.

On an even-dimensional orientable manifold Y, there is always a chirality operator $\overline{\gamma}$ (usually called γ_5 in four-dimensional particle physics) that anti-commutes with the self-adjoint Dirac operator $\mathcal{D}_Y$. Accordingly, the nonzero eigenvalues of $\mathcal{D}_Y$ are equal and opposite in pairs: if $\mathcal{D}_Y\Psi = \lambda\Psi$, then $\mathcal{D}_Y(\overline{\gamma}\Psi) = -\lambda\overline{\gamma}\Psi$. From the definition of the η-invariant

$$\eta_Y = \lim_{\epsilon\to 0}\sum_k \exp(-\epsilon|\lambda_k|)\mathrm{sign}(\lambda_k) \tag{12.103}$$

(where the function $\mathrm{sign}(x)$ was defined in Eq. (12.19)), we see at once that a pair of eigenvalues $\lambda, -\lambda$ do not contribute. Therefore, in this situation, η_Y simply equals the number of linearly independent zero-modes of $\mathcal{D}_Y$.

A zero-mode of $\mathcal{D}_Y$ can have either positive or negative chirality. In $d = 2$, complex conjugation exchanges the two types of mode. So the number of zero-modes of $\mathcal{D}_Y$ is two times the number of zero-modes of positive chirality.

It is physically sensible in two dimensions to consider a fermion field ψ of positive chirality only, transforming in a real representation of some symmetry group. For our case, the relevant real representation is the vector representation of $SO(n)$ or $O(n)$. Such a field can have an action

$$\int d^2x\sqrt{g}\,(\psi, \mathcal{D}_Y^+\psi), \tag{12.104}$$

where $\mathcal{D}_Y^+$ is the Dirac operator acting on a fermion field of positive chirality. By fermi statistics, we can here think of $\mathcal{D}_Y^+$ as an antisymmetric matrix. The canonical form of such a matrix is

$$\begin{pmatrix} 0 & a_1 & & & & & \\ -a_1 & 0 & & & & & \\ & & 0 & a_2 & & & \\ & & -a_2 & 0 & & & \\ & & & & \ddots & & \\ & & & & & 0 & \\ & & & & & & 0 \end{pmatrix}, \tag{12.105}$$

with skew "eigenvalues" a_i that appear in 2×2 blocks, and unpaired zero-modes. The only way that the number of zero-modes can change is that one of the a_i can become zero or nonzero. When that happens, the number of zero-modes jumps by ± 2. So if ζ is the number of zero-modes mod 2, ζ is invariant in any continuous deformation. ζ is called the mod 2 index of the chiral Dirac operator on Y. The derivation shows that any fermion system with an action consistent with fermi statistics has a mod 2 index (this invariant is not always interesting as for many fermion systems it identically vanishes or is the mod 2 reduction of a more familiar invariant). In our problem of the chiral fermion field ψ in the vector representation of $SO(n)$ or $O(n)$, this means that the number ζ of zero-modes of ψ mod 2 is a topological invariant: it is unchanged if one varies the metric of Y or the background gauge field.

Since η_Y is twice the number of zero-modes of ψ, we have $\eta_Y = 2\zeta$ mod 4. Hence $\exp(-i\pi\eta_Y/2) = (-1)^\zeta$. So the anomaly reduces to a sign ± 1, as we saw more directly in Eq. (12.101). To evaluate the anomaly more explicitly, first note that an $SO(n)$ bundle $E \to \overline{Y}$, for any closed two-manifold $\overline{Y}$, is classified topologically by its second Stieffel–Whitney class $w_2(E)$. Hence the anomaly can only depend on $w_2(E)$. Furthermore, the structure group of E can be reduced to an $SO(2)$ subgroup. (See the end of Sec. 12.4.4 for this argument.) We have $SO(2) \cong U(1)$. From the point of view of $U(1)$, the vector representation of $SO(n)$ consists of $n - 2$ neutral fermions and two components of charge ± 1. In the field of a $U(1)$ gauge field with first Chern class k, generically one of the charged components has $|k|$ zero-modes, and the other has none. Since we assume n even, the $n - 2$ neutral fermions do not contribute to the mod 2 index. So ζ is the mod 2 reduction of k. But the mod 2 reduction of k coincides with $w_2(E)$. So $\zeta = w_2(E)$ and the anomaly is $(-1)^{w_2(E)}$. In particular, the anomaly does not depend on the spin structure and could have arisen in a purely bosonic theory. Since $w_2(E)$ is the obstruction to lifting the structure group of E from $SO(n)$ to its double cover $\mathrm{Spin}(n)$, the anomaly would disappear if we view the original system (12.99) as a system with $\mathrm{Spin}(n)$ rather than $SO(n)$ symmetry. This is consistent with the observation that we made at the outset: the Hilbert space $\mathcal{H}$ of this theory furnishes a representation of $\mathrm{Spin}(n)$, not $SO(n)$.

Now let us consider the case of Ramond spin structure. A single circle W with Ramond spin structure is not the boundary of any spin manifold Y. But two such circles are the boundary of such a Y (we can take Y to be a cylinder). The cobordism group in this problem is $\mathbb{Z}_2$; for a generator, we can pick a circle W_0 with Ramond spin structure and with any chosen holonomy U_0 for the background gauge field. Since W_0 is not a boundary, our formalism gives no natural way to compute the sign $\mathrm{Tr}\,(-1)^F U_0$. But once we fix this sign, any other Ramond sector path integral is determined by the procedure explained at the end of Sec. 12.3.1. Thus, the Ramond sector path integral is uniquely determined up to an arbitrary overall sign, as we saw more directly in Eq. (12.102).

Finally, we briefly consider three generalizations that were mentioned at the outset.

(1) For odd n, consider the path integral of the theory (12.99) on $W = S^1$ with a spin structure of R-type. With such a spin structure, each of $\chi_1, \ldots, \chi_n$ has a zero-mode, so to get a nonzero path integral, we need to insert the product of all these fields. With this insertion, we get $\langle \mathrm{Tr} \, (-1)^\mathsf{F} \chi_1 \chi_2 \cdots \chi_n \rangle \neq 0$. But for odd n, the operator $\chi_1 \chi_2 \cdots \chi_n$ is odd under $(-1)^\mathsf{F}$. So we have found an anomaly in $(-1)^\mathsf{F}$.

Let us try to recover this anomaly from a $D = 2$ point of view. For odd n, the derivation showing that the anomaly is governed by $(-1)^\varsigma$ is still valid. The only difference is that $(-1)^\varsigma$ can be nontrivial even if we ignore the $O(n)$ symmetry and do not turn on any background $O(n)$ gauge field. For an example, take Y to be a two-torus with fermions periodic in both directions. Then the Dirac operator on Y for a one-component chiral fermion ψ has a single zero-mode (the "constant" mode of ψ), and so $(-1)^\varsigma = -1$. Taking n identical chiral fermions does not change $(-1)^\varsigma$ if n is odd. So there is an anomaly even if the only symmetry we consider is $(-1)^\mathsf{F}$. This symmetry is implicit whenever we discuss fermions and spin structures.

This anomaly in $(-1)^\mathsf{F}$ has a variety of applications in different areas of physics; for example, see [44–47]. An application to Type II superstring theory, closely related to the recent paper [48], was actually the subject of the lecture by one of us at the Shoucheng Zhang Memorial Workshop [3].

(2) Another generalization is to allow background gauge fields of $O(n)$ rather than $SO(n)$. In discussing this, for simplicity we take n even. Let W be a circle with NS spin structure and with a background $O(n)$ gauge field with monodromy $\mathsf{R}_1 = \mathrm{diag}(-1, 1, 1, \ldots, 1)$. In this background, χ_1 has a zero-mode and χ_r, $r > 1$, does not. So again we see an anomaly in $(-1)^\mathsf{F}$, this time a "mixed anomaly" between $(-1)^\mathsf{F}$ and the $O(n)$ symmetry, since to detect it we had to turn on a background $O(n)$ gauge field.

Replacing $O(n)$ by a double cover $\mathrm{Pin}^+(n)$ or $\mathrm{Pin}^-(n)$ will not eliminate this anomaly. R_1 can be lifted to $\mathrm{Pin}^+(n)$ or $\mathrm{Pin}^-(n)$, though not uniquely, and the same analysis leads to the same anomaly in $(-1)^\mathsf{F}$ in the presence of a background field.

From a two-dimensional point of view, the anomaly is still governed by $(-1)^\varsigma$. But in contrast to what we found in discussing $SO(n)$ for even n, $(-1)^\varsigma$ now depends on the spin structure. To see this, take Y to be a two-torus $S_A^1 \times S_B^1$, where the first circle S_A^1 has trivial $O(n)$ gauge field and the second circle S_B^1 has an $O(n)$ gauge field with monodromy R_1. Then $(-1)^\varsigma$ can again be computed just by counting "constant" fermion modes; it equals 1 if S_A^1 has NS spin structure, and equals -1 if S_A^1 has spin structure of R-type (regardless of the spin structure of S_B^1). Thus there is an anomaly, and, since $(-1)^\varsigma$ depends on the spin structure, the anomaly is not given by a cohomological formula and cannot be reproduced in a purely bosonic theory. That is consistent with the fact that in the boundary description, the anomaly involves $(-1)^\mathsf{F}$.

One can relate the two ways of seeing the anomaly by trying to compute $\langle \mathrm{Tr}\,(-1)^\mathsf{F}\mathsf{R}_1\chi_1\rangle^2$ via a path integral on the cylinder $Y = S^1 \times I$, where S^1 is a circle with Ramond spin structure and I is an interval. We assume that the gauge field is a pullback from S^1 and has monodromy R_1. Once we specify the spin structure on S^1, there are two possible spin structures on Y, for a reason explained at the end of Sec. 12.2.5. (After picking an isomorphism between the two boundary circles and their spin bundles, one can say that the two spin bundles on Y differ by the sign of parallel transport of a fermion from one end of the cylinder to the other.) As explained at the end of Sec. 12.2.5, the sign one gets for $\langle \mathrm{Tr}\,(-1)^\mathsf{F}\mathsf{R}_1\chi_1\rangle^2$ depends on the spin structure of Y, and this dependence reproduces the $(-1)^\mathsf{F}$ anomaly of the original $d = 1$ theory on a single boundary component.

The theory (12.99) can certainly be quantized, with the Hilbert space providing an irreducible representation of the Clifford algebra $\{\chi_i, \chi_j\} = \delta_{ij}$. But after quantization, the "internal symmetry" R_1 anti-commutes with the "spacetime symmetry" $(-1)^\mathsf{F}$, rather than commuting. As a result, the symmetry group after quantization does not fit the general framework introduced at the start of Sec. 12.2.4, where $(-1)^\mathsf{F}$ is supposed to be central, commuting with the full symmetry group (as noted in footnote m, there are more general possibilities in $d = 1$). The symmetry group of the quantized $d = 1$ theory does not extend naturally to $D = 2$ (in dimension $D \geq 2$, the element $(-1)^\mathsf{F} \in \mathrm{Spin}(D)$ is always central in any relativistic fermion theory). Hence it is not immediately obvious how to deduce the symmetry group of the quantized $d = 1$ theory from the $D = 2$ anomaly.

(3) To include time-reversal symmetry, we should consider W to be unoriented and allow unorientable Y. The obvious time-reversal symmetry of Eq. (12.99) that acts on the fermions by $\mathsf{T}\chi_j(t)\mathsf{T}^{-1} = \chi_j(-t)$ satisfies $\mathsf{T}^2 = 1$ at the classical level and corresponds to a pin^- structure in 1 dimension. To incorporate this symmetry in the boundary theory, one should allow Y to be an unorientable manifold endowed with a pin^- structure. In this case,

as there is no longer a chirality operator $\overline{\gamma}$ that anti-commutes with $\mathcal{D}_Y$, it is no longer true that $\exp(-i\pi\eta_{\overline{Y}}/2) = \pm1$ for a closed manifold $\overline{Y}$, and instead $\exp(-i\pi\eta_{\overline{Y}}/2)$, for a single Majorana fermion, can be an arbitrary eighth root of 1. (This can be proved along lines explained in Appendix C of [14] for the analogous problem in four dimensions.) Indeed, in the time-reversal invariant case, the theory (12.99) that we started with does have a mod 8 anomaly from a Hamiltonian point of view [49]; that is, it is anomalous unless n is a multiple of 8. The interpretation of this in terms of cobordism was originally suggested in [39]. If n is even, it is possible to define a time-reversal symmetry such that $\mathsf{T}^2 = (-1)^{\mathsf{F}}$ at the classical level. This leads to a slightly different analysis, in which the description by the η-invariant again reduces to a mod 2 index.

12.4.2 $d = 3$

Our example in $d = 3$ will actually involve one of the celebrated contributions of Shoucheng Zhang [18].

First consider in $d = 3$ a massless Dirac fermion χ coupled with charge 1 to a U(1) gauge field. Such a field could have a gauge-invariant bare mass, so there will be no anomaly spoiling the gauge symmetry. However, a bare mass of the χ field would explicitly violate the time-reversal and reflection symmetries of the massless theory. So there is a possibility of an anomaly that would spoil those symmetries. Indeed, this model is the original context for the "parity" anomaly [15]: the field χ cannot be quantized, purely in three-dimensional terms, in a way that preserves time-reversal and reflection symmetry. We essentially computed the anomaly in Sec. 12.2.2, when we explained that integrating out the χ field on a manifold W, with a particular regulator mass, generates a phase $\exp(-i\pi\eta_{D,W}/2)$ (as we are dealing with a Dirac fermion, we use $\eta_{D,W}$, the η-invariant on W for charge 1 modes only, rather than $\eta_W = 2\eta_{D,W}$, the η-invariant on W for all modes of charge ±1). The partition function of χ, including this phase, is

$$|\mathrm{Det}\,\mathcal{D}_\chi|\exp(-i\pi\eta_{D,W}/2), \qquad (12.106)$$

where $\mathcal{D}_\chi$ is the Dirac operator of the χ field. This phase is odd under time-reversal or reflection symmetry and cannot be removed by any counterterm. Reversing the sign of the regulator mass would give the opposite phase $\exp(i\pi\eta_{D,W}/2)$, still violating time-reversal and reflection symmetry. The anomaly is a mod 2 anomaly, because a pair of χ fields could be quantized with regulator masses of opposite signs, in which case the phases cancel out and all symmetries are preserved.

We are interested here in the anomalous case with a single χ field. In the theory of topological insulators, instead of trying to quantize χ by itself in purely three-dimensional terms, one views it as a field that propagates on the boundary of a four-manifold Y, the worldvolume of a topological insulator. The combined system can then be quantized in a way that preserves time-reversal and reflection symmetry. In this interpretation, the U(1) gauge field is just the usual gauge field of electromagnetism. The result of Shoucheng Zhang, together with Hughes and Qi [18], was to show that the electromagnetic θ-angle in the bulk of a topological insulator is equal to π. (A striking consequence of this can be explicitly demonstrated in a lattice model of a topological insulator [50]: a magnetic monopole of unit charge immersed in a $3 + 1$-dimensional topological insulator will acquire a half-integral electric charge.) Let us see how the result that $\theta = \pi$ may be understood from the point of view of the present paper.

The doubling procedure applied to the χ field produces a massive charge 1 Dirac fermion Ψ in dimension $d + 1 = 4$. With the couplings, boundary conditions, and regulator described in Secs. 12.2.1 and 12.2.2, Ψ provides a model of a topological insulator; in particular, χ can be interpreted as a boundary localized mode of Ψ, and the path integral of Ψ gives a consistent framework that describes χ together with massive degrees of freedom in the bulk. According to the general formalism, the universal part of the path integral of Ψ on a manifold Y with boundary W is

$$|\mathrm{Det}\, \mathcal{D}_W^+|\exp(-\mathrm{i}\pi\eta_{D,Y}). \tag{12.107}$$

We want to show that this formula is consistent with time-reversal and reflection symmetry, and moreover we would like to recover the result of Shoucheng Zhang and his colleagues showing that the electromagnetic θ-angle of the bulk theory is $\theta = \pi$. For simplicity, we work on oriented manifolds only. The self-adjoint Dirac operator $\mathcal{D}_Y$ anti-commutes with a chirality operator $\overline{\gamma}$, so just as in Sec. 12.4.1, its nonzero modes do not contribute to $\eta_{D,Y}$; $\eta_{D,Y}$ is simply equal to the number of linearly independent (charge 1) zero-modes of $\mathcal{D}_Y$. However, unlike the two-dimensional case, in four dimensions, complex conjugation does not exchange zero-modes of positive and negative chirality. On the contrary, if n_+ and n_- are the numbers of linearly independent zero-modes of $\mathcal{D}_Y$ (acting on fermions of charge 1) with positive or negative chirality, then the index of $\mathcal{D}_Y$ is $\mathcal{I} = n_+ - n_-$. On the other hand, $\eta_{D,Y}$ is the total number of zero-modes: $\eta_{D,Y} = n_+ + n_-$. We see that $\eta_{D,Y} \cong \mathcal{I}$ mod 2, and hence $\exp(-\mathrm{i}\pi\eta_{D,Y}) = (-1)^{\mathcal{I}}$. Thus the combined path integral of the bulk and boundary modes of the topological insulator — or more precisely its universal part — is

$$|\text{Det}\,\mathcal{D}_W|(-1)^{\mathcal{I}}. \tag{12.108}$$

This is real, and thus manifestly consistent with time-reversal and reflection symmetry.

To understand the result of Hughes, Qi, and Zhang concerning the θ-angle, we write $(-1)^{\mathcal{I}} = \exp(i\pi\mathcal{I})$ and think of $-i\pi\mathcal{I}$ as a contribution to the effective action. For discussing the bulk effective action, we can temporarily work on a closed manifold $\overline{Y}$. In that case, the Atiyah–Singer index formula gives $\mathcal{I} = \int_{\overline{Y}} \Phi$ with

$$\Phi = \widehat{A}(R) + \frac{1}{2}\frac{F \wedge F}{(2\pi)^2}, \tag{12.109}$$

where $\widehat{A}(R)$ is a certain quadratic polynomial in the Riemann tensor. Because of the electromagnetic contribution to Φ, the term $-i\pi\mathcal{I} = -i\pi\int\Phi$ in the action corresponds to an electromagnetic θ-angle $\theta = \pi$.

A key fact in this derivation was that on a closed manifold, $\exp(-i\pi\eta_{\overline{Y},D}) = \exp(i\pi\mathcal{I}) = \exp(i\pi\int_{\overline{Y}}\Phi)$ is the exponential of the integral of a characteristic class. Therefore, we are in the situation that was analyzed in detail in Sec. 12.3.3. As in Eq. (12.98), it is possible to give a purely three-dimensional formula for the partition function of χ by replacing $\exp(-i\pi\eta_{Y,D})$ with $\exp(-i\pi\eta_{Y,D})\exp(-i\pi\int_Y \Phi)$:

$$Z_W = |\text{Det}\,\mathcal{D}_W^+|\exp(-i\pi\eta_{D,Y})\exp\left(-i\pi\int_Y \Phi\right)$$

$$= |\text{Det}\,\mathcal{D}_W^+|\exp(i\pi\mathcal{I})\exp\left(-i\pi\int_Y \Phi\right). \tag{12.110}$$

The general formalism tells us that the right-hand side of Eq. (12.110) will make sense in purely three-dimensional terms. We can confirm this by using the APS index theorem:

$$\mathcal{I} = \int_Y \Phi - \frac{\eta_{D,W}}{2}. \tag{12.111}$$

So in fact Eq. (12.110) is equivalent to

$$Z_W = |\text{Det}\,\mathcal{D}_W^+|\exp(-i\pi\eta_{D,W}/2). \tag{12.112}$$

This is the purely three-dimensional, but time-reversal violating, formula that was explained more directly in Eq. (12.106), except that after doubling, we write $\mathcal{D}_W^+$ for $\mathcal{D}_\chi$. Time-reversal violation entered this derivation when we canceled the Y dependence of $(-1)^{\mathcal{I}} = \exp(\pm i\pi\mathcal{I})$ with a factor of $\exp(-i\pi \int_Y \Phi)$. Time-reversal would map this to a conjugate construction using $\exp(+i\pi \int_Y \Phi)$.

So we recover the familiar fact that this system can be quantized as a purely three-dimensional theory if we are willing to give up time-reversal and reflection symmetry. Alternatively, we can consider the χ field to propagate on the boundary of a four-manifold, and use the time-reversal invariant partition function (12.108). If we wish to define the χ field on unorientable three-manifolds, then time-reversal and reflection symmetry are essential and the purely three-dimensional quantization is not available. In this case, we have to consider the χ field as living on the surface of a topological insulator. The appropriate formula for the partition function is Eq. (12.107), in which the η-invariant no longer reduces to $\mathcal{I}$.

12.4.3 $d = 2$

Our remaining examples will be cases with nontrivial cancellation of perturbative anomalies.

In $d = 2$, we consider first a $U(1)$ gauge theory with positive chirality Dirac fermions of charges $n_1, n_2, \ldots, n_s$ and negative chirality Dirac fermions of charges $m_1, m_2, \ldots, m_s$. The perturbative gauge theory anomaly cancels if and only if

$$\sum_{i=1}^{s} n_i^2 = \sum_{i=1}^{s} m_i^2. \tag{12.113}$$

By taking equal numbers of positive and negative chirality fermions, we have ensured cancellation of perturbative gravitational anomalies. Therefore, when Eq. (12.113) holds, this theory is free of perturbative anomalies.

A possible global anomaly would be controlled as usual by an η-invariant in dimension $d + 1 = 3$. In the present case, let us write $\eta_{D,r}$ for the η-invariant of a charge r Dirac fermion on a three-manifold Y. (We use η_D as we are dealing with Dirac fermions.) The global anomaly of our given theory is controlled by $\Upsilon = \exp\left(-i\pi \left(\sum_{i=1}^{s} \eta_{D,n_i} - \sum_{j=1}^{s} \eta_{D,m_i}\right)\right)$. (As explained at the end of Sec. 12.2.2, reversing the sign of the fermion chirality in two dimensions reverses the sign with which the η-invariant appears in the exponent.) However, as is always the case when perturbative anomalies are absent, the factor Υ that controls the global anomaly is a cobordism invariant. The cobordism of a three-dimensional closed spin manifold $\overline{Y}$ with a $U(1)$ gauge field is trivial (any such $\overline{Y}$ is the boundary of some X over which the spin structure and $U(1)$ gauge field extend). So Υ always equals

1 on a closed manifold, and a theory of this kind that is free of perturbative anomalies is also always free of global or nonperturbative anomalies.[bb]

We can get an example in $d = 2$ that does have a nontrivial global anomaly if we replace U(1) by a $\mathbb{Z}_k$ subgroup, for some k. Then there is no perturbative anomaly as long as there are equally many positive and negative chirality fermions (to avoid a gravitational anomaly). In particular, there is no condition analogous to (12.113). For generic choices of the $\mathbb{Z}_k$ quantum numbers of the fermions, such a theory will have a global anomaly.

We will consider a special case in a moment, but first we explain the motivation to consider this special case. Spacetime supersymmetry in string theory was originally discovered by Gliozzi, Olive, and Scherk (GOS) [51]. Their original insight was that the partition function of 8 chiral fermions in two dimensions (in genus 1, where they computed explicitly) vanishes if it is summed over spin structures; this vanishing was a reflection of spacetime supersymmetry. (8 is the number of light cone oscillator modes in the Ramond–Neveu–Schwarz model.) Though this point was not made explicitly until later, it only makes sense to add together the partition functions with different spin structures if the anomaly does not depend on the spin structure. Later discoveries involved elaborations of the original GOS analysis. One always finds that spacetime supersymmetry in string theory depends on the fact that the anomaly of 8 chiral fermions in two dimensions is independent of the spin structure. It turns out that this is true for a system of k chiral fermions if and only if k is divisible by 8.

A closely related problem has been much studied in the context of condensed matter physics. In that context, one studies "symmetry protected topological" (SPT) states [19–21], which are states that are topologically nontrivial when some global symmetry is taken into account but become trivial if that global symmetry is explicitly broken. Such a system can have an anomalous boundary state in one dimension less. A much-studied special case is a fermionic system in spacetime dimension $D = 3$ with a global $\mathbb{Z}_2$ symmetry [52–54]. Those systems have a $\mathbb{Z}_8$ classification. An example

---

[bb]One could avoid relying on a knowledge of the cobordism group by making an argument similar to what we will make for the Standard Model in Sec. 12.4.4. A U(1) gauge field is classified topologically by its first Chern class. In three dimensions, this is dual to an embedded circle $C \subset \overline{Y}$. Using this, and cobordism invariance of Υ, one can reduce as in Sec. 12.4.4 to two special cases: (a) the gauge field is trivial, or (b) $\overline{Y} = S^2 \times S^1$ with a gauge field is a pullback from S^2. In case (a), trivially $\Upsilon = 1$ since the U(1) charges do not matter, and in case (b), arguing as in Sec. 12.4.4 one shows that $\Upsilon = 1$. This argument will also work for a theory with gauge group U(1)^n for any n: if such a theory in two dimensions has no perturbative anomaly, it also has no global anomaly.

associated to a nonzero element $k \in \mathbb{Z}_8$ has a boundary state in dimension $d = 2$ that carries the anomaly that one would find in the GOS calculation if the number of chiral fermions considered were k (or any integer congruent to k mod 8) rather than 8.

To put this problem in the context of gauge theory, we consider a $\mathbb{Z}_2$ gauge theory in two dimensions, with k positive chirality real fermions that are invariant under $\mathbb{Z}_2$, and k negative chirality real fermions that transform as -1 under the nontrivial element of $\mathbb{Z}_2$. This system has no perturbative anomaly, but an explicit genus 1 calculation shows that if k is not divisible by 8, it has a global anomaly. We will explore from the vantage point of the present paper the absence of anomalies when k is a multiple of 8. (For a previous analysis, see the discussion of Eq. (24) in [43]. An explicit computation of the η-invariant when k is not a multiple of 8 to show the $\mathbb{Z}_8$ classification was done in [55].)

Let W be a two-manifold with spin structure α and some background $\mathbb{Z}_2$ bundle, which we can think of as a real line bundle L with structure group $\mathbb{Z}_2 = \{\pm 1\}$. In the model just introduced, positive chirality fermions (being invariant under $\mathbb{Z}_2$) are coupled only to the spin structure α. But negative chirality fermions are coupled to the spin structure α and also to L. Effectively the negative chirality fermions are coupled to a new spin structure $\beta = \alpha \otimes L$. Thus for studying anomalies, we can forget about the $\mathbb{Z}_2$ gauge field and just say that fermions of positive or negative chirality are coupled to different spin structures α or β.

The spin cobordism problem for a two-manifold with two spin structures α, β (or even just with one spin structure) is not trivial. This is related to the fact that in string theory, the sign of the GOS projection in the Ramond sector is not uniquely determined and could be reversed. It is also related to the existence of two different Type II superstring theories.[cc] We will pass over such issues here and just ask if there is any consistent way to define the theory with 8 positive chirality fermions coupled to spin structure α and 8 negative chirality fermions coupled to β. As usual, the potential obstruction is a global anomaly that can be measured by an η-invariant.

In detail, let $\overline{Y}$ be a closed three-manifold with spin structures α, β. The global anomaly is then measured by $\exp\left(-\frac{\pi i}{2} 8(\eta_{\overline{Y},\alpha} - \eta_{\overline{Y},\beta})\right)$ where $\eta_{\overline{Y},\alpha}$ and $\eta_{\overline{Y},\beta}$ are η-invariants on $\overline{Y}$ for a Majorana fermion coupled to spin structure

---

[cc]The cobordism invariant of a Riemann surface with one spin structure is $(-1)^\varsigma$, which was already discussed, with references to various applications, at the end of Sec. 12.4.1. With two spin structures α, β, one has $(-1)^{\varsigma_\alpha}$ and $(-1)^{\varsigma_\beta}$.

α or β. We note that this is trivial if and only if one always has

$$\exp\left(-\frac{\pi i}{2} 8\eta_{\overline{Y},\alpha}\right) = \exp\left(-\frac{\pi i}{2} 8\eta_{\overline{Y},\beta}\right), \tag{12.114}$$

or in other words if and only if the anomaly for 8 positive chirality fermions in two dimensions does not depend on the spin structure. This is how we formulated the question initially.

In this form, it is not immediately obvious how to answer the question. But a more general question is easier to answer. Consider a two-dimensional theory with gauge group Spin(8). This group has two spinor representations — spinors of positive or negative Spin(8) chirality. Each of these is a real representation of dimension 8. Let us call the two representations S_+ and S_-. We consider a two-dimensional theory with gauge group Spin(8), with 8 positive chirality real fermions in the representation S_+, and 8 negative chirality fermions in the representation S_-. This theory is free of perturbative anomalies, because the representations S_+ and S_- have the same dimension and quadratic Casimir operator. We will show that the theory is also free of global anomalies.

To answer this question, we consider a closed three-manifold $\overline{Y}$ with spin structure α and some Spin(8) bundle. The global anomaly of the Spin(8) theory described in the last paragraph is measured by $\Upsilon = \exp\left(-\frac{\pi i}{2}\left(\eta_{\overline{Y},\alpha,S_+} - \eta_{\overline{Y},\alpha,S_-}\right)\right)$. The notation is hopefully clear; $\eta_{\overline{Y},\alpha,S_\pm}$ is the η-invariant on $\overline{Y}$ for a Majorana fermion in the representation $S_\pm$ coupled to some background Spin(8) gauge field, as well as to the spin structure α.

Since this theory has no perturbative anomaly, Υ is a cobordism invariant and in particular it is a topological invariant. But any Spin(8) bundle on a three-manifold is topologically trivial (since $\pi_i(\text{Spin}(8)) = 0$ for $i \leq 2$). So we can continuously deform to the case that the background Spin(8) gauge field is trivial, in which case trivially $\Upsilon = 1$. Thus the Spin(8) theory under consideration has no global anomaly, and for any background Spin(8) gauge field,

$$\exp\left(-\frac{\pi i}{2}\eta_{\overline{Y},\alpha,S_+}\right) = \exp\left(-\frac{\pi i}{2}\eta_{\overline{Y},\alpha,S_-}\right). \tag{12.115}$$

It follows from this, together with a judicious embedding of $\mathbb{Z}_2$ in Spin(8), that the $\mathbb{Z}_2$ theory with 8 positive chirality neutral fermions and 8 negative chirality charged fermions is also free of global anomaly. For this, we embed $\mathbb{Z}_2$ in SO(8) so that the nontrivial element $x \in \mathbb{Z}_2$ maps to the central element

$-1 \in \mathrm{SO}(8)$. The element $-1 \in \mathrm{SO}(8)$ can be lifted to $\mathrm{Spin}(8)$ in two ways. We can pick a lift so that x acts as $+1$ on S_+ and as -1 on S_-. (With the other lift, these signs are reversed.) Now we consider the identity (12.115), specialized to the case that the background $\mathrm{Spin}(8)$ gauge field actually has structure group $\mathbb{Z}_2$, embedded in $\mathrm{Spin}(8)$ as just described. With this choice, the $\mathrm{Spin}(8)$ identity (12.115) reduces to the identity (12.114), which says that the anomaly of 8 chiral fermions does not depend on the spin structure. Indeed, for a background $\mathrm{Spin}(8)$ gauge field that is induced from a $\mathbb{Z}_2$ bundle L by embedding $\mathbb{Z}_2$ in $\mathrm{Spin}(8)$ in the way that we have described, the vector bundle over $\overline{Y}$ corresponding to S_+ is a rank 8 trivial bundle, and the vector bundle over $\overline{Y}$ corresponding to S_- is the direct sum of 8 copies of L. On the left-hand side of Eq. (12.115), $\eta_{\overline{Y},\alpha,S_+}$ reduces in this example to $8\eta_{\overline{Y},\alpha}$ on the left-hand side of Eq. (12.114), while on the right-hand side of Eq. (12.115), $\eta_{\overline{Y},\alpha,S_-}$ similarly reduces to $8\eta_{\overline{Y},\beta}$ on the right-hand side of Eq. (12.114). So Eq. (12.115) does reduce to Eq. (12.114).

The only property of $\mathrm{Spin}(8)$ that we used was that it is connected and simply-connected; the only important property of the fermion representation was that it has no perturbative anomaly. So a two-dimensional theory with a connected and simply-connected gauge group and no perturbative anomaly also has no global anomaly.

12.4.4 $\;d=4$

For an example in $d = 4$ with nontrivial cancellation of perturbative anomalies, we can take the Standard Model of particle physics. Does it have any global anomaly?

The Standard Model can actually be embedded in the $\mathrm{SU}(5)$ grand unified theory [56] which is also free of perturbative anomalies. It turns out that the $\mathrm{SU}(5)$ grand unified model has no global anomaly. Thus, the phase of the fermion path integral of the $\mathrm{SU}(5)$ grand unified theory, coupled to an arbitrary background metric and $\mathrm{SU}(5)$ gauge field, can be defined in a consistent way. Specializing to the case that the structure group of the background gauge field reduces to the gauge group of the Standard Model, it follows that the Standard Model is also free of global anomalies. This was originally shown in [57, Example 3.4]. Here we will explain how to establish the result in the framework of the present paper. See also [58] for another discussion.

Since the Standard Model does not have a time-reversal or reflection symmetry, we formulate it only on oriented four-manifolds W, and in the anomaly inflow problem, we consider only oriented manifolds Y. Since the Standard Model has fermions whose definition requires a spin structure, both W and Y are endowed with spin structures.

We are not going to be able to get a unique answer for the phase of the path integral of the $SU(5)$ model in an arbitrary background gauge and gravitational field. The reason is that a four-dimensional spin manifold W with an $SU(5)$ bundle is not necessarily the boundary of a five-manifold Y over which the spin structure and $SU(5)$ bundle of W can be extended. The relevant cobordism group is $\mathbb{Z} \times \mathbb{Z}$. The two integer-valued invariants are $\sigma(W)/16$ (where σ is the signature) and the $SU(5)$ instanton number.

For generators of the cobordism group, we can use the following two manifolds W_1 and W_2. For W_1, we take a K3 surface, with some chosen metric and orientation, and with the background gauge field being $A = 0$. For W_2, we take a four-sphere S^4 with some chosen metric and orientation and with some chosen gauge field A_0 of instanton number 1.

We have no way to determine the phases of the path integral measure for those two examples, so we make arbitrary choices. One can think of those choices as representing a precise definition of what is meant by the gravitational θ-angle and the $SU(5)$ θ-angle including quantum effects of the fermions.

Any other W is cobordant, by some manifold Y over which the relevant structures extend, to a linear combination $n_1 W_1 + n_2 W_2$ for some integers n_1 and n_2. (By $n_1 W_1$ or $n_2 W_2$ with n_1 or n_2 negative, one means $|n_1|$ or $|n_2|$ copies of W_1 or W_2 with orientation reversed.) So once the gauge and gravitational θ-angles have been fixed, the procedure of Sec. 12.3.1 gives a definition of the path integral measure for any W. A priori, this definition might depend on Y.

To know that the phase does not depend on Y so that the $SU(5)$ grand unified theory has no global anomaly, we need to know that $\Upsilon_{\overline{Y}} = \exp(-i\pi\eta_{\overline{Y}}/2)$ is trivial for any closed five-manifold $\overline{Y}$ with $SU(5)$ gauge field. Here $\eta_{\overline{Y}}$ is the η-invariant of the operator $\mathcal{D}_{\overline{Y}}$, acting on a five-dimensional field Ψ that is obtained by combining the Standard Model fermions χ with dual fields $\widetilde{\chi}$. We will use cobordism invariance, plus an examination of some special cases, to show that Υ is always trivial.

We can proceed as follows. First of all, spin cobordism is trivial in five dimensions, so in case the $SU(5)$ gauge field is trivial, $\overline{Y}$ is the boundary of a spin manifold and $\Upsilon_{\overline{Y}} = 1$. Actually any fermion coupled only to gravity can have a bare mass in $d = 4$ and hence does not have any pure gravitational anomaly.[dd] Second, an SU(5) bundle E over a five-manifold $\overline{Y}$ is completely classified by its second Chern class $c_2(E)$. (This follows from the fact that homotopy groups $\pi_d(SU(5))$ vanish for $d \leq 4$ except for $\pi_3(SU(5)) = \mathbb{Z}$, which is related to the second Chern class.) Moreover, in five dimensions, $c_2(E)$ is dual to an embedded circle $C \subset \overline{Y}$. It means that, topologically, the SU(5) gauge field describes some integer number ν of instantons propagating along the circle C. In other words, we can deform the SU(5) gauge field so that it is trivial except very near some circle C, while in the normal plane to C the integral that defines the instanton number integrates to ν.

In this situation, by an elementary cobordism,[ee] $\overline{Y}$ is cobordant to a disjoint union $Y_1 + Y_2$, where Y_1 and Y_2 are as follows. Y_1 is a copy of $\overline{Y}$, but with $A = 0$. Y_2 is a copy of $S^4 \times C$ (S^4 being a four-sphere) with instanton number ν on S^4.

We already know that Y_1 has $\Upsilon_{Y_1} = 1$. We therefore only have to investigate Y_2.

There are two possible spin structures on $Y_2 = S^4 \times C$, since the spin structure around C may be of R or NS type. In the NS case (antiperiodic fermions), $S^4 \times C$ is the boundary of $S^4 \times D$, where D is a two-dimensional disk, and the SU(5) gauge field on $S^4 \times C$ extends over $S^4 \times D$. Therefore, $\Upsilon_{Y_2} = 1$ in this case.

In the R case (periodic fermions), there is no obvious six-dimensional spin manifold X with boundary $S^4 \times C$ over which the instanton bundle on S^4 extends. Such an X actually can be constructed,[ff] but without having to

---

[dd] Since a bare mass is possible, one suspects that there will be a simple direct proof, without knowing about the cobordism group, that, for a $d = 4$ Majorana fermion χ coupled only to gravity, in five dimensions one has $\Upsilon_{\overline{Y}} = 0$. Such a proof may be constructed as follows. A rank five Clifford algebra has an irreducible representation of dimension four, so there exists on a five-dimensional spin manifold $\overline{Y}$ a self-adjoint Dirac operator $\mathcal{D}_{\overline{Y}}^0 = i\slashed{D}$ acting on a four-component fermion field. It is not possible for the five gamma matrices to be all real, but it is possible for three of them to be real while two, say γ_1 and γ_2, are imaginary. Using this fact in a locally Euclidean frame, one can define an antilinear operator $\mathsf{C} = *\gamma_1\gamma_2$ that commutes with all gamma matrices and anti-commutes with $i\slashed{D}$. Therefore, the η-invariant of $\mathcal{D}_{\overline{Y}}^0$ reduces to the number of its zero-modes. That number is always even, since $\mathsf{C}^2 = -1$, so $\eta(\mathcal{D}_{\overline{Y}}^0)$ is an even integer. The doubling procedure applied to the four-dimensional Majorana fermion χ produces an operator $\mathcal{D}_{\overline{Y}}$ that is the direct sum of two copies of $\mathcal{D}_{\overline{Y}}^0$. ($\chi$ has four components, so after doubling we get an eight component fermion in five dimensions, whose Dirac operator is the direct sum of two copies of $\mathcal{D}_{\overline{Y}}^0$.) So $\eta_{\overline{Y}} = 2\eta(\mathcal{D}_{\overline{Y}}^0)$ is a multiple of 4 and $\Upsilon_{\overline{Y}} = \exp(-i\pi\eta_{\overline{Y}}/2) = 1$.

[ee] Instead of cobordism, one can use a cut and paste argument that is explained near the end of this section.

[ff] This is slightly technical. First, by an elementary cobordism or by a cut and paste argument (as described later), one can replace S^4 with another convenient manifold containing an instanton. A useful choice is $S^2 \times T^2$ where T^2 is a two-torus, which we take

know this, we can proceed as follows. We can deform the metric on $S^4 \times C$ to be a product and we can choose the SU(5) gauge field on $S^4 \times C$ to be a pullback from S^4. This condition means that if C is parameterized by an angle τ, then the gauge field on $S^4 \times C$ is independent of τ and has no component in the τ-direction. Concretely then, if the metric of C is chosen to be $d\tau^2$, the self-adjoint Dirac operator on Y_2 takes the form

$$\mathcal{D}_{Y_2} = i\gamma^\tau \frac{\partial}{\partial \tau} + \mathcal{D}_{S^4}, \qquad (12.116)$$

where $\mathcal{D}_{S^4}$ is the self-adjoint Dirac operator of S^4.

The chirality operator $\overline{\gamma}_{S^4}$ of S^4 (the product of the four gamma matrices of S^4) anti-commutes with $\mathcal{D}_{S^4}$ and commutes with γ^τ. If we combine $\overline{\gamma}_{S^4}$ with a reflection $\tau \to -\tau$, we get an operator that anti-commutes with $\mathcal{D}_{Y_2}$. Hence the nonzero eigenvalues of $\mathcal{D}_{Y_2}$ come in pairs $\lambda, -\lambda$. As usual, such pairs do not contribute to the η-invariant. Hence, $\eta_{\mathcal{D}_{Y_2}}$ is just the number of zero-modes of $\mathcal{D}_{Y_2}$.

Since

$$\mathcal{D}_{Y_2}^2 = -\frac{\partial^2}{\partial \tau^2} + \mathcal{D}_{S^4}^2, \qquad (12.117)$$

a zero-mode of $\mathcal{D}_{Y_2}$ is independent of τ and is a zero mode of $\mathcal{D}_{S^4}$. Conversely, any such mode is a zero-mode of $\mathcal{D}_{Y_2}$. So η_{Y_2} is the same as the number of zero-modes of $\mathcal{D}_{S^4}$.

Because of the doubling that is involved in going from W to Y in this construction, the operator $\mathcal{D}_{S^4}$ acts on the Standard Model fermions χ plus a dual or complex conjugate set of fermions $\widetilde{\chi}$. Let n be the number of zero modes of the Dirac operator on S^4 acting on the original Standard Model fermions χ. Complex conjugation exchanges zero-modes of χ with zero-modes of $\widetilde{\chi}$, so the number of zero-modes of $\mathcal{D}_{S^4}$ is $2n$. Hence $\eta_{Y_2} = 2n$.

As we will explain in a moment, n is always even. So η_{Y_2} is a multiple of 4, which implies that $\exp(-i\pi\eta_{Y_2}/2) = 1$.

---

with a spin structure periodic in both directions. Then $Y_2 = S^4 \times S^1$, with Ramond spin structure on S^1, is replaced by $Y_2' = S^2 \times T^2 \times S^1$, where now $T^2 \times S^1$ is a three-torus with spin structure periodic in all directions. $T^2 \times S^1$, with this spin structure, is the boundary of a "half-K3 surface," that is, a four-manifold Q that maps to a disk D with generic fiber an elliptic curve. In particular, the map $Q \to D$ has a section $s : D \to Q$. We can use $S^2 \times Q$ as the six-manifold of boundary Y_2'. Picking any point $p \in S^2$, the world-volume of the instanton in $S^2 \times Q$ can be taken to be the two-manifold $p \times s(D)$. We proceed with the argument in the text because it seems illuminating to know how to directly evaluate the η-invariant, rather than relying on a technical argument about cobordism.

To show that n is always even, let n_+ and n_- be the number of positive or negative chirality zero-modes of SU(5) model fermions, coupled to some background SU(5) gauge field on S^4. Then $n = n_+ + n_-$, and this is congruent mod 2 to the index $\mathcal{I} = n_+ - n_-$. The index theorem for SU(5) gauge theory shows that for the SU(5) model fermions, $\mathcal{I}$ is always even (in fact always a multiple of 4), so n is always even.

It is inevitable that in one way or another we were going to have to check that in the SU(5) grand unified model, the number of fermion zero-modes in an instanton field on S^4 is always even. If this number could be odd, it really would represent an anomaly [8, 59].

An alternative to cobordism in reducing to the two special cases of Y_1 and Y_2 is the following. Let $Y' = S^4 \times S^1$ with trivial gauge field. Given any closed five-manifold $\overline{Y}$, to show $\Upsilon_{\overline{Y}} = 1$, let $\widehat{Y} = \overline{Y} + Y'$. Since $\Upsilon_{Y'} = 1$, we have $\Upsilon_{\widehat{Y}} = \Upsilon_{\overline{Y}}$ and we want to show that $\Upsilon_{\widehat{Y}} = 1$. We will reduce to the special cases $\Upsilon_{Y_1} = \Upsilon_{Y_2} = 1$ by showing that $\Upsilon_{\widehat{Y}} = \Upsilon_{Y_1} \Upsilon_{Y_2}$. Since cutting S^4 on the equator decomposes it into two copies of a four-dimensional ball B_4, Y' can be cut to make two copies of $B_4 \times S^1$. The boundary of $B_4 \times S^1$ is $S^3 \times S^1$. Likewise by cutting along the boundary of a tubular neighborhood of $C \subset \overline{Y}$, we can remove from $\overline{Y}$ a copy of $B_4 \times S^1$, leaving behind another manifold $\widetilde{Y}$, also with boundary $S^3 \times S^1$. At this point we have decomposed $\widehat{Y}$ to the disjoint union of $\widetilde{Y}$ and three copies of $B_4 \times S^1$. Of these four manifolds, only one copy of $B_4 \times S^1$ has a nontrivial gauge field. Exchanging that copy of $B_4 \times S_1$ with one of the others and then gluing the pieces back together, we get $Y_1 + Y_2$. In other words, we have described a procedure of cutting and regluing that converts $\widehat{Y} = \overline{Y} + Y'$ into $Y_1 + Y_2$. The gluing law (12.88) for the η-invariant implies that Υ is invariant under cutting and regluing, so we conclude that $\Upsilon_{\widehat{Y}} = \Upsilon_{Y_1} \Upsilon_{Y_2}$. Since invariance under cobordism can essentially be deduced from invariance under cutting and regluing and vice versa [40, 41], it is inevitable that a cobordism argument can be expressed in terms of cutting and regluing.

What happens if (still assuming there is no time-reversal or reflection symmetry) we replace SU(5) by another compact gauge group G? Since $\pi_2(G) = 0$ for any G, we can repeat the above argument verbatim if $\pi_0(G) = \pi_1(G) = \pi_4(G) = 0$. For example, the grand unified theories based on Spin(10) or E_6 satisfy those conditions.[gg] We conclude that if

---

[gg] The Spin(10) theory has a refinement in which we take the symmetry group of the fermions to be $(\text{Spin}(4) \times \text{Spin}(10))/\mathbb{Z}_2$ rather than $\text{Spin}(4) \times \text{Spin}(10)$. Absence of a global anomaly in this case was argued in [60] based on cobordism invariance of the anomaly. It is possible to modify the following argument for this case. See also [61].

G satisfies those conditions and moreover the fermion representation is free of perturbative anomaly and the number of fermion zero-modes in a G-bundle on S^4 is always even, then the model is completely anomaly-free. With some additional knowledge of Lie group topology, one can omit the assumption that $\pi_4(G) = 0$. First of all, with $\pi_0(G) = \pi_1(G) = 0$ but without assuming $\pi_4(G) = 0$, the classification of G-bundles on a five-manifold $\overline{Y}$ is modified only in a very simple way. The only change is that we have to allow for the possibility that a G-bundle over $\overline{Y}$ can be modified in a small neighborhood of a point $p \in \overline{Y}$ by twisting by a nontrivial element of $\pi_4(G)$. This means that in the cobordism analysis, we have to allow a third special case Y_3, namely a five-sphere S^5 with some G-bundle. To proceed farther, we need some specific facts about Lie group topology. For simple G, one has $\pi_4(G) = 0$ except for $G = \mathrm{Sp}(2k)$, which satisfies $\pi_4(\mathrm{Sp}(2k))) = \mathbb{Z}_2$. If G is semi-simple rather than simple, then $\pi_4(G)$ is a product of copies of $\mathbb{Z}_2$, one for each $\mathrm{Sp}(2k)$ factor in G. A nontrivial $\mathrm{Sp}(2k)$-bundle over S^5 associated to the nonzero element of $\pi_4(\mathrm{Sp}(2k))$ can be constructed as follows. Think of an instanton as a particle in five dimensions. Consider an instanton propagating in S^5 in such a way that its worldline is an embedded circle C. To make from this a nontrivial $\mathrm{Sp}(2k)$ bundle, the instanton should undergo a 2π rotation as it propagates around C. Starting with this description of the nontrivial G-bundles on S^5, one can use a cobordism argument — or a cut and paste argument, as in the last paragraph — to show that Υ_{Y_3}, for a G-bundle on S^5, is equal to Υ_{Y_2}, for a corresponding G-bundle on $Y_2 = S^4 \times S^1$. So the condition $\Upsilon_{Y_3} = 1$ does not add anything new. In fact, the mod 2 cobordism invariant associated to $\pi_4(\mathrm{Sp}(2k)) = \mathbb{Z}_2$ is the mod 2 index of the Dirac operator with values in the fundamental representation of $\mathrm{Sp}(2k)$. This invariant is nonzero for $Y_2 = S^4 \times S^1$ with Ramond spin structure around S^1 [59], so in effect our analysis for Y_2 already incorporated the role of $\pi_4(G)$.

The discussion so far demonstrates the general utility of cobordism invariance and a cut and paste argument. It is possible to reorganize the argument in a perhaps more elementary way as a reduction from any connected and simply-connected G to the case $G = \mathrm{SU}(2)$. In doing so, we may assume that G is simple; otherwise, we make the following analysis for each simple factor of G. In this version of the argument, we de-emphasize the role of cobordism invariance and proceed as much as possible using only the fact that when a four-dimensional theory has no perturbative anomaly, the global anomaly is a topological invariant in five dimensions. To make the argument explicit, let us first consider gauge groups $G = \mathrm{Spin}(n), \mathrm{SU}(n)$, and $\mathrm{Sp}(2n)$. The bundle associated to the fundamental representation of G is a real, complex, or quaternionic vector bundle of rank n, respectively.

Let α be 1, 2 or 4 for Spin, SU or Sp, respectively. The real rank of the vector bundle is αn. If $\alpha n > D$, there is a section of the vector bundle which is nonzero everywhere. This is simply because a sufficiently generic section of a vector bundle of rank αn is always nonzero as a function of D variables $x^1, \ldots, x^D$ if $\alpha n > D$. (Locally, such a section is a collection of αn real-valued functions, and generically these functions have no common zero as a function of $D < \alpha n$ real variables.) By taking such a nonzero section, the structure group of a vector bundle can be reduced from Spin(n), SU(n), or Sp($2n$) to Spin($n-1$), SU($n-1$), and Sp($2n-2$), respectively. By repeating this process, the structure group in $D = 5$ dimensions is topologically reduced to Spin(5), SU(2), and Sp(2). But we have Spin(5) $\cong$ Sp(4) and this can be further reduced to Sp(2) $\cong$ SU(2), so we can always reduce the structure group from G to SU(2). Using "obstruction theory" and a knowledge of the homotopy groups $\pi_i(G)$, $i \leq 4$, one can show that this is also possible if G is a connected and simply-connected exceptional Lie group. Therefore, for any G, the global anomaly can always be captured for background fields valued in a subgroup SU(2) $\subset G$. An irreducible representation of SU(2) is either strictly real or pseudoreal. A strictly real representation does not contribute to the anomaly because a mass term is possible in $d = 4$. The exponentiated η-invariant of a pseudoreal representation is given by the mod 2 index in $D = 5$. This vanishes if the number of zero-modes in an instanton field on S^4 is always even. But to prove that this last statement holds for a general five-manifold (and therefore that the anomaly associated to the mod 2 index in $D = 5$ is entirely captured by a counting of zero-modes on S^4) appears to require cobordism or cut and paste arguments such as we have explained above.

At any rate, the conclusion is that a four-dimensional theory with connected and simply-connected gauge group has no anomalies beyond the familiar ones. If one drops the requirement for the gauge group to be connected, then, as in $d = 2$, there definitely are new anomalies.

Acknowledgments

We thank R. Mazzeo for some discussions. The work of KY is supported by JSPS KAKENHI Grant-in-Aid (Wakate-B), No. 17K14265. Research of EW is supported in part by NSF Grant PHY-1911298.

References

[1] S. L. Adler, Axial vector vertex in spinor electrodynamics, *Phys. Rev.* **177**, 2426 (1969); doi:10.1103/PhysRev.177.2426.

[2] J. S. Bell and R. Jackiw, A PCAC puzzle: $\pi^0 \to \gamma\gamma$ in the σ model, *Nuovo Cim. A* **60**, 47 (1969); doi:10.1007/BF02823296.

[3] E. Witten, Nonsupersymmetric D-branes and the kitaev fermion chain, *Lecture at the Shoucheng Zhang Memorial Workshop*, Available at https://glam.stanford.edu/sites/g/files/sbiybj10026/f/may3-1-2_witten.pdf.

[4] R. Jackiw, Topological investigations of quantized gauge theories, in B. S. DeWitt *et al.* (eds)., *Relativity, Groups, and Topology, II*, Les Houches 1983, Reprinted in updated form in S. B. Treiman *et al.* (eds)., *Current Algebra and Anomalies* (World Scientific, 1985).

[5] B. Zumino, Chiral anomalies in differential geometry, in B. S. DeWitt *et al.* (eds)., *Relativity, Groups, and Topology, II*, Les Houches 1983, Reprinted in S. B. Treiman *et al.* (eds)., *Current Algebra and Anomalies* (World Scientific, 1985).

[6] R. Stora, Algebraic structure and topological origin of anomalies, in *Progress in Gauge Field Theory* (Plenum, 1984).

[7] C. G. Callan, Jr. and J. A. Harvey, Anomalies and fermion zero-modes on strings and domain walls, *Nucl. Phys. B 250*, 427 (1985).

[8] E. Witten, An SU(2) Anomaly, *Phys. Lett. B* **117**, 324, (1982).

[9] E. Witten, Global gravitational anomalies, *Commun. Math. Phys.* **100**, 197 (1985); doi:10.1007/BF01212448.

[10] M. F. Atiyah, V. K. Patodi and I. M. Singer, Spectral asymmetry and Riemannian geometry 1, *Math. Proc. Cambridge Philos. Soc.* **77**, 43 (1975); doi:10.1017/S0305004100049410.

[11] X. Z. Dai and D. S. Freed, Eta invariants and determinant lines, *J. Math. Phys.* **35**, 5155 (1994) Erratum: [*J. Math. Phys.* **42**, 2343 (2001)]; doi:10.1063/1.530747 [hep-th/9405012].

[12] K. Yonekura, Dai–Freed theorem and topological phases of matter, *J. High Energy Phys.* **1609**, 022 (2016); doi:10.1007/JHEP09 (2016)022 [arXiv:1607.01873 [hep-th]].

[13] E. Witten, World-sheet corrections via *D*-instantons, preprint (2000), arXiv:hep-th/9907041.

[14] E. Witten, Fermion path integrals and topological phases, *Rev. Mod. Phys.* **88**(3), 035001 (2016); doi:10.1103/RevModPhys.88.035001, [arXiv:1508.04715 [cond-mat.mes-hall]].

[15] N. Redlich, Gauge noninvariance and parity violation of three-dimensional fermions, *Phys. Rev. Lett.* **52**, 18 (1984).

[16] L. Alvarez-Gaume, S. Della Pietra and G. W. Moore, Anomalies and odd dimensions, *Ann. Phys.* **163**, 288 (1985); doi:10.1016/0003-4916(85)90383-5.

[17] D. S. Freed and M. J. Hopkins, M-theory anomaly cancellation, preprint (2019); arXiv:1908.09916 [hep-th].

[18] X.-L. Qi, T. Hughes and S.-C. Zhang, Topological field theory Of time-reversal invariant insulators, *Phys. Rev. B 78*, 195424 (2008); arXiv:0802.3537.

[19] X. Chen, Z.-C. Gu, Z.-X. Liu and X.-G. Wen, Symmetry-protected topological orders in interacting bosonic systems, *Science* **338**, 1604 (2012).

[20] X. Chen, Z.-C. Gu, Z.-X. Liu and X.-G. Wen, Symmetry protected topological orders and the group cohomology of their symmetry group, *Phys. Rev. B 87*, 155114 (2013).

[21] T. Senthil, Symmetry protected topological phases of quantum matter, preprint (2014), arXiv:1405.4015.

[22] D. S. Freed, Anomalies and invertible field theories, *Proc. Symp. Pure Math.* **88**, 25 (2014); doi: 10.1090/pspum/088/01462[arXiv:1404.7224[hep-th]].

[23] S. Monnier, A modern point of view on anomalies, preprint (2019); arXiv:1903.02828.

[24] M. Kurkov and D. Vassilevich, Parity anomaly in four dimensions, *Phys. Rev. D* **96**(2), 025011 (2017); doi:10.1103/PhysRevD.96.025011 [arXiv:1704.06736[hep-th]].

[25] M. Kurkov and D. Vassilevich, Gravitational parity anomaly with and without boundaries, *J. High Energy Phys.* **1803**, 072 (2018); doi:10.1007/JHEP03(2018)072 [arXiv:1801.02049[hep-th]].

[26] I. Fialkovsky, M. Kurkov and D. Vassilevich, Quantum dirac fermions in a half-space and their interaction with an electromagnetic field, *Phys. Rev. D* **100**(4), 045026 (2019); doi:10.1103/PhysRevD.100.045026 [arXiv:1906.06704 [hep-th]].

[27] M. F. Atiyah and I. M. Singer, Dirac operators coupled to vector potentials, *Proc. Natl. Acad. Sci. USA* **81**, 2597 (1984); doi:10.1073/pnas.81.8.2597.

[28] M. F. Atiyah and I. M. Singer, The index of elliptic operators, I, *Ann. Math.* **87**, 484 (1968).

[29] H. Fukaya, T. Onogi and S. Yamaguchi, Atiyah–Patodi–Singer index from the domain-wall fermion dirac operator, *Phys. Rev. D* **96**(12), 125004 (2017); doi:10.1103/PhysRevD.96.125004 [arXiv:1710.03379 [hep-th]].

[30] A. Dabholkar, D. Jain and A. Rudra, APS η-invariant, path integrals, and mock modularity, preprint (2019); arXiv:1905.05207.

[31] M. Hortacsu, K. D. Rothe and B. Schroer, Zero energy eigenstates for the dirac boundary problem, *Nucl. Phys. B* **171**, 530 (1980); doi:10.1016/0550-3213(80)90384-3.

[32] D. Belov and G. W. Moore, Conformal blocks for AdS(5) singletons, preprint (2014); arXiv:hep-th/0412167.

[33] E. Witten, On *S*-duality in abelian gauge theory, *Selecta Math.* **1**, 383 (1995); doi:10.1007/BF01671570 [hep-th/9505186].

[34] N. Seiberg, Y. Tachikawa and K. Yonekura, Anomalies of duality groups and extended conformal manifolds, *Prog. Theor. Exp. Phys.* **2018**(7), 073B04 (2018); doi:10.1093/ptep/pty069 [arXiv:1803.07366 [hep-th]].

[35] C. T. Hsieh, Y. Tachikawa and K. Yonekura, On the anomaly of the electromagnetic duality of the maxwell theory, preprint (2019); arXiv:1905.08943 [hep-th].

[36] E. Witten, The 'Parity' anomaly on an unorientable manifold, *Phys. Rev. B* **94**(19), 195150 (2016); doi:10.1103/PhysRevB.94.195150 [arXiv:1605.02391 [hep-th]].

[37] D.-E. Diaconescu, G. W. Moore and E. Witten, E_8 gauge theory, and a derivation of K-theory from *M*-theory, *Adv. Theor. Math. Phys.* **6**, 1031 (2003).

[38] D. S. Freed and G. W. Moore, Setting the quantum integrand of *M*-theory, *Commun. Math. Phys.* **263**, 89 (2006); [hep-th/0409135].

[39] A. Kapustin, R. Thorngren, A. Turzillo and Z. Wang, Fermionic symmetry protected topological phases and cobordisms, *J. High Energy Phys.* **1512**, 052 (2015) [*J. High Energy Phys.* **1512**, 052 (2015)] doi:10.1007/JHEP12(2015)052 [arXiv:1406.7329 [cond-mat.str-el]].

[40] D. S. Freed and M. J. Hopkins, Reflection positivity and invertible topological phases, preprint (2016); arXiv:1604.06527 [hep-th].

[41] K. Yonekura, On the cobordism classification of symmetry protected topological phases, *Commun. Math. Phys.* **368**(3), 1121 (2019); doi:10.1007/s00220-019-03439-y [arXiv:1803.10796 [hep-th]].

[42] C. Closset, T. Dumitrescu, G. Festuccia, Z. Komargodski and N. Seiberg, Comments on chern–simons contact terms in three dimensions, *J. High Energy Phys.* **1209**, 091 (2012); arXiv:1206.5218.

[43] E. Witten, Global anomalies in string theory, in W. A. Bardeen and A. R. White (eds)., *Symposium on Anomalies, Geometry, Topology* (World Scientific, 1986), Available at https://www.sns.ias.edu/sites/default/files/files/global-anomalies-in-stringtheory-1985.pdf.

[44] A. Y. Kitaev, Unpaired majorana fermions in quantum wires, *Phys. Usp.* **44**(10S), 131 (2001); arXiv:cond-mat/0010440 [cond-mat.mes-hall].

[45] A. Kapustin and N. Seiberg, Coupling a QFT to a TQFT and duality, *J. High Energy Phys.* **04**, 001 (2014); arXiv:1401.0740 [hep-th].

[46] R. Dijkgraaf and E. Witten, Developments in topological gravity, preprint (2018); arXiv:1804.03275 [hep-th].

[47] D. Stanford and E. Witten, JT gravity and the ensembles of random matrix theory, preprint (2019); arXiv:1907.03363 [hep-th].

[48] J. Kaidi, J. Parro-Martinez and Y. Tachikawa, GSO projections via SPT phases, preprint (2019); arXiv:1908.04805 [hep-th].

[49] L. Fidkowski and A. Kitaev, The effects of interactions on the topological classification of free fermion systems, *Phys. Rev. B 81*, 134509, (2010); arXiv:0904.2197.

[50] C. Rosenberg and M. Franz, Witten effect in a crystalline topological insulator, preprint (2010); arXiv:1001.3179.

[51] F. Gliozzi, D. Olive and J. Scherk, Supersymmetry, supergravity theories, and the dual spinor model, *Nucl. Phys. B 122*, 253 (1977).

[52] X.-L. Qi, A new class of $(2 + 1)$-dimensional topological superconductors with $\mathbb{Z}_8$ topological classification, *New J. Phys.* **15**, 065002 (2013); [arXiv:1202.3983].

[53] H. Yao and S. Ryu, Interaction effect on topological classification of superconductors in two dimensions, *Phys. Rev. B 88*, 064507 (2013); [arXiv:1202.5805].

[54] Z.-C. Gu and M. Levin, Effect of interactions on two-dimensional fermionic symmetry-protected topological phases with $\mathbb{Z}_2$ symmetry, *Phys. Rev. B 89*, 201113 (2014); [arXiv:1304.4569].

[55] Y. Tachikawa and K. Yonekura, Why are fractional charges of orientifolds compatible with dirac quantization? preprint (2018); arXiv:1805.02772 [hep-th].

[56] H. Georgi and S. L. Glashow, Unity of all elementary particle forces, *Phys. Rev. Lett.* **32**, 438 (1974).

[57] D. S. Freed, Pions and generalized cohomology, *J. Differential Geom.* **80**, 45 (2008); *J. High Energy Phys.* arXiv:hep-th/0607134.

[58] I. Garcia-Etxebarria and M. Montero, Dai–Freed anomalies in particle physics, *J. High Energy Phys.* **1908**, 003 (2019); doi:10.1007/JHEP08(2019)003 [arXiv:1808.00009 [hep-th]].

[59] J. Wang, X. G. Wen and E. Witten, A new SU(2) anomaly, *J. Math. Phys.* **60**(5), 052301 (2019); doi:10.1063/1.5082852 [arXiv:1810.00844 [hep-th]].

[60] J. Wang and X. G. Wen, A nonperturbative definition of the standard models, preprint (2018); arXiv:1809.11171 [hep-th].

[61] Z. Wan and J. Wang, Higher anomalies, higher symmetries, and cobordisms I: Classification of higher-symmetry-protected topological states and their boundary fermionic/bosonic anomalies via a generalized cobordism theory, preprint (2018); arXiv:1812.11967 [hep-th].

Chapter 13

Detection of the Orbital Hall Effect by the Orbital–Spin Conversion

Jiewen Xiao, Yizhou Liu and Binghai Yan*

*Department of Condensed Matter Physics,
Weizmann Institute of Science, Rehovot 7610001, Israel*

*binghai.yan@weizmann.ac.il

The intrinsic orbital Hall effect (OHE), the orbital counterpart of the spin Hall effect, was predicted and studied theoretically for more than one decade, yet to be observed in experiments. Here we propose a strategy to convert the orbital current in OHE to the spin current via the spin–orbit coupling from the contact. Furthermore, we find that OHE can induce large nonreciprocal magnetoresistance when employing the magnetic contact. Both the generated spin current and the orbital Hall magnetoresistance can be applied to probe the OHE in experiments and design orbitronic devices.

13.1 Introduction

The intrinsic orbital Hall effect (OHE), where an electric field induces a transverse orbital current, was proposed by the Zhang group [1] soon after the prediction of the intrinsic spin Hall effect (SHE) [2, 3]. The SHE was soon observed [4,5], later applied for the spintronic devices [6] and references therein, and also led to the seminal discovery of the quantum SHE, i.e., the 2D topological insulator [7, 8]. Different from the SHE, the OHE does not rely on the spin–orbit coupling (SOC), and thus, it was predicted to exist in many materials [1, 9–18] with either weak or strong SOC, for example, in metals Al, Cu, Au, and Pt.

In an OHE device, the transverse orbital current leads to the orbital accumulation at transverse edges, similar to the spin accumulation in a SHE device. Zhang *et al* [1] proposed to measure the edge orbital accumulation by the Kerr effect. Recently, Ref. [19] predicted the orbital torque generated by the orbital current. However, the OHE is yet to be detected in experiments until today. The detection of the orbital is rather challenging, because the orbital is highly nonconserved compared to the spin, especially at the device boundary.

A very recent work by us proposed [20] that the longitudinal current through DNA-type chiral materials is orbital-polarized, and contacting DNA to a large-SOC material can transform the orbital current into the spin current. Thus, we are inspired to conceive a similar way to detect the transverse OHE by converting the orbital to the spin by the SOC proximity.

In this chapter, we propose two ways to probe the OHE, where the strong SOC from the contact transforms the orbitronic problem to the spintronic measurement. One way is to generate spin current or spin polarization from the transverse orbital current by connecting the edge to a third lead with the strong interfacial SOC. Then the edge spin polarization and spin current is promising to be measured by the Kerr effect [4] and the inverse SHE [21–23], respectively. The other way is to introduce a third magnetic lead and measure the magnetoresistance. We call it the orbital Hall magnetoresistance (OHMR), similar to the spin Hall magnetoresistance [24,25]. In our proposal, the OHE refers to orbitals that resemble atomic-like orbitals, which naturally couple to the spin via the atomic SOC. We first demonstrate detection principles in a lattice model by transport calculations. Then we incorporate these principles into the metal copper, which has negligible SOC and avoids the coexistence of the SHE, as a typical example of realistic materials. In the copper-based device, we demonstrate the resultant spin polarization/current and very large OHMR (0.3–1.3%), which are measurable by present experiment techniques.

13.2   Results and discussions

13.2.1   *Methods and general scenario*

To detect the OHE, we introduce an extra contact with the strong SOC on the boundary of the OHE material, as shown in Fig. 13.1. This device can act for both two-terminal (2T) and three-terminal (3T) measurements (or more terminals). In theoretical calculations, we completely exclude SOC from all leads so that we can well define the spin current. We also remove SOC in the OHE material, the device regime in the center, to avoid the existence of SHE. Only finite atomic SOC is placed in the interfacial region (highlighted by yellow in Fig. 13.1) between the OHE and the third lead.

We first prove the principle by a simple square-lattice model that hosts OHE. As shown in the inset of Fig. 13.2(a), a tight-binding spinless model is constructed, with three orbitals s, p_x and p_y assigned to each site. Under the

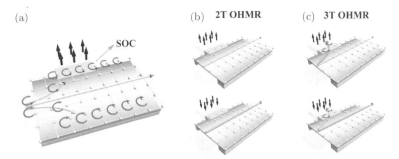

Fig. 13.1. Illustration of the orbital–spin conversion and the orbital Hall magnetoresistance (OHME). (a) The orbital Hall effect and the spin polarization/current generation. Opposite orbitals (red and blue circular arrows) from the left lead deflect into opposite boundaries. The red and blue backgrounds represent the orbital accumulation at two sides. Because of the SOC region (yellow) at one side, the orbital current is converted into the spin current (indicated by black arrows). (b) The two-terminal (2T) OHMR. The third lead is magnetized but open. (c) The three-terminal (3T) OHMR. The third lead is magnetized and conducts current. The 2T/3T conductance between different leads relies on the magnetization sensitively. The thickness of gray curves represent the relative magnitude of the conductance.

above basis, the atomic orbital angular momentum operator $\hat{L}_z$ is written as

$$\hat{L}_z = \hbar \begin{bmatrix} 0 & 0 & 0 \\ 0 & 0 & -i \\ 0 & i & 0 \end{bmatrix}, \tag{13.1}$$

and three eigenstates $p_\pm \equiv (p_x \pm ip_y)/\sqrt{2}, s$ correspond to eigenvalues $L_z = \pm 1, 0$, respectively. After considering the nearest-neighboring hopping, the Hamiltonian is written as

$$H(k_x, k_y) = \begin{pmatrix} E_s + 2t_s \cos k_x a & -2it_{sp} \sin k_x a & -2it_{sp} \sin k_y a \\ \quad + 2t_s \cos k_y a & & \\ 2it_{sp} \sin k_x a & E_{px} + 2t_{p\sigma} \cos k_x a & 0 \\ & \quad + 2t_{p\pi} \cos k_y a & \\ 2it_{sp} \sin k_y a & 0 & E_{py} + 2t_{p\pi} \cos k_x a \\ & & \quad + 2t_{p\sigma} \cos k_y a \end{pmatrix},$$

$$\tag{13.2}$$

where E_s, E_{px} and E_{py} are onsite energies of s-, p_x- and p_y-orbitals t_s, $t_{p\sigma}$, $t_{p\pi}$, t_{sp} are electron hopping integrals between s-orbitals, σ-type oriented p-orbitals, π-type oriented p-orbitals, and s- and p-orbitals, respectively. In the following calculations, their values are specified as $E_s = 1.3$, $E_{px} = E_{py} = -1.9$, $t_s = -0.3$, $t_{p\sigma} = 0.6$, $t_{p\pi} = 0.3$, and $t_{sp} = 0.5$, in the unit of eV.

To realize the OHE, it requires the inter-orbital hopping to induce the transverse L_z current. Since p_x- and p_y-orbitals are orthogonal under the square lattice geometry, the inter-orbital hopping t_{sp} becomes the critical parameter that controls the existence of the OHE. Then we introduce the atomic SOC on the boundary to demonstrate the OHE detection by $\lambda_{soc} \hat{S}_z \cdot \hat{L}_z$, where $\hat{S}_z$ is the spin operator.

We estimate the OHE conductivity (σ_{OH}) with the orbital Berry curvature in the Kubo formula [26, 27],

$$\sigma_{OH} = \frac{e}{\hbar} \sum_n \int \frac{d^3 k}{(2\pi)^3} f_{nk} \Omega_n^{L_z}(k), \tag{13.3}$$

$$\Omega_n^{L_z}(k) = 2\hbar^2 \sum_{m \neq n} \text{Im} \left[\frac{\langle u_{nk} | j_y^{L_z} | u_{mk} \rangle \langle u_{mk} | \hat{v}_x | u_{nk} \rangle}{(E_{nk} - E_{mk})^2} \right], \tag{13.4}$$

where $\Omega_n^{L_z}(k)$ is the "orbital" Berry curvature for the nth-band with Bloch state $|u_{nk}\rangle$ and energy eigenvalue E_{nk}. f_{nk} is the Fermi–Dirac distribution function. v_x is the x-component of the band velocity operator while $j_y^{L_z}$ is the orbital current operator in the y-direction, defined as $j_y^{L_z} = (\hat{L}_z \hat{v}_y + \hat{v}_y \hat{L}_z)/2$. Therefore, the above formula indicates that the interband perturbation induces the orbital Berry curvature, further reiterating the importance of inter-orbital hopping. We also note that, the orbital Berry curvature is even under the time-reversal symmetry or the spatial inversion symmetry, $\Omega_n^{L_z}(k) = \Omega_n^{L_z}(-k)$.

For the device schematically presented in Fig. 13.1, we calculated the conductance by the Landauer–Büttiker formula [28] with the scattering matrix from lead i to lead j,

$$G^{i \to j} = \frac{e^2}{h} \sum_{n \in j, m \in i} |S_{nm}|^2, \tag{13.5}$$

where S_{mn} is the scattering matrix element from the mth eigenstate in lead i to the nth eigenstate in lead j. In all three leads ($i, j = 1, 2, 3$), spin ($S_z = \uparrow\downarrow$) is a conserved quantity because of the lack of SOC. We turn off the inter-orbital hopping in leads so that L_z is also conserved, i.e., L_z commutes with the Hamiltonian (see Supplementary Materials). Therefore, with the spin and orbital conserved leads, we can specify the conductance in each S_z and L_z channel, and define the orbital- and spin-polarized conductance as:

$$G_{S_z}^{ij} = G^{i \to j\uparrow} - G^{j \to j\downarrow}, \tag{13.6}$$

$$G_{L_z}^{ij} = G^{i \to j+} - G^{i \to j-}, \tag{13.7}$$

where $G^{ij}_{S_z(L_z)}$ is the conductance from lead i to the $S_z(L_z)$ channel of lead j. $G^{i \to j0}$ is omitted here since $L_z = 0$ contributes no polarization. We performed the conductance calculations with the quantum transport package Kwant [29].

As illustrated in Fig. 13.1(a), electrons with the opposite orbital angular momentum deflect into transverse directions in the OHE region, resulting in the transverse orbital current. Therefore, orbital accumulates at two sides, and the orbital polarization emerges. To detect the orbital polarization, atomic SOC is added at one side, as highlighted by yellow in Fig. 13.1(a). After electron deflecting into the SOC region, the right-handed orbital (red circular arrows) is converted to the up spin polarization. If a third lead is further attached, the SOC region converts the orbital current into the spin current. If the third lead exhibits magnetization along z (M_z) (Figs. 13.1(b) and 13.1(c)), inversely, the OHE induces the OHMR, relying on whether M_z is parallel or anti-parallel to the generated spin polarization. In the 2T measurement (Fig. 13.1(b)), the conductance from lead 1 to lead 2 ($G_{1 \to 2}$) changes when the M_z-direction is reversed, and the changing direction of $G_{1 \to 2}$ depends sensitively on the size of the device, due to the complex orbital accumulation and reflection with an open lead. While for its spin counterpart, the SHE-induced magnetoresistance is commonly measured in a 2T setup [24,25]. In the 3T device (Fig. 13.1(c)), the situation is simpler since the transverse orbital current can flow into the third lead. If M_z and spin polarization is parallel (anti-parallel), the transverse orbital current matches (mismatches) the lead magnetization, resulting in the high (low) $G_{1 \to 3}$ and low (high) $G_{1 \to 2}$ accordingly. We point out that the 3T measurement is usually more favorable than 2T, since the 3T device avoids the 2T reciprocity constrain [30] and the conductance change [$\Delta G = G(M_z) - G(-M_z)$] is also relatively larger in the third lead, as discussed in the following.

13.2.2 *Spin polarization and spin current generated by the OHE*

The band structure weighted by the orbital Berry curvature for the square lattice is plotted in Fig. 13.2(a). The highest band corresponds to the s-orbital dispersion, while two lower bands are dominated by p-orbitals. The orbital Berry Curvature concentrates near the Γ-point, M-point and Γ–M line in the Brillouin zone, where band hybridization is strong. After integrating Ω_{Lz} in the Brillouin zone, the orbital Hall conductivity is derived and presented in Fig. 13.2(a). It shows that, due to the inter-orbital hopping t_{sp}, states below (p-orbitals) and above (s-orbital) Fermi level both exhibit

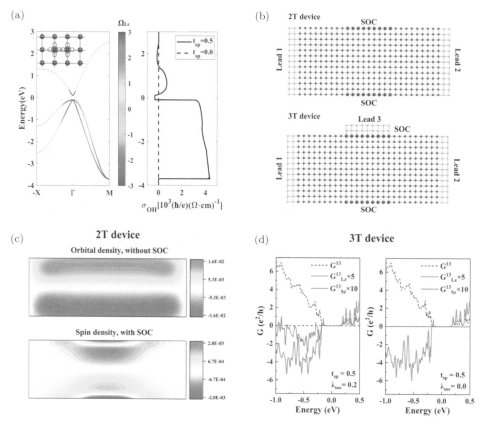

Fig. 13.2.   Orbital–spin conversion in the two terminal (2T) and three terminal (3T) device. (a) Band structure of the square lattice with $t_{sp} = 0.5$ eV (left) and the orbital Hall conductivity with $t_{sp} = 0.5$ eV and $t_{sp} = 0.0$ eV (right). In the inset, the tight-binding model of the square lattice is presented. (b) 2T and 3T detection devices, where larger spheres at two sides represent SOC regions. The yellow spheres at left, right and upper sides represent leads. (c) Orbital and spin density distribution in the 2T setup, at the energy level of 0.2 eV. (d) Total, orbital and spin conductance from lead 1 to lead 3 with (left) and without (right) SOC.

significant σ_{OH}. However, if t_{sp} is turned off so that L_z is conserved, both Ω_{Lz} and σ_{OH} vanish.

Based on the square lattice with finite t_{sp}, the 2T device is constructed, as shown in Fig. 13.2(b). Without SOC at two sides, the orbital density distribution is plotted in Fig. 13.2(c), which shows that opposite orbitals accumulate and polarize at two boundaries. With SOC turned on, spin density appears and largely concentrates on the local SOC atoms, which is promising to be detected by the Kerr effect [4]. Since SOC couples the p_+ (p_-) orbital to the $\uparrow$ ($\downarrow$) spin and forms the $\left| j_m = \frac{3}{2} \right\rangle$ ($\left| j_m = -\frac{3}{2} \right\rangle$) state, the spin density near the SOC region largely follows the orbital density pattern: positive at the upper side and negative at the lower side. To verify that the spin polarization is directly induced by the OHE rather than SOC, we turned off the OHE by setting $t_{sp} = 0$ eV and preserve the SOC at the interface. The supplementary Fig. S2 shows that both the orbital and spin polarization disappear.

On the basis of 2T device, a third lead is attached to the SOC side to form a 3T device, as shown in Fig. 13.2(b). Therefore, rather than the orbital accumulation, the orbital current will flow into the third lead and generate the spin current. Figure 13.2(d) shows that the orbital current from lead 1 to lead 3 ($G_{L_z}^{13}$) exists with and without SOC at the interface. For instance, for states above Fermi level, $L_z = +1$ states are more easily transported into lead 3 than $L_z = -1$ states, and thus polarizes the lead, being consistent with the positive orbital polarization at the upper side in Fig. 13.2(c). On the other hand, for the spin conductance, it only appears when turning on SOC, and the energy dependence of $G_{S_z}^{13}$ largely follows the orbital conductance, further demonstrating the spin generation process from the orbital. If we increase the SOC strength, $G_{S_z}^{13}$ increases accordingly, because of the higher orbital–spin conversion efficiency (see Fig. S3). We also test orbital nonconserved leads with nonzero t_{sp}, whose spin conductance remains the similar feature (see Fig. S4).

13.2.3 Orbital Hall magnetoresistance

As discussed above, the current injected into lead 3 is spin polarized. When lead 3 is magnetized along the z-axis, we expect the existence of magnetization-dependent conductance, i.e., $G^{13}(M_z) \neq G^{13}(-M_z)$. From the current conservation [30], we deduce the relation,

$$\Delta G^{13} = -\Delta G^{12}, \tag{13.8}$$

where $\Delta G^{ij} \equiv G^{ij}(M_z) - G^{ij}(-M_z)$. To demonstrate this, M_z is introduced to lead 3 as an exchange field to the spin, as shown in the 3T setup in Fig. 13.3(a). Results in Figs. 13.3(c) and 13.3(d) indicate that ΔG^{12} and ΔG^{13} can reach several percentage of the total conductance at some energies. We also confirm that ΔG^{12} and ΔG^{13} are proportional to the exchange field strength (see Fig. S5).

To understand the orbital-induced magnetoresistance, the spin and orbital conductances from lead 1 to lead 3 are calculated. As shown in Fig. 13.3(e), $G_{S_z}^{13}$ almost changes its sign when flipping M_z in lead 3, as expected, and $G_{S_z}^{13}$ now inversely affects $G_{L_z}^{13}$ because of the interfacial SOC.

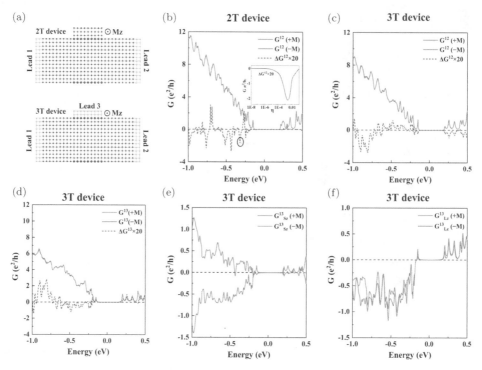

Fig. 13.3.   Orbital magnetoresistance with magnetic leads. (a) 2T and 3T detection devices with the exchange field $\pm M_z$ in the open and conducting lead 3. Larger spheres represent the interfacial SOC region. (b) Total conductance from lead 1 to lead 2 in $\pm M_z$ field in the 2T device, with the dephasing term η set to 0.001. In the inset, the dephasing dependent ΔG_{12} for the peak inside the circle is presented. (c) Total conductance from lead 1 to lead 2 in $\pm M_z$ field in the 3T device. (d) Total, (e) spin and (f) orbital conductance from lead 1 to lead 3 in $\pm M_z$ field in the 3T device. In all these calculations, λ_{SOC} and M_z is set to 0.2 eV and 0.4 eV, respectively.

When further comparing Figs. 13.3(d) and 13.3(f), we found that the change of the magnitude of $G_{L_z}^{13}$ is proportional to the change of total conductance ΔG^{13}. Therefore, it verifies the scenario in Fig. 13.1(c): when the orbital matches the spin in magnetic leads, $G_{L_z}^{13}$ and thus G^{13} is higher while G^{12} is accordingly lower. Thus, it indicates the essential role of the orbital in connecting charge and spin in the transport.

However, the 2T results exhibit qualitatively different features from the 3T results. According to the reciprocity relation [30], the 2T conductance obeys $G^{12}(M_z) = G^{12}(-M_z)$. Only when the current conservation is broken, one may obtain the 2T magentoresistance. Therefore, we introduce a dephasing term $i\eta$ to leak electrons into virtual leads [31] to release the above constrain. As shown in the inset of Fig. 13.3(b), the ΔG^{12} is zero at $\eta = 0$, first increases quickly and soon decreases as further increasing η.

In the large η limit, the system is totally out of coherence and thus, the conductance cannot remember the spin and orbital information. We note that the dephasing exists ubiquitously in experiments due to the dissipative scattering for example by electron–phonon interaction and impurities.

For the same M_z, the 2T ΔG^{12} (Fig. 13.3(b)) roughly exhibits the opposite sign compared to the 3T ΔG^{12} (Fig. 13.3(c)) in the energy window investigated. Unlike that ΔG^{13} follows the change of $G^{13}_{L_z}$ (Fig. 13.3(f)), the change direction of G_{12} depends on the geometry of the 2T device (see Fig. S6). The magnitude of the 2T ΔG^{12} also depends sensitively on the value of η. Its peak value (Fig. 13.3(b)), with η around 0.001, is comparable with the 3T value in the same parameter regime. However, the 3T conductance avoids the strict constrain of the 2T reciprocity, and the existence of 3T OHMR does not rely on the dephasing. Furthermore, in the 3T setup, the magnetoresistance ratio $\Delta G_{13}/G_{13}$ in the third lead is also larger than $\Delta G_{12}/G_{12}$, because of the lower total conductance of G_{13}. Therefore, we propose that the 3T setup may be more advantageous to detect the OHE.

13.2.4 *Realistic material Cu*

Based on the simple square lattice model, we demonstrate two main phenomena, the OHE-induced spin polarization/spin current current assisted by the atomic SOC on the boundary and the existence of OHMR. We further examine them in a realistic material Cu. This light noble metal is predicted to exhibit the strong OHE.

As shown in Fig. 13.4(a), the 2T (without lead 3) and 3T devices are composed of Cu (without SOC) in both the scattering region and leads, and the heavy metal Au (with SOC) at two boundaries. We adopted the tight-binding method to describe the Cu and leads, where 9 atomic orbitals (s, p_x, p_y, p_z, d_{z^2}, $d_{x^2-y^2}$, d_{xy}, d_{yz}, d_{zx}) are assigned to each site. The nearest-neighboring and the second-nearest-neighboring hoppings are considered with the Slater–Koster-type parameters from Ref. [32]. For the heavy metal Au at two sides, the SOC strength is set to 0.37 eV as suggested by Ref. [32]. With the tight-binding approach, the first-principles band structure of Cu is reproduced (see Fig. S7).

As shown in Fig. 13.4(b), the orbital Berry curvature concentrates on the d-orbital region (-4 eV to -2 eV) due to the orbital hybridization, consistent with previous works [12, 15]. After integrating Ω_{L_z}, Fig. 13.4(b) shows that the orbital Hall conductivity is around 6000 $(\hbar/e)(\Omega cm)^{-1}$ in the d-orbital region, even larger than the spin Hall conductivity of Pt. Near the Fermi level, the orbital Hall conductivity is determined by the s-orbital derived bands and reduces to around 1000 $(\hbar/e)(\Omega cm)^{-1}$.

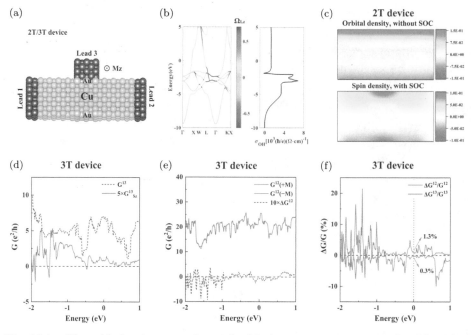

Fig. 13.4.　The orbital–spin conversion and orbital magnetoresistance in Cu. (a) 2T/3T device based on real materials, where the scattering region and leads are treated as Cu and the SOC region is treated as Au. (b) The band structure of Cu, weighted by the orbital Berry curvature Ω_{Lz}, and the corresponding orbital Hall conductivity. (c) Orbital and spin density distribution at the Fermi level in the 2T setup. (d) Spin conductance from lead 1 to lead 3 in nonmagnetic leads in the 3T setup. (e) Total conductance from lead 1 to lead 2 in $\pm M_z$ exchange field in the 3T setup. (f) Magnetoresistance ratio $\Delta G^{12}/G^{12}$ and $\Delta G^{13}/G^{13}$ in the 3T setup.

For the 2T device, the orbital and spin density at Fermi level are plotted in Fig. 13.4(c). The orbital polarization exists at two sides as a consequence of the OHE. With the heavy metal Au attached, spin polarization is generated, which concentrates on Au atoms and follows the orbital density pattern. To confirm that the spin polarization is induced by the OHE, we artificially turn off the inter-orbital hopping in Cu to eliminate the OHE, but still keep the SOC in the Au region. Result show that both the orbital and spin polarization disappear (see Fig. S8), in accordance with our prediction.

For the 3T device, we add a third Cu lead to one SOC side and calculate the spin conductance from lead 1 to lead 3 ($G_{S_z}^{13}$). As shown in Fig. 13.4(d), the generated spin conductance displays an energy-dependence similar to the bulk σ_{OH}. Near the Fermi level, the spin-polarization rate can reach 4%, and it is even around 20% in the d-orbital region. Therefore, a sizable spin current can also be generated from the OHE by adding an interfacial SOC layer. Similarly, when artificially switching off the OHE of Cu but keeping the Au part, the spin current disappears, eliminating the contribution of the SHE brought by the thin Au layer (see Fig. S9).

We also studied the OHMR by applying an exchange field M_z in the lead 3. We choose $M_z = 0.95$ eV according to the approximate spin splitting in the transition metal Co (see Fig. S10). As shown in Figs. 13.4(e) and 13.4(f), the 3T OHMR is rather large, where we find $\Delta G^{12}/G^{12} \approx 0.3\%$ and $\Delta G^{13}/G^{13} \approx 1.3\%$ at the Fermi level. In experiment, the SHE magentoresistance is around 0.05–0.5% (see Ref. [33] for example). Therefore, the sizable OHMR in copper can be fairly measurable by present experimental techniques. We should point out that similar effects can be generalized to other OHE materials like Li and Al [15].

13.3 Summary

In summary, we have proposed the OHE detection strategies by converting the orbital to spin by the interfacial SOC, and inducing the strong spin current/polarization. Inversely, the OHE can also generate the large nonreciprocal magnetoresistance when employing the magnetic contact. We point out that, compared to the two-terminal one, the three-terminal OHMR does not require the dephasing term, and may be more advantageous to detect the OHE. Using the device setup based on the metal Cu, we demonstrate that the generated spin polarization and OHMR are strong enough to be measured in the present experimental condition. Our work will pave a way to realize the OHE in experiment, and further design orbitronic or even orbitothermal devices for future applications.

Acknowledgments

We honor the memory of Prof. Shoucheng Zhang. This chapter follows his earlier works on the intrinsic orbital Hall effect and spin Hall effect. B.Y. acknowledges the financial support by the Willner Family Leadership Institute for the Weizmann Institute of Science, the Benoziyo Endowment Fund for the Advancement of Science, Ruth and Herman Albert Scholars Program for New Scientists, and the European Research Council (ERC) under the European Union's Horizon 2020 research and innovation programme (Grant No. 815869, NonlinearTopo).

References

[1] B. A. Bernevig, T. L. Hughes and S.-C. Zhang, *Phys. Rev. Lett.* **95**, 066601 (2005).
[2] S. Murakami, N. Nagaosa and S.-C. Zhang, *Science* **301**, 1348 (2003).

[3] J. Sinova, D. Culcer, Q. Niu, N. Sinitsyn, T. Jungwirth and A. H. MacDonald, *Phys. Rev. Lett.* **92**, 126603 (2004).

[4] Y. K. Kato, R. C. Myers, A. C. Gossard and D. D. Awschalom, *Science* **306**, 1910 (2004).

[5] J. Wunderlich, B. Kaestner, J. Sinova and T. Jungwirth, *Phys. Rev. Lett.* **94**, 047204 (2005).

[6] T. Jungwirth, J. Wunderlich and K. Olejník, *Nature Mater.* **11**, 382 (2012).

[7] C. L. Kane and E. J. Mele, *Phys. Rev. Lett.* **95**, 226801 (2005).

[8] B. A. Bernevig, T. L. Hughes and S.-C. Zhang, *Science* **314**, 1757 (2006).

[9] G. Y. Guo, Y. Yao and Q. Niu, *Phys. Rev. Lett.* **94**, 226601 (2005).

[10] H. Kontani, T. Tanaka, D. S. Hirashima, K. Yamada and J. Inoue, *Phys. Rev. Lett.* **100**, 096601 (2008).

[11] H. Kontani, T. Tanaka, D. S. Hirashima, K. Yamada and J. Inoue, *Phys. Rev. Lett.* **102**, 016601 (2008), 0806.0210.

[12] T. Tanaka, H. Kontani, M. Naito, T. Naito, D. S. Hirashima, K. Yamada and J. Inoue, *Phys. Rev. B* **77** (2008), 10.1103/physrevb.77.165117.

[13] I. V. Tokatly, *Phys. Rev. B* **82**, 161404 (2010).

[14] D. Go, D. Jo, C. Kim and H.-W. Lee, *Phys. Rev. Lett.* **121**, 086602 (2018).

[15] D. Jo, D. Go and H.-W. Lee, *Phys. Rev. B* **98**, 214405 (2018).

[16] V. T. Phong, Z. Addison, S. Ahn, H. Min, R. Agarwal and E. J. Mele, *Phys. Rev. Lett.* **123**, 236403 (2019).

[17] L. M. Canonico, T. P. Cysne, A. Molina-Sanchez, R. Muniz and T. G. Rappoport, arXiv:2001.03592 (2020).

[18] S. Bhowal and S. Satpathy, *Phys. Rev. B* **101**, 121112 (2020).

[19] D. Go and H.-W. Lee, *Phys. Rev. Res.* **2**, 013177 (2020).

[20] Y. Liu, J. Xiao, J. Koo and B. Yan, arXiv:2008.08881 (2020).

[21] E. Saitoh, M. Ueda, H. Miyajima and G. Tatara, *Appl. Phys. Lett.* **88**, 182509 (2006).

[22] S. O. Valenzuela and M. Tinkham, *Nature* **442**, 176 (2006).

[23] H. Zhao, E. J. Loren, H. M. van Driel and A. L. Smirl, *Phys. Rev. Lett.* **96**, 246601 (2006).

[24] S. Y. Huang, X. Fan, D. Qu, Y. P. Chen, W. G. Wang, J. Wu, T. Y. Chen, J. Q. Xiao and C. L. Chien, *Phys. Rev. Lett.* **109**, 107204 (2012).

[25] M. Weiler, M. Althammer, F. D. Czeschka, H. Huebl, M. S. Wagner, M. Opel, I.-M. Imort, G. Reiss, A. Thomas, R. Gross and S. T. B. Goennenwein, *Phys. Rev. Lett.* **108**, 106602 (2012).

[26] D. Xiao, M.-C. Chang and Q. Niu, *Rev. Mod. Phys.* **82**, 1959 (2010).

[27] N. Nagaosa, J. Sinova, S. Onoda, A. H. MacDonald and N. P. Ong, *Rev. Mod. Phys.* **82**, 1539 (2010).

[28] M. Büttiker, *Phys. Rev. Lett.* **57**, 1761 (1986).

[29] C. W. Groth, M. Wimmer, A. R. Akhmerov and X. Waintal, *New J. Phys.* **16**, 063065 (2014).

[30] M. Büttiker, *IBM J. Res. Develop.* **32**, 317 (1988).

[31] M. Büttiker, *Phys. Rev. B* **33**, 3020 (1986).

[32] D. A. Papaconstantopoulos *et al.*, *Handbook of the Band Structure of Elemental Solids* (Springer, 1986).

[33] C. O. Avci, K. Garello, A. Ghosh, M. Gabureac, S. F. Alvarado and P. Gambardella, *Nature Phys.* **11**, 570 (2015), 1502.06898.

Chapter 14

Non-Bloch Band Theory and Beyond

Zhong Wang

Institute for Advanced Study, Tsinghua University,
Beijing, 100084, China
wangzhongemail@tsinghua.edu.cn

Non-Hermitian physics commonly arises in open systems. A non-Hermitian band theory is a necessary groundwork for understanding a wide range of non-Hermitian phenomena. To meet this need, the non-Bloch band theory has been formulated recently. This chapter presents an intuitive introduction to this theory from our perspective.

14.1 Introduction

Open systems can be seen everywhere. Strictly speaking, every physical system is an open system because complete isolation from the rest of the world is impossible. When the interaction between the system and the environment is negligibly weak, the system can be safely regarded as closed. When the system–environment interaction is not negligible, however, the system should be seriously treated as open. One may still attempt to focus on the total system including the environment to avoid studying open systems. However, it is often the case that the total system is too complicated to solve. Even if the total system could be fully solved, it remains a nontrivial task to extract the relevant physics of the system that interests us. Therefore, a more viable approach is to focus on the degrees of freedom of the relevant system itself, at the price that the effective Hamiltonian becomes non-Hermitian. Thus, non-Hermitian Hamiltonians naturally arise in many systems.

In view of the significance of band theory in closed systems with Hermitian Hamiltonians, a natural question is to generalize it to open systems with non-Hermitian Hamiltonians. This question has led to the non-Bloch band theory, which is the focus of this chapter. Although this theory has some seemingly wild features, such as a revision of the basic concept of Brillouin zone (BZ), it has proved natural and useful for understanding and computing various aspects of non-Hermitian systems. As far as we can see, these unfamiliar features are intrinsic to non-Hermitian physics.[a]

We will present an intuitive introduction to the non-Bloch band theory. First, we will briefly introduce and clarify several relevant aspects of non-Hermitian physics, without assuming that the readers have prior knowledge of non-Hermitian systems. We then introduce the key idea and picture of non-Bloch band theory, followed by its applications in non-Hermitian topology, nonreciprocal amplification, etc. We end with a brief outlook. This chapter is not intended to be a comprehensive review of the subject, and the reference list is far from complete. Instead, it focuses on several crucial yet accessible aspects of the non-Bloch band theory from our own perspective.

14.2   Band theory and topology

One of the cornerstones of condensed matter physics is the energy band theory. Among other achievements, it successfully explains the dramatic differences between insulators, metals, and semiconductors. According to the band theory, electrons in a periodic lattice exist in the form of Bloch waves, which are plane waves modulated by the periodic potential (Bloch's theorem). As the eigenstates of Hamiltonian of a periodic structure, Bloch waves are the starting points for understanding innumerable phenomena in condensed matter physics. The space of all distinct wave vectors ($\mathbf{k}$) is the Brillouin zone (BZ), which is a basic stage of band theory. These band-theoretical concepts are not only indispensable in condensed matters, but also widely used in other fields, for example, photonic lattices and metamaterials.

Exploiting the translational symmetry, the Hamiltonian can be conveniently written in the BZ, which is called the Bloch Hamiltonian and denoted by $H(\mathbf{k})$. As a matrix, it can be diagonalized as

$$H(\mathbf{k})|u_n(\mathbf{k})\rangle = E_n(\mathbf{k})|u_n(\mathbf{k})\rangle, \tag{14.1}$$

where the eigenvalues $E_n(\mathbf{k})$ form the band structures and contain the information of the band gap and numerous other physical quantities. Many physical properties of solids can be derived in the Bloch band framework. While it is not our purpose to recall the standard band theory, we mention

---

[a]While the aforementioned "wild features" had at first generated considerable doubts from some friends, I was encouraged by the positive view of my great advisor, Shoucheng Zhang. In fact, Shoucheng was among the earliest to appreciate this theory. I was so delighted to hear that he liked it.

the following point that will become subtle in non-Hermitian physics. Strictly speaking, realistic materials have open boundary condition (OBC) and the wave vector **k** is not a good quantum number. Nevertheless, it is generally assumed that the boundary condition plays a negligible role in a large system and, for the sake of convenience, periodic boundary condition (PBC) is commonly used. Thus, translational symmetry is restored and **k** retains its role as a good quantum number.

Although the band theory had been regarded as a mature subject for many years, topology entered as a surprise during the last several decades. The breakthrough was the discovery of topological insulators protected by time-reversal symmetry, which has generated tremendous interests in topological bands during the last fifteen years [1–6]. In contrast to the conventional focuses on local band structures in the BZ, band topology is determined by the global topological features in the entire BZ. The topological properties can be expressed in terms of topological invariants that are insensitive (i.e., robust) to small variations of physical parameters. Among the most well-known examples of topological invariants are the Chern numbers, Z_2 invariants,[b] and winding numbers, etc., which are useful in different contexts. Topological bands are the simplest and most tractable ingredients of topological states of matter, and they serve as the departure point to understand the more intricate topological phases in strongly correlated systems.

A guiding principle in topological phases of matter is the bulk–boundary correspondence, which states that a nontrivial topological invariant of a bulk material guarantees the existence of robust boundary states. For example, a nonzero Chern number of a two-dimensional insulator means that there must be chiral edge states inside the bulk band gap, and a nontrivial Z_2 invariant implies the existence of Dirac-cone surface states. To experimentalists, the robust boundary states are the most prominent features of topological phases, and the bulk–boundary correspondence is important because it relates these physical boundary states to the underlying topological concepts. As we will see shortly, this principle should be revised in an unexpected way in non-Hermitian systems.

---

[b]I was fortunate to have the opportunity to do my doctoral thesis with Shoucheng in 2010. Stimulated by him, I proved that the Fu–Kane–Mele Z_2 Pfaffian topological invariant and the Qi–Hughes–Zhang Chern–Simons topological invariant are precisely equivalent [7], bridging these two pioneering approaches to topological insulators. That was my first scientific paper. Shoucheng has been teaching me how to do physics since then, from which I have been benefiting immensely. He remains my guide in physics.

14.3 Non-Hermitian physics

In the standard quantum mechanics, the Hermiticity of Hamiltonian is an axiom. It is so far unclear whether non-Hermitian Hamiltonians play a fundamental role at the level of quantum foundation, which is beyond the scope of this chapter. However, effective non-Hermitian Hamiltonians have found widespread applications in many branches of physics, which are collectively known as "non-Hermitian physics". A basic equation in non-Hermitian physics is

$$i\frac{d|\psi(t)\rangle}{dt} = \mathcal{H}|\psi(t)\rangle, \qquad (14.2)$$

where $\mathcal{H}$ is a non-Hermitian operator or matrix, and $|\psi\rangle$ is a state vector. This equation is not in conflict with the standard quantum mechanics, because $|\psi\rangle$ does not have the usual interpretation of the standard quantum mechanics. In fact, $|\psi\rangle$ can stand for a classical field configuration, a probability distribution, or the density matrix of a quantum mechanical system, etc. In one of the interpretations, $|\psi\rangle$ is indeed the usual quantum-mechanical state ket, but Eq. (14.2) describes time evolution under postselection, which is again not in contradiction with quantum mechanics (more on this shortly).

Many physical platforms of non-Hermitian physics can be put into the two major classes: Classical wave systems and quantum mechanical systems. In both cases, non-Hermiticity stems from the systems being open. Due to the richness of non-Hermitian physics, this is not a complete classification; for example, certain problems in statistical mechanics can be mapped to non-Hermitian Hamiltonian problems [8], whose origin is neither classical waves nor quantum mechanics.

Classical waves share many similar features as the quantum mechanical waves. They both exhibit wave superpositions and interferences. Owing to their similarity, several interesting phenomena initially studied in the quantum mechanical context, such as the Anderson localization caused by disorders, have later been seen in classical wave systems. It is therefore natural that the non-Hermiticity of Hamiltonian also has counterparts in the classical wave context [9–11]. Among the most popular platforms are photonic crystals and metamaterials [11]. The Maxwell equations governing the behaviors of light or microwaves in these materials can be recast in the form of Hamiltonian evolution akin to Eq. (14.2), with $|\psi\rangle$ being the classical field configuration, and the gain and loss naturally enters the effective Hamiltonian $\mathcal{H}$ as non-Hermitian terms. The gain can be added by external pumping [12,13], and loss by material absorption or radiation [14].

The presence of either gain or loss implies that the system is open, i.e., it exchanges energy or matter with an environment. Since the Maxwell equations are classical wave equations, the issue of consistency with standard quantum mechanics simply does not arise. This is a classical-level non-Hermiticity.

In the quantum mechanical context, non-Hermitian physics is also ubiquitous. Nevertheless, the physical meanings of non-Hermiticity are richer than the classical-level counterpart, and there are certain ambiguities and even confusions in some literatures. In its essence, non-Hermitian quantum physics arises from the openness of a quantum system. The interaction with environment means that the time evolution of the studied system, with the environment ignored, cannot be described as a unitary evolution under a Hermitian Hamiltonian. However, simply using an effective non-Hermitian Hamiltonian to describe the nonunitary time evolution of a state ket, which is called the "Hamiltonian-level non-Hermiticity" hereafter, is not suitable under the usual interpretation of quantum mechanics. To see when and how the Hamiltonian-level non-Hermiticity is meaningful, we have to take a closer look.

Intuitively, the environment constantly "observes" the system, causing "quantum jumps" from time to time. These quantum jumps, in addition to the Hamiltonian evolution, have to be included to provide a coherent picture. To this end, a solid and unambiguous formulation is based on the Markovian quantum master equation that governs the time evolution of the density matrix ρ [15],

$$\frac{d\rho(t)}{dt} = \mathcal{L}\rho = -i[H, \rho] + \sum_{\mu} \left(L_{\mu}\rho L_{\mu}^{\dagger} - \frac{1}{2}\{L_{\mu}^{\dagger}L_{\mu}, \rho\} \right), \qquad (14.3)$$

where H is the Hamiltonian, and L_{μ} are the "dissipators" that describe the quantum jump induced by the system–environment interaction. The Liouvillian $\mathcal{L}$ is called a superoperator because it operates on the density matrix, which itself is an operator. This equation can be rewritten as

$$\frac{d\rho(t)}{dt} = -i(H_{\text{eff}}\rho - \rho H_{\text{eff}}^{\dagger}) + \sum_{\mu} L_{\mu}\rho L_{\mu}^{\dagger}, \qquad (14.4)$$

$$H_{\text{eff}} = H - \frac{i}{2}\sum_{\mu} L_{\mu}^{\dagger}L_{\mu}, \qquad (14.5)$$

whose physical meaning is intuitive. Suppose that the system starts from a pure state $|\psi(t)\rangle$ at time t, with density matrix $\rho(t) = |\psi(t)\rangle\langle\psi(t)|$. After a short duration Δt, the density matrix becomes $\rho(t + \Delta t) = |\psi(t + \Delta t)\rangle \langle\psi(t + \Delta t)| + \Delta t \sum_\mu L_\mu |\psi(t)\rangle \langle\psi(t)| L_\mu^\dagger$, in which

$$|\psi(t + \Delta t)\rangle = \exp(-iH_{\text{eff}}\Delta t)|\psi(t)\rangle. \tag{14.6}$$

Thus, the wavefunction evolves under an effective non-Hermitian Hamiltonian H_{eff}, occasionally interrupted by quantum jumps to $\{L_\mu|\psi\rangle\}$. In this quantum trajectory picture [16, 17], it is clear that the presence of environment not only causes stochastic quantum jumps, but also generates non-Hermitian evolution of the system between successive jumps.

Therefore, we can see that the Hamiltonian-level non-Hermiticity [e.g., Eq. (14.6)] is physical and useful in the following sense. It describes the no-jump time evolution between two successive quantum jumps. In other words, the non-Hermitian Hamiltonian H_{eff} describes the time evolution of an open quantum system under postselection. If no quantum jump is observed for some time, then the evolution is governed by H_{eff} in this duration. This is experimentally relevant and testable, as the quantum jump can often be monitored and therefore the no-jump trajectory can be followed (e.g., certain quantum jump is accompanied by a photon emission, which can be detected). Recently, effective non-Hermitian Hamiltonians in strongly correlated systems have been advocated and used to make interesting predictions [18, 19], which can also be understood in the framework of open quantum systems. The effective non-Hermitian Hamiltonian associated with the Green's function of strongly correlated systems is shown to be identical to the effective non-Hermitian Hamiltonian of an open quantum system under postselection [20]. This open system consists of one single-particle state, with all other states treated as the environment.

Finally, we can also view the density matrix ρ as the "state" of the system, and the Liouvillian $\mathcal{L}$ as a "Hamiltonian" that governs the time evolution of state. In fact, if we identify $\mathcal{L}$ as $-i\mathcal{H}$, and ρ as $|\psi\rangle$, Eq. (14.3) can be viewed as a special case of Eq. (14.2). The Liouvillian is apparently always non-Hermitian. Thus, this is called the Liouvillian-level non-Hermiticity.

Since the density matrix plays a ubiquitous role in open quantum systems, the Liouvillian is also a most essential object therein. It indeed controls many aspects of the open quantum system. For example, its spectral gap determines how fast the system relaxes to the steady state when the initial state itself is not steady. In recent studies of non-Hermitian physics, the Liouvillian-level non-Hermiticity seems less emphasized than the Hamiltonian-level non-Hermiticity, though we believe that the Liouvillian-level one offers even richer physics for further investigations.

To summarize, we have emphasized three levels of non-Hermitian physics:

(i) Classical-level non-Hermiticity;
(ii) Hamiltonian-level non-Hermiticity;
(iii) Liouvillian-level non-Hermiticity.

Apparently, (i) belongs to classical wave systems, while both (ii) and (iii) are relevant to quantum systems. The common feature is that the non-Hermitian physics originates from the openness of the system, irrespective of the system being classical or quantum. Nevertheless, the important distinctions between different levels of non-Hermiticity should also be kept in mind.

14.4   Non-Bloch band theory

14.4.1   *Non-Hermitian skin effect*

Periodic structures have the same significance in non-Hermitian systems as in Hermitian systems; it is therefore a natural and important question to generalize the band theory from Hermitian systems to non-Hermitian cases. At first sight, it might look straightforward to do so. For example, it appears that nothing is wrong in using Eq. (14.1) when the Hamiltonian is non-Hermitian. As to non-Hermitian band topology, it appears natural to define topological invariants in the same way as in Hermitian bands, for example, as integrals in the BZ. These topological invariants remain quantized as integers, and it seems natural to expect that their values correspond to the number of topological edge modes, embodying the bulk–boundary correspondence as wished. This picture was indeed assumed in some early works on non-Hermitian topology [21–23]. For example, the Chern number has been defined as an integral of Berry curvature in the BZ, which was supposed to predict the number of chiral edge modes [21, 22].

Although this picture looks tempting, it has proved problematic in recent studies, which was a surprise. In fact, it overlooks a most essential and unexpected difference between Hermitian and non-Hermitian bands. For a Hermitian system, taking OBC has significant consequences only near the boundary, with the extensive bulk states unaffected. For a non-Hermitian system, however, the bulk states are highly sensitive to the boundary condition, and the OBC bulk states can be dramatically different from the PBC bulk states. Although the conventional band theory based on the BZ correctly describes the PBC bulk states, it fails to capture the OBC bulk states, though it is the latter that enters the bulk–boundary correspondence, because a PBC system simply has no boundary. In Hermitian systems, this

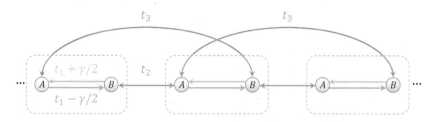

Fig. 14.1.   The non-Hermitian SSH model. The rectangles indicate the unit cells.

is not an issue because the PBC and OBC bulk states are the same, and therefore the BZ-based Bloch band theory works well for OBC systems and correctly describes the band topology. For non-Hermitian systems, the needed band theory fulfilling bulk–boundary correspondence should be based on the OBC bulk states, which differ from the PBC bulk states.

   This general argument may look somewhat strange and abstract. To be transparent, let us explain the physics in a concrete model, the non-Hermitian Su–Schrieffer–Heeger (SSH) model [Fig. 14.1] [24]. Each unit cell contains two sites, A and B, and the Bloch Hamiltonian reads

$$H(k) = d_x(k)\sigma_x + \left[d_y(k) + i\frac{\gamma}{2}\right]\sigma_y, \qquad (14.7)$$

where $d_x(k) = t_1 + (t_2 + t_3)\cos k$, $d_y(k) = (t_2 - t_3)\sin k$, and $\sigma_{x,y}$ are the standard Pauli matrices. The basis is chosen such that $\sigma_z = +1 \,(-1)$ corresponds to the $A \,(B)$ site. The non-Hermitian term proportional to γ causes a left-right asymmetry in the hopping. This simple Hamiltonian is not merely a mathematical toy; it appears in many different physical contexts. For example, this model itself or its close relatives underlie some realistic photonic systems [25], electric circuits [26] (classical-level non-Hermiticity), and quantum walks [27] (Hamiltonian-level non-Hermiticity). Furthermore, such non-Hermitian band models play key roles in the master equation approach to some open quantum systems [28, 29] (Liouvillian-level non-Hermiticity).

   We now calculate the energy spectrums with PBC and OBC, and show the results in Fig. 14.2. The entire PBC and OBC spectrums are quite different [24, 30–32]. This is in sharp contrast to the Hermitian cases (e.g., when $\gamma = 0$), in which the PBC and OBC spectrums differ only in a few discrete topological or trivial edge modes. The PBC–OBC difference is even more prominent when the energy spectrums are shown in the complex energy plane [Fig. 14.3]. While the PBC eigenenergies form loops, the OBC eigenenergies form lines or arcs.

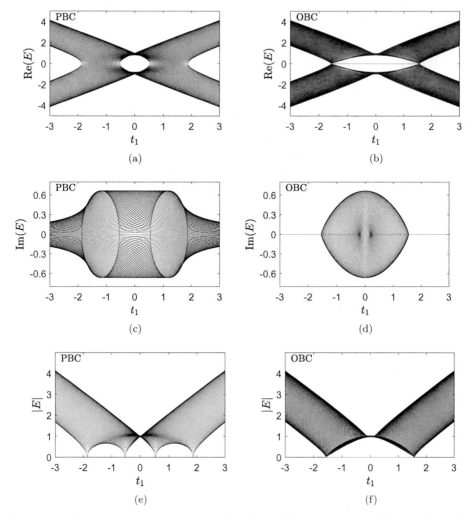

Fig. 14.2.   The energy spectrums of non-Hermitian SSH model under PBC and OBC. The real parts $\mathrm{Re}(E)$, imaginary parts $\mathrm{Im}(E)$, and the absolute values $|E|$ of eigenenergies E are all shown. The parameter t_1 is varied, with fixed parameters $t_2 = 1$, $t_3 = 1/5$, and $\gamma = 4/3$. The red lines represent the zero modes.

The mechanism of this dramatic PBC–OBC differences is the "non-Hermitian skin effect" (NHSE) [24, 31, 32]. It means that the nominal bulk eigenstates of the Hamiltonian, corresponding to the continuous spectrums, are all exponentially localized at the boundary. These nominal bulk modes are called the "skin modes". The number of skin modes is proportional to the size of the system, which clearly distinguishes them from the conventional boundary modes that is proportional to the boundary area. (For a one-dimensional system, the boundary consists of several discrete points, which do not change with the system size.) Note that the NHSE cannot occur

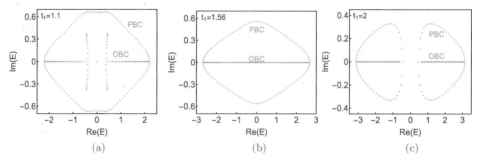

Fig. 14.3.   The energy spectrums of non-Hermitian SSH model in the complex plane. The PBC and OBC eigenenergy spectrums are shown in blue and red, respectively. The value of t_1 is shown in each figure; $t_2 = 1$, $t_3 = 1/5$, and $\gamma = 4/3$.

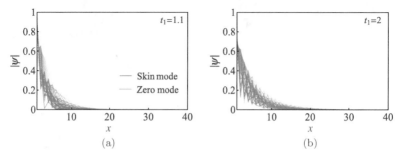

Fig. 14.4.   Energy eigenstates of a non-Hermitian SSH chain with OBC, exhibiting the non-Hermitian skin effect. The horizontal axis stands for the spatial coordinate. The nominal bulk eigenstates are localized near the boundary as the skin modes. The value of t_1 is shown in each figure; $t_2 = 1$, $t_3 = 1/5$, and $\gamma = 4/3$.

in Hermitian systems because the eigenstates are orthogonal, preventing them from accumulation. Two typical examples of skin modes are shown in Fig. 14.4. By observing this non-Hermitian SSH model, one may have the impression that the asymmetric hoppings drive all the eigenstates towards the direction favored by the larger hopping. This picture is somewhat naive. In fact, there are very simple models whose eigenstates are localized at both the left and right ends of an open chain, which means that some of the eigenstates are localized in the direction unfavored by the larger hopping. This is called the bipolar NHSE [33].

Seeing the NHSE partially answers the puzzling PBC–OBC difference. The PBC and OBC spectrums are attributed to the extended Bloch waves and the skin modes, respectively. Since the OBC skin modes are far from the Bloch waves, their energy spectrums should not coincide. Nevertheless, deeper questions remain. Why should the NHSE occur? Why do the OBC spectrums take the line or arc shape? How to calculate the OBC eigenenergies without diagonalizing the large Hamiltonian matrix in real space?

14.4.2 *Generalized Brillouin zone*

The answers to these questions and deeper insights into the physics should be obtained from a non-Hermitian band theory, which should contain the analytical formulas of the energy spectrums and eigenstates in the large-size limit. For Hermitian bands, these formulas are well known; they are contained in Eq. (14.1). In view of the NHSE, however, the non-Hermitian generalization of band theory should not be straightforward. The initial efforts were made in [24, 34], which laid the groundwork for the non-Bloch band theory of non-Hermitian systems. In this chapter, we do not plan to follow the technical aspects of this theory. Instead, we would like to present some of the main results with intuitive "derivations" and perspectives. The purpose is to understand.

For a PBC system, Bloch's theorem remains valid even though the Hamiltonian is non-Hermitian, and therefore the energy eigenstates are Bloch waves as usual. For an OBC system, we do not have the assistance of Bloch's theorem, and therefore finding the exact OBC eigenstates is a quite complicated task. Fortunately, the problem simplifies in the large size limit. Following an intuitive path (briefly mentioned in [24]), we view the OBC eigenstates as standing waves, and our goal is to find out the conditions for their formation. Let us consider a long chain with length L, and send an initial wave packet ψ_i from the left boundary, which propagates rightward [see Fig. 14.5]. Suppose that the energy of this wave packet is E, then its wave vector k is one of the roots of the characteristic equation $\det[H(k) - E] = 0$. A crucial feature of the following theory is not requiring k to be real-valued, which is reasonable in view of the NHSE, as a nonzero imaginary part of k indicates the exponential localization of eigenstates at the boundary. Thus, it will be convenient to change the variable to $\beta = e^{ik}$, which is called the "generalized phase factor" because $|\beta| = 1$ is not assumed. Accordingly, the wave e^{ikx} becomes β^x, where x is the spatial coordinate. Imaginary part of k being nonzero means $|\beta| \neq 1$. We introduce the notation $H(\beta) = H(k = -i \ln \beta)$ or, equivalently, $H(\beta) = H(k)|_{e^{ik} \to \beta}$. For Hermitian bands, one commonly focuses on the BZ, which is the unit circle $|\beta| = 1$. In our non-Hermitian problem, however, we view $H(\beta)$ as defined on the complex β-plane. Given the energy E, the initial wave packet ψ_i has a generalized phase factor $\beta_i(E)$, which should be one of the roots of the characteristic equation $\det[H(\beta) - E] = 0$. When arriving at the right boundary, the wave packet acquires a factor β_i^L and becomes $\psi_i \beta_i^L$. It is then reflected back by the right boundary, and propagates leftward with another generalized phase factor $\beta_j(E)$, acquiring a factor β_j^{-L} along the left-moving

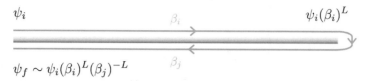

Fig. 14.5.   An intuitive picture that leads to the GBZ equation $|\beta_i(E)| = |\beta_j(E)|$. The initial wave packet ψ_i propagates rightward, and gets reflected by the right boundary. After coming back to the initial location, it acquires a factor $\beta_i^L \beta_j^{-L}$. In the large-L limit, the formation of standing wave requires that $\beta_i^L \beta_j^{-L} \sim 1$, and therefore $|\beta_i| = |\beta_j|$.

journey. When the wave packet comes back to the left boundary, it becomes $\psi_f \sim \psi_i \beta_i^L \beta_j^{-L}$ [see Fig. 14.5].

Now we are approaching the key point. To form a standing wave, ψ_f and ψ_i must have the same order of magnitude, $\psi_f \sim \psi_i$, which means that $\beta_i^L \beta_j^{-L} \sim 1$. In general, this is a complicated equation. However, when $L \to \infty$, this approximate equation can be greatly simplified and promoted to an exact equation $|\frac{\beta_i(E)}{\beta_j(E)}| = 1$ or, equivalently [24]

$$|\beta_i(E)| = |\beta_j(E)|. \tag{14.8}$$

In this equation, the i, j indices are not specified. A more rigorous calculation by Yokomizo and Murakami [34] shows that not all i, j should enter the above equation. In fact, if we order the roots $\beta_j(E)$'s of $\det[H(\beta) - E] = 0$ such that $|\beta_1(E)| \leq |\beta_2(E)| \leq \cdots \leq |\beta_{2M}(E)|$ ($M = 2$ for our present model), the only relevant i, j are the middle two, M and $M + 1$, and the refined version of Eq. (14.8) is

$$|\beta_M(E)| = |\beta_{M+1}(E)|. \tag{14.9}$$

While we do not want to include the technical steps of derivation, we note that taking the middle two is more reasonable than taking any other pair, because inverting the coordinate $x \to L - x$ leads to $\beta_i \to 1/\beta_i$, and the ordering of β's is therefore reversed. The middle two remain the middle two under this inversion, indicating that the choice is consistent.

Notably, in the presence of symplectic symmetry, Eq. (14.9) gives way to a different equation [35], which still takes the form of Eq. (14.8).

In essence, Eq. (14.9) contains only one complex variable, as E and β's are not independent variables; they are related by the energy dispersion $E(\beta)$. Since Eq. (14.9) provides a single real-valued constraint, the solution space is a one-dimensional object. We can take the complex variable as E, or β_M, or β_{M+1}. If Eq. (14.9) is viewed as an equation for E, the solutions are the OBC continuous energy spectrums. If Eq. (14.9) is viewed as an

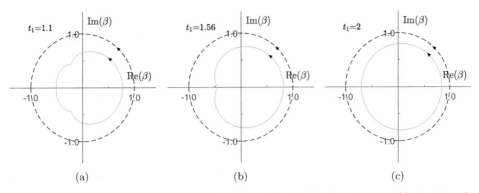

Fig. 14.6.   Generalized Brillouin zone (solid loop) and Brillouin zone (dashed loop). Parameter values are $t_2 = 1$, $t_3 = 1/5$, $\gamma = 4/3$, and t_1 is marked in each figure.

equation for β_M or β_{M+1}, the solutions form a one-dimensional loop in the complex β-plane. This loop is called the "generalized Brillouin zone" (GBZ) [see Fig. 14.6], which is a central concept in the non-Bloch band theory. In many senses, it is a natural generalization of the BZ to the non-Hermitian systems. First, the GBZ generates the OBC energy spectrum. Similar to Eq. (14.1), we consider the eigenvalue equation.

$$H(\beta)|u_n(\beta)\rangle = E_n(\beta)|u_n(\beta)\rangle, \tag{14.10}$$

with $\beta \in$ GBZ. Notably, the eigenvalues $E_n(\beta)$ belong to the OBC continuous spectrums when $\beta \in$ GBZ. For Hermitian bands, it is well known that the eigenvalues of $H(\beta)$ belong to the OBC continuous spectrum when $\beta \in$ BZ, i.e., $|\beta| = 1$. Thus, for Hermitian bands, the GBZ reduces to the BZ and Eq. (14.10) reduces to Eq. (14.1). Second, the eigenstate profiles are determined by the GBZ. For example, if the GBZ is entirely inside (outside) the unit circle, it means that all the eigenstates are exponentially localized at the left (right) boundary. For Hermitian bands, GBZ being the unit circle corresponds to that OBC eigenstates are superpositions of Bloch waves.

Within the GBZ picture, we can now understand why NHSE should occur in our model. Assuming that the eigenstates are Bloch waves would mean that the GBZ is a unit circle, namely that it coincides with the BZ. However, the energy spectrums of $H(\beta)$ with $|\beta| = 1$, i.e., the BZ energy spectrums, form loops in the complex plane [see Fig. 14.3]. These loop structures reflect the circle topology of the BZ, and there is a one-to-one correspondence between $E_\pm(\beta)$ and β in the BZ. Therefore, for a given

β_i with $|\beta_i| = 1$, we cannot find its wanted partner $\beta_j(\neq \beta_i)$ such that Eq. (14.8) is satisfied. Thus, to find solutions to Eq. (14.8), we have to remove the constraint $|\beta| = 1$. This is also the reason why k should be treated as complex-valued. The above argument is not specific to the present model. It is a quite general observation that whenever the PBC energy spectrums (BZ energy spectrums) form loops, there should be NHSE. In fact, this has been proved in a rigorous manner in recent studies [36, 37].

We can also see why the OBC eigenenergies form lines or arcs in the complex plane. Equation (14.8) means that at least two β's must exist in the GBZ for a given E in the OBC spectrums. As β varies in the GBZ loop, the OBC eigenenergies $E(\beta)$ varies in the complex energy plane, depicting a one-dimensional trajectory. This trajectory has to turn around at some point and then repeat itself in the reversed direction, to ensure the existence of another β to pair a given β such that Eq. (14.8) can be satisfied. Thus, the OBC energy spectrums should form lines or arcs, with the turning points being the ends of the lines and arcs. The line and arc shapes of OBC spectrums have interesting consequences. For example, it enables non-Bloch parity-time (PT) symmetry and its breaking, namely the transition between purely real and complex spectrums. In contrast, the PBC spectrums in the presence of NHSE form loops, and therefore can never be entirely real-valued.

Although we focus on the OBC geometry, the GBZ theory is applicable not only to the OBC case; it can be applied to arbitrary geometries, e.g., domain-wall configurations [38]. For PBC, the GBZ simply reduces to the BZ, namely the unit circle.

At this point, we may rethink a question: What justifies the Bloch band theory for Hermitian bands? In other word, for Hermitian bands, why does Eq. (14.1) produce the correct continuous energy spectrums under arbitrary boundary condition? A common answer is Bloch's theorem. However, Bloch's theorem is just a statement about PBC systems. In fact, Bloch's theorem also applies to non-Hermitian bands under PBC, yet the Bloch band theory fails to predict the energy spectrums for OBC, which indicates that Bloch's theorem is insufficient to fully justify the Bloch band picture. A further understanding is provided by the following consideration. For a non-Hermitian Hamiltonian, the eigenstates do not have to be orthogonal to each other, and therefore they can all accumulate to the boundary. For a Hermitian Hamiltonian, the orthogonality of eigenstates prevents them from crowding together. Thus, we may say that the orthogonality of eigenstates helps to protect the validity of Bloch band picture in Hermitian systems.

14.5   Applications of non-Bloch band theory

14.5.1   *Non-Bloch topological invariants and non-Bloch bulk–boundary correspondence*

The initial motivation of the non-Bloch band theory was to understand the band topology and bulk–boundary correspondence in non-Hermitian systems. As explained above, the Bloch band theory is not an appropriate description of the OBC skin modes (or non-Bloch modes), and therefore the conventional BZ-based topological invariants should not be expected to predict the topological edge modes. In view of the non-Bloch band theory, the natural bulk–boundary correspondence should be that between the OBC skin modes and the topological boundary modes, rather than that between the PBC Bloch modes and the OBC topological boundary modes. Since GBZ underlies the skin modes, and it is the natural generalization of the conventional BZ for arbitrary boundary conditions, we are justified to define non-Bloch topological invariants in GBZ.

Let us continue to take the non-Hermitian SSH as the example. This model has a chiral (sublattice) symmetry $\sigma_z^{-1}H(k)\sigma_z = -H(k)$, which protects the topological zero modes at the boundary, if the bulk is topologically nontrivial [3]. This symmetry protection is not destroyed by non-Hermiticity, and the zero modes can be robust. In the Hermitian limit $\gamma = 0$, it is well known that a winding number determines the number of zero modes. There are two bands in the SSH model, corresponding to $n = 1,2$ in Eq. (14.1). In the Hermitian limit, one can define a Q-matrix $Q(k) = |u_1(k)\rangle\langle u_1(k)| - |u_2(k)\rangle\langle u_2(k)|$, which is off-diagonal because of the chiral symmetry $\sigma_z^{-1}Q\sigma_z = -Q$. As such, we can write $Q = \begin{pmatrix} & q \\ q^{-1} & \end{pmatrix}$. The conventional winding number is then defined as $W = \frac{i}{2\pi}\int_{\text{BZ}} q^{-1}dq$.

Now for the non-Hermitian cases, we should consider Eq. (14.10) instead of Eq. (14.1), and define $Q(\beta) = |u_{R1}(\beta)\rangle\langle u_{L1}(\beta)| - |u_{R2}(\beta)\rangle\langle u_{L2}(\beta)|$, where the subscript "$R$" stands for the eigenvector of $H(\beta)$ while "L" stands for that of $H^\dagger(\beta)$, i.e., the right and left eigenstates of $H(\beta)$, respectively. The $Q(\beta)$-matrix and $q(\beta)$ can be defined akin to the Hermitian case, and the non-Bloch winding number is [24]

$$W = \frac{i}{2\pi}\int_{\text{GBZ}} q^{-1}dq, \tag{14.11}$$

which is an integral in the GBZ. This non-Bloch topological invariant correctly predicts the topological zero modes [24].

Recently, the NHSE has been confirmed in a number of experiments based on various physical platforms [25–27,39,40]. In particular, in addition to seeing the NHSE, the photon quantum walk experiment has also confirmed that the non-Bloch winding number of Eq. (14.11) indeed predicts the number of topological zero modes [27].

14.5.2   Non-Bloch band theory of bulk physics

14.5.2.1   Long-time bulk dynamics and the generalized Brillouin zone

Although the non-Bloch band theory was initially motivated by topological thoughts, it turns out to have unexpected applications beyond topology. Notably, interesting physics deep in the bulk (far from the boundary) is also naturally described in the framework of GBZ.

A prominent example showing that the bulk physics can be described by the GBZ is presented in Longhi's recent work [41]. First, this work rigorously proves that the wave packet dynamics in a bulk non-Hermitian system (PBC or OBC) can display qualitative difference between the presence and absence of NHSE. In the presence of NHSE, the wave packet exhibits a nonzero drifting velocity. This is independent of the boundary condition as long as the wave packet remains far from the boundary (or if there is no boundary). It indicates that the NHSE is not merely a boundary effect that only affects the regions near boundary. Second, and even more remarkably, Ref. [41] shows that the GBZ quantities can be directly measured by bulk physics. If a wave packet is created at a certain location x deep in the bulk, the long-time growth or decay behavior at the same x is precisely determined by special points in the GBZ. In fact, the wavefunction at x for a later time t can be written as [41] $\psi_x(t) = -i \sum_n \int_{\mathrm{BZ}} \frac{d\beta}{\beta} C_n(\beta) e^{-iE_n(\beta)t}$, where the coefficients $C_n(\beta)$ are determined by the initial wave packet. Taking $t \to \infty$, the saddle point method leads to

$$\psi_x(t) \sim e^{-iE_n(\beta_s)t} \sim e^{\mathrm{Im}[E_n(\beta_s)]t} \tag{14.12}$$

with a dominant n, where β_s is the saddle point determined by $\frac{dE_n(\beta)}{d\beta} = 0$. Remarkably, this saddle point always belongs to the GBZ, though it does not belong to the BZ in the presence of NHSE [41]. Although it is possible to take the BZ as a calculational assistance, the GBZ naturally emerges in the final results for the physical observables. This fact indicates that the GBZ is the natural language for non-Hermitian bands.

14.5.2.2   Non-Bloch band theory of nonreciprocal transmission and amplification

Another interesting and useful subject, to which the non-Bloch band theory has natural applications, is the nonreciprocal amplification (or directional amplification). Non-Bloch band theory provides highly concise formulas for the gain and directionality of nonreciprocal amplifiers [29].

Nonreciprocal amplification means that a signal is amplified towards a preferred direction, while the reverse propagation is prohibited or strongly suppressed. This feature is important to protecting the signal source from harmful extraneous noises. The amplifier has to be an open system to supply the energy gain during the amplification process. A simple model for nonreciprocal amplifier is a one-dimensional chain of bosonic modes [29], as shown in Fig. 14.7(a). As an open quantum system, the basic equation is the quantum master equation in Eq. (14.3). For our specific model, the Hamiltonian $H = \sum_i [(t_1 a_i^\dagger a_{i+1} + t_2 a_i^\dagger a_{i+2} + \text{H.c.}) + \omega_0 a_i^\dagger a_i + \epsilon_i(t) a_i^\dagger + \epsilon_i^*(t) a_i]$, where $\epsilon_i(t)$ represents the signal entering the site i. The dissipators are $\{L_\mu\} = \{L_i^g, L_i^l\}$, which contains the single-boson gain $L_i^g = \sqrt{\gamma'} a_i^\dagger$ and loss $L_i^l = \sqrt{\gamma}(a_i - i a_{i+2})$.

The most important quantity in this problem is the field expectation value $\phi_i(t) = \text{Tr}[a_i \rho(t)]$, which is directly measurable in experiments and applications. Its time evolution follows from the master equation: $\frac{d\phi_i}{dt} = -i\text{Tr}([a_i, H]\rho(t)) + \frac{1}{2}\sum_\mu \text{Tr}([L_\mu^\dagger, a_i]L_\mu \rho(t) + L_\mu^\dagger[a_i, L_\mu]\rho(t))$, from which we obtain

$$\frac{d\phi_i(t)}{dt} = -i\sum_j h_{ij}\phi_j(t) - i\epsilon_i(t), \qquad (14.13)$$

where the effective non-Hermitian Hamiltonian h is pictorially illustrated in Fig. 14.7(b). The corresponding Bloch Hamiltonian is

$$h(\beta) = \left(t_2 + \frac{\gamma}{2}\right)\beta^{-2} + t_1\beta^{-1} + \omega_0 + i\kappa + t_1\beta + \left(t_2 - \frac{\gamma}{2}\right)\beta^2, \quad (14.14)$$

where $\beta = e^{ik}$ and $\kappa = \frac{\gamma'}{2} - \gamma$.

For an input signal $\epsilon_j(t)$ with frequency ω at site j, i.e., $\epsilon_j(t) = \epsilon_j(\omega)e^{-i\omega t}$, the generated output at site i is $\phi_i(t) = \phi_i(\omega)e^{-i\omega t}$, with $\phi_i(\omega) = G_{ij}(\omega)\epsilon_j(\omega)$, where the Green's function matrix

$$G(\omega) = \frac{1}{\omega - h}. \qquad (14.15)$$

We get nonreciprocal amplification when $|G_{ij}| \gg 1$ while $|G_{ji}| \ll 1$, meaning that signals can propagate in the preferred direction. Apparently, $|G_{ij}|$ is the gain, and $|G_{ji}/G_{ij}|$ quantifies the directionality. Our goal is to find their explicit formulas. For our model, taking $i, j = 0, L$, we indeed see that $|G_{L0}| \sim \alpha_\rightarrow^L$ and $|G_{0L}| \sim \alpha_\leftarrow^L$ with $\alpha_\rightarrow > 1$ and $\alpha_\leftarrow < 1$ [Fig. 14.7(c)], which guarantee nonreciprocal amplification for L sufficiently large. Notably,

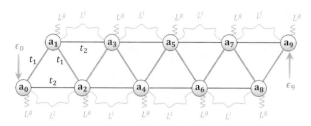

(a) A nonreciprocal amplifier.

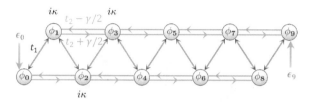

(b) Effective hopping Hamiltonian.

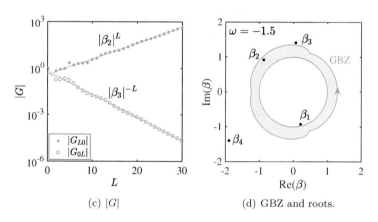

(c) $|G|$ (d) GBZ and roots.

Fig. 14.7.   Model of nonreciprocal amplifier. (a) The model. The length $L = 9$ in this illustration. (b) The effective Hamiltonian for ϕ derived from the master equation. (c) The L dependence of G_{L0} and G_{0L}. The solid lines are the results of GBZ theory in Eq. (14.17), which predicts $|G_{L0}| \sim |\beta_2|^L$ and $|G_{0L}| \sim |\beta_3|^{-L}$. (d) GBZ and the roots of $\omega - h(\beta) = 0$. The blue circle is the BZ (unit circle). For (c) and (d), $t_1 = 1$, $t_2 = 1$, $\gamma = 1.2$, $\kappa = -0.7$, and $\omega = -1.5$. For convenience, we have redefined $\omega - \omega_0$ as ω, and therefore a negative ω means a frequency below the bare frequency ω_0.

the boundary condition is crucial. To have nonreciprocal amplification, we must take OBC rather than PBC [42, 43].

What are the formulas for $\alpha_\rightarrow$ and $\alpha_\leftarrow$? For Hermitian bands, we have the following suggestive formula for Green's function:

$$G_{ij}(\omega) = \int_0^{2\pi} \frac{dk}{2\pi} \frac{e^{ik(i-j)}}{\omega - h(k)}, \tag{14.16}$$

which can be rewritten as an integral in the unit circle $|\beta| = 1$ with $\beta = e^{ik}$. For non-Hermitian systems with OBC, however, this formula turns out to be incorrect. In fact, the non-Bloch band theory advocates the GBZ as the natural language. The correct formula is an integral in the GBZ [29]:

$$G_{ij}(\omega) = \int_{\text{GBZ}} \frac{d\beta}{2\pi i \beta} \frac{\beta^{i-j}}{\omega - h(\beta)}. \tag{14.17}$$

This formula can be further simplified by the residue theorem, which leads to $\alpha_\rightarrow = |\beta_M(\omega)|$ and $\alpha_\leftarrow = |\beta_{M+1}(\omega)|^{-1}$, where $\beta_1, \beta_2, \ldots, \beta_{2M}$ are the roots of $\omega - h(\beta) = 0$, ordered as $|\beta_1| \leq |\beta_2| \leq \cdots \leq |\beta_{2M}|$. For our model, we have $M = 2$ [Fig. 14.7(d)]. An important geometrical fact in the above application of residue theorem is that the GBZ encloses only the first M roots, $\beta_1, \ldots, \beta_M$ [29, 37], and therefore the residue theorem selects β_M and β_{M+1}. The BZ, however, does not enjoy any nice property similar to this. For our specific model with $M = 2$, $\alpha_\rightarrow = |\beta_2(\omega)|$, and $\alpha_\leftarrow = |\beta_3(\omega)|^{-1}$. This is confirmed by numerical calculations [see Fig. 14.7(c)].

These formulas are natural and concise only in the GBZ. As far as we can see, approaching them within the BZ framework is difficult, if not impossible.

14.6 Outlooks

The non-Bloch band theory is still quite young. While it provides answers to several interesting questions of non-Hermitian systems, it also generates a new set of questions, many of which have not been answered yet. Perhaps many interesting questions remain to be raised yet.

Even at the level of band theory, there are quite a few open questions. One of the important questions is the generalization of GBZ to higher dimensions. For example, in two spatial dimensions, we should have (β_x, β_y) in the place of a single complex variable β for one spatial dimension. As has been explained, in the one-dimensional case, we have obtained a simple equation [Eq. (14.8) or Eq. (14.9)] that reduces the two independent real variables $(\text{Re}(\beta), \text{Im}(\beta))$ to one, which leads to the one-dimensional GBZ. In two spatial dimensions, the GBZ should be a two-dimensional subspace of the four-dimensional space $(\text{Re}(\beta_x), \text{Im}(\beta_x), \text{Re}(\beta_y), \text{Im}(\beta_y))$. However, it is not straightforward to find a generalization of Eq. (14.8). The counterpart of the intuitive picture in Fig. 14.5, which hints Eq. (14.8), is not easy to see in two spatial dimensions.

In view of the challenge to systematically generalize the GBZ to higher dimensions, there have been several alternatives in recent works. One approach is to take OBC only in the direction perpendicular to the boundary, and keep PBC in all the other directions. Thus, the GBZ can be calculated using the one-dimensional equation. Another approach is to calculate the GBZ under certain approximations, e.g., in the low-energy continuum model [44, 45]. However, a general equation that can generate higher-dimensional GBZs under arbitrary boundary conditions remains highly desirable.

Another interesting direction is to study the transport phenomena in the framework of non-Bloch band theory. As has been explained, even the bulk dynamics far from the boundary can feel the non-Bloch features of non-Hermitian bands. However, the investigations so far have been at a preliminary stage. A complete transport theory based on the non-Bloch band theory has not been established yet. Such a theory should be able to make interesting predictions in transport phenomena. For example, the cusps in the GBZ, which is discernable in Fig. 14.6, may have interesting physical consequences in transports. Moreover, finding unique features of transport via topological boundary modes, even in the presence of skin modes, in also an intriguing question.

In addition, the GBZ has certain provocative mathematical properties. It seems that this geometrical object may have some deep connections to the algebraic geometry, which is a vigorous branch of mathematics. It is not impossible that certain concepts from algebraic geometry may have interesting physical consequences in non-Hermitian physics.

Perhaps the most intriguing direction is the many-body non-Hermitian physics beyond the framework of band theory. As far as we can see, the Liouvillian-level non-Hermitian physics is most amenable to nontrivial many-body generalizations. Hopefully, the non-Bloch physics initially seen in the band theory framework remains a useful guide or hint for the new physics beyond band theory.

Acknowledgments

I would like to take this opportunity to express my deepest gratitude to Shoucheng Zhang, my fantastic advisor and friend, who has inspired me in the deepest way.

References

[1] X.-L. Qi and S.-C. Zhang, Topological insulators and superconductors, *Rev. Mod. Phys.* **83**, 1057 (2011); doi: 10.1103/RevModPhys.83.1057.
[2] M. Z. Hasan and C. L. Kane, *Colloquium.* Topological insulators, *Rev. Mod. Phys.* **82**, 3045 (2010); doi: 10.1103/RevModPhys.82.3045.

[3] C.-K. Chiu, J. C. Y. Teo, A. P. Schnyder and S. Ryu, Classification of topological quantum matter with symmetries, *Rev. Mod. Phys.* **88**, 035005 (2016); doi: 10.1103/RevModPhys.88.035005.

[4] C. L. Kane and E. J. Mele, Z_2 topological order and the quantum spin Hall effect, *Phys. Rev. Lett.* **95**, 146802 (2005).

[5] B. A. Bernevig, T. L. Hughes and S. C. Zhang, Quantum spin Hall effect and topological phase transition in HgTe quantum wells, *Science* **314**, 1757 (2006).

[6] M. König, S. Wiedmann, C. Brüne, A. Roth, H. Buhmann, L. Molenkamp, X.-L. Qi and S.-C. Zhang, Quantum spin Hall insulator state in HgTe quantum wells, *Science* **318**, 766 (2007).

[7] Z. Wang, X.-L. Qi and S.-C. Zhang, Equivalent topological invariants of topological insulators, *New J. Phys.* **12**(6): 065007 (2010); doi: 10.1088/1367-2630/12/6/065007.

[8] N. Hatano and D. R. Nelson, Localization transitions in non-hermitian quantum mechanics, *Phys. Rev. Lett.* **77**, 570 (1996); doi: 10.1103/PhysRevLett.77.570. https://link.aps.org/doi/10.1103/PhysRevLett.77.570.

[9] R. El-Ganainy, K. G. Makris, M. Khajavikhan, Z. H. Musslimani, S. Rotter and D. N. Christodoulides, Non-hermitian physics and pt symmetry, *Nature Phys.* **14**(1), 11 (2018);

[10] L. Feng, R. El-Ganainy and L. Ge, Non-hermitian photonics based on parity-time symmetry, *Nature Photonics* **11**(12), 752 (2017).

[11] T. Ozawa, H. M. Price, A. Amo, N. Goldman, M. Hafezi, L. Lu, M. C. Rechtsman, D. Schuster, J. Simon, O. Zilberberg and I. Carusotto, Topological photonics, *Rev. Mod. Phys.* **91**, 015006 (2019); doi: 10.1103/RevModPhys.91.015006; https://link.aps.org/doi/10.1103/RevModPhys.91.015006.

[12] G. Harari, M. A. Bandres, Y. Lumer, M. C. Rechtsman, Y. Chong, M. Khajavikhan, D. N. Christodoulides and M. Segev, Topological insulator laser: Theory, *Science* p. eaar4003 (2018).

[13] M. A. Bandres, S. Wittek, G. Harari, M. Parto, J. Ren, M. Segev, D. N. Christodoulides and M. Khajavikhan, Topological insulator laser: Experiments, *Science* p. eaar4005 (2018).

[14] H. Zhou, C. Peng, Y. Yoon, C. W. Hsu, K. A. Nelson, L. Fu, J. D. Joannopoulos, M. Soljačić and B. Zhen, Observation of bulk fermi arc and polarization half charge from paired exceptional points. (2018); doi: 10.1126/science.aap9859.

[15] H.-P. Breuer, *et al.*, *The Theory of Open Quantum Systems.* (Oxford University Press on Demand, 2002).

[16] J. Dalibard, Y. Castin and K. Mølmer, Wave-function approach to dissipative processes in quantum optics, *Phys. Rev. Lett.* **68**, 580 (1992); doi: 10.1103/PhysRevLett.68.580. https://link.aps.org/doi/10.1103/PhysRevLett.68.580.

[17] H. J. Carmichael, Quantum trajectory theory for cascaded open systems, *Phys. Rev. Lett.* **70**, 2273 (1993); doi: 10.1103/PhysRevLett.70.2273. https://link.aps.org/doi/10.1103/PhysRevLett.70.2273.

[18] V. Kozii and L. Fu, Non-Hermitian Topological Theory of Finite-Lifetime Quasiparticles: Prediction of Bulk Fermi Arc Due to Exceptional Point, Preprint (2017); arXiv:1708.05841.

[19] T. Yoshida, R. Peters and N. Kawakami, Non-hermitian perspective of the band structure in heavy-fermion systems, *Phys. Rev. B.* **98**, 035141 (2018); doi: 10.1103/PhysRevB.98.035141. https://link.aps.org/doi/10.1103/PhysRevB.98.035141.

[20] Y. Michishita and R. Peters, Equivalence of effective non-hermitian hamiltonians in the context of open quantum systems and strongly correlated electron systems, *Phys. Rev. Lett.* **124**, 196401 (2020); doi: 10.1103/PhysRevLett.124.196401. https://link.aps.org/doi/10.1103/PhysRevLett.124.196401.

[21] K. Esaki, M. Sato, K. Hasebe and M. Kohmoto, Edge states and topological phases in non-hermitian systems, *Phys. Rev. B* **84**, 205128 (2011); doi: 10.1103/PhysRevB. 84.205128. https://link.aps.org/doi/10.1103/PhysRevB.84.205128.

[22] H. Shen, B. Zhen and L. Fu, Topological band theory for non-hermitian hamiltonians, *Phys. Rev. Lett.* **120**, 146402 (2018); doi: 10.1103/PhysRevLett.120.146402. https://link.aps.org/doi/10.1103/PhysRevLett.120.146402.

[23] D. Leykam, K. Y. Bliokh, C. Huang, Y. D. Chong and F. Nori, Edge modes, degeneracies and topological numbers in non-hermitian systems, *Phys. Rev. Lett.* **118**, 040401 (2017); doi: 10.1103/PhysRevLett.118.040401. https://link.aps.org/doi/10.1103/PhysRevLett.118.040401.

[24] S. Yao and Z. Wang, Edge states and topological invariants of non-hermitian systems, *Phys. Rev. Lett.* **121**, 086803 (2018); doi: 10.1103/PhysRevLett.121.086803. https://link.aps.org/doi/10.1103/PhysRevLett.121.086803.

[25] S. Weidemann, M. Kremer, T. Helbig, T. Hofmann, A. Stegmaier, M. Greiter, R. Thomale and A. Szameit, Topological funneling of light, *Science* **368** (6488), 311 (2020); doi: 10.1126/science.aaz8727. https://science.sciencemag.org/content/368/6488/311.

[26] T. Helbig, T. Hofmann, S. Imhof, M. Abdelghany, T. Kiessling, L. Molenkamp, C. Lee, A. Szameit, M. Greiter and R. Thomale, Generalized bulk–boundary correspondence in non-hermitian topolectrical circuits, *Nature Phys.* **16**, 747 (2020).

[27] L. Xiao, T. Deng, K. Wang, G. Zhu, Z. Wang, W. Yi and P. Xue, Non-Hermitian bulk-boundary correspondence in quantum dynamics, *Nature Phys.* **16**, 761 (2020).

[28] F. Song, S. Yao and Z. Wang, Non-hermitian skin effect and chiral damping in open quantum systems, *Phys. Rev. Lett.* **123**, 170401 (2019); doi: 10.1103/PhysRevLett. 123.170401. https://link.aps.org/doi/10.1103/PhysRevLett.123.170401.

[29] W.-T. Xue, M.-R. Li, Y.-M. Hu, F. Song and Z. Wang, Non-Hermitian band theory of directional amplification, Preprint (2020); arXiv:2004.09529.

[30] T. E. Lee, Anomalous edge state in a non-hermitian lattice, *Phys. Rev. Lett.* **116**, 133903 (2016); doi: 10.1103/PhysRevLett.116.133903. https://link.aps.org/doi/10.1103/PhysRevLett.116.133903.

[31] F. K. Kunst, E. Edvardsson, J. C. Budich and E. J. Bergholtz, Biorthogonal bulk-boundary correspondence in non-hermitian systems, *Phys. Rev. Lett.* **121**, 026808 (2018); doi: 10.1103/PhysRevLett.121.026808. https://link.aps.org/doi/10.1103/PhysRevLett.121.026808.

[32] C. H. Lee and R. Thomale, Anatomy of skin modes and topology in non-hermitian systems, *Phys. Rev. B.* **99**, 201103 (2019); doi: 10.1103/PhysRevB.99.201103. https://link.aps.org/doi/10.1103/PhysRevB.99.201103.

[33] F. Song, S. Yao and Z. Wang, Non-hermitian topological invariants in real space, *Phys. Rev. Lett.* **123**, 246801 (2019); doi: 10.1103/PhysRevLett.123.246801. https://link.aps.org/doi/10.1103/PhysRevLett.123.246801.

[34] K. Yokomizo and S. Murakami, Non-bloch band theory of non-hermitian systems, *Phys. Rev. Lett.* **123**, 066404 (2019); doi: 10.1103/PhysRevLett.123.066404. https://link.aps.org/doi/10.1103/PhysRevLett.123.066404.

[35] K. Kawabata, N. Okuma and M. Sato, Non-bloch band theory of non-hermitian hamiltonians in the symplectic class, *Phys. Rev. B.* **101**, 195147 (2020); doi: 10.1103/PhysRevB.101.195147. https://link.aps.org/doi/10.1103/PhysRevB.101.195147.

[36] N. Okuma, K. Kawabata, K. Shiozaki and M. Sato, Topological origin of non-hermitian skin effects, *Phys. Rev. Lett.* **124**, 086801 (2020); doi: 10.1103/PhysRevLett. 124.086801. https://link.aps.org/doi/10.1103/PhysRevLett.124.086801.

[37] K. Zhang, Z. Yang and C. Fang, Correspondence between winding numbers and skin modes in non-hermitian systems, Preprint (2019); arXiv:1910.01131.

[38] T.-S. Deng and W. Yi, Non-bloch topological invariants in a non-hermitian domain wall system, *Phys. Rev. B.* **100**, 035102 (2019); doi: 10.1103/PhysRevB.100.035102. https://link.aps.org/doi/10.1103/PhysRevB.100.035102.

[39] A. Ghatak, M. Brandenbourger, J. van Wezel and C. Coulais, Observation of non-Hermitian topology and its bulk-edge correspondence, Preprint (2019); arXiv: 1907.11619.

[40] T. Hofmann, T. Helbig, F. Schindler, N. Salgo, M. Brzezińska, M. Greiter, T. Kiessling, D. Wolf, A. Vollhardt, A. Kabaši, C. H. Lee, A. Bilušić, R. Thomale and T. Neupert, Reciprocal skin effect and its realization in a topolectrical circuit, *Phys. Rev. Res.* **2**, 023265 (2020); doi: 10.1103/PhysRevResearch.2.023265. https://link.aps.org/doi/10.1103/PhysRevResearch.2.023265.

[41] S. Longhi, Probing non-hermitian skin effect and non-bloch phase transitions, *Phys. Rev. Res.* **1**, 023013 (2019); doi: 10.1103/PhysRevResearch.1.023013. https://link.aps.org/doi/10.1103/PhysRevResearch.1.023013.

[42] A. McDonald, T. Pereg-Barnea and A. A. Clerk, Phase-dependent chiral transport and effective non-hermitian dynamics in a bosonic Kitaev-Majorana chain, *Phys. Rev. X.* **8**, 041031 (2018); doi: 10.1103/PhysRevX.8.041031. https://link.aps.org/doi/10.1103/PhysRevX.8.041031.

[43] C. C. Wanjura, M. Brunelli and A. Nunnenkamp, Topological framework for directional amplification in driven-dissipative cavity arrays, Preprint (2019); arXiv: 1909.11647.

[44] S. Yao, F. Song and Z. Wang, Non-hermitian chern bands, *Phys. Rev. Lett.* **121**, 136802 (2018); doi: 10.1103/PhysRevLett.121.136802. https://link.aps.org/doi/10.1103/PhysRevLett.121.136802.

[45] T. Liu, Y.-R. Zhang, Q. Ai, Z. Gong, K. Kawabata, M. Ueda and F. Nori, Second-order topological phases in non-hermitian systems, *Phys. Rev. Lett.* **122**, 076801 (2019); doi: 10.1103/PhysRevLett.122.076801. https://link.aps.org/doi/10.1103/PhysRevLett.122.076801.

Chapter 15

Quantum Anomalous Hall Effect in Magnetic Topological Insulators

Yayu Wang, Ke He and Qikun Xue

Department of Physics, Tsinghua University,
Beijing 100084, China

The quantum anomalous Hall effect refers to the quantized version of the anomalous Hall effect originated from spontaneous ferromagnetic order. It represents an important member of the quantum Hall effect family, and remained elusive for decades. It was first proposed by Shoucheng Zhang and coworkers that magnetic topological insulators are ideal material systems for realizing this exotic effect. In 2013, the quantum anomalous Hall effect was discovered in Cr-doped $(Bi,Sb)_2Te_3$ ultrathin films in the absence of external magnetic field. In this chapter, we will first introduce the theoretical background about the quantum anomalous Hall effect. We then review its experimental realization in magnetic topological insulators step by step. In the end, we will discuss recent progresses and future perspectives about the quantum anomalous Hall effect.

15.1 Introduction

The discovery of quantum Hall effect in two-dimensional (2D) electron gas at strong magnetic field represents a remarkable breakthrough in modern physics [1]. The precise quantization of the Hall resistivity with wide plateaus at $\rho_{yx} = h/ie^2$ (here h is the Planck constant, e is the electron charge, and i is an integer) and the vanishing longitudinal resistivity ($\rho_{xx} = 0$) are astonishing features that are totally indifferent to material details. It was revealed by the TKNN formalism that such peculiar behavior reflects the topological nature of the electronic structure of the Landau levels induced by strong magnetic field in 2D electron systems [2]. The quantum Hall effect not only directly illustrates the existence of topological quantum matter, but may also find applications in metrology and next generation electronic devices.

One obvious obstacle for the practical application of quantum Hall effect is the requirement of strong magnetic field, or the existence of Landau levels.

An important conceptual breakthrough was made by Haldane in 1988, who proposed a theoretical model for quantum Hall effect without Landau levels [3]. The Haldane model is basically a graphene lattice with the time-reversal symmetry (TRS) broken by a periodic magnetic field, but the net magnetic flux is zero. It was the first theoretical model for realizing the quantum anomalous Hall effect (QAHE), the quantized version of the anomalous Hall effect due to spontaneous magnetization in the absence of external magnetic field. The Haldane model has been highly influential in latter theoretical developments in topological phases of matter, but it was too abstract to realize experimentally.

The discovery of topological insulators (TI) in recent years brought new hope to the realization of QAHE [4–8]. The topological property of a TI is induced by strong spin–orbit coupling (SOC), which causes the inversion of bulk conduction and valence bands. The bulk band structure then becomes topologically nontrivial and can be characterized by a Z_2 number. A 2D TI is expected to exhibit the quantum spin Hall effect (QSHE), in which a pair of spin-polarized edge states counter-propagates along the sample edges. According to the Landauer–Buttiker formalism for ballistic transport, the helical edge states will lead to a quantized resistance $h/2e^2$ in six-terminal longitudinal transport (with standard Hall-bar measurement geometry). A landmark contribution of Shoucheng to this field is the theoretical prediction that the HgTe/CdTe quantum well is a 2D TI that should exhibit the QSHE [9], which connects this exotic effect with a realistic material. This prediction was soon confirmed experimentally by Molenkamp's group [10].

Together with Chaoxing Liu *et al.*, Shoucheng quickly realized that breaking the TRS of the HgTe/CdTe quantum well may realize a ferromagnetic (FM) TI, giving rise to the QAHE [11]. If the FM exchange splitting of the lowest-order quantum well sub-band in the 2D TI is large enough, one set of spin sub-band will be driven back to the topologically trivial phase. As a consequence, only one spin channel remains topologically protected, which transforms the QSHE with a pair of helical edge states into the QAHE with one chiral edge state. They proposed that doping Mn ions into the HgTe/CdTe quantum well may generate FM order, which unfortunately has not been realized so far because it is rare for an insulating material to obtain long-range FM order. After the discovery of the Bi_2Se_3 family 3D TIs [12, 13], Xi Dai, Zhong Fang, and Shoucheng *et al.* proposed that these compounds could be even better materials for realizing the QAHE [14]. A major advantage of this system is the existence of the so-called van Vleck mechanism for FM ordering that, unlike the conventional RKKY mechanism due to itinerant carriers, is valid even when the samples are truly insulating.

These series of theoretical works, mainly owing to Shoucheng's deep insight, laid out the road map for realizing the QAHE in magnetic TIs.

15.2 Experimental realization of the quantum anomalous Hall effect

Although the route towards the experimental realization of QAHE seems to be rather clear, it is by no means an easy task. The QAHE actually puts very stringent requirements on the material properties because the ultimate goal is to find a *2D Ferromagnetic Topological Insulator*. The material must have topologically nontrivial bulk band structure so that there are topologically protected boundary states. Moreover, the sample must obtain long-range FM order with its magnetization direction perpendicular to the sample plane so that the anomalous Hall effect (AHE) is present. Finally, the sample must be truly insulating both in the bulk and surface so that only the chiral edge states contribute to charge transport. In Fig. 15.1, we illustrate the schematic band structure required for realizing the QAHE.

One of the biggest challenges facing the transport studies of TI is the existence of bulk carriers. Most as-grown TIs are not truly insulating due to the defects that donate charges into the bulk. The large bulk conduction can easily dominate the transport process, and the AHE cannot be quantized. When the TI becomes truly insulating, another challenge for QAHE emerges, namely the establishment of FM order. Most FM materials are metallic because itinerant carriers are needed in the RKKY mechanism to mediate FM exchange between local moments. It is very rare to find FM insulators in nature, and the tendency for FM order is further suppressed when the system is reduced to two-dimensionality. Moreover, it is unclear if the TI remains

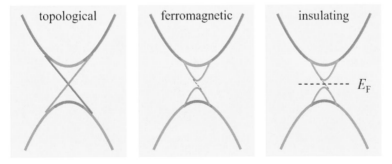

Fig. 15.1.   (a) The band structure of a pure topological insulator. (b) Ferromagnetic order breaks the time-reversal symmetry and opens a gap at the Dirac point. (c) The Fermi level must lie in the gap of the surface state so that only the 1D edge state contributes to transport.

topologically nontrivial when a significant number of magnetic dopants are introduced. Therefore, finding a 2D material system that simultaneously possesses ferromagnetism, nontrivial topology, and insulating bulk is a formidable task for experimentalist.

Staring from 2009, the Tsinghua and IOP joint research team led by Qikun Xue started the adventure towards the experimental realization of QAHE in magnetic TIs. Throughout this remarkable journey, Shoucheng has constantly been the main source of intellectual support and encouragement. His theoretical insight and unflappable confidence were crucial for the eventual success of the collaboration. Below we will present the key steps that led to the experimental observation of the QAHE.

The general strategy we employed is to combine a variety of complementary techniques including molecular beam epitaxy (MBE), scanning tunneling microscopy (STM), angle-resolved photoemission spectroscopy (ARPES), and transport measurements. The state-of-art MBE is used for the growth of high-quality TI films, in which the film thickness and chemical compositions can be accurately controlled. The STM and ARPES instruments are installed in the same ultrahigh vacuum chamber as the MBE, which allows *in situ* characterization of the surface morphology and electronic structure of the TI films. For transport measurements, the TI film is taken out of the UHV chamber and put into a low temperature strong magnetic field system to obtain the information about magnetic order, anomalous Hall resistance, and charge carrier density and mobility. The comprehensive information obtained by the various probes and the quick feedback between growth and characterization processes are the main reasons for the continuous improvement of our understanding and control of the magnetic TI films.

In order to realize the QAHE, ultrathin TI films with thickness in the order of a few quintuple layers (QLs) are necessary. Decreasing film thickness can reduce the bulk conduction and parallel conduction along the side surfaces. The growth and electronic structure studies of Bi_2Se_3 family TIs with well-controlled thickness down to 1 QL has been reported [15, 16]. Although MBE-grown thin films presumably have higher quality than single crystals, there are still unavoidable defects that donate charge carriers into the bulk. Shown in Fig. 15.2 are two early examples of magnetic TI films, Cr-doped Bi_2Te_3 and Sb_2Te_3, in which the anomalous Hall (AH) resistances are in the order of 50–150 Ω in zero magnetic field at 1.5 K due to the existence of significant bulk conduction.

In order to further eliminate the bulk carriers and create intrinsic TI, band structure engineering technique has been applied to tune the electronic

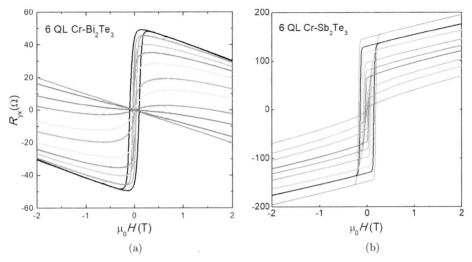

Fig. 15.2. AH effect measured in 6-QL Cr-doped Bi_2Te_3 (a) and Sb_2Te_3 films (b) at varied temperatures. The AH resistances are in the order of 50–150 Ω in zero magnetic field at 1.5 K due to the existence of significant bulk conduction.

structure of the TIs in the form of $(Bi_{1-x}Sb_x)_2Te_3$ ternary compounds [17]. Bi_2Te_3 and Sb_2Te_3 are both V–VI group TIs with the same crystal structure and similar lattice constants, making it ideal to form $(Bi_{1-x}Sb_x)_2Te_3$ ternary compounds with arbitrary mixing ratio. The advantages of mixing the two TIs can be anticipated from their different electronic properties. For Bi_2Te_3, the Fermi level E_F lies in the bulk conduction band due to the electron-type bulk carriers induced by Te vacancies, which can be seen from the negative slope of the ordinary Hall effect in Fig. 15.2(a). In contrast, for Sb_2Te_3 the E_F lies in the bulk valence band due to the hole-type bulk carriers induced by Sb–Te antisite defects, which leads to the positive ordinary Hall slope in Fig. 15.2(b). By mixing the two compounds one can achieve charge compensation, which may lead to an ideal TI with truly insulating bulk.

The ARPES band maps of eight $(Bi_{1-x}Sb_x)_2Te_3$ films with $0 \leq x \leq 1$ are shown in Fig. 15.3(a). Pure Bi_2Te_3 film shows well-defined surface states with linear dispersion but E_F also cuts through the bulk conduction band. With the addition of Sb, E_F moves downwards from the bulk conduction band, indicating the reduction of the electron-type bulk carriers. Moreover, the Dirac point moves upwards relative to the bulk valence band due to the increasing weight of the Sb_2Te_3 band structure. When the Sb content is increased to $x = 0.88$, both the Dirac point and E_F lie within the bulk energy gap. The system is now an ideal TI with a truly insulating bulk and a nearly symmetric surface Dirac cone with exposed Dirac point. When x increases from $x = 0.94$ to $x = 0.96$, E_F moves from above the Dirac point

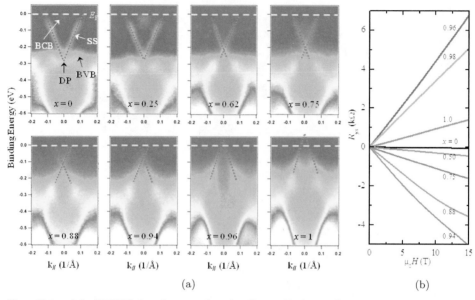

Fig. 15.3.   (a) ARPES bandmaps of eight $(Bi_{1-x}Sb_x)_2Te_3$ films with $0 \leq x \leq 1$.
(b) The Hall effect measured on the same films show consistent evolution of the electronic structure.

to below it, indicating a crossover from electron- to hole-type Dirac fermion gas. The charge neutrality point where E_F meets Dirac point can thus be identified to be located between $x = 0.94$ and $x = 0.96$.

Hall effect measurements on the $(Bi_{1-x}Sb_x)_2Te_3$ films show consistent evolutions. The variation of R_{yx} with magnetic fields is shown in Fig. 15.3(b). For films with $x \leq 0.94$, R_{yx} is always negative due to the existence of electron-type carriers. The Hall coefficient R_H increases with x in this regime due to the decrease of electron-type carrier density with Sb doping (recall that $R_H = 1/n_{2D}e$, n_{2D} is density of charge carrier). As x increases slightly from 0.94 to 0.96, the Hall curve suddenly jumps to the positive side with a large slope, indicating the reversal to hole-type Dirac fermions with a small carrier density. At even higher x, the slope of the positive curves decreases systematically due to the increase of hole-type carrier density.

The band structure engineering method allows us to reach the intrinsic insulating regime of TI, but then the question is whether the system can still establish long-range FM order when all the bulk carriers are depleted. In order to check this issue, we have grown a series of Cr-doped $(Bi_{1-x}Sb_x)_2Te_3$ films on sapphire (0001) substrates with the same Cr-doping level but different Bi:Sb ratio, and carried out Hall effect and ARPES measurements [18].

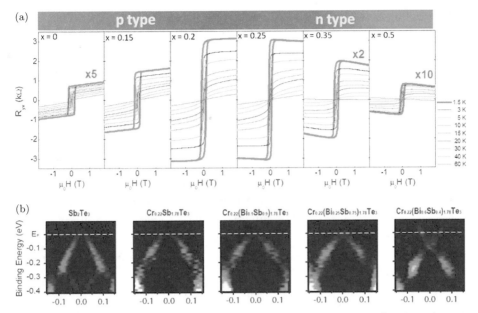

Fig. 15.4. (a) Magnetic field-dependent Hall resistance ρ_{yx} of $Cr_{0.22}(Bi_xSb_{1-x})_{1.78}Te_3$ films with $x = 0$, 0.15, 0.2, 0.25, 0.35 and 0.5. (b) ARPES measurements on four selected samples show that the surface states with linear dispersion are still present.

Figure 15.4(a) shows the ρ_{yx} vs. H curves of 5 QL $Cr_{0.22}(Bi_xSb_{1-x})_{1.78}Te_3$ films from $x = 0$ to $x = 0.5$. At the base temperature $T = 1.5$ K (the thicker lines), the Hall traces in all the films show nearly square-shaped hysteresis loops at low field, suggesting long-range ferromagnetic order with the easy magnetization axis perpendicular to the sample plane. With increasing Bi content, the ordinary Hall effect evolves from positive to negative, indicating the change of the dominating carriers from p- to n-type. So the ferromagnetism of Cr-doped $(Bi_xSb_{1-x})_2Te_3$ always exists despite the significant change in carrier type and density. The T_c of the films are in the range between 30 K to 35 K, showing weak dependence on carrier type and density even in the rather insulating samples around the crossover region. The carrier independent ferromagnetism supports the van Vleck mechanism, suggesting that FM insulator phase indeed exists in this material. The anomalous Hall resistance is significantly enhanced up to ~3 kΩ in insulating samples, much larger than the AHE observed in most ferromagnetic metals. Moreover, ARPES measurements on these samples in Fig. 15.4(b) can still resolve the linearly dispersed surface states, confirming that they are still TIs. Therefore, the Cr-doped $(Bi_{1-x}Sb_x)_2Te_3$ thin films are ferromagnetic, topological and insulating, which fulfill all the requirements for realizing the QAHE.

Band structure engineering by varied Bi/Sb ratio can tune the E_F into the bulk energy gap. However, to realize QAHE the E_F must precisely lie in the small energy gap at the Dirac point of the surface state opened by the ferromagnetism. If the Curie temperature of the FM TI is in the order of 10 K, a rough estimate suggests that the energy gap due to broken-TRS is only in the order of 1 meV. In order to control the E_F position to such precision, an external electrical gate is necessary.

A very convenient way of achieving gating is by growing the TI thin film on a SrTiO$_3$ substrate, which has a huge low temperature dielectric constant. By using SrTiO$_3$ as the gate dielectric material, one can realize carrier density variation of $\sim 3 \times 10^{13}$ cm^{-2} with a back-gate voltage (V_g) between ± 200 V for a typical substrate thickness of 0.5 mm [19]. The schematic drawing for measurement setup with SrTiO$_3$ substrate as gate dielectric is shown in Fig. 15.5(a). In Fig. 15.5(b), we display the gate-tuned AHE curves of a 5 QL Cr$_{0.15}$(Bi$_{0.1}$Sb$_{0.9}$)$_{1.85}$Te$_3$ film grown on SrTiO$_3$ (111) measured with different V_gs at $T = 1.5$ K. The coercivity and shape of the hysteresis loops are nearly unchanged with V_g, reconfirming the carrier independent ferromagnetism due to the van Vleck mechanism. At the same time, the AH resistance at $H = 0$ changes significantly with V_g. It increases from less than 5 kΩ at $V_g = 0$ to the maximum of close to 20 kΩ at $V_g = 80$ V, which is a significant fraction of the quantum resistance.

Up to now, all the requirements for realizing the QAHE can in principle be fulfilled in ultrathin Cr-doped (Bi$_{1-x}$Sb$_x$)$_2$Te$_3$ TI films grown on an STO

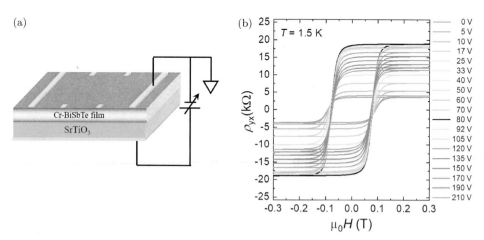

Fig. 15.5.   (a) Schematic setup of the bottom gate device using the STO substrate as dielectric material. (b) Gate-voltage dependence of the anomalous Hall effect in a Cr$_{0.22}$(Bi$_{0.2}$Sb$_{0.8}$)$_{1.78}$Te$_3$ film. The anomalous Hall resistance can be enhanced to close to 20 kΩ at $T = 1.5$ K.

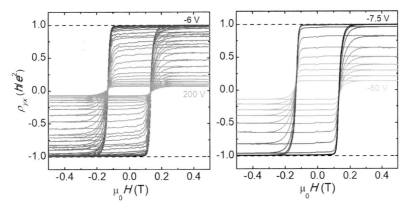

Fig. 15.6. The QAHE measured at $T = 30$ mK. Magnetic field dependence of ρ_{yx} at different V_gs show perfect quantization of the AH resistivity for a range of V_g near the charge neutrality point.

substrate. All we need to do is to further lower the measurement temperature so that the thermally activated conduction channels can be suppressed. This is particularly important due to the small energy gap ($\sim$1 meV) at the Dirac point of the surface state opened by the magnetic order. Figure 15.6 shows the magnetic field dependence of ρ_{yx} of a 5 QL $Cr_{0.15}(Bi_{0.1}Sb_{0.9})_{1.85}Te_3$ film at different V_gs measured at $T = 30$ mK in a dilution fridge [20]. Staring from $V_g = 200$ V, the AH resistance at zero magnetic field is merely $\sim$0.1 h/e^2 when the sample is strongly electron doped by a large positive gate voltage. With decreasing V_g, the AH resistivity increases rapidly, and for a range of V_g around the charge neutrality point $V_g^0 = -1.5$ V, the ρ_{yx} value is perfectly quantized and forms a wide plateau, which is the landmark signature of QAHE. With further decreasing of V_g to a large negative value, the ρ_{yx} value decreases again when too much holes are injected to the sample. Accompanying the formation of AH plateau, the longitudinal resistivity of the magnetic TI film decreases rapidly and approaches zero, indicating the dissipationless transport that is expected for QAHE.

Figures 15.7(a) and 15.7(b) display the magnetic field dependence of ρ_{yx} and ρ_{xx} to a magnetic field up to 18 T. Except for the large MR at H_c, increasing the field further suppresses ρ_{xx} towards zero. Above 10 T, ρ_{xx} vanishes completely, indicating a perfect QHE state. Since the increase in ρ_{xx} from zero (above 10 T) to 0.098 h/e^2 (at zero field) is very smooth and ρ_{yx} remains at the quantized value h/e^2, no quantum phase transition occurs and the sample stays in the same quantum Hall phase as the field sweeps from 10 T to zero field. Therefore, the complete quantization above 10 T can only be attributed to the same QAHE state at zero field.

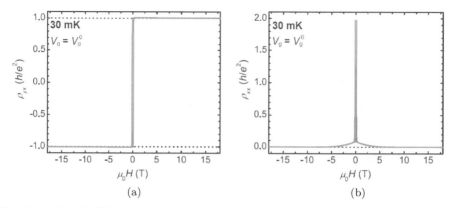

Fig. 15.7.   The QAHE under strong magnetic field measured at $T = 30$ mK. (a) Magnetic field dependence of ρ_{yx} at V_g^0. (b) Magnetic field dependence of ρ_{xx} at V_g^0. The blue and red lines in (a) and (b) indicate the data taken with increasing and decreasing fields, respectively.

These experimental observations concluded a decade-long search for the QAHE [21]. So far, the results have been confirmed by several groups around the world in magnetic TI thin films with different chemical compositions, on different substrates, and with different gating method [22–27]. Therefore, the QAHE in magnetic TI thus becomes a well-established phenomenon and represents a new class of topological quantum effect.

15.3   Recent progresses on the quantum anomalous Hall effect

Magnetically doped TIs are notorious "dirty" materials, and is partially responsible for the ultralow temperature that is needed to realize the QAHE. An ideal QAH insulator should be an intrinsic one, namely a stoichiometric compound with orderly arranged and exchange-coupled magnetic ions without extrinsic dopants. An important recent progress in the studies of QAHE is the finding of $MnBi_2Te_4$, a new stoichiometric magnetic TI compound. $MnBi_2Te_4$ has a layered ternary tetradymite structure composed of septuple-layers (SLs) bonded by van der Waals force, and each SL consists of seven elemental monolayers stacking in the sequence Te-Bi-Te-Mn-Te-Bi-Te. Interestingly, an SL of $MnBi_2Te_4$ on $Bi_2Te(Se)_3$ was reported to be able to open a large magnetic gap at the topological surface states of the latter [28, 29].

Gong *et al.* obtained single crystal thin films of $MnBi_2Te_4$ by repeatedly growing one Bi_2Te_3 QL and one MnTe bilayer with MBE [30]. With *in situ* ARPES, they found that the films thicker than 1 SL show Dirac-cone-shaped

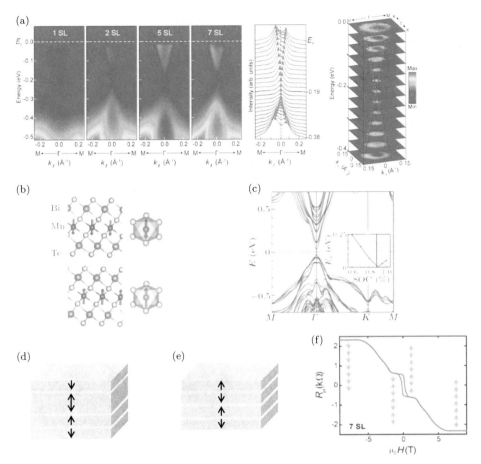

Fig. 15.8. MnBi$_2$Te$_4$ as an intrinsic magnetic topological insulator. (a) Topological surface states measured with ARPES. (b) Schematic lattice and magnetic structures of MnBi$_2$Te$_4$. (c) Band structure of MnBi$_2$Te$_4$ film obtained by first principles calculations. The inset displays the evolution of the gap size with the strength of SOC. The dip represents a topological phase transition from normal insulator to QAH insulator. (d, e) Schematic magnetic configurations of MnBi$_2$Te$_4$ films of odd-SL (d) and even-SL (e). (f) Step-like hysteresis loop of Hall resistance of a 7 SL MnBi$_2$Te$_4$ film. The arrows show the magnetic configuration at each layer.

linearly dispersed bands in the bandgap around E_F, which suggests that MnBi$_2$Te$_4$ is a 3D TI [Fig. 15.8(a)]. First-principles calculations reveal rich topological phases in this material resulting from the interplay between the SOC and magnetic order [31–33]. The ground state of bulk MnBi$_2$Te$_4$ is an 3D antiferromagnetic (AFM) TI with an A-type AFM configuration: neighboring ferromagnetic SLs with out-of-plane easy magnetic axis are AFM coupled [Fig. 15.8(b)]. Its topological property is protected by a combination of TRS and translational symmetry. An odd-SL MnBi$_2$Te$_4$

thin film is a QAH insulator with its top and bottom surfaces having the same magnetization direction but different topological characteristics [Figs. 15.8(c) and 15.8d)]. An even-SL $MnBi_2Te_4$ thin film, on the other hand, is an axion insulator since the top and bottom surfaces have magnetization vectors opposite to each other and thus topologically identical [Fig. 15.8(e)]. A magnetic field of several tesla can overcome the weak inter-SL AFM coupling of $MnBi_2Te_4$ and drive it into FM configuration. Most of the properties and topological phases of $MnBi_2Te_4$ have been verified in experiment. In flake samples exfoliated from $MnBi_2Te_4$ single crystals with the thickness from 3 SL to 9 SL, quantized anomalous Hall resistance and vanishing longitudinal resistance was observed by several groups at above 1.5 K, a temperature that is much higher than the QAHE in magnetically doped TIs and can be easily reached with common He-4 refrigerators [34–36].

In the 6-SL $MnBi_2Te_4$ flake, the magnetic field-driven quantum phase transition from axion insulator to Chern insulator has been observed [35]. This work demonstrates the intimate relationship between topology and magnetism in TI, which can be used to toggle between two distinct topological phases. The realization of such exotic states of matter at relatively high temperature in a simple stoichiometric compound makes the 6-SL $MnBi_2Te_4$ a perfect material platform for exploring other emergent topological quantum phenomena. In particular, the robust zero-magnetic-field axion insulator state in an easily accessible material significantly facilitates the search for quantized topological magnetoelectric effects and axion electrodynamics in condensed matter systems, another important theoretical prediction made by Shoucheng and his collaborators [37]. The field of topological quantum matter, exemplified by the QAHE, will continue to be an exciting field of research for the decades to come, thanks to the series of pioneering works by Shoucheng.

References

[1] K. V. Klitzing, G. Dorda and M. Pepper, *Phys. Rev. Lett.* **45**, 494 (1980).
[2] D. J. Thouless, M. Kohmoto, M. P. Nightingale and M. den Nijs, *Phys. Rev. Lett.* **49**, 405 (1982).
[3] F. D. M. Haldane, *Phys. Rev. Lett.* **61**, 2015 (1988).
[4] C. L. Kane and E. J. Mele, *Phys. Rev. Lett.* **95**, 226801 (2005).
[5] B. A. Bernevig and S. C. Zhang, *Phys. Rev. Lett.* **96**, 106802 (2006).
[6] L. Fu, C. L. Kane and E. J. Mele, *Phys. Rev. Lett.* **98**, 106803 (2007).
[7] M. Z. Hasan and C. L. Kane, *Rev. Mod. Phys.* **82**, 3045 (2010).
[8] X.-L. Qi and S.-C. Zhang, *Rev. Mod. Phys.* **83**, 1057 (2011).
[9] B. A. Bernevig, T. L. Hughes and S. C. Zhang, *Science* **314**, 1757 (2006).
[10] M. Konig, S. Wiedmann, C. Brune, A. Roth, H. Buhmann, L. W. Molenkamp, X. L. Qi and S. C. Zhang, *Science* **318**, 766 (2007).

[11] C.-X. Liu, X.-L. Qi, X. Dai, Z. Fang and S.-C. Zhang, *Phys. Rev. Lett.* **101**, 146802 (2008).
[12] H. J. Zhang, C. X. Liu, X. L. Qi, X. Dai, Z. Fang and S. C. Zhang, *Nat. Phys.* **5**, 438 (2009).
[13] Y. Xia *et al.*, *Nat. Phys.* **5**, 398 (2009).
[14] R. Yu, W. Zhang, H. J. Zhang, S. C. Zhang, X. Dai and Z. Fang, *Science* **329**, 61 (2010).
[15] Y. Zhang *et al.*, *Nat. Phys.* **6**, 584 (2010).
[16] M. H. Liu *et al.*, *Phys. Rev. B* **83**, 165440 (2011).
[17] J. S. Zhang *et al.*, *Nature Comm.* **2**, 574 (2011).
[18] C.-Z. Chang *et al.*, *Adv. Mater.* **25**, 1065 (2013).
[19] J. Chen *et al.*, *Phys. Rev. Lett.* **105**, 176602 (2010).
[20] C.-Z. Chang *et al.*, *Science* **340**, 167 (2013).
[21] S. Oh, *Science* **340**, 153 (2013).
[22] J. G. Checkelsky *et al.*, *Nat. Phys.* **10**, 731 (2014).
[23] X. Kou *et al.*, *Phys. Rev. Lett.* **113**, 137201 (2014).
[24] A. J. Bestwick *et al.*, *Phys. Rev. Lett.* **114**, 187201 (2015).
[25] C.-Z. Chang *et al.*, *Nat. Mater.* **14**, 473 (2015).
[26] M. H. Liu *et al.*, *Sci. Adv.* **2**, e1600167 (2016).
[27] S. Grauer *et al.*, *Phys. Rev. Lett.* **118**, 246801 (2017).
[28] T. Hirahara *et al.*, *Nano Lett.* **17**, 3493 (2017).
[29] M. Otrokov *et al.*, *2D Mater.* **4**, 025082 (2017).
[30] Y. Gong *et al.*, *Chin. Phys. Lett.* **36**, 076801 (2019).
[31] J. Li *et al.*, *Sci. Adv.* **5**, eaaw5685 (2019).
[32] D. Zhang *et al.*, *Phys. Rev. Lett.* **122**, 206401 (2019).
[33] M. M. Otrokov *et al.*, *Phys. Rev. Lett.* **122**, 107202 (2019).
[34] Y. Deng *et al.*, arXiv: 1904.11468.
[35] C. Liu *et al.*, arXiv: 1905.00715.
[36] J. Ge *et al.*, arXiv: 1907.09947.
[37] J. Wang, J. B. Lian, X. L. Qi and S. C. Zhang, *Phys. Rev. B* **92**, 081107 (2015).

Chapter 16

SciviK: A Versatile Framework for Specifying and Verifying Smart Contracts

Shaokai Lin[*,†], Xinyuan Sun[*,‡], Jianan Yao[†], and Ronghui Gu[†,‡]

†*Department of Computer Science, Columbia University*
‡*CertiK*

The growing adoption of smart contracts on blockchains poses new security risks that can lead to significant monetary loss, while existing approaches either provide no (or partial) security guarantees for smart contracts or require huge proof effort. To address this challenge, we present SciviK, a versatile framework for specifying and verifying industrial-grade smart contracts. SciviK's versatile approach extends previous efforts with three key contributions: (i) an expressive annotation system enabling built-in directives for vulnerability pattern checking, neural-based loop invariant inference, and the verification of rich properties of real-world smart contracts (ii) a fine-grained model for the Ethereum Virtual Machine (EVM) that provides low-level execution semantics, (iii) an IR-level verification framework integrating both SMT solvers and the Coq proof assistant.

We use SciviK to specify and verify security properties for 12 benchmark contracts and a real-world Decentralized Finance (DeFi) smart contract. Among all 158 specified security properties (in six types), 151 properties can be automatically verified within 2 seconds, 5 properties can be automatically verified after moderate modifications, and two properties are manually proved with around 200 lines of Coq code.

16.1 Introduction

While Shoucheng Zhang is known for his many outstanding contributions to physics, his contributions to the development and adoption of blockchain technology are also indispensable. Blockchain and other distributed ledger technologies enable consensus over global computation to be applied to situations where decentralization and security are critical [1]. Smart contracts are decentralized programs on the blockchain that encode the logic of transactions and businesses, enabling a new form of collaboration — rather than requiring a trusted third party, users only need to trust that

---

*Equal contribution.

the smart contracts faithfully encode the transaction logic [2]. Shoucheng Zhang viewed the blockchain and smart contract technology as a key to reach a new era where trust is built upon math [3]. However, the smart contract implementations are not trustworthy due to program errors and, by design, are difficult to change once deployed [1, 4], posing new security risks. Million dollars' worth of digital assets have been stolen every week due to security vulnerabilities in smart contracts [5].

Techniques such as static analysis and formal verification have been applied to improve smart contracts' security and ensure that given contracts satisfy desired properties [6–10]. While promising, existing efforts still suffer fundamental limitations.

There is a long line of work to improve smart contracts' security using highly automated techniques, such as security pattern matching [6], symbolic execution [7], and model checking [8,9]. These efforts can only deal with pre-defined vulnerability patterns and do not support contract-specific properties. For example, a smart contract implementing a decentralized exchange may need to ensure that traders are always provided with the optimal price. Such a property is related to the underlying financial model and cannot be verified using existing automated approaches. Furthermore, even for pre-defined patterns, these highly automated techniques may still fail to provide any security guarantee. For example, the accuracy rates of Securify [6] and Oyente [7] are 62.05% and 54.68%, respectively, on various benchmarks [11], making them ineffective and impractical in securing complex smart contracts.

Previous efforts focusing on developing mechanized proofs for smart contracts can provide security guarantees but often require substantial manual effort to write boilerplate specifications, infer invariants for every loop or recursive procedure, and implement proofs with limited automation support [12, 13], limiting their application to industry-grade smart contracts.

Most of the existing techniques on securing smart contracts, regardless of their approach, work either at the source code level or at the bytecode level. However, each of these two target levels has its own drawbacks. On the one hand, techniques working at the source code level are hard to adapt to the rapid development of smart contract's toolchain. Programming languages for writing smart contracts are still under active and iterative development and usually do not have a formal (or even informal) semantics definition. Solidity, the official language for Ethereum smart contracts, has released 84 versions from Aug 2015 to Jan 2021 and still does not have a stable formal semantics. On the other hand, due to a lack of program structure and the exposure of low-level EVM details, techniques working at the bytecode level usually only support a limited set of security properties while incapable of writing manual proofs for the properties that cannot be handled by automated techniques, such as the ones related to financial models [7, 10, 12, 14].

To address the above challenges, we present SciviK, a versatile framework enabling the formal verification of smart contracts at an intermediate representation (IR) level with respect to a rich set of pre-defined and user-defined properties expressed by lightweight source-level annotations.

SciviK introduces a source-level annotation system, with which users can write expressive specifications by directly annotating the smart contract source code. The annotation system supports a rich set of built-in directives, including vulnerability pattern checking (`@check`), neural-based loop invariant inference (`@learn`), and porting proof obligations to the Coq proof assistant (`@coq`). The annotation system offers the flexibility between using automation techniques to reduce the proof burden and using an interactive proof assistant to develop complex manual proofs.

SciviK parses and compiles annotated source programs into annotated programs in Yul [15], a stable intermediate representation for all programming languages used by the Ethereum ecosystem. Such a design choice unifies the verification of smart contracts written in various languages and various versions.

We develop an IR-level verification engine on top of WhyML [16], an intermediate verification language, to encode the semantics of annotated Yul IR programs. To make sure that the verified guarantees hold on EVM, SciviK formalizes a high-granularity EVM execution model in WhyML, with which verification conditions (VCs) that respect EVM behaviors can be generated from annotated IR programs and discharged to a suite of SMT solvers (Z3 [17], Alt-Ergo [18], CVC4 [19], etc.) and the Coq proof assistant [20].

To evaluate SciviK, we study the smart contracts of 167 real-world projects audited by CertiK [21], a blockchain security company invested by Shoucheng Zhang, and characterize common security properties into six types. We then use SciviK to specify and verify 12 benchmark contracts and a real-world Decentralized Finance (DeFi) smart contract with respect to these six types of security properties. Among all 158 security properties (in six types) specified for the evaluated smart contracts, 151 properties can be automatically verified within 2 seconds, 5 properties can be automatically verified after moderate retrofitting to the generated WhyML programs, and two properties are manually proved with around 200 lines of Coq code.

In summary, this paper makes the following technical contributions:

- An expressive and lightweight annotation system for specifying built-in and user-defined properties of smart contracts;
- A high-granularity EVM model implemented in WhyML;
- An IR-level verification framework combining both SMT solvers and the Coq proof assistant to maximize the verification ability and flexibility;
- Evaluations showing that SciviK is effective and practical to specify and verify all six types of security properties for real-world smart contracts.

16.2   Overview

To give an overview of SciviK, we use a simplified version of the staked voting smart contract, which is commonly used in DeFi, as a running example (see Fig. 16.1 for the Solidity contract). This contract allows participants on the blockchain to collectively vote for proposals by staking their funds. Contracts like this, such as the Band Protocol [22] staked voting application, have managed more than $150 million worth of cryptocurrencies. However, many bugs such as Frozen Funds [5] were frequently discovered in such voting-related multi-party contracts due to their complex business logic. Here, we intentionally insert three common vulnerabilities to this voting contract, respectively, on lines 12-13, 15, and 31, which cover vulnerabilities including reentrancy attacks [23], calculation mistakes, and incorrect data structure operations.

To demonstrate the workflow of SciviK and its architecture (see Fig. 16.2), we verify the correctness of the toy voting contract shown in Fig. 16.1 by finding the three inserted vulnerabilities and proving that the following theorems hold:

- T_1 (*On-Chain Data Correctness*): The data structure storing voted results will always reflect actual voter operations.
- T_2 (*Access Control*): Only authenticated users that have neither voted nor collected lottery rewards before can vote.
- T_3 (*Correctness of Voting Procedure*): The voting procedure itself is correct, i.e., stake accounts and user decisions are correctly updated after each vote.

Step 1: Writing annotations. SciviK requires users to provide the desired properties in the comments using its annotation system, which supports arithmetic operators, comparison operators, first-order logic operators, and built-in directives, such that a rich set of properties can be easily defined. For example, T_2, which states that people who have voted cannot vote again, can be formally specified by the @post tag on line 5 that if the user does not

```
1   /* @meta forall i: address, stakers[i] != 0x0 */
2   contract StakedVoting {
3     // ... variable declarations and functions ...
4
5     /* @post userVoted[msg.sender] -> revert
6      * @post ! old userVoted[msg.sender] ->
7      *   stakes[msg.sender] = old stakes[msg.sender] + stake
8      * @check reentrancy */
9     function vote(uint256 choice, uint256 stake, address token) public {
10      if (userVoted[msg.sender]) { revert(); }
11      rewards[msg.sender] += _lotteryReward(random(100));
12      bool transferSuccess = ERC777(token).
13                          transferFrom(msg.sender,address(this), stake);
14      /* @assume transferSuccess */
15      stakes[msg.sender] *= stake;
16      // ... the voting procedure and adding stakes to the stake pool ...
17      userVoted[msg.sender] = true; past_stakes.push(stake);
18      rebalanceStakers(); // sort the updated list of voters
19    }
20
21    /* @coq @pre sorted_doubly_linked_list stakers
22     * @coq @post sorted_doubly_linked_list stakers */
23    function rebalanceStakers() internal {
24      if (stakers[msg.sender] == address(0)) {
25        /* @assert (prevStakers[msg.sender] == address(0)); */
26        address prevStaker = HEAD;
27        for (address x; x != END; prevStaker = x) {
28          // ... add msg.sender node to right position in stakers
29        }
30      } else {
31        // delete old msg.sender node in stakers
32        delete stakers[msg.sender];
33        delete prevStakers[msg.sender];
34        rebalanceStakers();
35      }
36    }
37
38    function _lotteryReward(uint256 n) pure internal returns (uint) {
39      uint x = 0; uint y = 0;
40      // The user can also use "@inv" to manually provide a loop invariant.
41      /* @learn x y
42       * @pre n < = 15
43       * @post y < 1500
44       * @post x >= n */
45      while (x < n) { x += 1; y += x ** 2; }
46      return y;
47    }
48  }
```

Fig. 16.1. A simplified staked voting contract in Solidity with three vulnerabilities.

meet the requirement, the function reverts. This is an example of a function-level annotation that only holds for the annotated function call. Further, the @meta specification on line 1 asserts a contract-level invariant that the array of stakers will never contain the address 0x0 (an invalid address for msg.sender) at any observable point in the contract execution. The @post condition on line 6-7 ensures that if a qualified voter votes successfully, the voter's stakes increase by the number of staked tokens after the vote.

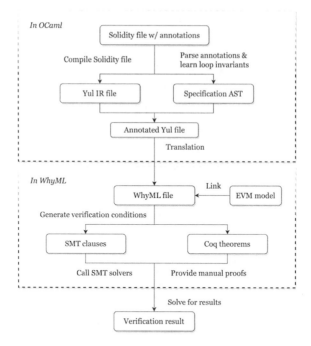

Fig. 16.2.   Architecture of SciviK verification framework.

The @check directive on line 8 instantiates a vulnerability pattern check that checks for any reentrancy attacks in the function. These specifications together contribute to the validity of $\mathcal{T}_3$.

The SciviK annotation system provides users the flexibility to reason about different properties at different abstraction levels. For example, one can choose to abstract the hashmap operators in Yul to functional data structures and inductive predicates (see §16.4 for details). Another example is the specifications annotated with @coq prefix on line 21-22, where it aims to prove $\mathcal{T}_1$ by asserting that, if stakers is a sorted doubly linked list, it remains a sorted doubly linked list after the insertion of a new voter node. This annotation enables SciviK to port the verification condition of rebalanceStakers to Coq.

Step 2: Parsing annotated programs into annotated Yul IR. SciviK parses the program with user-provided annotations and then uses the Continuous Logic Networks (CLN) [24] to automatically generate numeric loop invariants based on @learn directive. For example, the annotation on line 41 indicates that we expect SciviK to automatically infer a loop invariant for our pseudo-random _lotteryReward function.

After parsing annotated Solidity, SciviK compiles the smart contract using the solc compiler and outputs annotated Yul IR of the contract with embedded EVM opcodes. Based on the generated Yul IR, SciviK compiles annotations and learned invariants to IR-level annotations and insert them into the generated Yul IR. Since Yul is intermixed with low-level EVM opcodes and manipulates non-local variables via pointer arithmetic and hashing, at this step, SciviK also needs to map all non-local variables to corresponding memory and storage operations for later data type abstraction. For example, the Yul IR snippet that contains the IR-level annotation corresponding to the `@post` tag on line 6-7 is shown below:

```
/* ...
 * @post
 *    !old read_from_storage_split_offset_0_t_bool(
 *         mapping_index_access_t_mapping_t_address_t_bool_of_t_address(
 *            0x05, caller()))
 *    → read_from_storage_split_offset_0_t_uint256(
 *         mapping_index_access_t_mapping_t_address_t_uint256_of_t_address(
 *            0x06, caller()))
 *       = old read_from_storage_split_offset_0_t_uint256(
 *            mapping_index_access_t_mapping_t_address_t_uint256_of_t_address(
 *               0x06, caller()))
 *         + vloc_stake_226
 * ... */
function fun_vote_277(vloc_choice_224, vloc_stake_226, vloc_token_228) {
    ... the voting procedure ...
}
```

Step 3: Translating annotated Yul IR to WhyML. In this step, SciviK translates the compiled Yul IR into WhyML [16] with specifications inserted at designated locations. During translation, each Yul IR function, with the IR-level annotations, is mapped to a semantically equivalent definition in WhyML. The patterns specified by `@check` are also expanded to concrete specifications in WhyML. SciviK further links the generated WhyML to an EVM model that contains the semantic definitions of the EVM opcodes implemented in WhyML. A WhyML snippet of the translated example contract is shown below:

```
let ghost function fun_vote_277 (st_c: evmState) (vloc_choice_224: int)
  (vloc_stake_226: int) (vloc_token_228: int) : evmState
  ...
  ensures { !(Map.get st_c.map_0x05 st_c.msg.sender)
              → (Map.get result.map_0x06 result.msg.sender)
                = (Map.get st_c.map_0x06 st_c.msg.sender) + vloc_stake_226 }
  = let _r: ref int = ref 0 in
    let ghost st_g: ref evmState = ref st_c in
    try
      begin
        ... the voting procedure ...
      end;
      st_g := (setRet !st_g !_r);
      raise Ret
    with Ret → (!st_g)
    end
```

Step 4: Generating VCs and proofs. SciviK generates the VCs using the WhyML programs and the EVM model, and then discharges the VCs to SMT solvers. In case SMT solvers fail to solve the VCs automatically, the user can choose to port these proof obligations to the Coq proof assistant with SciviK and prove them manually in Coq's interactive proof mode.

We now dive into the three vulnerabilities in the example above. The first one is the reentrancy bug on line 12-13 that can be detected by SciviK's `@check reentrancy` directive on line 8. The reentrancy attack could happen as follows. On line 17, the state variable `userVoted` is updated after an invocation to an external function, namely, the user-provided ERC777 [25] token contract's function `transferFrom`. This function's semantics is to transfer `stake` amount of token from the voter to the contract. However, since the ERC777 token contract's address is provided by the user, the implementation of its `transferFrom` function could be malicious. For example, `transferFrom` could call the `vote` function again and claim reward multiple times on line 12-13, bypassing the check of `userVoted` at line 10. This kind of exploits on ERC777-related reentrancy has caused tens of millions of capital loss on DeFi protocol Lendf.me [26]. The exploit can be fixed if we place the call to the ERC777 token immediately after line 18. Furthermore, SciviK's discharge of the VC generated by annotation on line 6-7 does not pass and the SMT solver gives a counter-example, where we find that the stakes of a voter should be added rather than multiplied. During the process of manually proving the annotations on line 21 and 22, we find that Coq always fails to reason the case where an old voter node already existed in the doubly linked list `stakers`. After inspecting the implementation of `rebalanceStakers`, we find that there should have been a line of `stakers[prevStakers[msg.sender]] = stakers[msg.sender];` immediately after line 31 to connect the original voter node's predecessor and successor before their deletions.

Limitations and assumptions. SciviK does not support the generation of loop invariants and intra-procedural assertions for Coq. Among 158 security properties verified for evaluated smart contracts, 5 properties require manual retrofitting for the generated WhyML programs. This retrofitting step can be automated and is left for future work. SciviK trusts the correctness of the (1) compiler backend that compiles Yul IR programs to EVM bytecode, (2) the translation from source-level annotations to IR-level annotations, (3) the translation from annotated Yul IR programs to annotated WhyML programs, (4) the generation of verification conditions from WhyML programs, (5) SMT solvers, and (6) the Coq proof checker. We do not need to trust the compiler front-end that parses the source programs to IR programs.

16.3 The annotation system

As shown in Fig. 16.3, the annotation system of SciviK consists of a large set of directives for constructing specifications and verification conditions. SciviK supports different types of annotations, including pre-condition (@pre), post-condition (@post), loop invariant (@inv), global invariant that holds true at all observable states (@meta), assumption (@assume), and assertion (@assert). We illustrate the use of the above directives in the overview example (Fig. 16.1).

In Fig. 16.3, quantifiers and forms have their standard meanings. A pattern, denoted as pat, is an idiom corresponding to a well-established security vulnerability in smart contracts. These patterns are identified by analyzing existing attacks and can be easily extended. For example, the reentrancy pattern on line 8 in Fig. 16.1 checks for the classic reentrancy vulnerability which caused the infamous DAO hack [5]. Even though reentrancy attacks have been largely addressed in recent Solidity updates by limiting the gas for send and transfer functions [23], the threat of other forms of reentrancy still persists as there are no gas limits for regular functions, like the transferFrom function in ERC777 contract on line 12-13 of Fig. 16.1. Pattern overflow checks for integer overflow and underflow. As a word in the EVM has 256 bits, an unsigned integer faces the danger of overflow or underflow when an arithmetic operation results in a value

$$
\begin{array}{lll}
(\textit{IdPrefix}) & \textit{idp} & ::= \; \texttt{old} \mid \epsilon \\
(\textit{Ident}) & \textit{idnt} & ::= \; \texttt{x} \mid \texttt{result} \mid \textit{idp} \; \texttt{x} \\
(\textit{Quant}) & \textit{qunt} & ::= \; \texttt{forall} \mid \texttt{exists} \\
(\textit{Pattern}) & \textit{pat} & ::= \; \texttt{overflow} \mid \texttt{re-entrancy} \\
& & \quad \mid \texttt{timestamp} \\
(\textit{Status}) & \textit{stat} & ::= \; \texttt{return} \mid \texttt{revert} \\
(\textit{Prefix}) & p & ::= \; \texttt{@coq} \mid \texttt{old} \mid \epsilon \\
(\textit{Form}) & \textit{form} & ::= \; e \mid \textit{stat} \mid \; ! \; \textit{form} \mid \textit{form} \; \Rightarrow \; \textit{form} \\
& & \quad \mid \textit{form} \; \wedge \; \textit{form} \mid \textit{form} \; \vee \; \textit{form} \\
& & \quad \mid \textit{qunt} \; \overline{\textit{idnt} : t} \, , \; \textit{form} \\
(\textit{Spec}) & \textit{spec} & ::= \; p \; \texttt{@pre} \; \{ \; \textit{form} \; \} \mid p \; \texttt{@post} \; \{ \; \textit{form} \; \} \\
& & \quad \mid p \; \texttt{@meta} \; \{ \; \textit{form} \; \} \mid \texttt{@inv} \; \{ \; \textit{form} \; \} \\
& & \quad \mid \texttt{@assume} \; \{ \; \textit{form} \; \} \mid \texttt{@assert} \; \{ \; \textit{form} \; \} \mid \\
& & \quad \mid \texttt{@check} \; \textit{pat} \\
& & \quad \mid \texttt{@learn} \; \overline{\texttt{x}}
\end{array}
$$

Fig. 16.3. The syntax of SciviK's annotation system.

greater than 2^{256} or less than 0. Pattern `timestamp` check for timestamp dependencies. It is a kind of vulnerability in which the program logic depends on block timestamp, an attribute that can be manipulated by the block miner and therefore susceptible to consensus-level attacks. The notation `@check pat` checks if the given pattern `pat` is satisfied at any point of the program. The annotation system also supports built-in predicates for the termination status. If a function returns without errors, the `return` predicate evaluates to true and false otherwise. If a function terminates by invoking the `REVERT` opcode, the `revert` predicate evaluates to true. We describe in detail how vulnerability patterns are checked in §16.6.

When the `@coq` prefix is applied to certain annotation, SciviK ports these proof obligations to Coq [20] and launches an interactive proving session, a workflow enabled by Why3's Coq driver. In the example above, a property of interest would be that the mapping `stakers` always behaves as a sorted doubly linked list. This kind of verification condition is hard to prove using SMT solvers and can be designated to Coq with the `@coq` prefix. Note that assumptions, assertions, and loop invariants cannot be annotated for abstraction to Coq because they involve Hoare triples inside function bodies, which means they use intermediate variables that are mapped to compiler allocated temporaries in Yul, so for the simplicity of implementation we do not support them now.

16.4   Generating annotated IR

Smart contract intermediate representation.   Yul [15] is an intermediate representation (IR) for Ethereum smart contracts. It is designed to be the middle layer between source languages (Solidity, Vyper, etc.) and compilation targets (EVM bytecode [27], eWASM [28], etc.). Yul uses the EVM instruction set, including `ADD`, `MLOAD`, and `SSTORE`, while offering native support for common programming constructs such as function calls, control flow statements, and switch statements. These built-in programming constructs abstract away obscure low-level instructions such as `SWAP` and `JUMP` which are difficult to reason about. As such, Yul bridges the high-level semantics written in the contract's source code and the low-level execution semantics determined by the EVM backend. Because of these advantages, we use Yul IR as the verification target for SciviK. The parsing process from source code to Yul is completed by the compiler front-end (e.g., `solc` for Solidity) and does not need to be trusted. In fact, SciviK can be used to detect bugs in the compiler front-end (see §16.7).

Parsing source annotation to IR annotation Since SciviK reasons about the contract program at the Yul level, it generates Yul-level specifications from the source-level annotations. Most of the annotations describing functional correctness can be straightforwardly translated (by mapping procedures and variables to their Yul-level counterparts) once we encoded the memory allocation mechanism of Yul. The challenging part is how to deal with state variables that live in EVM storage and operate according to EVM storage rules, and data structures that live in EVM memory/ storage.

State variables in a contract are variables defined explicitly and accessible in that contract's scope. Recall from Fig. 16.4 that Yul, as an intermediate representation, does not have the notion of a state variable. According to the Ethereum blockchain specification [27], state variables are stored in `storage`. In the backend, `storage` is manipulated by opcodes such as `sstore` and `sload`. For example, line 26 in Fig. 16.1 refers to variable `HEAD`, which, in the translated Yul file, is a pointer to a hashed location in the storage. We observe that all state variables such as `HEAD` exhibit similar behaviors at the Yul level: they are mapped to storage segments and operations on them are abstracted to the following three functions: read ($\overline{\Psi}_r$), write ($\overline{\Psi}_w$), and metadata ($\overline{\Psi}_m$). Each operation function takes an initial identifier *id* differentiating the location of state variables in `storage`. All three operation functions first hash the parameters and then manage EVM storage by calling opcodes `sstore` and `sload` on the hashed values. In reality, precise modeling of these hash operations and storage management is infeasible for automated SMT solvers. Therefore, we model a state variable as a tuple $\langle id, \overline{\Psi}_r, \overline{\Psi}_w, \overline{\Psi}_m \rangle$. More generically, since variable operations depend on their type (mapping, array, struct, uint256, etc.), location (storage, memory), and declaration (dynamic or static), SciviK provide predefined templates of $\langle id, \overline{\Psi}_r, \overline{\Psi}_w, \overline{\Psi}_m \rangle$ tuples for each class of variables. For example, variable

$$
\begin{array}{ll}
(Lit) & l \in Nat \\
(Type) & t ::= uint256 \\
(Block) & b ::= \{\ \overline{s}\ \} \\
(Expr) & e ::= \mathtt{x} \mid \mathtt{f}(\overline{e}) \mid l \\
(Stmt) & s ::= b \mid \mathtt{break} \mid \mathtt{leave} \mid e \mid o \\
& \quad\ \mid \mathtt{if}\ e\ b \mid \mathtt{let}\ \mathtt{x} := e \\
& \quad\ \mid \mathtt{function}\ \mathtt{f}\ \overline{\mathtt{x}}\ \rightarrow\ \mathtt{r}\ b \\
& \quad\ \mid \mathtt{switch}\ e\ \overline{\mathtt{case}(l, s)}\ \mathtt{default}\ \overline{s} \\
& \quad\ \mid \mathtt{for}\ b\ e\ b\ b
\end{array}
$$

Fig. 16.4. The syntax of a restricted subset of Yul.

past_stakes from line 17 in Fig. 16.1 is a dynamic array. Its corresponding operation tuple is shown below (here, field *id* is 0x00 for past_stakes since it is the first state variable defined in the contract):

$$id \quad 0x00$$

$$\overline{\Psi}_r \text{ fun p i} \rightarrow \text{read\_from\_storage\_dynamic}$$
$$\quad \text{storage\_array\_index\_access\_t\_array\_storage p i}$$
$$\overline{\Psi}_w \text{ fun p i v} \rightarrow \text{update\_storage\_value p i v}$$
$$\overline{\Psi}_m \text{ fun p} \rightarrow \text{array\_length\_t\_array\_storage p}$$

When translating the annotations mentioning past_stakes, SciviK automatically expands abstract operators (e.g., past_stakes.length) into concrete Yul function calls like Ψ_m past_stakes, which corresponds to ((fun p → array_length_t_array_storage) past_stakes). Similarly, source-level specifications like past_stakes[0] = 0 are translated into Ψ_w *id* 0 0. Besides modeling storage and memory variables with abstracted functions, SciviK provides additional abstraction layer refinement theorems to the proof engine to reduce the reasoning effort, while also maintaining enough detail so that layout-related specification can still be expressed and checked. For a state variable in EVM storage like past_stakes, we take a cut through theorems about Ψ_w and Ψ_r, like the following one on direct storage reduction:

Theorem 16.1 (Storage Reduce).

$$\forall \phi : \texttt{evm\_state}, \; i : \texttt{int}, \; v : \texttt{int}. \; v = \texttt{pop} \; (\Psi_r(\Psi_w \; \phi \; i \; v))$$

Loop invariant learning. To verify the functional correctness of programs with loops, loop invariants must be provided. A loop invariant is a formula that holds true before and after each iteration of the loop. Decentralized gaming and finance applications, due to heavy numerical operations, often involve loops in their execution logic and require developers to generate loop invariants for verification, which is a non-trivial task. In principle, SciviK can plug-in any language-independent data-driven loop invariant inference tool. Currently, SciviK integrates Continuous Logic Networks (CLN) [24] into its verification workflow to automatically infer numeric loop variants based on simple user annotations. This step happens after we parsed source-level annotations and before we generate IR annotations.

Consider the _lotteryReward function in Fig. 16.1, which calculates the reward amount and transfers the reward to the recipient. In order to infer the desired invariant formula, SciviK requires the user to label the variable of interest using the @learn directive. SciviK then parses the loop source

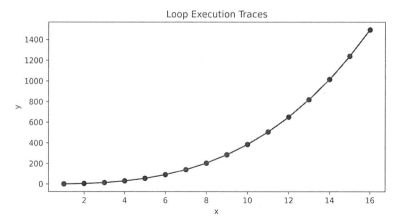

Fig. 16.5. Traces of the while loop in the _lotteryReward function.

code, reads the user-provided annotations, and keeps the specified variables on a list of monitored variables.

CLN infers invariant formulas based on loop execution traces. To obtain the traces of the **reward** function, SciviK first receives a WhyML version of the smart contract IR from the translator module and injects variable monitor code which keeps track of the intermediary values of the watched variables during each iteration of the loop. SciviK then executes the **reward** function based on the built-in EVM execution semantics, capturing the intermediary outputs of x and y (see Fig. 16.5).

The execution traces are then fed into CLN to learn the parameters and logical connectives of the formula [24]. For the while loop shown on line 45 in Fig. 16.1, CLN learns the following invariant:

$$6y - 2x^3 - 3x^2 - x = 0. \tag{16.1}$$

The learned invariant in Eq.(16.1) is then checked by Z3 against the specifications detailed on line 42-44 in Fig. 16.1 and proved to be valid.

When there are sequential or nested loops in a single function, CLN learns their invariants independently by logging execution traces at each loop and training separate neural models. When CLN fails to infer the correct invariant, SciviK prompts the user to manually provide a loop invariant.

16.5 Translating annotated IR into WhyML

Intermediate verification language. After generating the annotated Yul IR from a smart contract source code, SciviK translates the Yul IR into an intermediate verification language called WhyML (syntax in Fig. 16.6) [16].

$$
\begin{array}{lll}
(\mathit{Lit}) & l \ \in \ \mathit{Nat} \\
(\mathit{Type}) & t \ ::= \ \texttt{int} \\
(\mathit{Spec}) & \mathit{wsp} \ ::= \ \texttt{assume} \ \{ \ e \ \} \ | \ \texttt{assert} \ \{ \ e \ \} \\
& \qquad\quad | \ \texttt{requires} \ \{ \ e \ \} \ | \ \texttt{ensures} \ \{ \ e \ \} \\
(\mathit{FuncDef}) & \mathit{fd} \ ::= \ \texttt{let function f} \ \mathit{spec}^* \ = \ \mathit{spec}^* e \\
(\mathit{Expr}) & e \ ::= \ \texttt{x} \ | \ l \ | \ e \ \oplus \ e \ | \ (\ e \ ) \ | \ e \ ; \ e \ | \ \mathit{fd} \\
& \qquad\quad | \ \texttt{if} \ e \ \texttt{then} \ e \\
& \qquad\quad | \ \texttt{while} \ e \ \texttt{do} \ \mathit{wsp}^* \ e \ \texttt{done} \\
& \qquad\quad | \ \texttt{let x} \ = \ e \ \texttt{in} \ e \\
& \qquad\quad | \ \texttt{x} \ := \ e \\
& \qquad\quad | \ \texttt{match} \ e \ \texttt{with} \ \overline{(| \ \mathit{pattern} \ \to \ e)}^{\,+} \ \texttt{end} \\
& \qquad\quad | \ e \ e^{+} \ | \ \{ \ \overline{\texttt{x}\,=\,\texttt{e}} \ \} \ | \ e.\texttt{x} \ | \ \texttt{raise x} \\
& \qquad\quad | \ e \ : \ t \ | \ e \ \mathit{wsp}^{+} \\
& \qquad\quad | \ \texttt{try} \ e \ \texttt{with x} \to e \ \texttt{end} \\
& \qquad\quad | \ \texttt{begin} \ e \ \texttt{end} \\
& \qquad\quad | \ \texttt{function} \ \overline{\texttt{x}} \ \to \ e \\
(\mathit{Exception}) & \mathit{exp} \ ::= \ \texttt{exception x} \\
(\mathit{Decl}) & \mathit{wsp} \ ::= \ \mathit{exp} \ | \ \mathit{fd} \\
(\mathit{Module}) & \mathit{module} \ ::= \ \texttt{module m} \ \mathit{decl}^* \ \texttt{end} \\
(\mathit{Prog}) & p \ ::= \ \overline{(\mathit{theory} \ | \ \mathit{module})}^{\,*}
\end{array}
$$

Fig. 16.6.   The syntax of a restricted subset of WhyML.

An intermediate verification language combines a software program with specifications and serves as a middle layer between programs and theorem provers. WhyML is a functional programming language and a part of a larger deductive verification framework Why3 [29].

EVM execution semantics modeling. Since Yul embeds EVM opcodes, we provide an EVM formalization as a WhyML library. Our library formalizes opcode semantics as a state machine using function primitives. Specifically, EVM operations in Yul can either be pure functions (i.e., functions that do not generate side-effects, such as ADD, which simply returns the sum of the two input values) or impure (such as SSTORE, which modifies the underlying storage of the EVM). The presence of both pure and impure functions in Yul not only makes it difficult to translate Yul to WhyML, but also poses a challenge in analyzing EVM state during the execution of a function. To address these two issues, we make our EVM model functional, which means that every EVM instruction takes an input state and returns an output state. Impure functions can directly return a state with updated attributes (e.g., SSTORE returns a new EVM state with updated storage) and pure functions simply return a state with an updated stack, which holds the results of these computations. We extend prior EVM formalizations, including a Lem formalization [30] and Imandra EVM [31], and present the first implementation in WhyML. In our execution semantics model, we specify the semantics of the EVM state, message, memory, storage, and instructions, according to the Ethereum Yellow Paper [27], which outlines standard EVM behaviors.

We now walk through two components of SciviK's EVM execution semantics: the EVM state and the memory model. Since EVM is a stack-based machine with a word size of 256-bit, we introduce the **word** type using WhyML's built-in integer theory and place a bound $0 \leqslant i < 2^{256}$ to model the valid range of the bit array.

We define the EVM state in Fig. 16.7. EVM's memory is a storage of temporary data that persists during smart contract execution and is deleted at the end of the contract call. Since EVM memory is a word-addressed byte array, we implement the **data** field of the memory with a list of words using the built-in list theory in WhyML. The **size** field stores the current number of bytes in the memory.

In Yul IR, EVM operations can either be pure functions (i.e., functions that do not generate side-effects, such as **ADD**, which simply returns the sum of the two input values) or impure (such as **SSTORE**, which modifies the underlying storage of the EVM). The presence of both pure and impure operations in Yul not only makes it difficult to translate Yul to WhyML, but also poses a challenge in analyzing EVM state during the execution of a function. To address this issue, we make our EVM model functional, which means that every EVM instruction takes an input state and returns an output state. Impure functions can directly return a state with updated attributes (for example, **SSTORE** returns a new EVM state with an updated storage), and pure functions simply return a state with an updated **stack**, which holds the results of these computations.

We present an example of SciviK's EVM instruction semantics modeling in Fig. 16.8. The **MLOAD** instruction takes an EVM state as input and outputs a new EVM state. We implement functions such as **get_mem** and **pad_mem**

```
1   type array_t   = { data      : list word;
2                       size      : word      }
3
4   type message_t = { recipient  : word;
5                       sender     : word;
6                       value      : word;
7                       gas        : word      }
8
9   type state_t   = { stack      : list word; (* stores return values *)
10                      calldata   : list word; (* stores input params *)
11                      memory     : array_t;
12                      storage    : array_t;
13                      message    : message_t;
14                      pc         : word;      }
```

Fig. 16.7.   EVM state definition.

```
1   (* Extract param from state *)
2   let ghost function param (s : State.t) (i : int) : word
3   = let params = s.cd in nth i params
4
5   (* EVM internal operations *)
6   let function ceiling (x : word) : word
7   = (x + byte_size) / byte_size
8
9   let function round_up (max_index i : word) : word
10  = let end_idx = i + byte_size in max max_index (ceiling end_idx)
11
12  let rec function k_zeroes (k : word) : list word
13  variant { k }
14  = if k < 0 then (Nil : list word)
15    else (Cons 0 (Nil : list word)) ++ (k_zeroes (k - 1))
16
17  let function pad_mem (data : list word) (cur_size new_size : word) : list word
18  = let d = data in let zeros = k_zeroes (new_size - cur_size) in d ++ zeros
19
20  let ghost function get_mem (byte_idx : word) (mem : t) : t
21  = let size_w = mem.cur_size in
22    if byte_idx < (size_w * 32) then
23      let d = nth (byte_idx / 32) mem.data in { mem with peek = d }
24    else
25      let new_size = round_up mem.cur_size byte_idx in
26      let new_data = pad_mem mem.data mem.cur_size new_size in
27      { data = new_data; cur_size = new_size; peek = 0 }
28
29  (* MLOAD instruction *)
30  let ghost function mload (s : state_t) : state_t
31  = let mem' = (get_mem (param s 0) s.m) in
32    { s with m = mem' ; sk = Cons mem'.peek Nil }
```

Fig. 16.8.   Execution semantics of the MLOAD instruction in WhyML.

which simulate the internal behaviors of the memory. According to the Ethereum Yellow Paper, EVM memory automatically adjusts its size when the memory location being accessed is greater than the current memory size, under which circumstance it resizes itself to the queried location and dynamically increases the gas fee.

To ensure our formalized EVM model is correct, we write specifications for the opcode instructions and then verify their correctness against our implementation in WhyML. Take the MLOAD function for example, the Yellow Paper describes the following behaviors of MLOAD:

- The gas fee associated with MLOAD is proportional to the smallest multiple of 32 bytes that can include all memory indices in its range;
- If the queried memory address exceeds the current bound of active memory, memory is extended to include the queried memory address;
- The content of the memory, excluding extended sections, does not change after each MLOAD.

```
1    (* s0 is the input state. s1 is the output state. *)
2    (* Gas cost of MLOAD is proportional to memory resource consumption. *)
3    assert { let cost = (mem_cost s1.m - mem_cost (s0.m) + gas_verylow) in
4             s1.gas = (s0.gas) - cost }
5
6    (* If queried memory address exceeds memory bound, extend the address size such
             that the address is included. *)
7    assert { (byte_idx < (s0.m.size) ∧ s1.m.size = (s0.m.size))
8             ∨  (byte_idx ≥ (s0.m.size)
9                ∧ s1.m.size = (word_ceiling (s0.m.size) byte_idx)) }
10
11   (* Content of memory remains the same after MLOAD. *)
12   assert { forall i. 0 ≤ i ≤ s1.m.size → s1.m.data[i] = old s1.m.data[i] }
```

Fig. 16.9.   Specifications of `MLOAD` according to Ethereum Yellow Paper.

The specifications of `MLOAD` are discharged by the translator as `assert` statements in the translated WhyML contract, as shown in Fig. 16.9. With the opcode instructions checked against the expected behaviors outlined in the Yellow Paper, we ensure that the EVM model conforms to the formal semantics and provides the capability to reason about the correctness of smart contracts from a low-level view.

Translating Yul IR to WhyML.   With the EVM model implemented in WhyML, we can now translate the Yul program into WhyML. The translation rules are defined in Fig. 16.10. We use $\bar{e}$ to denote a sequence of expressions, with the expanded form of $e_1, e_2, ..., e_i$. We write $e \triangleright \kappa$ to denote that expression e can be translated into κ in SciviK. Σ, the exception identifier, and its return value symbol, σ of type *unit*, are used to mimic imperative control flow. π denotes the variable to be returned (corresponding to r in the Yul syntax in Fig. 16.4), and ϕ denotes the EVM state variable that we explicitly pass throughout instructions.

We present the inference rule for translating Yul IR to WhyML in Fig. 16.10. We now expand on how SciviK treats certain Yul language features. Ethereum smart contracts provide globally available variables about the current transaction and block, accessible by direct reference at the source level (front-end languages). For example, `msg.sender` returns the address of the function caller, `msg.value` returns the amount of wei (the smallest denomination of ether) wired to current contract.

When translating the contract program, SciviK treats these opcodes just like function calls. This is enabled by the aforementioned EVM model, which defines concrete behaviors for the opcodes. As shown in Fig. 16.1, the `msg.sender` on line 10 is translated into `caller()`. On the other hand, the translation of certain opcode calls written in user-provided annotations is treated differently. For example, the `msg.sender` on line 5, which is

$$\frac{}{x} \; \text{(Var)} \qquad \frac{\vdash IsReturnVar(\text{x})}{\vdash \text{x} \triangleright \pi} \; \text{(VarRet)} \qquad \frac{\text{leave}}{\text{raise } \Sigma_l} \; \text{(Leave)}$$

$$\frac{\vdash \overline{e} \triangleright \overline{\kappa} \qquad \vdash \text{f} \triangleright \Psi}{\vdash \text{f} \, (\, \overline{e} \,) \; \triangleright \; (\; \phi := \text{push } \overline{\kappa} \, ; \phi := \Psi \, \phi \, ; \text{ pop } \phi \,)} \; \text{(Call)}$$

$$\frac{\vdash e \triangleright \kappa \qquad \vdash b \triangleright \overline{\tau}}{\vdash \text{if } e \, b \; \triangleright \; (\text{ if } \kappa \text{ then } \overline{\tau} \,) \, ;} \; \text{(If)}$$

$$\frac{\vdash \text{x} \triangleright \text{x} \qquad \vdash e \triangleright \kappa}{\vdash \text{let x} := e \; \triangleright \; \text{let x } = \kappa \, ;} \; \text{(Asg)} \qquad \frac{\vdash \text{x} \triangleright \text{x} \qquad \vdash e_1 \triangleright \kappa_1 \qquad \vdash e_2 \triangleright \kappa_2}{\vdash \text{let x} := e_1 \, , \; e_2 \; \triangleright \; \text{let x } = \kappa_1 \text{ in } \kappa_2} \; \text{(Seq)}$$

$$\frac{\vdash e \triangleright \kappa \qquad \vdash \overline{(l,s)} \triangleright \overline{(\ell, \tau)} \qquad \vdash \overline{s}_i \triangleright \overline{\tau}_i}{\vdash \text{switch } e \; \overline{\text{case } (l,s)} \text{ default } \overline{s}_i \; \triangleright \; \text{match } \kappa \text{ with } \overline{(\, | \, \ell \, \rightarrow \, \tau \,)} \, \overline{\tau}_i \text{ end}} \; \text{(Switch)}$$

$$\frac{\vdash e \triangleright \kappa \qquad \vdash b_1 \triangleright \overline{\tau}_1 \qquad \vdash b_2 \triangleright \overline{\tau}_2 \qquad \vdash b_3 \triangleright \overline{\tau}_3}{\vdash \text{for } b_1 \; e \; b_2 \; b_3 \; \triangleright \; \overline{\tau}_1 \, ; \text{ while } \kappa \text{ do } (\text{ try } \overline{\tau}_2 \, ; \; \overline{\tau}_3 \text{ with } \Sigma \rightarrow \sigma \text{ end }) \text{ done}} \; \text{(Loop)}$$

$$\frac{\vdash \text{f} \triangleright \Psi \qquad \vdash \text{r} \triangleright \pi \qquad \vdash \overline{\text{x}} \triangleright \overline{\kappa} \qquad \vdash b \triangleright \overline{\tau}}{\begin{array}{c} \vdash \text{function f } \overline{\text{x}} \; \rightarrow \text{r } b \triangleright \\ \text{let function } \Psi \; \overline{\kappa} \; = \; (\text{ try } \overline{\tau} \text{ with } \Sigma_l \rightarrow \text{push } \pi \text{ end }) \,) \end{array}} \; \text{(Def)}$$

Fig. 16.10.   The translation inference rule of SciviK.

being used in the annotation rather than in the program, is translated into ϕ.message.sender.

Dealing with the complex semantics of inter-contract calls is also challenging. In particular, we specify functions being Effectively Callback Free (ECF) [32]. ECF is a property that avoids most blockchain-level bugs by restricting R/W on transaction states (e.g., globally available variables and storage) after calling external contract functions. Existing algorithms for detecting ECF either are online or inspect deployed bytecode and transaction histories. Since it is impossible to know what other contracts do before deployment, SciviK requires input from the developer to secure the ECF property. It asks if a function f changes the EVM state. If it is, SciviK proceeds by adding an axiom specifying that the return state of f satisfies the same predicate as if there were no callback altering the original ϕ. For example, if we have an external contract call like transferFrom on line 12-13 of Fig. 16.1, SciviK inserts and axiom shown in Fig. 16.11:

where push is used to load the two arguments of sendEtherTo into state st.

For state variables and data structures, SciviK maps them to abstract data types in ϕ. For example, operations on the userVoted state variable

```
1  axiom transferFromECF : forall st: evmState, st' : evmState, sender: address,
      receiver: address, amount: uint256.
2    st = push sender receiver amount
3    → st' = trnasferFrom st
4    → st' = st
```

Fig. 16.11. WhyML axiom of `transferFrom`'s ECF condition.

on line 17 in Fig. 16.1, which was translated to a set of Ψ_w, Ψ_r, and Ψ_m on function ids in the last phase, are further translated into abstract map operations, namely, uninterpreted functions. As shown by step 2 and step 3 in §16.2, Ψ_r 0x05 i, reading element at i of mapping `userVoted`, will be translated into `Map.get` ϕ.`map_0x05 i`.

Yul has the traditional C-style implicit type conversions between values of type `uint256` and `bool`. Since WhyML does not support such features, SciviK simulates this implicit conversion by restricting the type of Yul to only `int` and treats all operations on `bool` with `iszero`, with explicit conversion functions. The reason we do not model our types using 256-bit bitvectors is that its theory is not scalable with certain SMT solvers. With integers, reasoning can be much simplified.

There exists a gap (scoping, continuation, and return type) between Yul, an imperative IR, and WhyML, a functional language. SciviK fills this gap by mimicking imperative control flow constructs (e.g., `while`, `for`, `return`, and `break`) by exceptions. Since Yul only has well-structured control flow operators (i.e., no `goto`), this modeling is sound in the sense that it creates the exact same CFG structure.

We also experimented with using a fully monadic translation style, where we model all the imperative continuations and mutations through monad transformers, but our experiments showed that this representation actually made the SMT clauses much more complex (through the higher-order reasoning of `unit` and `bind`) and therefore cannot be efficiently solved. More importantly, a monadic representation of the program breaks the specification written at the front-end level. For example, loop invariants cannot be easily translated to their counterparts, and often we have to manually rewrite the specifications at the IR (Yul) level.

16.6 Generating verification conditions

SciviK generates a variety of verification conditions at different abstraction levels from WhyML. With various drivers provided by Why3, SciviK can leverage SMT solvers as well as the Coq proof assistant to verify the correctness specifications.

Checking vulnerability patterns. The Ethereum community has long observed security attack patterns and set up community guidelines to avoid pitfalls during contract development. Recall that SciviK allows the user to specify a @check directive at the source level to detect certain pre-defined vulnerability patterns (see Fig. 16.3). SciviK implements the verification for three vulnerability patterns: integer overflow, reentrancy, and timestamp dependence. SciviK uses WhyML and static analysis to perform pattern checking. For the overflow pattern, SciviK generates specifications in WhyML to be verified. For the reentrancy and timestamp patterns, SciviK performs static analysis to detect any potential vulnerabilities.

If the @check overflow annotation is specified, SciviK automatically adds assertions to all the integer variables that check whether the program handles integer overflow and underflow. For an unsigned integer variable, v, that has b bits in size, WhyML inserts the following assertion at the end of the function:

$$\texttt{assert}\,\{\,\neg(0 \leqslant v < 2^b) \Rightarrow \texttt{revert}\,\} \quad \text{(unsigned integer)}$$

Similarly, for a signed integer variable, WhyML inserts the following assertion at the end of the function:

$$\texttt{assert}\,\{\,\neg(-2^{b/2} \leqslant v < 2^{b/2}) \Rightarrow \texttt{revert}\,\} \quad \text{(signed integer)}$$

where revert is the termination status in the annotation system. We choose to place the assertions at the end of a function, instead of placing right after arithmetic operations. This allows for correct checking of the following Ethereum smart contract development common practice: developers first intentionally perform an unsound arithmetic operation and then check if the result is larger (in case of underflow) or smaller (in case of overflow) than the two operands. If it is, revert the EVM state (i.e., exit the program and rollback to previous memory and storage state), otherwise continue program execution. In this case, using the naive overflow check (placing assertion immediately after variable definition) will give a false warning on the intentionally overflowed (underflowed) variable. However, with our approach, intentionally overflowed variables will have enough time to propagate their results to where smart contract developers manually check overflow. We notice that solc-verify [33] also adopt this approach. However, since their annotation system and framework is at the source (Solidity) level, they introduce extra false negatives by delaying the assertion. Consider a overflowed variable x, the delayed assertion at the source code-level permits an edge case where x is modified by operations before the delayed assertions such that it still passes certain overflow checks. Now the approach using delayed assertions at the source-code level does not detect the overflow incident. In our approach, since SciviK discharges VCs at the IR level (where every variable is in SSA form) and Yul create temporary variables for assignments, the reassignment of x at the end of the function can detect the overflow incident happened to a snapshot of the x variable.

This approach also has its limitation when certain SMT solvers cannot handle 256-bit integers; however, smart contract development has focused increasingly on gas efficiency, which promotes the use of cheaper data types[34] such as uint64, which can be handled by most SMT solvers.

To detect time dependency vulnerabilities, SciviK adds specifications requiring that variables within the contract do not depend on the current timestamp, which can be manipulated by the miner. [35] To detect the reentrancy pattern, SciviK performs a static analysis on the Yul IR level to ensure that all storage related-operations are placed before external contract calls. If this property is violated, SciviK produces a warning signaling a potential reentrancy error.

Discharging VCs from SMT solvers. For each of the verification condition pair (pre-condition and post-condition), SciviK discharges VCs for a SMT solver chosen by the user, a process enabled by Why3's drivers. Currently, the supported solvers are Z3 [17], Alt-Ergo [18], and CVC4 [19].

As mentioned in the prior section, some SMT solvers are not capable of handling 256-bit unsigned integers. Without loss of generality, we reduce the word size to 64 bits in Fig. 16.15 and Fig. 16.16 when demonstrating the vulnerability.

Discharging VCs from Coq. For annotations written with @coq, SciviK discharges the Coq proof assistant by directly calling WhyML's porting mechanism, which shallowly embeds annotated WhyML program as Gallina, Coq's specification language. For example, the ported Coq proof obligation for annotation on line 21 and 22 in Fig. 16.1 is as follows:

```
Theorem rebalanceStakers'vc :
  sorted_doubly_linked_list (map_0x03 st_c)
  -> let st_c2 := rebalanceStakers st_c in
    sorted_doubly_linked_list (map_0x03 st_c2).
```

In the above theorem, st_c is the EVM state when user is calling the function rebalanceStakers. It has a Coq Record type with a field map_0x03 defined as an uninterpreted function from addresses to addresses. map_0x03 is an abstraction of the source-level variable stakers, which stores the addresses of users who participated in the staked voting. The st_c variable is generated as a *Parameter* (bound to its type in the global environment) in the Coq file. The sorted_doubly_linked_list is a user defined property which asserts that the mapping forms a sorted doubly linked list. `rebalanceStakers` is also defined as an uninterpreted function with its behavior bounded by an axiom in the environment called `rebalanceStakers'def` which embeds the function body in Gallina.

16.7 Evaluation

Environment. All the experiments are conducted on a Ubuntu 18.04 system with Intel Core i7-6500U @ 4x 3.1GHz and 8GB RAM. The specific versions of tools used are Solidity 0.8.1, Why3 1.3.1, Alt-ergo 2.3.2, Z3 4.8.6, and CVC4 version 1.7. We set the timeout of SMT solvers in SciviK to 5 seconds.

Benchmarks. We compiled a set of smart contracts, a part from prior work on VeriSol [36], that reflects common uses of Solidity, and a real-world Decentralized Finance (DeFi) [37] protocol contract that mimics Uniswap [38]. We show the details of those benchmarks in Table 16.1. The "Scenario" column briefly explains each contract's purpose. For example, the contract `BazaarItemListing` allows users to list items publicly on the blockchain and then trade in a decentralized way; therefore its scenario is *Market*. The *Governance* scenario means that the contract is used for managing interactions between multiple on-chain parties. Other scenarios like *Game* and *Supply Chain* means the contract encodes application-specific business logic. The "LOC" column is defined as the total lines of Solidity program, and the length of the generated Yul program is shown in the "LOC (Yul)" column.

Table 16.1. SciviK benchmark statistics.

Contract	Scenario	LOC	LOC (Yul)
AssetTransfer [36]	Banking	218	1,789
BasicProvenance [36]	Governance	53	686
BazaarItemListing [36]	Market	130	3,653
DefectiveComponentCounter [36]	Utility	44	1,026
DigitalLocker [36]	Utility	152	1,636
FrequentFlyerRewardsCalculator [36]	Utility	60	1,072
HelloBlockchain [36]	Education	45	1,013
PingPongGame [36]	Game	95	2,964
RefrigeratedTransportation [36]	Supply Chain	145	1,865
RefrigeratedTransportationWithTime [36]	Supply Chain	110	1,912
RoomThermostat [36]	Utility	50	755
SimpleMarketplace [36]	Exchange	74	1,055
UniswapStyleMarketMaker [38]	Exchange	135	2,197

Properties of interest. To understand the common types of desired security properties in real-world smart contracts, we conduct a thorough analysis of our smart contract benchmark and the smart contracts of 167 real-world projects audited by CertiK from the year of 2018 to 2020 [21], including 19 DeFi protocols whose TVL (total value locked) are larger than 20M USD, and categorize six types of desired security properties for smart contracts.

T1: *User-defined functional properties* represent whether smart contract implementations satisfy the developer's intent.

T2: *Access control requirements* define proper authorization of users to execute certain functions in smart contracts.

T3: *Contract-level invariants* define properties of the global state of smart contracts that hold at all times.

T4: *Virtual-machine-level properties* specify the low-level behaviors of state machine to ensure the execution environment functions as expected.

T5: *Security-pattern-based properties* require that the smart contract does not contain insecure patterns at the IR level.

T6: *Financial model properties* define the soundness of economic models in decentralized financial systems.

Results. SciviK successfully verifies a total of 153/158 security properties listed in Table 16.2. The security properties are categorized into the six types above. In the table, T1 to T6 correspond to the six types of security properties. The "Time(s)" column shows the total time for SciviK to prove all six types of security properties. Out of the 153 verified properties, 136 are verified at the WhyML level via SMT solvers, two are verified by porting definitions to Coq, and 15 are verified using static analysis.

Now we investigate the reasons why certain checks have failed. Verification conditions in SciviK fail when SMT solvers either timeout or return a counterexample.

In our experiment, out of the 5 properties that SciviK failed to verify, all were due to timeout. After inspecting the generated VCs, we found that there are two reasons for solver timeout. First, some of the translated contract representations in WhyML have too many, often redundant, verification conditions. In these cases, we manually compress the WhyML program by removing temporary variables to reduce redundancy. Second, certain VCs are inherently complex due to the logic of the contract as well as user annotations. We find that adding intermediary assertions and using Why3's goal-splitting transformations can help the SMT solvers terminate. SciviK

Table 16.2.   Experimental results of SciviK

Contract	T1	T2	T3	T4	T5	T6	Total	Solved	Time(s)
AssetTransfer [36]	10	19	0	2	3	0	34	32	8.86
BasicProvenance [36]	2	2	1	0	0	0	5	5	4.84
BazaarItemListing [36]	5	3	0	1	3	0	12	12	8.95
DefectiveComponentCounter [36]	2	1	0	2	2	0	7	7	4.09
DigitalLocker [36]	10	10	1	7	0	0	28	27	13.57
FrequentFlyerRewardsCalculator [36]	2	1	0	2	2	0	7	5	3.69*
HelloBlockchain [36]	2	1	1	0	0	0	4	4	2.25
PingPongGame [36]	5	0	0	1	3	0	9	9	5.26
RefrigeratedTransportation [36]	3	10	0	0	3	0	16	16	6.29
RefrigeratedTransportationWithTime [36]	3	3	0	0	6	0	12	12	7.70
RoomThermostat [36]	3	3	1	0	1	0	8	8	2.95
SimpleMarketplace [36]	3	6	1	1	1	0	12	12	5.89
UniswapStyleMarketMaker [38]	0	0	0	1	0	3	4	4	1.25*
Total	50	59	5	17	24	3	158	153	75.59
Average time for each property (s)	0.75	0.18	1.54	1.11	0.05	0.13	—	—	—

*The total verification time shown here is only for properties that can be automatically solved by Z3. The verification of both contracts also include a manual proof of around 200 lines of Coq, whose time is not included in the table.

successfully verifies all 5 of the manually retrofitted version of the WhyML contracts. Given that Yul has a stable formal semantics on which the Solidity compiler is implemented, we plan to automate such a retrofitting process while respecting the official Yul semantics in future work.

Case study on T6 properties. To demonstrate that SciviK can verify the correctness of decentralized financial models for smart contracts, we study a Uniswap-style [39] constant product market maker contract (135 LOC in Solidity). It represents the standard practice adopted by the constant product market makers in the current DeFi ecosystem. Uniswap and its variants are the biggest liquidity source for decentralized cryptocurrency trading (at the time of writing, they collectively carry 6B USD token market capital and a 5B USD total value locked). These contracts have various desirable properties, e.g., resistance to price manipulation. However, these properties were only proven on paper, and in fact some of the underlying assumptions could be easily violated in practice, leading to high-profile price manipulation attacks that stole millions worth of cryptocurrencies from the contract users (e.g., the bZx protocol hack on February 18, 2020 [40]). SciviK is able to prove the financial security properties against the contract implementation and avoid such attacks. We show an example contract in Fig. 16.12 where SciviK proves its soundness by verifying three lemmas:

- $\mathcal{L}_1$ (*Swap correctness*): if one calls the `swap0to1` function, he will lose some `token0` in return for `token1`, and vice-versa.
- $\mathcal{L}_2$ (*Reserve synchronization*): the `reserve` variable in the market maker contract always equals the actual liquidity it provides, i.e., its token balances.
- $\mathcal{L}_3$ (*Increasing product*): constant product market makers are named because the product of the contract's reserves is always the same before and after each trade. But in our example, due to the 0.3% transaction fee, the product no longer stays constant but increases slightly after each trade.

The WhyML program generated from the Yul IR consists of 2,197 LOC. We first tried to prove the relatively simple property $\mathcal{L}_1$, but, due to the length of the program, the solvers timed out. After a manual inspection into the code, we discovered that it has too many low-level typecasts and temporary variables that overburdened the SMT solvers. Then we tried to manually prove these theorems, but found that the generated raw Coq file was too large (345K LOC) to reason about. In short, the complexity of DeFi has rendered reasoning about low-level programs difficult.

```
1    /* @meta _token0.balanceOf(address(this)) = reserve0
2     * @meta _token1.balanceOf(address(this)) = reserve1 */
3    contract AMM {
4        ...
5        uint reserve0;
6        uint reserve1;
7
8        /* @pre (token0.allowance(fromA, address(this)) == 1);
9         * @coq @post (reserve0 * reserve1 > old reserve0 * old reserve1)
10        * @post (token0.balanceOf(fromA) < old token0.balanceOf(fromA))
11        * @post (token1.balanceOf(fromA) > old token1.balanceOf(fromA)) */
12       function trade (address fromA, uint amount0) returns (uint) {
13           require (amount0 > 0);
14           require (fromA != address(this));
15           require (reserve0 > 0);
16           require (reserve1 > 0);
17           bool success = _token0.transferFrom (fromA, address(this), amount0);
18           assert (success);
19           uint swapped = swap(fromA);
20           return swapped;
21       }
22
23       function swap0to1 (address toA) returns (uint) {
24           uint balance0 = _token0.balanceOf(address(this));
25           uint amount0In = balance0 - reserve0;
26           assert (toA != token0 && toA != token1);
27           assert (amount0In > 0);
28           assert (reserve0 > 0 && reserve1 > 0);
29           uint amountInWithFee = amount0In * 997;
30           uint numerator = amountInWithFee * reserve1;
31           uint denominator = reserve0 * 1000 + amountInWithFee;
32           uint result = numerator / denominator;
33           bool success = _token1.transfer(toA, result);
34           assert (success);
35           reserve0 = _token0.balanceOf(address(this));
36           reserve1 = _token1.balanceOf(address(this));
37           return result;
38       }
39       ...
40   }
```

Fig. 16.12.   Simplified automated market maker contract.

As a result, we retrofitted the translated WhyML program by removing low-level helper functions, temporary variables, and typecasts. The retrofitting work took 1 person-day. The final WhyML program consists of 320 LOC and generates a Coq program of 403 LOC. An example of a retrofitted function is shown in Fig. 16.13. We leave the automated retrofitting for WhyML programs as future work.

SciviK successfully verifies the retrofitted version with respect to $\mathcal{L}_1$ and $\mathcal{L}_2$ using the automated SMT solver Z3 with a total time of 1.25 seconds. And $\mathcal{L}_3$ was able to be verified in about 200 lines of Coq.

During the experiments, we also found that most industry-grade contracts that are hard to reason about are not necessarily long, but involve

```
1   let ghost function fun_swap0to1_552 (st_c: evmState)  (vloc_toA_444: int)
2       : evmState
3       requires { st_c.var_0x07
4               = pop (fun_balanceOf_33 (push st_c st_c.this.address)) }
5       ...
6   = let _r: ref int = ref 0 in
7     let ghost st_g: ref evmState = ref st_c in
8     let bvconstzero = 0 in
9     try
10      begin
11        ... the trading procedure ...
12        st_g := { st_g with var_0x07 = pop (!st_g) };
13        let _51 = vloc_result_512 in
14        let expr_549 = _51 in
15        _r := expr_549;
16        leave
17      end;
18      st_g := (setRet !st_g !_r);
19      raise Ret
20    with Ret → (!st_g)
21    end
```

Fig. 16.13.   Snippet of the retrofitted `swap0to1` function in WhyML

```
1   contract C {
2       function f(uint length) public {
3           uint[] memory x = new uint[](length);
4           /* other operations */
5       }
6   }
```

Fig. 16.14.   A contract with a bug of array creation overflow

complex data structures and financial models. The complexity is needed to ensure either the integrity of the free and fair market against all possible trader behavior, or the correctness of crucial data structures that involve frequent re-structuring, e.g., deletion, insertion, sorting in a linked list, etc.

Compiler front-end bugs. In §16.4, we mention that SciviK can also be used to detect bugs or any semantic changes due to versioning [41] in the compiler front-end. Here, we describe how we use SciviK to detect a bug in `Solc`, the Solidity compiler, before the 0.6.5 version. Consider the code snippet in Fig. 16.14 in which function `f` creates a dynamically-sized array in memory based on a `length` parameter. A simplified Yul IR compiled by a faulty compiler is shown in Fig. 16.15.

Inspecting the compiled `create_memory_array` function, we can see that since it does not check the length of the dynamically allocated array, overflow could occur by invoking `f`. SciviK detects this compiler bug with the `overflow` pattern annotation. The translated WhyML function (simplified)

```
1   function allocate(size) -> memPtr {
2       memPtr := mload(64)
3       let newFreePtr := add(memPtr, size)
4       // protect against overflow
5       if or(gt(newFreePtr, 0xffffffffffffffff), lt(newFreePtr, memPtr))
6           { revert(0, 0) }
7       mstore(64, newFreePtr)
8   }
9
10  function create_memory_array(length) -> memPtr {
11      size := mul(length, 0x20)
12      size := add(size, 0x20)
13      memPtr := allocate(size)
14      mstore(memPtr, length)
15  }
16
17  // main function
18  function f(length)  {
19      let vloc_x_10_mpos := create_memory_array(length)
20  }
```

Fig. 16.15.   Yul IR (simplified) compiled by a faulty compiler

```
1   let function create_memory_array (st_c: evmState) (length : int) : (evmState)
2   (* Compiler comes with overflow check for parameters; therefore length variable
         does not over/underflow. *)
3   require { 0 ≤ length < 0xffff_ffff_ffff_ffff }
4   = let st_g: ref evmState = ref st_c in
5     let ret: ref int = ref 0 in
6     let st_g := push (!st_g) (Cons length (Cons 0x20 Nil)) in
7     let st_g := mul (!st_g) in
8     let ret := pop (!st_g) in
9     (* ensure the size variable is within bound after MUL *)
10    assert { 0 ≤ (!ret) < 0xffff_ffff_ffff_ffff }
11    let st_g := push (!st_g) (Cons (!ret) (Cons 0x20 Nil)) in
12    let st_g := add (!st_g) in
13    let ret := pop (!st_g) in
14    (* ensure the size variable is within bound after ADD *)
15    assert { 0 ≤ (!ret) < 0xffff_ffff_ffff_ffff }
16    (* ... other operations ... *)
```

Fig. 16.16.   The translated WhyML function of create_memory_array

is shown in Fig. 16.16. With these fine-grained specifications embedded, SciviK returns counterexamples that violate the two conditions on line 10 and 15 in Fig. 16.16, such as $length = 2^{64}/32$. This indicates that although the input parameter length is within a maximum bound b, overflow can occur after the MUL and ADD instructions (line 11 and 12 in Fig. 16.15). The overflow checks embedded by SciviK thus allow us to detect the anomalies after MUL and ADD. To eliminate this bug, a potential solution is to limit the size of the length parameter such that its value after MUL and ADD operations are still kept in bound. This solution is implemented in Solc 0.6.5. [41] To reflect this fix in WhyML, we can modify the pre-condition on line 3 with the following

```
1    requires { 0 ≤ length < 0xffff }
```

and the verification conditions pass successfully. By inspecting fine-grained EVM intermediary states, SciviK allows compiler front-end developers to peek into the execution of EVM, enabling the development of a more robust front-end.

Cross-comparison. We also compare SciviK with Securify [6], Solc-verify[33], and VeriSol [36], in terms of the verification techniques and capabilities of each framework, shown in Table 16.3. Securify performs static analysis on Solidity contracts to check against pre-defined patterns. Securify supports many useful patterns and we found in our experiment that it could handle more T5 properties than SciviK; however, performing security patterns alone is insufficient when many crucial properties are user-defined and contract-specific. Due to time limits and engineering challenges, we did not experiment with all of the tools available. However at a high level, while VeriSol can verify most of the properties in the benchmark, it does not provide an expressive way for the user to specify custom properties. As for Solc-verify, although it provides an annotation system, its expressiveness is limited and the source-level approach relies on the correctness of the `solc` compiler. Compared to these tools, SciviK is the most versatile framework that can verify all six types of properties. Besides an expressive annotation

Table 16.3. Comparison of verification tools based on techniques and capabilities.

Techniques	Securify	VeriSol	Solc-verify	SciviK
Static analysis				
Symbolic execution				
Semantic formalization				
Model checking				
Mechanized proof				
IVL				
Machine learning				
Capabilities				
T1				
T2				
T3				
T4				
T5				
T6				
Detecting compiler bugs				

system, SciviK allows the user to use security patterns and loop invariant inference to enhance automaton, reducing the proof burden on the user side. In addition, SciviK's IR-level approach and the fine-grained EVM model ensure that the verified guarantees still hold on EVM and enables the detection of errors in compiler front-end. Finally, SciviK's use of the Why3 IVL enables porting the generated VCs to multiple SMT solvers as well as the Coq proof assistant, facilitating flexible and compositional proofs. Overall, SciviK is a capable verification framework when dealing with various proof tasks from real-world contracts.

16.8   Related work

Pattern-based static analysis over smart contracts.   Pattern-based techniques have been applied to detect various pre-defined vulnerability patterns in smart contracts. Vandal [42] and Mythril [10] provide a static analysis framework for semantic inference. They transform the original program into custom IRs, which enable the extraction of dataflow facts and dependency relations. Datalog-based approaches [43], such as Securify [6], decompile EVM bytecode and use optimizations on semantic database queries to match vulnerability patterns. MadMax [44] identifies gas-related vulnerabilities and provides a control-flow-based decompiler for declarative pattern programming. However, pattern-based techniques may not provide any security guarantees due to false positives and cannot express complex patterns such as financial model related properties, which are both supported by SciviK.

Model checking for smart contracts.   Model checking techniques [8,9,45, 46] can be applied to verify access control and temporal-related properties in smart contracts. However, these model checking-based approaches may not be sound as they tend to abstract away the contract source code and low-level execution semantics. In addition, these approaches often require contracts to be written using explicit state definitions and transitions, which can be non-intuitive to developers. On the contrary, SciviK's EVM model ensures that the verified guarantees hold on the EVM, and that the source-level annotation system can seamlessly embed into the source code of smart contracts and does not require any change to the program's logic.

Automated verification frameworks for smart contracts.   Existing verification frameworks use SMT solvers to encode and prove the security properties of smart contracts. Oyente [7] employs symbolic execution to check user-defined specifications of each function on its custom IR. VerX [47] provides refinement-based strategies to contract-level invariant verification. Solc-verify [13] also provides an encoding of Solidity data structures for reasoning and modeling of its semantics. Solidity compiler's SMT solver [48] enables native keywords to be used for specification and symbolic execution of functions. Although these frameworks achieve great automation, their capabilities are constrained by SMT solvers, which cannot express and solve clauses beyond first-order logic. On the contrary, SciviK combines SMT-based verification techniques with manual verification through the Coq proof assistant. This combination enables SciviK to automatically prove VCs expressed in first-order logic and port complex and contract-specific VCs to the Coq proof assistant.

Semantic formalization and mechanized proofs for smart contracts. There have been extensive works on the formalization of EVM semantics [49–51], which can be used to reason about EVM bytecode but have not been applied to verify real-world smart contracts. The K Framework [12] introduces an executable EVM formal semantics and has been shown to enable the verification of smart contracts. This work, however, restricts the specifications and the reasoning to the bytecode level, which discards certain human-readable information, such as variable names and explicit control flows, making it hard for developers to write specifications and proofs. Li *et al.* [52] provide a formalization of smart contracts in Yul and prove the correctness of a set of benchmark contracts in Isabelle; however, the methodology presented mostly relies on manual efforts to formulate specifications and construct proofs, rendering the verification approach unscalable to complex smart contracts. On the contrary, SciviK allows users to express properties using our annotation system at the source level while, at the same time, offering a highly automated verification engine, as well as a formal model of the EVM semantics, ensuring that the verified guarantees still hold over the EVM.

16.9 Conclusion

SciviK is a versatile framework for specifying and verifying security properties for real-world smart contracts. SciviK provides an annotation system for writing specifications at the source code level, translates annotated smart contracts into annotated Yul IR programs, and then generates VCs using an EVM model implemented in WhyML. Generated VCs can be discharged to SMT solvers and the Coq proof assistant. To further reduce the proof burden, SciviK uses automation techniques, including loop invariant inference and static analysis for security patterns. We have successfully verified 158 security properties in six types for 12 benchmark contracts and a real-world DeFi contract. We expect that SciviK will become a critical building block for developing secure and trustworthy smart contracts in the future.

Acknowledgments

We would like to thank Shoucheng Zhang for his tremendous support of our work on building trustworthy blockchain systems and smart contracts. We also thank Zhong Shao, Vilhelm Sjöberg, Zhaozhong Ni, and Justin Wong for their helpful comments and suggestions that improved this paper and the implemented tools. This research is based on work supported in part by NSF grants CCF-1918400, a Columbia-IBM Center Seed Grant Award, and an Arm Research Gift. Any opinions, findings, conclusions, or recommendations expressed herein are those of the authors and do not necessarily reflect those of the US Government, NSF, IBM, or Arm.

References

[1] Z. Zheng, S. Xie, H.-N. Dai, X. Chen, and H. Wang, Blockchain challenges and opportunities: A survey, *International Journal of Web and Grid Services.* **14**(4), 352–375 (2018).

[2] Ethereum Foundation. Ethereum whitepaper, (2020). URL https://github.com/ethereum/wiki.

[3] S. Zhang, In math we trust, *Stanford Blockchain Salon.* (2018).

[4] S. Nakamoto, Bitcoin: A peer-to-peer electronic cash system, *Cryptography Mailing list at https://metzdowd.com* (03. 2009).

[5] N. Atzei, M. Bartoletti and T. Cimoli. A survey of attacks on ethereum smart contracts (sok). In *International conference on principles of security and trust*, pp. 164–186. Springer, (2017).

[6] P. Tsankov, A. Dan, D. Drachsler-Cohen, A. Gervais, F. Buenzli and M. Vechev, Securify: Practical security analysis of smart contracts. In *Proceedings of the 2018 ACM SIGSAC Conference on Computer and Communications Security*, pp. 67–82. ACM, (2018).

[7] L. Luu, D.-H. Chu, H. Olickel, P. Saxena, and A. Hobor, Making smart contracts smarter. In *Proceedings of the 2016 ACM SIGSAC conference on computer and communications security*, pp. 254–269. ACM, (2016).

[8] Z. Nehai, P.-Y. Piriou, and F. Daumas, Model-checking of smart contracts. In *2018 IEEE International Conference on Internet of Things (iThings) and IEEE Green Computing and Communications (GreenCom) and IEEE Cyber, Physical and Social Computing (CPSCom) and IEEE Smart Data (SmartData)*, pp. 980–987. IEEE, (2018).

[9] I. Sergey, A. Kumar, and A. Hobor, Temporal properties of smart contracts. In *International Symposium on Leveraging Applications of Formal Methods*, pp. 323–338. Springer, (2018).

[10] J. Feist, G. Grieco, and A. Groce, Slither: a static analysis framework for smart contracts. In *2019 IEEE/ACM 2nd International Workshop on Emerging Trends in Software Engineering for Blockchain (WETSEB)*, pp. 8–15. IEEE, (2019).

[11] R. M. Parizi, A. Dehghantanha, K.-K. R. Choo, and A. Singh, Empirical vulnerability analysis of automated smart contracts security testing on blockchains. In *Proceedings of the 28th Annual International Conference on Computer Science and Software Engineering*, pp. 103–113, (2018).

[12] E. Hildenbrandt, M. Saxena, N. Rodrigues, X. Zhu, P. Daian, D. Guth, B. Moore, D. Park, Y. Zhang, A. Stefanescu, *et al.* Kevm: A complete formal semantics of the ethereum virtual machine. In *2018 IEEE 31st Computer Security Foundations Symposium (CSF)*, pp. 204–217. IEEE, (2018).

[13] Á. Hajdu and D. Jovanovic, Smt-friendly formalization of the solidity memory model. In *ESOP*, pp. 224–250, (2020).

[14] B. Mueller. Mythril-reversing and bug hunting framework for the Ethereum blockchain, (2017).

[15] The Yul development team. Yul — Solidity 0.6.2 documentation, (2020). URL https ://solidity.readthedocs.io/en/v0.6.2/yul.html.

[16] F. Bobot, J.-C. Filliâtre, C. Marché, and A. Paskevich, Why3: Shepherd your herd of provers. (2011).

[17] L. de Moura and N. Bjørner, Z3: An efficient SMT solver. In *Tools and Algorithms for the Construction and Analysis of Systems (TACAS)*, (2008).

[18] F. Bobot, S. Conchon, E. Contejean, M. Iguernelala, S. Lescuyer, and A. Mebsout, The alt-ergo automated theorem prover, *URL: http://alt-ergo.lri.fr.* (2008).

[19] C. Barrett, C. L. Conway, M. Deters, L. Hadarean, D. Jovanović, T. King, A. Reynolds, and C. Tinelli. Cvc4. In *International Conference on Computer Aided Verification*, pp. 171–177. Springer, (2011).

[20] B. Barras, S. Boutin, C. Cornes, J. Courant, J.-C. Filliatre, E. Gimenez, H. Herbelin, G. Huet, C. Munoz, C. Murthy, *et al.*, The coq proof assistant reference manual: Version 6.1. (1997).

[21] CertiK Security Leaderboard, (2021). URL https://certik.org.

[22] S. Srinawakoon, Band protocol white paper v3. (2019). URL https://bandprotocol.c om/whitepaper-3.0.1.pdf.

[23] The Solidity development team. Security Considerations — Solidity 0.6.5 documen-tation, (2020).URL https://solidity.readthedocs.io/en/v0.6.5/security-consideration s.html?highlight=reentrancy.

[24] J. Yao, G. Ryan, J. Wong, S. Jana, and R. Gu. Learning nonlinear loop invariants with gated continuous logic networks. In *Proceedings of the 41st ACM SIGPLAN Conference on Programming Language Design and Implementation*, pp. 106–120, (2020).

[25] Jacques, Dafflon and Jordi, Baylina and Thomas, Shababi. Eip-777: Erc777 token standard, (2020). URL https://eips.ethereum.org/EIPS/eip-777.

[26] Duncan Riley. $25m in cryptocurrency stolen in hack of lendf.me and uniswap, (2020).URL https://siliconangle.com/2020/04/19/25m-cryptocurrency-stolen-hack-lendf-uniswap/.

[27] G. Wood *et al.* Ethereum: A secure decentralised generalised transaction ledger. ethereum project yellow paper, (2014).

[28] The eWasm development team. ewasm/design (Feb., 2020). URL https://github.co m/ewasm/design. original-date: 2016-03-05T02:21:36Z.

[29] J.-C. Filliâtre and A. Paskevich. Why3 — where programs meet provers. In *European symposium on programming*, pp. 125–128. Springer, (2013).

[30] Y. Hirai. pirapira/eth-isabelle (Feb., 2020). URL https://github.com/pirapira/eth-i sabelle.

[31] AestheticIntegration/contracts. URL https://github.com/AestheticIntegration/cont racts.

[32] S. Grossman, I. Abraham, G. Golan-Gueta, Y. Michalevsky, N. Rinetzky, M. Sagiv, and Y. Zohar, Online detection of effectively callback free objects with applications to smart contracts, *Proceedings of the ACM on Programming Languages.* **2**(POPL), 1–28, (2017).

[33] Á. Hajdu and D. Jovanović. solc-verify: A modular verifier for solidity smart contracts. In *Working Conference on Verified Software: Theories, Tools, and Experiments*, pp. 161–179. Springer, (2019).

[34] Scott Chipolina. Ethereum transaction fees soar amid high demand, (2020). URL https://decrypt.co/53095/ethereum-gas-fees-soar-as-price-peaks-at-1150.

[35] Known Attacks - Ethereum Smart Contract Best Practices. URL https://consensys.github.io/smart-contract-best-practices/known_attacks/.

[36] Y. Wang, S. Lahiri, S. Chen, R. Pan, I. Dillig, C. Born, I. Naseer and K. Ferles, Formal verification of workflow policies for smart contracts in azure blockchain. In *Verified Software: Theories, Tools and Experiments*. Springer (September, 2019).

[37] Y. Chen and C. Bellavitis, Blockchain disruption and decentralized finance: The rise of decentralized business models, *Journal of Business Venturing Insights.* **13**, e00151, (2020).

[38] H. Adams, N. Zinsmeister, and D. Robinson. Uniswap v2 core, (2020). URL https://uniswap.org/whitepaper.pdf.

[39] G. Angeris, H.-T. Kao, R. Chiang, C. Noyes, and T. Chitra, An analysis of uniswap markets, *arXiv preprint arXiv:1911.03380.* (2019).

[40] Yogita Khatri. bzx attacked again, $645k in eth estimated to be lost, (2020). URL https://www.theblockcrypto.com/post/56207/bzx-attacked-again-645k-in-eth-estimated-to-be-lost.

[41] S. Team. Solidity Memory Array Creation Overflow Bug, (2020). URL https://solidity.ethereum.org/2020/04/06/memory-creation-overflow-bug/.

[42] L. Brent, A. Jurisevic, M. Kong, E. Liu, F. Gauthier, V. Gramoli, R. Holz, and B. Scholz, Vandal: A scalable security analysis framework for smart contracts, *arXiv preprint arXiv:1809.03981.* (2018).

[43] P. Tsankov. Security analysis of smart contracts in datalog. In *International Symposium on Leveraging Applications of Formal Methods*, pp. 316–322. Springer, (2018).

[44] N. Grech, M. Kong, A. Jurisevic, L. Brent, B. Scholz, and Y. Smaragdakis, Madmax: Surviving out-of-gas conditions in ethereum smart contracts, *Proceedings of the ACM on Programming Languages.* **2** (OOPSLA), 1–27, (2018).

[45] W. Ahrendt, R. Bubel, J. Ellul, G. J. Pace, R. Pardo, V. Rebiscoul, and G. Schneider. Verification of smart contract business logic. In *International Conference on Fundamentals of Software Engineering*, pp. 228–243. Springer, (2019).

[46] A. Mavridou and A. Laszka. Tool demonstration: FSolidM for designing secure Ethereum smart contracts. In *International Conference on Principles of Security and Trust*, pp. 270–277. Springer, (2018).

[47] A. Permenev, D. Dimitrov, P. Tsankov, D. Drachsler-Cohen, and M. Vechev. Verx: Safety verification of smart contracts. In *2020 IEEE Symposium on Security and Privacy (SP)*, pp. 1661–1677. IEEE, (2020).

[48] L. Alt and C. Reitwießner. SMT-based verification of solidity smart contracts. In *International Symposium on Leveraging Applications of Formal Methods*, pp. 376–388. Springer, (2018).

[49] Y. Hirai. Defining the ethereum virtual machine for interactive theorem provers. In *International Conference on Financial Cryptography and Data Security*, pp. 520–535. Springer, (2017).

[50] I. Grishchenko, M. Maffei, and C. Schneidewind. A semantic framework for the security analysis of ethereum smart contracts. In *International Conference on Principles of Security and Trust*, pp. 243–269. Springer, (2018).

[51] S. Amani, M. Bégel, M. Bortin, and M. Staples. Towards verifying ethereum smart contract bytecode in Isabelle/HOL. In *Proceedings of the 7th ACM SIGPLAN International Conference on Certified Programs and Proofs*, pp. 66–77, (2018).

[52] X. Li, Z. Shi, Q. Zhang, G. Wang, Y. Guan, and N. Han. Towards verifying Ethereum smart contracts at intermediate language level. In *International Conference on Formal Engineering Methods*, pp. 121–137. Springer, (2019).

Appendix

Schedule of the Shoucheng Zhang
Memorial Workshop

May 2nd, 2019

9:00-9:10AM	Opening Remarks	Barbara Zhang
9:10-9:40AM	An Overview of Shoucheng's Scientific Life	Xiao-Liang Qi
Chair: Steven Kivelson	**Topological materials and topological effects**	
9:40-10:10AM	Topology in HgTe	Laurens Molenkamp
10:10-10:40AM	Topological superconductor and Majorana zero mode	Jinfeng Jia
Chair: Akira Furusaki	**Topological materials and topological effects**	
11:00-11:30AM	Magnetic topological materials	Claudia Felser
11:30AM-12:00PM	Noncommutative geometry at boundaries of topological insulators	Shinsei Ryu
12:00-12:30PM	First principle calculation of the effective Zeeman's couplings in topological materials	Xi Dai

Chair: Chyh-Hong Chern	**Spintronics**	
2:00-2:30PM	Persistent spin helix	Joe Orenstein
2:30-3:00PM	How spin currents defy our high-school intuition	Yaroslaw Bazaliy
Chair: Jing Wang	**Spintronics**	
3:30-4:00PM	Anti-skyrmion	Stuart Parkin
4:00-4:30PM	Twisted Bilayer Graphene: Fractional Quantum Hall Effect Reprise?	Allan MacDonald
4:30-5:00PM	Topological phase transitions in topological insulators	Shuichi Murakami
5:00-6:30PM	Reception	
6:30-9:30PM	**Banquet dinner**	Peter van Nieuwenhuizen Robert Laughlin Duncan Haldane Laurens Molenkamp

May 3rd, 2019

Chair: Shinsei Ryu	**Fundamental physics and mathematics**	
9:00-9:30AM	Topological Chern-Simons Matter Theories	Cumrun Vafa
9:30-10:00AM	Anomalies and Nonsupersymmetric D-Branes	Edward Witten
10:00-10:30AM	From Mathematics to a Secure Digital World	Shafi Goldwasser
Chair: Hong Yao	**Topological states of matter**	
11:00-11:30AM	Topological Insulators from Beginning to Today	Andrei Bernevig
11:30AM-12:00PM	Higher Order Topological Phases: Multipole Insulators and Metamaterials	Taylor Hughes
12:00-12:30PM	topological entanglement entropy without locality	Leonid Pryadko

Chair: Eun-Ah Kim	**Spintronics**	
2:00-2:30PM	Quantum Nucleation of Skyrmions	Daniel Arovas
2:30-3:00PM	Cooper pair spin current in ferromagnet / spin-triplet superconductor heterostructure	Suk Bum Chung
Chair: Chao-Xing Liu	**Topological materials and topological effects**	
3:30-4:00PM	Quantum Spin Hall Effect and Beyond in InAs/GaSb Double Layers	Rui-Rui Du
4:00-4:30PM	Visualizing Topological Electronic Structures	Yulin Chen
4:30-5:00PM	Ordered and maybe topological states in flat band 2D moire systems	David Goldhaber-Gordon

May 4th, 2019

Chair: Cenke Xu	**Strongly correlated electrons**	
9:00-9:30AM	Theory of a Planckian metal with a remnant Fermi surface	Subir Sachdev
9:30-10:00AM	Genes of unconventional high temperature superconductor	Jiangping Hu
10:00-10:30AM	Symmetry protected gapless states and their regularizability	Dung-Hai Lee
Chair: Casten Honerkamp	**Strongly correlated electrons**	
11:00-11:30AM	Disordered and deconfined: exotic quantum criticality with Dirac fermions	Joseph Maciejko
11:30AM-12:00PM	Application of the symmetry principle to condensed matter physics	Congjun Wu
12:00-12:30PM	A chiral SYK model	Biao Lian

Chair: Xiao-Liang Qi	**Algorithms for the future**	
2:00-2:30PM	Robustness in many-body systems: from topological insulators to mechanism design	Brian Zhang
2:30-3:00PM	Learning atoms to discover new materials	Quan Zhou
Chair: Peizhe Tang	**Algorithms for the future**	
3:30-4:00PM	Towards building trustworthy blockchain ecosystems	Ronghui Gu
4:00-4:30PM	Engineering the physics of genetics to empower disease treatment	Stanley Qi
4:30-4:45PM	Conclusion remarks	Eugene Demler
4:45-4:50PM	Closing remarks	Xiao-Liang Qi Barbara Zhang